SpringerWienNewYork

Texts and Monographs in Symbolic Computation

A Series of the Research Institute for Symbolic Computation, Johannes Kepler University, Linz, Austria

For further volumes:
http://www.springer.com/series/3073

Ulrich Langer
Peter Paule
Editors

Numerical and Symbolic Scientific Computing

Progress and Prospects

SpringerWienNewYork

Prof.Dr. Ulrich Langer
Johannes Kepler University Linz
Institute of Computational Mathematics
Altenbergerstr. 69
4040 Linz
Austria
ulanger@numa.uni-linz.ac.at

Prof.Dr. Peter Paule
Johannes Kepler University Linz
Research Institute for Symbolic Computation
Altenbergerstr. 69
4040 Linz
Austria
peter.paule@risc.jku.at

SpringerWienNewYork is part of Springer Science+Business Media
springer.at

Typesetting: SPi, Pondicherry, India

Printed on acid-free and chlorine-free bleached paper
SPIN 80044234

With 50 Figures

Library of Congress Control Number: 2011942415

ISSN 0943-853X
ISBN 978-3-7091-0793-5 e-ISBN 978-3-7091-0794-2
DOI 10.1007/978-3-7091-0794-2
SpringerWienNewYork

Preface

For more than 10 years, the numerical analysis and symbolic computation groups at the Johannes Kepler University Linz (JKU) have made serious efforts to combine two different worlds of scientific computing, numerics and symbolics. This work has been carried out in the frame of two excellence programs of the Austrian Science Funds (FWF), a special research program (SFB, 1998–2008) and a doctoral program (DK, 2008–). In addition to the JKU institutes for Applied Geometry, Computational Mathematics, Industrial Mathematics, and the Research Institute for Symbolic Computation (RISC), the Radon Institute for Computational and Applied Mathematics (RICAM), a branch of the Austrian Academy of Sciences, have been partners in this enterprise.

This book presents an offspring of this initiative. It contains surveys of the state of the art and of results achieved after more than 10 years of SFB/DK work. In addition, we included chapters that go beyond, this means, which set pointers for future developments. All of the chapters have been carefully refereed. Most of them center around the theme of partial differential equations. Major aspects are: fast solvers in elastoplasticity, symbolic analysis for boundary problems (from rewriting to parametrized Groebner bases), symbolic treatment of operators, use of computer algebra in the finite element method for the construction of recurrence relations in special high-order Maxwell solvers and for the construction of sparsity optimized high-order finite element basis functions on simplices, a symbolic approach to finite difference schemes, cylindrical algebraic decomposition and symbolic local Fourier analysis of multigrid methods, and white noise analysis for stochastic PDEs. The scope of other numerical-symbolic topics range from applied and computational geometry (approximate implicitization of space curves, symbolic–numeric genus computation, automated discovery in geometry) to computer algebra methods used for total variation energy minimization. One chapter deals with verification conditions in connection with functional recursive programs.

The contributions to this book are arranged in alphabetical order.

The editors express their sincerest thanks to all authors for their interesting contributions, the anonymous referees for their valuable work, Silvia Schilgerius and Wolfgang Dollhäubl from Springer-Verlag Wien and Gabriela Hahn from the DK.

Last but not least, we would like to acknowledge the moral and financial support by the Austrian Science Fund (FWF), the Johannes Kepler University Linz (JKU), the Government of Upper Austria, and the City of Linz.

Linz, June 2011

Ulrich Langer
Peter Paule

Contents

Approximate Implicitization of Space Curves

Martin Aigner, Bert Jüttler, and Adrien Poteaux

Abstract The process of implicitization generates an implicit representation of a curve or surface from a given parametric one. This process is potentially interesting for applications in Computer Aided Design, where the robustness and efficiency of intersection algorithm can be improved by simultaneously considering implicit and parametric representations. This paper gives an brief survey of the existing techniques for approximate implicitization of hyper surfaces. In addition it describes a framework for the approximate implicitization of space curves.

1 Introduction

There exist two main representations of curves and surfaces in Computer Aided Geometric Design: the implicit and the parametric form. In both cases, the functions which describe the curve or surface are almost always chosen as polynomial or rational functions or, more generally, as polynomial or rational spline functions [15]. Consequently, one deals with segments and patches of algebraic curves and surfaces.

Each of the two different representation is particularly well suited for certain applications. Parametric representations are well suited to generate points, e.g., for displaying curves and surfaces, and to apply the results of the classical differential geometry of curves and surfaces, e.g., for shape interrogation. Implicit representations encompass a larger class of shapes and are more powerful for certain geometric queries. Moreover, the class of algebraic curves and surfaces is closed

B. Jüttler (✉) · M. Aigner
Institute of Applied Geometry, Johannes Kepler University, Altenbergerstr. 69, 4040 Linz, Austria
e-mail: bert.juettler@jku.at; martin.aigner@jku.at

A. Poteaux
University of Lille, France

U. Langer and P. Paule (eds.), *Numerical and Symbolic Scientific Computing*, Texts and Monographs in Symbolic Computation, DOI 10.1007/978-3-7091-0794-2_1,

under certain geometric operations, such as offsetting, while the class of rational parametric curves and surfaces is not.

Consequently, it is often desirable to change from one representation to the other one. For instance, the implicitization of a planar curve reduces the computation of the intersection of two curves given in the parametric form to a root finding problem for a single polynomial [23].

The exact conversion procedures, implicitization and parameterization, have been studied in classical algebraic geometry and in symbolic computation. Their practical application in Computer Aided Design is rather limited, due to the feasibility reasons outlined below. As an alternative, approximate techniques have emerged recently. These alternatives contribute to the use of symbolic-numerical techniques in Computer Aided Geometric Design.

The remainder of this paper consists of four parts. First we introduce the notation. Section 3 then presents a survey of related techniques for the approximate implicitization of hypersurfaces. The following section describes a new framework for the approximate implicitizaton of space curves. Finally we conclude this paper.

2 Preliminaries

We start by introducing a few notations. A parametric representation of a curve segment or a surface patch is a mapping

$$\mathbf{p} : \Omega \to \mathbb{R}^d : \mathbf{t} \mapsto \mathbf{p}(\mathbf{t}) \tag{1}$$

where $\Omega \subset \mathbb{R}^k$ is the parameter domain (typically a closed interval in $\mathbb{R}$ or a box in $\mathbb{R}^2$). A curve or surface is described for $k = 1$ and $k = 2$, respectively. In many applications, e.g. in Computer-Aided Design, the mapping $\mathbf{p}$ is represented by piecewise rational functions (rational spline functions), see [15].

An implicitly defined hypersurface $\mathcal{F}$ in $\mathbb{R}^d$ is the zero-set of a function $f_\mathbf{s} : \mathbb{R}^d \to \mathbb{R}$,

$$\mathcal{F} = \{\mathbf{x} \in \mathbb{R}^d : f_\mathbf{s}(\mathbf{x}) = 0\}. \tag{2}$$

If $d = 3$ or $d = 2$, then it is called an implicitly defined *surface* or *planar curve*, respectively.

The subscript represents a vector $\mathbf{s} \in \mathbb{R}^N$ which collects the parameters which characterize the function $f_\mathbf{s}(\mathbf{x})$. They are called the *shape parameters*, since they control the shape of the curve or surface. For instance, if $f_\mathbf{s}$ is a polynomial of some finite degree,

$$f_\mathbf{s}(\mathbf{x}) = \sum_{i=1}^{N} s_i \, \phi_i(\mathbf{x}), \tag{3}$$

then $\mathbf{s} = (s_1, \ldots, s_N)$ contains the coefficients with respect to a suitable basis $(\phi_i)_{i=1}^N$ of the space of polynomials.

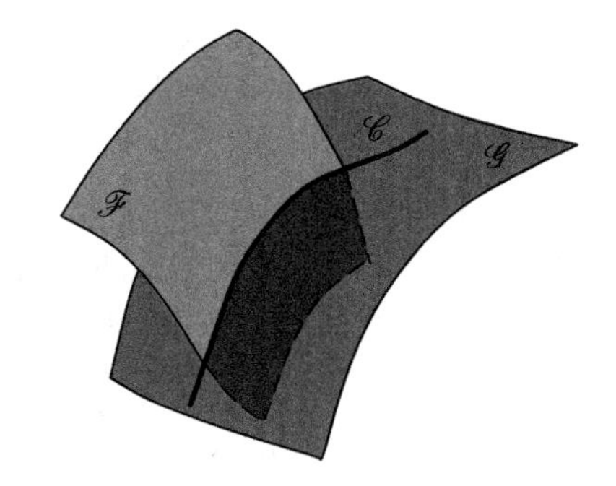

Fig. 1 An implicitly defined space curve

An implicitly defined space curve

$$\mathscr{C} = \{\mathbf{x} \in \mathbb{R}^3 : f_{\mathbf{s}}(\mathbf{x}) = 0 \wedge g_{\mathbf{s}}(\mathbf{x}) = 0\} \tag{4}$$

is defined by two intersecting implicitly defined surfaces $\mathscr{F}$ and $\mathscr{G}$, see Fig. 1. Clearly, $f_{\mathbf{s}}$ and $g_{\mathbf{s}}$ are not unique. This space curve is said to be *regular* at the point $\mathbf{x} \in \mathscr{F} \cap \mathscr{G}$, if there exists a representation (4) such that the two gradient vectors $\nabla_{\mathbf{x}} f_{\mathbf{s}}(\mathbf{x})$ and $\nabla_{\mathbf{x}} g_{\mathbf{s}}(\mathbf{x})$ with $\nabla_{\mathbf{x}} = (\frac{\partial}{\partial x}, \frac{\partial}{\partial y}, \frac{\partial}{\partial z})$ are linearly independent.

Typically, the two functions defining $\mathscr{F}$ and $\mathscr{G}$ are characterized by two independent sets of shape parameters, say $\mathbf{s}_f$ and $\mathbf{s}_g$. In order to simplify the notation, we shall use the convention that both functions depend on the union of these two sets, hence on $\mathbf{s} = \mathbf{s}_f \cup \mathbf{s}_g$. If the two functions $f_{\mathbf{s}}(\mathbf{x})$ and $g_{\mathbf{s}}(\mathbf{x})$ are polynomials, then $\mathscr{C}$ is said to be an *algebraic space curve*.

3 Approximate Implicitization

Exact techniques for the implicitization of curves and surfaces have been studied for a long time. In 1862, Salmon [20] noted that the surface implicitization can be performed by eliminating the parameters. This was improved by Dixon in 1908 [8], who published a more compact resultant for eliminating two variables from three polynomials. In 1983, Sederberg [21] considered the implicitization of surface patches for Computer Aided Geometric Design.

From a theoretical point of view, the problem of the implicitization of a given rational curve or surface is always solvable. However, there remains a number of challenging computational difficulties. As described in [15, chapter 12], while the 2D case can be handled satisfactorily by building the Bezout resultant, the 3D case is more complicated: for instance, a tensor product surface of degree (m, n) leads to an implicit formula of degree $2mn$. Then, in the simple case $m = n = 3$, we already have an algebraic representation of degree 18. After expanding this polynomial in monomial basis this would lead to 1330 terms.

Practical problems associated with the exact implicitization of curves and surfaces are addressed in [22] and [5]. Gröbner bases can also be used [7]. For more details on resultant based methods, the reader may also consult [6].

To conclude, as shown in [15, 22], exact implicitization has many associated difficulties, in particular in the case of surfaces. Moreover, the computed implicit form of a curve or surface can be difficult to use, since the degree of the polynomial is often too high. On the other hand, CAD (Computer-Aided Design) systems are based on floating point computations, and so all quantities are represented with a rounding error. Therefore, if we apply any of the exact implicitization method in this context, the result is not exact.

The existing techniques for approximate implicitization can be classified as direct ones, where the result is found in a single step, and evolution-based techniques, where an iterative process is needed to find the result.

3.1 Direct Techniques

We describe three approaches to approximate implicitization. The first two approaches are due to Dokken, who also coined the notion of AI. The third approach comprises various fitting-based techniques.

3.1.1 Dokken's method

In order to adapt implicitization to the need for approximate computation in CAD, and to achieve more practical algorithms, Dokken introduced the approximate implicitization of a curve or surface [9,10]. In the sequel we recall Dokken's method to compute the approximate implicitization of a curve or surface. See also [12] for a survey of these and related techniques.

Given a parametric curve or surface $\mathbf{p}(\mathbf{t})$, $\mathbf{t} \in \Omega$, a polynomial $f_{\mathbf{s}}(\mathbf{x})$ is called approximate implicitization of $\mathbf{p}(\mathbf{t})$ with tolerance $\epsilon > 0$ if we can find a continuous direction function $\mathbf{g}(\mathbf{t})$ and a continuous error function $\eta(\mathbf{t})$ such that

$$f_{\mathbf{s}}(\mathbf{p}(\mathbf{t}) + \eta(\mathbf{t})\mathbf{g}(\mathbf{t})) = 0, \tag{5}$$

with $\|\mathbf{g}(\mathbf{t})\|_2 = 1$ and $|\eta(\mathbf{t})| \leq \epsilon$ (see [9, Definition 35]). We denote by n the degree of the parametrization $\mathbf{p}$ and by m the degree of $f_{\mathbf{s}}$.

Dokken observes that the composition $f_{\mathbf{s}} \circ \mathbf{p}$ can be factorized as

$$f_{\mathbf{s}}(\mathbf{p}(\mathbf{t})) = (\mathbf{D}\mathbf{s})^T \alpha(\mathbf{t}), \tag{6}$$

where $\mathbf{D}$ is a matrix build from certain products of the coordinate functions of $\mathbf{p}(\mathbf{t})$, $\mathbf{s}$ is the vector of parameters that characterize the function $f_{\mathbf{s}}(\mathbf{x})$. Furthermore, $\alpha(\mathbf{t}) = (\alpha_1(\mathbf{t}), \ldots, \alpha_N(\mathbf{t}))^T$ is the basis of the space of polynomials of degree mn, which is used to describe $f_{\mathbf{s}}(\mathbf{p}(\mathbf{t}))$ and N is the dimension of polynomial space.

This basis is assumed to form a partition of unity,

$$\sum_{i=1}^{N} \alpha_i = 1$$

and in addition, the basis $\alpha(\mathbf{t})$ is assumed to be non-negative for $\mathbf{t} \in \Omega$:

$$\alpha_i(\mathbf{t}) \geq 0, \quad \forall i, \forall \mathbf{t} \in \Omega.$$

For instance, one may use the Bernstein-Bézier basis with respect to the interval Ω or with respect to a triangle which contains Ω in the case of curves and surfaces, respectively.

Consequently we obtain that

$$|f_{\mathbf{s}}(\mathbf{p}(\mathbf{t}))| = |(\mathbf{Ds})^T \alpha(\mathbf{t})| \leq \|\mathbf{Ds}\|_2 \|\alpha(\mathbf{t})\|_2 \leq \|\mathbf{Ds}\|_2, \tag{7}$$

hence we are led to find a vector $\mathbf{s}$ which makes $\|\mathbf{Ds}\|_2$ small. Using the Singular Value Decomposition (SVD) of the matrix $\mathbf{D}$, one can show that $\| f_{\mathbf{s}_1}(\mathbf{p}(\mathbf{t}))\|_\infty \leq \sqrt{\sigma_1}$, where σ_1 is the smallest singular value, and $\mathbf{s}_1$ is the corresponding singular vector. This strategy enables the use of Linear Algebra tools to solve the problem of approximate implicitization. Moreover, this approach provides high convergence rates, see [12, Table 1 and 2].

3.1.2 Dokken's weak method

Dokken's original method has several limitations: for instance, it is relatively costly to build the matrix $\mathbf{D}$. Moreover, it is impossible to use spline functions for describing $f_{\mathbf{s}}$, since no suitable basis for the composition $f_{\mathbf{s}} \circ \mathbf{p}$ can be found.

This problem can be avoided by using the *weak form* of approximate implicitization which was introduced in [11], see also [12, section 10]. For a given curve or surface $\mathbf{p}$ with parameter domain Ω, we now find the approximate implicitization by minimizing

$$\int_\Omega (f_{\mathbf{s}}(\mathbf{p}(\mathbf{t})))^2 \mathrm{d}\mathbf{t} = \mathbf{s}^T \mathbf{A} s \tag{8}$$

where

$$\mathbf{A} = \mathbf{D}^T \left(\int_\Omega \alpha(\mathbf{t}) \alpha(\mathbf{t})^T \mathrm{d}\mathbf{t} \right) \mathbf{D}. \tag{9}$$

The matrix $\mathbf{A}$ can be analyzed by eigenvalue decomposition, similar to the original approach, where the matrix $\mathbf{D}$ was analyzed with singular value decomposition. Note that one can apply this strategy even if no explicit expression is available: one only needs to be able to evaluate points on the curve or surface. The integrals can then be approximately evaluated by numerical integration.

Choosing the eigenvector which is associated with the smallest eigenvalue of the matrix **A** is equivalent to minimizing the objective function defined in (8) subject to the constraint $\|\mathbf{s}\| = 1$. This can be seen as a special case of fitting, see next section.

3.1.3 Algebraic curve and surface fitting

Given a number of points $(\mathbf{p}_i)_{i=1}^N$, which have been sampled from a given curve or surface, one may fit a curve or surface by minimizing the sum of the squared residuals (also called algebraic distances),

$$\sum_{i=1}^{N} (f_{\mathbf{s}}(\mathbf{p}_i))^2. \tag{10}$$

This objective function can be obtained by applying a simple numerical integration to (8).

If the algebraic curve or surface is given as in (3), then this objective function has the trivial minimum $\mathbf{s} = \mathbf{0}$. In order to obtain a meaningful result by minimizing (10), several additional constraints have been introduced.

Pratt [19] picks one of the coefficients and restricts it to 1, e.g.

$$s_1 = 1. \tag{11}$$

For instance, if $f_{\mathbf{s}}$ is a polynomial which is represented with respect to the usual power basis, then one may consider the absolute term. This constraint is clearly not geometrically invariant, since the curve and surface cannot pass through the origin of the system of coordinates.

Geometrically invariant constraints can be obtained by considering quadratic functions of the unknown coefficients **s**. An interesting normalization has been suggested by Taubin [24], who proposed to use the norm of the squared gradient vectors at the given data,

$$\sum_{i=1}^{N} \|\nabla_{\mathbf{x}} f_{\mathbf{s}}(\mathbf{p}_i)\|^2 = 1. \tag{12}$$

Adding this constraint leads to a generalized eigenvalue problem. Taubin's method gives results which are independent of the choice of the coordinate system.

Finally, Dokken's weak method – when combined with numerical integration for evaluating the objective function (8) – uses the constraint

$$\|\mathbf{s}\|^2 = \sum_{i=1}^{N} s_i^2 = 1. \tag{13}$$

These three approaches are able to provide meaningful solutions which minimize the squared algebraic distances (10). However, they may still lead to fairly unexpected results. Additional branches and isolated singular points may be present, even for data which are sampled from regular curves or surfaces.

If a method for approximate implicitization is to reproduce the exact results for sufficiently high degrees, then this unpleasant phenomenon is always present. For instance, consider a cubic planar curve with a double point. Even if we take sample points only from one of the two branches which pass through the singular point, any of the above-mentioned methods will generate the cubic curve with the double point, provided that the degree of $f_{\mathbf{s}}$ is at least 3.

These difficulties can be avoided by using additional normal (or gradient) information. More precisely, a nontrivial solution of the minimization problem can be found by considering a convex combination of the two objective functions (8) and

$$\sum_{i=1}^{N} \| \nabla_{\mathbf{x}} f_{\mathbf{s}}(\mathbf{p}_i) - \mathbf{n}_i \|^2, \tag{14}$$

where the vectors $(\mathbf{n}_i)_{i=1}^{N}$ represent additional normal vector information at the given points.

This gives a quadratic function of the unknown coefficients $\mathbf{s}$, hence the minimum is found by solving a system of linear equations. This approach has been introduced in [16], and it has later been extended in [17, 25, 26]. Among other topics, these papers also consider the case of curves which contain singular points, where a globally consistent propagation of the normals is needed.

3.2 Iterative (Evolution-Based) Techniques

Iterative (evolution-based) methods have been considered for several reasons. First, they lead to a uniform framework for handling various representations of curves and surfaces, which can handle implicitly defined curves and surfaces as well as parametric ones [1, 13]. Second, they make it possible to include various conditions, such as constraints on the gradient field, volume constraints or range constraints [14, 27, 28]. Finally, the sequence of curves or surfaces generated by an iterative method can be seen as discrete instances of a continuous evolution process, which links this approach to the level set method and to active curves and surfaces in Computer Vision [4, 18].

We recall the evolution-based framework for fitting point data $(\mathbf{p}_j)_{j=1,\ldots,M}$ with implicitly defined hypersurfaces, which was described in [1]. In this framework, the approximate solutions which are generated by an iterative algorithm are seen as discrete instances of a continuous movement of an initial curve or surface towards the target points (the given point data).

More precisely, we assume that the shape parameters $\mathbf{s}$ depend on a time-like parameter t, and consider the evolution of the hypersurface described by the parameters $\mathbf{s}(t)$ for $t \to \infty$. Each data point $\mathbf{p}_j$ attracts a certain point $\mathbf{f}_j$ on the hypersurface $\mathscr{F}$ which is associated with it. Usually $\mathbf{f}_j$ is chosen to be the closest point on $\mathscr{F}$, i.e.

$$\mathbf{f}_j = \arg\min_{\mathbf{p}\in\mathscr{F}} \|\mathbf{p}-\mathbf{p}_j\|. \tag{15}$$

These attracting forces push the time-dependent hypersurface towards the data. This is realized by assigning certain velocities to the points on the hypersurface. For a point lying on a time-dependent implicitly defined curve or surface, which is described by a function $f_{\mathbf{s}}$, the normal velocity is given by

$$\mathbf{v} = -\frac{\partial f_{\mathbf{s}}}{\partial t}\frac{\nabla_{\mathbf{x}} f_{\mathbf{s}}^{\top}}{\|\nabla_{\mathbf{x}} f_{\mathbf{s}}\|^2} = -\nabla_{\mathbf{s}} f_{\mathbf{s}}\,\dot{\mathbf{s}}\frac{\nabla_{\mathbf{x}} f_{\mathbf{s}}^{\top}}{\|\nabla_{\mathbf{x}} f_{\mathbf{s}}\|^2}, \tag{16}$$

where the dot indicates the derivative with respect to t and the gradient operator

$$\nabla_{\mathbf{s}} = \left(\frac{\partial}{\partial s_1}, \ldots, \frac{\partial}{\partial s_N}\right) \tag{17}$$

gives the row vector of the first partial derivatives. Note that we omitted the time dependency of $\mathbf{s}$ in (16), in order to simplify the notation.

The first term $-\nabla_{\mathbf{s}} f_{\mathbf{s}}\,\dot{\mathbf{s}}$ in (16) specifies the absolute value of the normal velocity. The second term is the unit normal vector of the curve, which identifies the direction of the velocity vector.

As the number of data points exceeds in general the degrees of freedom of the hypersurface, the velocities are found as the least squares solution of

$$\sum_{j=1}^{M}((\mathbf{v}_j-\mathbf{d}_j)^{\top}\mathbf{n}_j)^2 \rightarrow \min_{\dot{\mathbf{s}}}, \tag{18}$$

where $\mathbf{d}_j = \mathbf{f}_j - \mathbf{p}_j$ is the residual vector from a data point to its associated point on the hypersurface, $\mathbf{n}_j = \frac{\nabla_{\mathbf{x}} f_{\mathbf{s}}}{\|\nabla_{\mathbf{x}} f_{\mathbf{s}}\|}$ is the unit normal in this point and $\mathbf{v}_j$ is the velocity computed via (16) at $\mathbf{f}_j$. More precisely, this leads to the minimization problem

$$\sum_{j=1}^{M}\left(\left(\frac{(\nabla_{\mathbf{s}} f_{\mathbf{s}})(\mathbf{p}_j)\,\dot{\mathbf{s}}\,(\nabla_{\mathbf{x}} f_{\mathbf{s}})(\mathbf{p}_j)}{\|(\nabla_{\mathbf{x}} f_{\mathbf{s}})(\mathbf{p}_j)\|^2} - (\mathbf{f}_j-\mathbf{p}_j)^{\top}\right)\frac{(\nabla_{\mathbf{x}} f_{\mathbf{s}})(\mathbf{p}_j)^{\top}}{\|(\nabla_{\mathbf{x}} f_{\mathbf{s}})(\mathbf{p}_j)\|}\right)^2 \rightarrow \min_{\dot{\mathbf{s}}}. \tag{19}$$

We use Tikhonov regularization in order to obtain a unique solution. In addition, we apply a distance field constraint, in order to avoid the trivial solution, cf. [27].

The geometric interpretation of this approach is as follows: The bigger the distance to the associated data point, the greater is the velocity that causes the movement of the hypersurface at the corresponding point. Note that (18) takes only the normal component of the velocity into account, as a tangential motion does not change the distance to the data.

The objective function in (19) depends on $\mathbf{s}$ as well as on $\dot{\mathbf{s}}$. For a given value of $\mathbf{s}$, we can find $\dot{\mathbf{s}}$ by solving a system of linear equations. Consequently, (19) leads to an ordinary differential equation for the vector of shape parameters. We can solve it by using Euler steps with a suitable stepsize control, see [1] for details.

The solution converges to a stationary point, which defines the solution of the fitting problem. It can be shown that this evolution-based approach is equivalent to a Gauss-Newton method for the implicit fitting problem, and the stationary point of the ODE is a (generally only) local minimum of the objective function

$$\sum_{j=1}^{M} ||\mathbf{p}_j - \mathbf{f}_j||^2, \tag{20}$$

where $\mathbf{f}_j$ has been defined in (15), see [2].

The evolution viewpoint has several advantages. It provides a geometric interpretation of the initial solution, which is now seen as the starting point of an evolution that drives the hypersurface towards the data. It also provides a geometrically motivated stepsize control, which is based on the velocity of the points during the evolution (see [1]). Finally, the framework makes it possible to introduce various other constraints on the shape of the hypersurface, see [13, 14].

In the remainder of this paper we will apply the evolution framework to the approximate implicitization of space curves. In this situation we need to generate two surfaces which intersect in the given space curve. Moreover, these two surfaces should intersect transversely, in order to obtain a robustly defined intersection curve.

4 Approximate Implicitization of Space Curves

Now we consider a point cloud $(\mathbf{p}_j)_{j=1,\ldots,M}$ which has been sampled from a space curve. Recall that a point $\mathbf{p}_j$ lies on an implicitly defined space curve $\mathscr{C}$ if it is contained in both surfaces defining the curve. Consequently we fit the spatial data with two surfaces $\mathscr{F}$ and $\mathscr{G}$. The desired solution $\mathscr{C}$ is then contained in the intersection of $\mathscr{F}$ and $\mathscr{G}$. We need to couple the fitting of the two surfaces, in order to obtain a well-defined intersection curve.

4.1 Fitting Two Implicitly Defined Surfaces

Following the idea in [2] we use an approximation of the exact geometric distance from a data point to a space curve. More precisely, we use the Sampson distance which was originally introduced for the case of hypersurfaces [24]. The oriented distance from a point $\mathbf{p}_j$ to a curve or surface which is defined implicitly as the zero set of some function $f_{\mathbf{s}}$ can be approximated by

$$\frac{f_{\mathbf{s}}(\mathbf{p}_j)}{\|\nabla_{\mathbf{x}} f_{\mathbf{s}}(\mathbf{p}_j)\|}. \tag{21}$$

Geometrically speaking, the equation of the surface is linearized in the point $\mathbf{p}_j$ and the distance from this point to the zero-set of the linearization is taken as an approximation of the exact distance. Consequently, this measure is exact for planes, as they coincide with their linearization. The Sampson distance is not defined at points with vanishing gradients, which have to be excluded.

A natural extension of this distance to two surfaces defining a space curve is

$$d_j = \sqrt{\frac{f_{\mathbf{s}}(\mathbf{p}_j)^2}{\|\nabla_{\mathbf{x}} \mathbf{s}(\mathbf{p}_j)\|^2} + \frac{g_{\mathbf{s}}(\mathbf{p}_j)^2}{\|\nabla_{\mathbf{x}} g_{\mathbf{s}}(\mathbf{p}_j)\|^2}}. \tag{22}$$

If both surfaces intersect each other orthogonally in the data points, i.e.

$$(\nabla_{\mathbf{x}} f_{\mathbf{s}} \nabla_{\mathbf{x}} g_{\mathbf{s}}^{\top})\big|_{\mathbf{p}_j} = 0, \tag{23}$$

then this expression approximates the distance to the implicitly defined space curve.

In order to approximate a set of points which has been sampled from a space curve, we minimize the sum of the squared distances, which leads to the objective function

$$\sum_{j=1}^{M} d_j^2 = \sum_{j=1}^{M} \frac{f_{\mathbf{s}}(\mathbf{p}_j)^2}{\|\nabla_{\mathbf{x}} f_{\mathbf{s}}(\mathbf{p}_j)\|^2} + \frac{g_{\mathbf{s}}(\mathbf{p}_j)^2}{\|\nabla_{\mathbf{x}} g_{\mathbf{s}}(\mathbf{p}_j)\|^2} \to \min_{\mathbf{s}}. \tag{24}$$

Note that both functions $f_{\mathbf{s}}$ and $g_{\mathbf{s}}$ depend formally on the same vector $\mathbf{s}$ of shape parameters. Typically, each shape parameter s_i is uniquely associated with either $f_{\mathbf{s}}$ or $g_{\mathbf{s}}$. Consequently, (24) minimizes the Sampson distances from a point $\mathbf{p}_j$ to each of the surfaces $\mathscr{F}$ and $\mathscr{G}$ independently.

We adapt the evolution based-framework [2] in order to deal with the objective function (24). We consider the combination of the two evolutions for $\mathscr{F}$ and $\mathscr{G}$ which is defined by the minimization problem $E \to \min_{\dot{\mathbf{s}}}$, where

$$E(f,g) = \sum \left(\frac{f_{\mathbf{s}}}{\|\nabla_{\mathbf{x}} f_{\mathbf{s}}\|} + \frac{\nabla_{\mathbf{s}} f_{\mathbf{s}}}{\|\nabla_{\mathbf{x}} f_{\mathbf{s}}\|} \dot{\mathbf{s}} \right)^2 + \left(\frac{g_{\mathbf{s}}}{\|\nabla_{\mathbf{x}} g_{\mathbf{s}}\|} + \frac{\nabla_{\mathbf{s}} g_{\mathbf{s}}}{\|\nabla_{\mathbf{x}} g_{\mathbf{s}}\|} \dot{\mathbf{s}} \right)^2. \tag{25}$$

In order to simplify the notation, we omit the argument $\mathbf{p}_j$ from now on and omit the range of the sum, which is taken over all sampled points $(\mathbf{p}_j)_{j=1,\dots,M}$. This sum can also be seen as simple numerical integration along the given space curve.

The geometric meaning of this objective function is as follows: The normal velocity (cf. 16) of the level set of $f_{\mathbf{s}}$ (and analogously for $g_{\mathbf{s}}$) which passes through the given point $\mathbf{p}_j$ is to be equal to the estimated oriented distance, see (21), to the surface. Later we will provide another interpretation of this evolution as a Gauss-Newton-type method.

Similar to (19), the objective function in (25) depends on $\mathbf{s}$ and on $\dot{\mathbf{s}}$. For a given value of $\mathbf{s}$, we find $\dot{\mathbf{s}}$ by solving a system of linear equations. Consequently, (25) leads to an ordinary differential equation for the vector of shape parameters. We can again solve it simply by using Euler steps with a suitable stepsize control.

As a necessary condition for a minimum of (25), the first derivatives with respect to the vector $\dot{\mathbf{s}}$ have to vanish. This yields the linear system

$$\sum \left[\frac{\nabla_{\mathbf{s}} f_{\mathbf{s}}^{\top}}{\|\nabla_{\mathbf{x}} f_{\mathbf{s}}\|} \frac{\nabla_{\mathbf{s}} f_{\mathbf{s}}}{\|\nabla_{\mathbf{x}} f_{\mathbf{s}}\|} + \frac{\nabla_{\mathbf{s}} g_{\mathbf{s}}^{\top}}{\|\nabla_{\mathbf{x}} g_{\mathbf{s}}\|} \frac{\nabla_{\mathbf{s}} g_{\mathbf{s}}}{\|\nabla_{\mathbf{x}} g_{\mathbf{s}}\|} \right] \dot{\mathbf{s}} = - \sum \frac{f_{\mathbf{s}} \nabla_{\mathbf{s}} f_{\mathbf{s}}^{\top}}{\|\nabla_{\mathbf{x}} f_{\mathbf{s}}\|^2} + \frac{g_{\mathbf{s}} \nabla_{\mathbf{s}} g_{\mathbf{s}}^{\top}}{\|\nabla_{\mathbf{x}} g_{\mathbf{s}}\|^2}. \quad (26)$$

If there exists a zero-residual solution, then the right hand side vanishes, as $f_{\mathbf{s}}(\mathbf{p}_j) = g_{\mathbf{s}}(\mathbf{p}_j) = 0$ for all j. Hence $\dot{\mathbf{s}} = 0$ is a solution for the problem and we have reached a stationary point of the evolution. However, the solution may not be unique.

First, the trivial (and unwanted) functions $f_{\mathbf{s}} \equiv 0$ and $g_{\mathbf{s}} \equiv 0$ solve always the minimum problem (24) for all data sets $(\mathbf{p}_j)_{j=1\ldots M}$. Of course these solutions have to be avoided.

Second, the evolution defined via (25) pushes both surfaces *independently* towards the data points $\mathbf{p}_j$. This may lead to the unsatisfying result $f_{\mathbf{s}} \equiv g_{\mathbf{s}}$ (where the two functions are identical up to a factor λ). Consequently, we need to introduce additional terms which guarantee that $f_{\mathbf{s}}$ and $g_{\mathbf{s}}$ do not vanish and that they intersect orthogonally in the data points.

4.2 *Regularization*

So far, the implicitization problem is not well–posed. If $f_{\mathbf{s}}$ is a solution to the problem, then $\lambda f_{\mathbf{s}}$ is a solution as well. In this section we discuss several strategies that shall prevent the functions $f_{\mathbf{s}}$ and $g_{\mathbf{s}}$ from vanishing and that shall guarantee a unique solution to the individual fitting problems for the two defining surfaces $\mathscr{F}$ and $\mathscr{G}$. Additionally, we propose a coupling term that ensures a well-defined intersection curve of the surfaces $\mathscr{F}$ and $\mathscr{G}$.

Distance field constraint. In order to avoid the unwanted solutions $f_{\mathbf{s}} \equiv 0$ and $g_{\mathbf{s}} \equiv 0$ we use the distance field constraint which was described in [27]. Consider the term

$$D(f) = \left(\frac{d}{dt} \|\nabla_{\mathbf{x}} f_{\mathbf{s}}(\mathbf{x})\| + \|\nabla_{\mathbf{x}} f_{\mathbf{s}}(\mathbf{x})\| - 1 \right)^2. \quad (27)$$

It pushes the function $f_{\mathbf{s}}$ in a point $\mathbf{x}$ closer to a unit distance field, hence

$$\|\nabla_{\mathbf{x}} f_{\mathbf{s}}(\mathbf{x})\| = 1. \quad (28)$$

If the length of the gradient in (27) equals 1, it is expected to remain unchanged. Consequently, its derivative shall be 0. Otherwise (27) modifies $f_{\mathbf{s}}$ such that the norm of its gradient gets closer to 1.

We apply this penalty term to both functions f_s and g_s.

This side condition has also an important influence on the robustness of the implicit representation of the two surfaces $\mathcal{F}$ and $\mathcal{G}$, cf. [3]. Roughly speaking, the closer the defining functions f_s and g_s are to a unit gradient field, the less sensible is the representation to potential errors in its coefficients.

Theoretically, this condition can be integrated over the entire domain of interest. In order to obtain a robust representation of the implicit space curve, the robustness of the two generating surfaces is mainly required along their intersection, i.e. near the data points. This leads to the idea of imposing the distance field constraint only in the data points $\mathbf{p}_j$.

We note two more observations. First, the term is quadratic in the unknowns $\dot{\mathbf{s}}$ which follows directly from expanding the derivative in (27),

$$\frac{d}{dt}\|\nabla_{\mathbf{x}} f_s(\mathbf{x}_j)\| = \frac{\nabla_{\mathbf{x}} f_s}{\|\nabla_{\mathbf{x}} f_s\|}\nabla_{\mathbf{s}}\nabla_{\mathbf{x}} f_s\, \dot{\mathbf{s}}. \tag{29}$$

Consequently, the objective function with the distance field constraint is still quadratic in the unknowns, and we can compute the derivative vector $\dot{\mathbf{s}}$ of the shape parameters by solving a system of linear equations.

Second, the constrained problem does in general not reproduce exact solutions which would be available without any constraints. For instance, if the data were sampled from a low degree algebraic space curve, then the approximation technique would not provide an exact equation of this curve. Only if that solution possesses a unit gradient field along the data, then it can be recovered. In the next section we introduce another regularization term which makes it possible to reproduce the exact solution.

Averaged gradient constraint. This technique is related to a method that was introduced by Taubin [24]. The core idea is to restrict the sum of the norms of the gradients. Hence, not all the gradient lengths are expected to be uniform, but the average gradient length

$$\frac{1}{M}\sum \|\nabla_{\mathbf{x}} f_s(\mathbf{p}_j)\| = 1. \tag{30}$$

This can be dealt with by adding the term

$$A(f) = \left(\sum \frac{d}{dt}\|\nabla_{\mathbf{x}} f_s(\mathbf{p}_j)\| + \|\nabla_{\mathbf{x}} f_s(\mathbf{p}_j)\| - 1\right)^2 \tag{31}$$

to our framework.

Although (28) and (31) look quite similar, their effects on the solution are rather different. Note that (30) is only one constraint, whereas (28) is a set of constraints, which depends on the number of points.

Consequently, the condition on the average norm of the gradient can only handle the singularity that is due to the scalability of implicit representations.

If the ambiguity of the solution arises from an incorrectly chosen degree of the polynomial, then Taubin's method and the term (31) do not provide a unique solution.

For instance, when fitting a straight line with two quadratic surfaces, the obtained linear system is singular as the number of unknowns exceeds the number of linearly independent equations provided by the data points. On the other hand, if we use the distance field constraint (27), then we will obtain a unique solution.

Orthogonality constraint. The distance field constraint leads to a robust representation of each of the two surfaces which define the curve. Now we introduce an additional term which provides a robust representation of the curve itself.

Ideally, the two surfaces would intersect orthogonally along the space curve $\mathscr{C}$, i.e. (23) holds.

In this case, small displacements in the two surfaces cause only small errors in the curve. Moreover, the term (22) then approximates the distance to the space curve very well. On the other hand, if the two surfaces intersect tangentially

$$(\nabla_{\mathbf{x}} f_{\mathbf{s}}^{\top} \times \nabla_{\mathbf{x}} g_{\mathbf{s}}^{\top})\big|_{\mathscr{C}} = \mathbf{0} \tag{32}$$

even small perturbations may cause big changes of the curve.

In order to obtain two surfaces that intersect each other approximately orthogonally, we add the term

$$O(f,g) = \sum \left(\frac{d}{dt} \left(\frac{\nabla_{\mathbf{x}} f_{\mathbf{s}}}{\|\nabla_{\mathbf{x}} f_{\mathbf{s}}\|} \frac{\nabla_{\mathbf{x}} g_{\mathbf{s}}^{\top}}{\|\nabla_{\mathbf{x}} g_{\mathbf{s}}\|} \right) + \frac{\nabla_{\mathbf{x}} f_{\mathbf{s}}}{\|\nabla_{\mathbf{x}} f_{\mathbf{s}}\|} \frac{\nabla_{\mathbf{x}} g_{\mathbf{s}}^{\top}}{\|\nabla_{\mathbf{x}} g_{\mathbf{s}}\|} \right)^2 \tag{33}$$

to the objective function. This term penalizes deviations from the optimal case $\nabla_{\mathbf{x}} f_{\mathbf{s}} \nabla_{\mathbf{x}} g_{\mathbf{s}}^{\top} = 0$. More precisely, if the gradients of the surfaces are not orthogonal in a point where (33) is applied to, then the time derivative of the product of the unit gradients forces the surfaces to restore this property. Theoretically, this term should be imposed along the intersection of the surfaces $\mathscr{F}$ and $\mathscr{G}$. As the exact intersection curve is not known, we apply (33) to the data points $\mathbf{p}_j$.

We analyze the structure of this term in more detail. The time derivative of the first product in (33) gives

$$\begin{aligned} \frac{d}{dt} \frac{\nabla_{\mathbf{x}} f_{\mathbf{s}}}{\|\nabla_{\mathbf{x}} f_{\mathbf{s}}\|} \frac{\nabla_{\mathbf{x}} g_{\mathbf{s}}^{\top}}{\|\nabla_{\mathbf{x}} g_{\mathbf{s}}\|} &= \frac{\nabla_{\mathbf{x}} \dot{f}_{\mathbf{s}} \nabla_{\mathbf{x}} g_{\mathbf{s}}^{\top} + \nabla_{\mathbf{x}} f_{\mathbf{s}} \nabla_{\mathbf{x}} \dot{g}_{\mathbf{s}}^{\top}}{\|\nabla_{\mathbf{x}} f_{\mathbf{s}}\| \|\nabla_{\mathbf{x}} g_{\mathbf{s}}\|} \\ &\quad - \nabla_{\mathbf{x}} f_{\mathbf{s}} \nabla_{\mathbf{x}} g_{\mathbf{s}}^{\top} \left(\frac{\nabla_{\mathbf{x}} f_{\mathbf{s}} \nabla_{\mathbf{x}} \dot{\mathbf{s}}^{\top}}{\|\nabla_{\mathbf{x}} f_{\mathbf{s}}\|^3 \|\nabla_{\mathbf{x}} g_{\mathbf{s}}\|} + \frac{\nabla_{\mathbf{x}} g_{\mathbf{s}} \nabla_{\mathbf{x}} \dot{g}_{\mathbf{s}}^{\top}}{\|\nabla_{\mathbf{x}} f_{\mathbf{s}}\| \|\nabla_{\mathbf{x}} g_{\mathbf{s}}\|^3} \right) \end{aligned} \tag{34}$$

Since $\nabla_{\mathbf{x}} \dot{f}_{\mathbf{s}} = \nabla_{\mathbf{x}} \nabla_{\mathbf{s}} f_{\mathbf{s}} \dot{\mathbf{s}}$ and $\nabla_{\mathbf{x}} \dot{g}_{\mathbf{s}} = \nabla_{\mathbf{x}} \nabla_{\mathbf{s}} g_{\mathbf{s}} \dot{\mathbf{s}}$, the term (33) is quadratic in $\dot{\mathbf{s}}$.

4.3 Putting Things Together

Summing up, we obtain the minimization problem

$$F(\dot{\mathbf{s}}, \mathbf{s}) \to \min_{\dot{\mathbf{s}}} \tag{35}$$

where

$$F = E(f,g) + \omega_1(D(f) + D(g)) + \omega_2 O(f,g) + \omega_3(A(f) + A(g)) + \omega_4 \dot{\mathbf{s}}^2. \tag{36}$$

The non-negative weights ω_1, ω_2, ω_3 and ω_4 control the influence of the distance field constraint, the orthogonality constraint, the averaged gradient constraint and the Tikhonov regularization, respectively. Due to the special structure (36) is quadratic in the vector $\dot{\mathbf{s}}$. Hence, for a given vector $\mathbf{s}$ of shape parameters, we can find $\dot{\mathbf{s}}$ by solving a system of linear equations. The evolution of the implicit representation of the space curve can then be traced using explicit Euler steps with a suitable stepsize control (cf. [1]).

We conclude this section by discussing the coupled evolution from the optimization viewpoint. We show that the constrained optimization is in fact a Gauss-Newton method for a particular fitting problem.

Consider the optimization problem

$$\begin{aligned} C = \sum \Big(\frac{f_\mathbf{s}}{\|\nabla_\mathbf{x} f_\mathbf{s}\|} \Big)^2 + \Big(\frac{g_\mathbf{s}}{\|\nabla_\mathbf{x} g_\mathbf{s}\|} \Big)^2 + \omega_1 \left((\|\nabla_\mathbf{x} f_\mathbf{s}\| - 1)^2 + (\|\nabla_\mathbf{x} g_\mathbf{s}\| - 1)^2 \right) \\ + \omega_2 \left(\frac{\nabla_\mathbf{x} f_\mathbf{s}}{\|\nabla_\mathbf{x} g_\mathbf{s}\|} \frac{\nabla_\mathbf{x} g_\mathbf{s}^\top}{\|\nabla_\mathbf{x} g_\mathbf{s}\|} \right)^2 + \omega_3 \Big(\big(\sum \|\nabla_\mathbf{x} f_\mathbf{s}(\mathbf{p}_j)\| - 1 \big)^2 \\ + \big(\sum \|\nabla_\mathbf{x} g_\mathbf{s}(\mathbf{p}_j)\| - 1 \big)^2 \Big) \to \min_\mathbf{s}. \end{aligned} \tag{37}$$

Obviously, a solution of (37) minimizes simultaneously the Sampson distances from the data points to the space curve (term 1 and 2) the distance field constraint (term 3), the orthogonality constraint (term 4) and the averaged gradient constraint (term 5 and 6). Hence a zero residual solution of (37) interpolates all data points, the defining surfaces have slope one in the data points and furthermore, the surfaces intersect orthogonally.

Since (37) is non-linear in the vector of unknowns $\mathbf{s}$, we consider an iterative solution technique. A Gauss-Newton approach for (37) solves iteratively the linearized version of (37),

$$C^* \to \min_{\Delta \mathbf{s}} \tag{38}$$

where

$$C^* = \sum \left(\frac{f_s}{\|\nabla_x f_s\|} + \frac{\nabla_s f_s}{\|\nabla_x f_s\|} \Delta s \right)^2 + \left(\frac{g_s}{\|\nabla_x g_s\|} + \frac{\nabla_s g_s}{\|\nabla_x g_s\|} \Delta s \right)^2$$
$$+ \omega_1 \left[(\|\nabla_x f_s\| - 1 + \nabla_s(\|\nabla_x f_s\| - 1) \Delta s)^2 + (\|\nabla_x g_s\| - 1 + \nabla_s(\|\nabla_x g_s\| - 1) \Delta s)^2 \right]$$
$$+ \omega_2 \left(\frac{\nabla_x f_s}{\|\nabla_x f_s\|} \frac{\nabla_x g_s^\top}{\|\nabla_x g_s\|} + \nabla_s \left(\frac{\nabla_x f_s}{\|\nabla_x f_s\|} \frac{\nabla_x g_s^\top}{\|\nabla_x g_s\|} \right) \Delta s \right)^2$$
$$+ \omega_3 \left(\left(\sum \|\nabla_x f_s\| - 1 + \nabla_s \|\nabla_x f_s\| \Delta s \right)^2 + \left(\sum \|\nabla_x g_s\| - 1 + \nabla_s \|\nabla_x g_s\| \Delta s \right)^2 \right) \tag{39}$$

and computes an update of the previous solution via $\mathbf{s}^+ = \mathbf{s} + \Delta\mathbf{s}$. By comparing (36) and (39) we arrive at the following observation.

An explicit Euler step for the evolution equation (36) with stepsize 1 *is equivalent to the Gauss-Newton update (39) for the optimization problem (37).*

Indeed, if we use that for any function $h(\mathbf{s}(t))$,

$$\frac{d}{dt} h(\mathbf{s}(t)) = \nabla_s h(\mathbf{s}(t)) \dot{\mathbf{s}}, \tag{40}$$

then we can replace the time derivatives in (36). Substituting $\dot{\mathbf{s}}$ for $\Delta\mathbf{s}$ then gives the desired result.

4.4 Examples

Finally we present some examples.

Example 1. We sampled 50 points from a parametric space curve of degree 6. The two implicit patches that represent the implicit space curve are of degree 2. As initial configuration we have chosen two surfaces deviating from each other slightly, see Fig. 2(a).

The obtained result after 15 iterations is shown in Fig. 2(b). In order to demonstrate the robustness of the representation we note that the norm of the gradients of the two surfaces in the data points varies between 0.94 and 1.94. The maximal deviation of the gradients from orthogonality at the data points is 0.49 degrees.

Example 2. We choose again the same data set, but modify the various weights in order to demonstrate their influence. First we omit the orthogonality constraint. That is, the evolution is not coupled, and both surfaces move independently towards the data. The result is obvious, both surfaces converge towards the same result, as the initial values are quite similar, cf. Fig. 3(a).

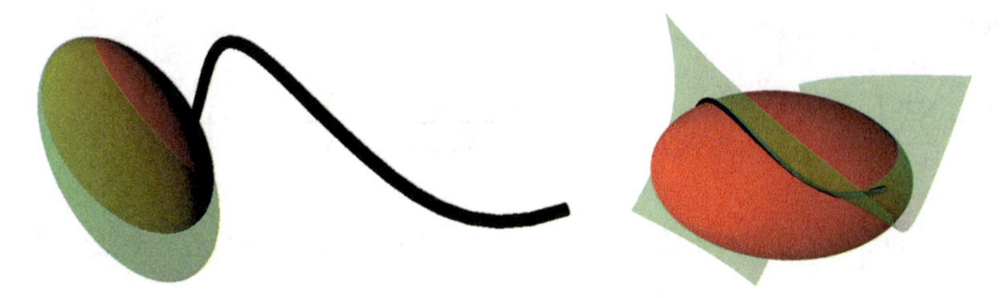

Fig. 2 Implicitization of a space curve represented by data points sampled from a parametric curve. *Left*: Initial surfaces, *right*: Final result

Fig. 3 Result with omitted orthogonality constraint (left) and omitted distance field constraint (right)

Fig. 4 Implicit description of a curve represented by perturbed data. *Left*: Initial surfaces, *right*: Final result

Alternatively, we omit the distance field constraint. The results can be seen in Fig. 3(b).

As one can verify, the two surfaces match still the data. However, one of the surfaces has a singularity. This is due to the fact that the averaged gradient constraint allows also vanishing gradients. For the distance field constraint this is not true, as the norm of the gradients in the data points is forced to be close to one, hence singular points are unlikely to appear.

Example 3. For this example we added a random error of maximal magnitude 0.05% of the diameter of the bounding box to the data points from the previous example. The fitted space curve is represented in Fig. 4.

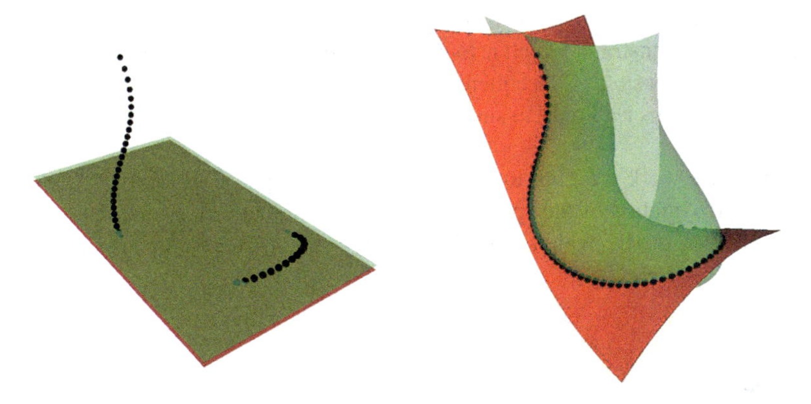

Fig. 5 Implicit representation of a curve described by exact point data. *Left*: Initial surfaces, *right*: Final result

Example 4. In a fourth example we consider a parametric curve of degree 8. The two surfaces were chosen to have degree 3. This example shall illustrate again the good convergence behavior, as the two initial surfaces are far away from the final result (Fig. 5).

5 Conclusion

In the first part of the paper we reviewed some of the existing techniques for approximate implicitization of hypersurfaces. Starting with Dokken's approach, which relies on the use of singular value decomposition, we observed that the weak version of Dokken's method can be seen as a special instance of a fitting method. Finally we described a general framework for evolution based fitting techniques.

The second part of the paper extended the existing evolution framework to the implicitization of space curves, by coupling the evolution of two implicitly defined surfaces. As the implicit representation of a curve or surface is not unique, additional regularization terms have to be added in order to achieve the uniqueness of the solution. We discussed two possibilities.

The first, called the distance field constraint, tries to achieve a unit gradient field along the intersecting surfaces. Hence a unique solution to the fitting problem is always guaranteed. Furthermore, it can even cope with an incorrectly chosen degree, that is when the degrees of the defining polynomials have been chosen too high. However, this approach prevents the evolution from finding the exact solution.

The second proposed regularization eliminates only the redundancy which is caused by the scalability of the underlying functions. As an advantage, it allows to find the exact solution, provided that the degrees of the implicitly defined surfaces are sufficiently high.

In order to obtain also a robust representation of the intersection curve we introduced another constraint which is to guarantee that the defining surfaces intersect as orthogonal as possible. Consequently, small perturbations of the coefficients of the defining functions lead only to small deviations of the intersection points of the two surfaces.

For future work we plan to use adaptive spline spaces to improve the quality of the approximation of the space curves. Furthermore a theoretical analysis of the approximation order, (which is until now only available for hypersurfaces) is under investigation.

References

1. Aigner, M., Jüttler, B.: Hybrid curve fitting. Computing **79**, 237–247 (2007)
2. Aigner, M., Jüttler, B.: Robust fitting of implicitly defined surfaces using Gauss–Newton–type techniques. Visual Comput. **25**, 731–741 (2009)
3. Aigner, M., Jüttler, B., Kim, M.-S.: Analyzing and enhancing the robustness of implicit representations. In Geometric Modelling and Processing, pp. 131–142. IEEE Press (2004)
4. Blake, A., Isard, M.: Active Contours: The Application of Techniques from Graphics, Vision, Control Theory and Statistics to Visual Tracking of Shapes in Motion. Springer, Secaucus (1998)
5. Cox, D., Goldman, R., Zhang, M.: On the validity of implicitization by moving quadrics for rational surfaces with no base points. J. Symb. Comput. **29**(3), 419–440 (2000)
6. Cox, D.A., Little, J., O'Shea, D.: Using Algebraic Geometry. Springer, Secaucus, NJ, USA (2005)
7. Cox, D.A., Little, J., O'Shea, D.: Ideals, Varieties, and Algorithms: An Introduction to Computational Algebraic Geometry and Commutative Algebra, 3/e (Undergraduate Texts in Mathematics). Springer, Secaucus, NJ, USA (2007)
8. Dixon, A.L.: The eliminant of three quantics in two independents variables. Proc. London Math. Soc. **6**, 49–69 (1908)
9. Dokken, T.: Aspects of Intersection Algorithms and Approximation. PhD thesis, University of Oslo (1997)
10. Dokken, T.: Approximate implicitization. In Mathematical Methods for Curves and Surfaces, pp. 81–102. Vanderbilt University Press, Nashville, TN (2001)
11. Dokken, T., Kellerman, H.K., Tegnander, C.: An approach to weak approximate implicitization. In *Mathematical Methods for Curves and Surfaces: Oslo 2000*, pp. 103–112. Vanderbilt University, Nashville, TN, USA (2001)
12. Dokken, T., Thomassen, J.: Overview of approximate implicitization. In *Topics in Algebraic Geometry and Geometric Modeling*, vol. 334, pp. 169–184. Amer. Math. Soc., Providence, RI (2003)
13. Feichtinger, R., Fuchs, M., Jüttler, B., Scherzer, O., Yang, H.: Dual evolution of planar parametric spline curves and T–spline level sets. Comput. Aided Des. **40**, 13–24 (2008)
14. Feichtinger, R., Jüttler, B., Yang, H.: Particle-based T-spline level set evolution for 3D object reconstruction with range and volume constraints. In: Cunningham, S., Skala, V. (eds.) Proc. WSCG, pp. 49–56. University of Plzen, Union Press (2008)
15. Hoschek, J., Lasser, D.: Fundamentals of Computer Aided Geometric Design. A K Peters Wellesley, MA (1993)
16. Jüttler, B., Felis, A.: Least–squares fitting of algebraic spline surfaces. Adv. Comput. Math. **17**, 135–152 (2002)

17. Jüttler, B., Wurm, E.: Approximate implicitization via curve fitting. In: Kobbelt, L., Schröder, P., Hoppe, H. (eds.) Symposium on Geometry Processing, pp. 240–247. New York, Eurographics/ACM Press (2003)
18. Osher, S., Fedkiw, R.: Level set methods and dynamic implicit surfaces, Applied Mathematical Sciences. vol. 153, Springer, New York (2003)
19. Pratt, V.: Direct least-squares fitting of algebraic surfaces. SIGGRAPH Comput. Graph. **21**(4), 145–152 (1987)
20. Salmon, G.: A treatise on the analytic geometry of three dimensions. Hodges, Figgis and Co., 4th ed. (1882)
21. Sederberg, T.W.: Implicit and parametric curves and surfaces for computer aided geometric design. PhD thesis, Purdue University, West Lafayette, IN, USA (1983)
22. Sederberg, T.W., Chen, F.: Implicitization using moving curves and surfaces. Comput. Graph. **29**(Annual Conference Series), 301–308 (1995)
23. Sederberg, T.W., Parry, S.R.: Comparison of three curve intersection algorithms. Comput. Aided Des. **18**(1), 58–64 (1986)
24. Taubin, G.: Estimation of planar curves, surfaces, and nonplanar space curves defined by implicit equations with applications to edge and range image segmentation. IEEE Trans. Pattern Anal. Mach. Intell. **13**(11), 1115–1138 (1991)
25. Wurm, E.: Approximate Techniques for the Implicitisation and Parameterisation of Surfaces. PhD thesis, Johannes Kepler University, Linz, Austria (2005)
26. Wurm, E., Thomassen, J.B., Jüttler, B., Dokken, T.: Comparative benchmarking of methods for approximate implicitization. In: Neamtu, M., Lucian, M. (eds.) Geometric Modeling and Computing: Seattle 2003, pp. 537–548. Nashboro Press, Brentwood (2004)
27. Yang, H., Fuchs, M., Jüttler, B., Scherzer, O.: Evolution of T-spline level sets with distance field constraints for geometry reconstruction and image segmentation. In Shape Modeling International, pp. 247–252. IEEE Press (2006)
28. Yang, H., Jüttler, B.: Evolution of T-spline level sets for meshing non–uniformly sampled and incomplete data. Visual Comput. **24**, 435–448 (2008)

Sparsity Optimized High Order Finite Element Functions on Simplices

Sven Beuchler, Veronika Pillwein, Joachim Schöberl, and Sabine Zaglmayr

Abstract This article reports several results on sparsity optimized basis functions for hp-FEM on triangular and tetrahedral finite element meshes obtained within the Special Research Program "Numerical and Symbolic Scientific Computing" and within the Doctoral Program "Computational Mathematics" both supported by the Austrian Science Fund FWF under the grants SFB F013 and DK W1214, respectively. We give an overview on the sparsity pattern for mass and stiffness matrix in the spaces L_2, H^1, $H(\mathrm{div})$ and $H(\mathrm{curl})$. The construction relies on a tensor-product based construction with properly weighted Jacobi polynomials.

1 Introduction

Finite element methods (FEM) are among the most powerful tools for the approximate solution of elliptic boundary value problems of the form: Find $u \in \mathbb{V}$ such that

S. Beuchler (✉)
Institute for Numerical Simulation, University of Bonn, Wegelerstr. 6, 53115 Bonn, Germany
e-mail: beuchler@ins.uni-bonn.de

V. Pillwein
Research Institute for Symbolic Computation, Johannes Kepler University Linz, 4040 Linz, Austria
e-mail: veronika.pillwein@risc.jku.at

J. Schöberl
Institute for Analysis and Scientific Computing, TU Wien, Wiedner Hauptstr. 8–10, 1040 Wien, Austria
e-mail: joachim.schoeberl@tuwien.ac.at

S. Zaglmayr
Computer Simulation Technology, 64289 Darmstadt, Germany
e-mail: sabine.zaglmayr@cst.com

U. Langer and P. Paule (eds.), *Numerical and Symbolic Scientific Computing*, Texts and Monographs in Symbolic Computation, DOI 10.1007/978-3-7091-0794-2_2,

$$a(u, v) = F(v) \quad \forall v \in \mathbb{V}, \tag{1}$$

where $\mathbb{V}$ is an infinite dimensional Sobolev space of functions on a bounded Lipschitz domain $\Omega \subset \mathbb{R}^d$, $d = 2, 3$, $a(\cdot, \cdot) : \mathbb{V} \times \mathbb{V} \mapsto \mathbb{R}$ is an elliptic and bounded bilinear form and $F(\cdot) : \mathbb{V} \mapsto \mathbb{R}$ is a bounded linear functional. Examples for the choice of $a(\cdot, \cdot)$ and $\mathbb{V}$ are:

1. The L_2 case, where $\mathbb{V} = L_2(\Omega)$ and $a(u, v) = \int_\Omega uv$
2. The H^1 case, where $\mathbb{V} = H^1(\Omega)$ and $a(u, v) = \int_\Omega \nabla u \cdot \nabla v + uv$
3. The $H(\mathrm{div})$ case, where $\mathbb{V} = H(\mathrm{div}, \Omega)$ and $a(u, v) = \int_\Omega \nabla \cdot u \, \nabla \cdot v + u \cdot v$
4. The $H(\mathrm{curl})$ case, where $\mathbb{V} = H(\mathrm{curl}, \Omega)$ and $a(u, v) = \int_\Omega \nabla \times u \cdot \nabla \times v + u \cdot v$

where the space $\mathbb{V}$ coincides with $\{v \in L_2(\Omega) : a(v, v) < \infty\}$. For a general overview of the involved spaces including their finite element approximation we refer to [48]. In all examples, the computation of an approximate solution u_N to u of (1) requires the solution of a linear system of algebraic equations

$$\mathscr{A} \underline{u} = \underline{f} \quad \text{with} \quad \mathscr{A} = \left[a(\psi_j, \psi_i)\right]_{i,j=1}^N \tag{2}$$

where $\psi = [\psi_1, \ldots, \psi_N]$ is a basis of a finite dimensional subspace $\mathbb{V}_N$ of $\mathbb{V}$, see e.g. [21, 26, 53].

In order to obtain a good approximation u_N to u for a fixed space dimension N of $\mathbb{V}_N$, finite elements with higher polynomial degrees p, e.g. the p and hp-version of the FEM, are preferred if the solution is piecewise smooth, see e.g. [8, 28, 31, 42, 56, 58] and the references therein. The fast solution of (2) with an iterative solution method like the preconditioned conjugate gradient method requires two main ingredients:

- A fast matrix vector multiplication $\mathscr{A} \underline{u}$,
- The choice of a good preconditioner in order to accelerate the iteration process.

Preconditioners based on domain decomposition methods (DD) for hp-FEM are extensively investigated in the literature, see e.g. [2, 5, 7, 12, 13, 18, 33, 38–41, 44, 46, 51] for the construction of DD-preconditioners and see [4, 10, 11, 27, 29, 30, 49] for extension operators which are required as one ingredient of the DD-preconditioners. The matrix vector multiplication becomes fast if $\mathscr{A}$ is a matrix that has as many non-zero entries as possible, i.e., it is a sparse matrix. Since the global stiffness matrix $\mathscr{A}$ in finite element methods is the result of assembling of local stiffness matrices, it is sufficient to consider the matrices on the element level.

In this survey, we will summarize the choice of sparsity optimized basis functions and the results for the above defined bilinear forms on triangular and tetrahedral finite elements. The results and their proofs have been presented in [14–16, 19], see also [3, 9, 32, 34, 54, 57] for the construction of scalar- and vector-valued high-order finite elements. For fast integration techniques we refer to [36, 42, 47].

For proving the sparsity pattern of the various system matrices we use a symbolic rewriting procedure to evaluate the integrals that determine the matrix entries explicitly. For this rewriting procedure several identities relating several orthogonal polynomials are necessary. Over the past decades algorithms for proving and finding such identities have been developed such as Zeilberger's algorithm [61, 63–65] or Chyzak's approach [23–25]. For a general overview on this type of algorithms see, e.g., [52].

The outline of this overview is as follows. Section 2 comprises several results about Jacobi and integrated Jacobi polynomials which are crucial for the sparsity of the system matrices. Some general basics for the definition of tensor product based shape functions on simplicial finite elements are presented in Sect. 3. The Sects. 4-7 include a summary of the definition of the basis functions and the sparsity results for mass and main term in L_2, H^1, $H(\nabla\cdot)$, and $H(\text{curl})$, respectively. Section 8 gives a brief overview on the algorithm applied for symbolic computation of the matrix entries.

2 Properties of Jacobi Polynomials with Weight $(1-x)^\alpha$

Sparsity optimization of high-order basis functions on simplices relies on using Jacobi-type polynomials and their basic properties which will be introduced in this section.

For $n \geq 0, \alpha, \beta > -1$ and $x \in [-1, 1]$ let

$$P_n^{(\alpha,\beta)}(x) = \frac{(-1)^n}{2^n n!(1-x)^\alpha (1+x)^\beta} \frac{\mathrm{d}^n}{\mathrm{d}x^n} \left((1-x)^{n+\alpha}(1+x)^{n+\beta}\right) \tag{3}$$

be the nth Jacobi polynomial with respect to the weight function $(1-x)^\alpha(1+x)^\beta$. The function $P_n^{(\alpha,\beta)}(x)$ is a polynomial of degree n, i.e. $P_n^{(\alpha,\beta)}(x) \in \mathbb{P}_n((-1,1))$, where $\mathbb{P}_n(I)$ is the space of all polynomials of degree n on the interval I. In the special case $\alpha = \beta = 0$, the functions $P_n^{(0,0)}(x)$ are called Legendre polynomials. Mainly, we will use Jacobi polynomials with $\beta = 0$. For sake of simple notation we therefore omit the second index in (3) and write $p_n^\alpha(x) := P_n^{(\alpha,0)}(x)$.

These polynomials are orthogonal with respect to the weight $(1-x)^\alpha$, i.e. there holds

$$\int_{-1}^{1} (1-x)^\alpha p_j^\alpha(x) p_l^\alpha(x)\,\mathrm{d}x = \rho_j^\alpha \delta_{jl}, \quad \text{where} \quad \rho_j^\alpha = \frac{2^{\alpha+1}}{2j+\alpha+1}. \tag{4}$$

This relation will be heavily used in computing the entries of the different mass and stiffness matrices. Moreover for $n \geq 1$, let

$$\hat{p}_n^\alpha(x) = \int_{-1}^{x} p_{n-1}^\alpha(y)\,\mathrm{d}y, \quad \text{with} \quad \hat{p}_0^\alpha(x) = 1, \tag{5}$$

be the nth integrated Jacobi polynomial. Obviously, $\hat{p}_n^\alpha(-1) = 0$ for $n \geq 1$. Integrated Legendre polynomials, by the orthogonality relation (4), vanish at both endpoints of the interval. Summarizing, one obtains

$$\hat{p}_n^\alpha(-1) = 0, \quad \hat{p}_n^0(1) = 0 \quad \text{for } n \geq 2. \tag{6}$$

Factoring out these roots, integrated Jacobi polynomials (5) can be expressed in terms of Jacobi polynomials (3) with modified weights, i.e.,

$$\hat{p}_n^\alpha(x) = \frac{1+x}{n} P_{n-1}^{(\alpha-1,1)}(x), \qquad n \geq 1, \tag{7}$$

$$\hat{p}_n^0(x) = \frac{1-x^2}{2n-2} P_{n-2}^{(1,1)}(x), \qquad n \geq 2. \tag{8}$$

There are several further identities relating Jacobi polynomials $p_n^\alpha(x)$ and integrated Jacobi polynomials (5) that have been proven in [19], [14] and [15]. These include three term recurrences for fast evaluation as well as identities necessary for proving the sparsity pattern of the mass and stiffness matrices below. We give a summary of all necessary identities in Sect. 8. For more details on Jacobi polynomials we refer the interested reader to the books of Abramowitz and Stegun [1], Szegö [59], and Tricomi [60].

3 Preliminary Definitions

We assume a conforming affine simplicial mesh. Although the basis functions are defined on arbitrary simplices, the analysis of the basis functions can be performed only on the reference elements $\hat{T}$ as defined in Fig. 1. The sparsity result on affine meshes then follows by the mapping principle. An arbitrary simplex can be mapped by an affine transformation to these reference elements. We mention that affine transformations guarantee that polynomials are mapped to polynomials of the same degree. The basis functions will be defined by means of barycentric coordinates λ_i that are functions depending on x, y (and z). For our reference triangle they are given as

$$\lambda_1(x,y) = \frac{1-2x-y}{4}, \quad \lambda_2(x,y) = \frac{1+2x-y}{4}, \quad \text{and} \quad \lambda_3(x,y) = \frac{1+y}{2},$$

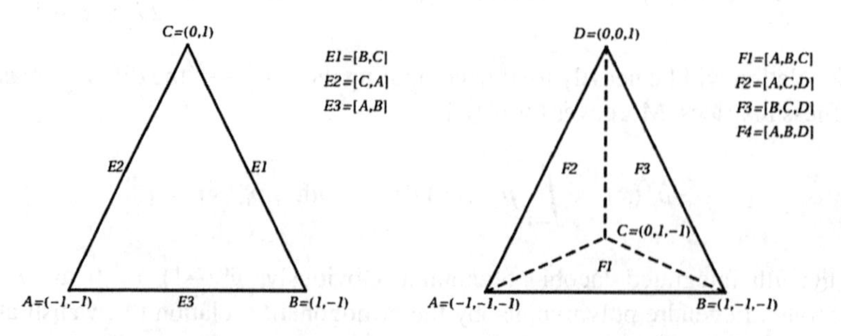

Fig. 1 Notation of the vertices and edges/faces on the reference element $\hat{T}$ for 2d and 3d

and for the reference tetrahedron they are defined as

$$\lambda_{1/2}(x,y,z) = \frac{1 \mp 4x - 2y - z}{8}, \quad \lambda_3(x,y,z) = \frac{1+2y-z}{4},$$

$$\text{and} \quad \lambda_4(x,y,z) = \frac{1+z}{2}.$$

We mention that the barycentric coordinates add up to 1.

By viewing the triangle (tetrahedron) as a collapsed quadrilateral (hexahedron) as suggested by Dubiner [34] and Karniadakis, Sherwin [42], we can construct a tensorial-type basis also for simplices. For this purpose, we need the Duffy transformation that maps the tensorial element to the simplicial element.

In two dimensions the Duffy transformation $\mathscr{D}$ mapping the unit square to the reference triangle is defined as

$$\begin{array}{rl} \mathscr{D}: \hat{Q} = [-1,1]^2 \to & \hat{T} \\ (\xi,\eta) & \to (x,y) \end{array} \quad \text{with} \quad \begin{array}{l} x = \frac{\xi}{2}(1-\eta), \\ y = \eta. \end{array} \tag{9}$$

Using the inverse of the Duffy transformation, we can parameterize the triangle $\hat{\Delta}$ by

$$\xi = \frac{2x}{1-y} = \frac{\lambda_2(x,y) - \lambda_1(x,y)}{\lambda_2(x,y) + \lambda_1(x,y)}, \quad \text{and} \quad \eta = y = 2\lambda_3(x,y) - 1.$$

Besides the Duffy transformation, polynomial basis functions which vanish on some or all edges of the triangle are required. Therefore, we introduce several auxiliary bubble functions, which are important for the definition of our basis functions. More precisely, the authors introduce the edge based function

$$g_i^E(x,y) := \hat{p}_i^0\left(\tfrac{\lambda_{e_2}-\lambda_{e_1}}{\lambda_{e_1}+\lambda_{e_2}}\right)(\lambda_{e_1}+\lambda_{e_2})^i \tag{10}$$

on the edge $E = [e_1, e_2]$, running from vertex V_{e_1} to V_{e_2} and the bubbles

$$g_i(x,y) := \hat{p}_i^0\left(\tfrac{\lambda_2-\lambda_1}{\lambda_1+\lambda_2}\right)(\lambda_1+\lambda_2)^i \quad \text{and} \quad h_{ij}(x,y) := \hat{p}_j^{2i-1}(2\lambda_3 - 1), \tag{11}$$

where the barycentric coordinates depend on x and y. Note that the functions in (10) and (11) are polynomial functions of degrees i, i and j, respectively. Using (6), one observes that the functions g_i^E as defined in (10) vanish at the endpoints of the edge E. In the same way, the functions $g_i(x,y)$ vanish at the edges $E2 = [1,3]$ and $E3 = [2,3]$, whereas h_{ij} vanishes at the edge $E1 = [1,2]$.

In three dimensions the Duffy transformation mapping the unit cube to the reference tetrahedron is defined as

$$\mathscr{D}: \hat{Q} = [-1,1]^3 \to \hat{T} \quad (\xi,\eta,\zeta) \to (x,y,z) \quad \text{with} \quad \begin{aligned} x &= \tfrac{\xi}{4}(1-\eta)(1-\zeta), \\ y &= \tfrac{\eta}{2}(1-\zeta), \\ z &= \zeta. \end{aligned}$$

Using the inverse of the Duffy transformation we can parameterize the triangle $\hat{\Delta}$ by

$$\begin{aligned} \xi &= \frac{4x}{1-2y-z} = \frac{\lambda_2(x,y,z)-\lambda_1(x,y,z)}{\lambda_2(x,y,z)+\lambda_1(x,y,z)}, \\ \eta &= \frac{2y}{1-z} = \frac{\lambda_3(x,y,z)-\lambda_2(x,y,z)-\lambda_1(x,y,z)}{\lambda_3(x,y,z)+\lambda_2(x,y,z)+\lambda_1(x,y,z)}, \\ \zeta &= z = 2\lambda_4(x,y,z)-1. \end{aligned}$$

Here, the edge-based functions

$$u_i^E(x,y,z) := \hat{p}_i^0\left(\tfrac{\lambda_{e_2}-\lambda_{e_1}}{\lambda_{e_1}+\lambda_{e_2}}\right)(\lambda_{e_1}+\lambda_{e_2})^i \tag{12}$$

are introduced on the edge $E = [e_1, e_2]$, running from vertex V_{e_1} to V_{e_2}. The face based functions

$$u_i^F := \hat{p}_i^0\left(\frac{\lambda_{f_2}-\lambda_{f_1}}{\lambda_{f_2}+\lambda_{f_1}}\right)(\lambda_{f_2}+\lambda_{f_1})^i, \quad v_{ij}^F := \hat{p}_j^{2i-1}(\lambda_{f_3}-\lambda_{f_2}-\lambda_{f_1}) \tag{13}$$

are defined on the face $F = [f_1, f_2, f_3]$ characterized by the vertices V_{f_1}, V_{f_2} and V_{f_3}. The functions

$$\begin{aligned} u_i(x,y,z) &:= \hat{p}_i^0\left(\frac{\lambda_2-\lambda_1}{\lambda_2+\lambda_1}\right)(\lambda_2+\lambda_1)^i, \\ v_{ij}(x,y,z) &:= \hat{p}_j^{2i-1}\left(\frac{2\lambda_3-(1-\lambda_4)}{1-\lambda_4}\right)(1-\lambda_4)^j, \\ \text{and} \quad w_{ijk}(x,y,z) &:= \hat{p}_k^{2i+2j-2}(2\lambda_4-1) \end{aligned} \tag{14}$$

will be central in the definition of the interior bubble functions. Again, the barycentric coordinates depend on x, y and z. For vector valued problems, the lowest-order Nédélec function [50] corresponding to the edge $E = [e_1, e_2]$ and the lowest order Raviart-Thomas function, [20, 50], corresponding to $F = [f_1, f_2, f_3]$, characterized by the vertices V_{f_1}, V_{f_2} and V_{f_3} are defined by

$$\varphi_{1,E} := \nabla\lambda_{e_1}\lambda_{e_2} - \lambda_{e_1}\nabla\lambda_{e_2} \quad \text{and} \tag{15}$$

$$\psi_0^F = \psi_0^{[f_1,f_2,f_3]} := \lambda_{f_1}\nabla\lambda_{f_2}\times\nabla\lambda_{f_3} + \lambda_{f_2}\nabla\lambda_{f_3}\times\nabla\lambda_{f_1} + \lambda_{f_3}\nabla\lambda_{f_1}\times\nabla\lambda_{f_2}, \tag{16}$$

respectively.

The functions (10)–(14) and the choice of the weights for the Jacobi polynomials are pivotal for obtaining the sparsity results in mass and stiffness matrices.

4 The L_2 Orthogonal Basis Functions of Dubiner

These basis functions have been introduced by [34], see also [42]. Another possible construction principle is based on Appell polynomials, [6, 22, 35].

Let $\triangle_s$ be a triangle with its baryzentrical coordinates $\lambda_m(x, y)$, $m = 1, 2, 3$. Instead of (11), we introduce the auxiliary functions

$$\tilde{g}_i(x,y) := p_i^0\left(\frac{\lambda_2 - \lambda_1}{\lambda_1 + \lambda_2}\right)(\lambda_1 + \lambda_2)^i \quad \text{and} \quad \tilde{h}_{ij}(x,y) := p_j^{2i+1}(2\lambda_3 - 1),$$

and define the L_2 orthogonal functions

$$\psi_{ij}(x,y) = \tilde{g}_i(x,y)\tilde{h}_{ij}(x,y), \quad 0 \le i, j, \; i + j \le p.$$

We prove this orthogonality for the reference triangle given in Fig. 1. The computations are straight forward: after using the Duffy transformation the integrals can be evaluated by a mere application of the orthogonality relation (4) for Jacobi polynomials:

$$\begin{aligned}
&\int_{\hat{T}} p_i^0\left(\frac{2x}{1-y}\right) p_k^0\left(\frac{2x}{1-y}\right)\left(\frac{1-y}{2}\right)^{i+k} p_j^{2i+1}(y) p_l^{2k+1}(y)\,\mathrm{d}(x,y)\\
&= \int_{-1}^{1} p_i^0(\xi) p_k^0(\xi)\,\mathrm{d}\xi \int_{-1}^{1}\left(\frac{1-\eta}{2}\right)^{i+k+1} p_j^{2i+1}(\eta) p_l^{2k+1}(\eta)\,\mathrm{d}\eta\\
&= \frac{2}{2i+1}\delta_{ik}\int_{-1}^{1}\left(\frac{1-\eta}{2}\right)^{2i+1} p_j^{2i+1}(\eta) p_l^{2i+1}(\eta)\,\mathrm{d}\eta\\
&= \frac{2\delta_{ik}\delta_{jl}}{(2i+1)(i+j+1)}.
\end{aligned}$$

Now, let $\triangle_s$ be a tetrahedron with its baryzentrical coordinates $\lambda_m(x, y)$, $m = 1, 2, 3, 4$. With the auxiliary functions

$$\begin{aligned}
\tilde{u}_i(x,y,z) &:= p_i^0\left(\frac{\lambda_2 - \lambda_1}{\lambda_2 + \lambda_1}\right)(\lambda_2 + \lambda_1)^i,\\
\tilde{v}_{ij}(x,y,z) &:= p_j^{2i+1}\left(\frac{\lambda_3 - \lambda_2 - \lambda_1}{\lambda_3 + \lambda_2 + \lambda_1}\right)(\lambda_3 + \lambda_2 + \lambda_1)^j,\\
\text{and} \quad \tilde{w}_{ijk}(x,y,z) &:= p_k^{2i+2j+2}(\lambda_4 - \lambda_1 - \lambda_2 - \lambda_3)
\end{aligned}$$

the basis functions read as

$$\psi_{ijk}(x,y,z) := \tilde{u}_i(x,y,z)\tilde{v}_{ij}(x,y,z)\tilde{w}_{ijk}(x,y,z), \quad i+j+k \le p, i,j,k \ge 0.$$

The evaluation of the L_2-inner product is completely analogous to the triangular case. For the reference tetrahedron as defined in Fig. 1, the final result is

$$\int_{\hat{T}} \psi_{ijk}(x,y,z)\psi_{lmn}(x,y,z)\,\mathrm{d}(x,y,z) = \frac{4\delta_{il}\delta_{jm}\delta_{kn}}{(2i+1)(i+j+1)(2i+2j+2k+3)}.$$

Also the sparsity results for the basis functions for H^1, $H(\mathrm{div})$ and $H(\mathrm{curl})$ are proved by evaluation that proceeds by rewriting until the orthogonality relation (4) for Jacobi polynomials can be exploited. These computations, however, become much more evolved as indicated in the sections below and ultimately this task is handed over to an algorithm, see Sect. 8.

5 Sparsity Optimized H^1-Conforming Basis Functions

The construction of the basis functions in this section follows [14, 15, 19]. Throughout we assume a uniform polynomial degree p.

In order to obtain H^1-conforming functions, the global basis functions have to be globally continuous. In 2D, the functions are split into three different groups, the vertex based functions, the edge bubble functions and the interior bubbles. In order to guarantee a simple continuous extension to the neighboring element, the interior bubbles are defined to vanish at all element edges, the edge bubbles vanish on two of the three edges whereas the vertex functions are chosen as the usual hat functions. In 3D, there additionally exist face bubble functions.

5.1 Sparse H^1-Conforming Basis Functions on the Triangle

Using the integrated Jacobi polynomials (5), we define the shape functions on the affine triangle Δ_s with baryzentrical coordinates $\lambda_m(x,y)$, $m = 1,2,3$.

- The vertex functions are chosen as the usual linear hat functions

$$\psi_{V,m}(x,y) := \lambda_m(x,y), \quad m = 1,2,3.$$

  Let $\Psi_V^2 := [\psi_{V,1}, \psi_{V,2}, \psi_{V,3}]$ be the basis of the vertex functions.
- For each edge $E = [e_1, e_2]$, running from vertex V_{e_1} to V_{e_2}, we define

$$\psi_{[e_1,e_2],i}(x,y) = g_i^E(x,y)$$

with the integrated Legendre type functions (10). By $\Psi_{[e_1,e_2]} = \left[\psi_{[e_1,e_2],i}\right]_{i=2}^{p}$, we denote the basis of the edge bubble functions on the edge $[e_1, e_2]$. $\Psi_E^2 = \left[\Psi_{[1,2]}, \Psi_{[2,3]}, \Psi_{[3,1]}\right]$ is the basis of all edge bubble functions.

- The interior bubbles are defined as

$$\psi_{ij}(x,y) := g_i(x,y)h_{ij}(x,y), \quad i+j \le p, i \ge 2, j \ge 1, \tag{17}$$

where the auxiliary bubble functions g_i and h_{ij} are given in (11). Moreover, $\Psi_I^2 = \left[\psi_{ij}\right]_{i \ge 2, j \ge 1}^{i+j \le p}$ denotes the basis of all interior bubbles.

Finally, let $\Psi_{\nabla,2} = \left[\Psi_V^2, \Psi_E^2, \Psi_I^2\right]$ be the set of all shape functions on $\triangle_s$.

The interior block of the mass and stiffness matrix on the triangle $\triangle_s$ are denoted by

$$M_{II,s,\nabla_2} = \int_{\triangle_s} [\Psi_I^2]^\top [\Psi_I^2] := \left[\mu_{ij;kl}^{s,2}\right]_{i,k=2;j,l=1}^{i+j \le p; k+l \le p}, \quad \text{and} \tag{18}$$

$$K_{II,s,\nabla_2} = \int_{\triangle_s} [\nabla \Psi_I^2]^\top \cdot [\nabla \Psi_I^2] := \left[a_{ij;kl}^{s,2}\right]_{i,k=2;j,l=1}^{i+j \le p; k+l \le p}, \tag{19}$$

respectively.

Theorem 5.1. *Let M_{II,s,∇_2} be defined via* (18)*, then the matrix has $\mathcal{O}(p^2)$ nonzero matrix entries. More precisely, $\mu_{ij;kl}^{s,2} = 0$ if $|i-k| \notin \{0,2\}$ or $|i-k+j-l| > 4$.*

Let K_{II,s,∇_2} be defined via (19)*, then the matrix has $\mathcal{O}(p^2)$ nonzero matrix entries. More precisely, $a_{ij;kl}^{s,2} = 0$ if $|i-k| > 2$ or $|i-k+j-l| > 2$.*

Proof. This sparsity result is proven by explicit evaluation of the matrix entries using the algorithm described in Sect. 8, see also [14, 15]. However, we will give the interested reader a short impression of the proofs. After the affine linear mapping of the element $\triangle_s$ to the reference element $\hat{\triangle}$ it suffices to prove the results there. We start with sketching the result for the mass matrix.

On the reference element $\hat{T}$, we have

$$\hat{\mu}_{ij;kl}^{(2)} = \int_{\hat{T}} \hat{p}_i^0\left(\frac{2x}{1-y}\right)\left(\frac{1-y}{2}\right)^i \hat{p}_j^{2i-1}(y)\hat{p}_k^0\left(\frac{2x}{1-y}\right)\left(\frac{1-y}{2}\right)^k \times \hat{p}_l^{2k-1}(y)\, \mathrm{d}(x,y)$$

by (11) and (17). With the substitution $\xi = \frac{2x}{1-y}$ and $\eta = y$, cf. (9), the integral simplifies to

$$\hat{\mu}_{ij;kl}^{(2)} = \int_{-1}^{1} \hat{p}_i^0(\xi)\hat{p}_k^0(\xi)\, \mathrm{d}\xi \int_{-1}^{1} \left(\frac{1-\eta}{2}\right)^{i+k+1} \hat{p}_j^{2i-1}(\eta)\hat{p}_l^{2k-1}(\eta)\, \mathrm{d}\eta.$$

Using (35) for $\alpha = 0$, the integrated Legendre polynomials can be expressed as the sum of two Legendre polynomials. The orthogonality relation (4) implies that the first integral is zero if $|i-k| \notin \{0,2\}$.

For $i = k$, we obtain

$$\hat{\mu}^{(2)}_{ij;il} = c_i \int_{-1}^{1} \left(\frac{1-\eta}{2}\right)^{2i+1} \hat{p}_j^{2i-1}(\eta)\hat{p}_l^{2i-1}(\eta)\, d\eta$$

with some constants c_i. Now, relation (36) is applied for $\hat{p}_j^{2i-1}(\eta)$ and $\hat{p}_l^{2i-1}(\eta)$. This gives

$$\hat{\mu}^{(2)}_{ij;il} = c_{i,j,l} \int_{-1}^{1} \left(\frac{1-\eta}{2}\right)^{2i+1} (p_j^{2i-1}(\eta) + p_{j-1}^{2i-1}(\eta))(p_l^{2i-1}(\eta) + p_{l-1}^{2i-1}(\eta))\, d\eta.$$

By the orthogonality relation (4), the term $(p_j^{2i-1}(\eta) + p_{j-1}^{2i-1}(\eta))$ is orthogonal to all polynomials of maximal degree $j-2$ with respect to the weight $\left(\frac{1-\eta}{2}\right)^{2i-1}$, e.g., is orthogonal to $\left(\frac{1-\eta}{2}\right)^{2} (p_l^{2i-1}(\eta) + p_{l-1}^{2i-1}(\eta)) \in \mathbb{P}_l$. Therefore, $\hat{\mu}^{(2)}_{ij;il} = 0$ for $j - l > 4$. By symmetry, we obtain $\hat{\mu}^{(2)}_{ij;il} = 0$ for $|j-l| > 4$. For $k = i-2$, one obtains

$$\hat{\mu}^{(2)}_{ij;i-2l} = c_i \int_{-1}^{1} \left(\frac{1-\eta}{2}\right)^{2i-1} \hat{p}_j^{2i-1}(\eta)\hat{p}_l^{2i-5}(\eta)\, d\eta.$$

Again, by (36) and (4), the result $\hat{\mu}_{ij;i-2l} = 0$ for $|j+2-l| > 4$ follows.

For the stiffness matrix, the proof is similar. Starting point is the computation of the gradient on the reference element, which is given by

$$\nabla\psi_{ij} = \begin{bmatrix} p_{i-1}^0\left(\frac{2x}{1-y}\right)\left(\frac{1-y}{2}\right)^{i-1} \hat{p}_j^{2i-1}(y) \\ \frac{1}{2}p_{i-2}^0\left(\frac{2x}{1-y}\right)\left(\frac{1-y}{2}\right)^{i-1} \hat{p}_j^{2i-1}(y) + \hat{p}_i^0\left(\frac{2x}{1-y}\right)\left(\frac{1-y}{2}\right)^{i} p_{j-1}^{2i-1}(y) \end{bmatrix}.$$

With this closed form representation at hand the computations follow the same pattern as outlined for the mass matrix. □

Remark 5.2. The family of basis functions defined by the auxiliary functions

$$g_i(x,y) := \hat{p}_i^0\left(\frac{\lambda_2 - \lambda_1}{\lambda_1 + \lambda_2}\right)(\lambda_1 + \lambda_2)^i \quad \text{and} \quad h_{ij}(x,y) := \hat{p}_j^{2i-a}(2\lambda_3 - 1), \quad (20)$$

for $0 \le a \le 4$ have been considered in [15]. For $a = 1$, the functions coincide with the functions given in (11). The sparsity optimal basis for H^1 for both mass and stiffness matrix is given by the choice $a = 0$ which also yields the best condition numbers for the system matrix.

The nonzero pattern obtained by Theorem 5.1 is displayed in Fig. 2 for the interior basis functions (17) obtained by (20) with $a = 0$ and $a = 1$. The best sparsity results are obtained for $a = 0$ with a maximum of nine nonzero entries per row for the element stiffness matrix on the reference element $\hat{\Delta}$. Because of this

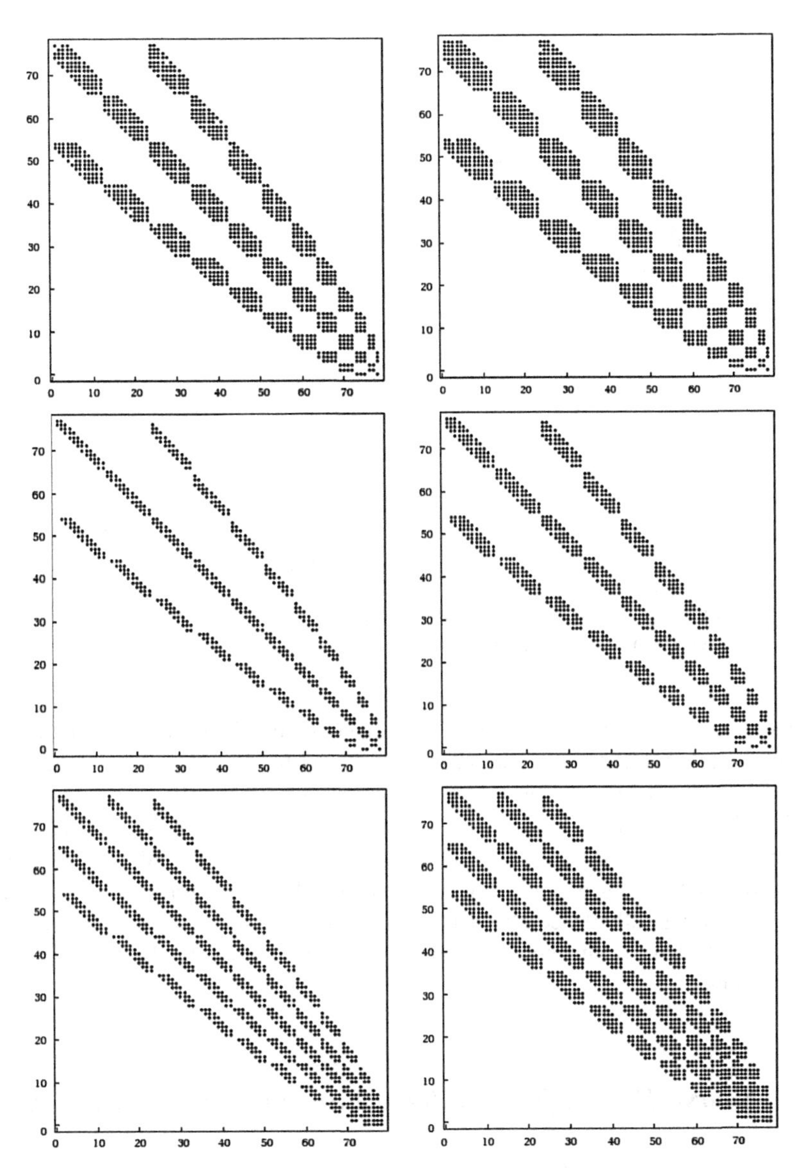

Fig. 2 Nonzero pattern for $p = 14$: mass matrix M_{II,s,∇_2} (*above*), stiffness matrix $\hat{K}_{II,\nabla_2}$ on $\hat{T}$ (*middle*), stiffness matrix K_{II,s,∇_2} on general element (*below*) for the interior bubbles based on the functions (20) with $a = 0$ (*left*) and $a = 1$ (*right*)

change of the weights in (20), the bandwidths of the nonzero blocks become larger for $a = 1$.

This nonzero pattern has a stencil like structure which makes it simpler to solve systems with linear combinations of M_{II,s,∇_2} and K_{II,s,∇_2} using sparse direct solvers

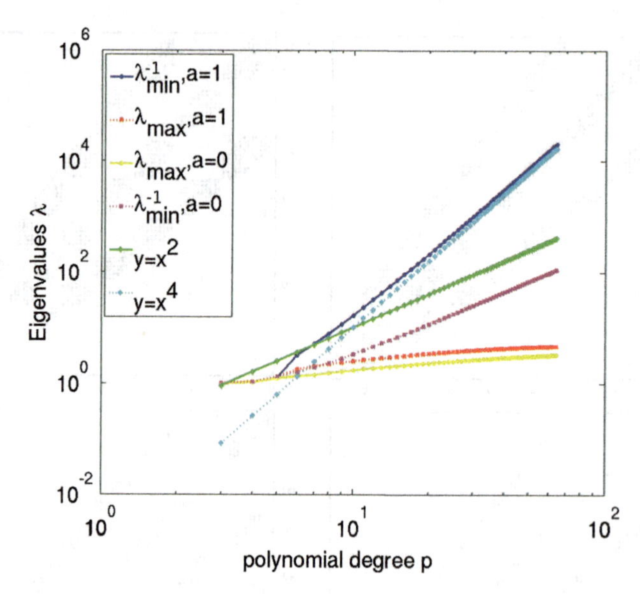

Fig. 3 Maximal and reciprocals of the minimal eigenvalues for the stiffness matrix $\hat{K}_{II,\nabla_2}$ (19) on the reference element $\hat{T}$ for the basis functions based on (20) with $a = 0$ and $a = 1$

as the method of nested dissection, [37], embedded in a DD-preconditioner. This is an important tool if static condensation is used in order to solve the system (2). We refer the interested reader for a more detailed discussion to [19].

Besides the sparsity, also the condition numbers of the local matrices are important. Figure 3 displays the diagonally preconditioned condition numbers of the stiffness matrix $\hat{K}_{II,\nabla_2}$ (19) on the reference element $\hat{T}$ for several polynomial degrees. Numerically the condition number grows at least as $\mathcal{O}(p^2)$ for the functions with $a=0$. This is the best possible choice for interior bubbles in two space dimensions.

5.2 *Sparse H^1-Conforming Basis Functions on the Tetrahedron*

The construction principle follows [14].

- The vertex functions are defined as the usual hat functions, i.e.

$$\psi_{V,m}(x,y,z) = \lambda_m(x,y,z), \quad m = 1,2,3,4.$$

  Let $\Psi_V^3 = [\psi_{V,m}]_{m=1}^4$ denote the basis of the hat functions.
- With (12), the edge bubbles are defined as

$$\psi_i^{[e_1,e_2]}(x,y) := u_i^E(x,y), \quad \text{for } 2 \le i \le p$$

for an edge $E = [e_1, e_2]$, running from vertex V_{e_1} to V_{e_2}. We denote the basis of all edge bubble functions by

$$\Psi_E^3 = \Big[\big[\psi_i^{[1,2]}\big]_{i=2}^p, \big[\psi_i^{[2,3]}\big]_{i=2}^p, \big[\psi_i^{[3,1]}\big]_{i=2}^p, \big[\psi_i^{[1,4]}\big]_{i=2}^p, \big[\psi_i^{[2,4]}\big]_{i=2}^p, \big[\psi_i^{[3,4]}\big]_{i=2}^p\Big].$$

- For each face $F = [f_1, f_2, f_3]$, characterized by the vertices V_{f_1}, V_{f_2} and V_{f_3}, the face bubbles are defined as

$$\psi_{j,k}^f(x,y,z) := u_i^F(x,y,z)\, v_{ij}^F(x,y,z), \quad i \ge 2, j \ge 1, i+j \le p$$

using the functions (13). We denote the basis of all face bubble functions by

$$\Psi_F^3 := \Big[\big[\psi_{i,j}^{[1,2,3]}\big]_{i=2,j=1}^{i+j=p} \big[\psi_{i,j}^{[2,3,4]}\big]_{i=2,j=1}^{i+j=p}, \big[\psi_{i,j}^{[3,4,1]}\big]_{i=2,j=1}^{i+j=p}, \big[\psi_{i,j}^{[4,1,2]}\big]_{i=2,j=1}^{i+j=p}\Big].$$

- With the functions (14), the interior bubbles read as

$$\psi_{ijk}(x,y,z) := u_i(x,y,z) v_{ij}(x,y,z) w_{ijk}(x,y,z), \qquad i+j+k \le p, i \ge 2, j,k \ge 1.$$

Moreover, $\Psi_I^3 = \big[\psi_{ijk}\big]_{i\ge 2, j\ge 1, k\ge 1}^{i+j+k\le p}$ denotes the basis of the interior bubbles.

Let $\Psi_{\nabla,3} = \big[\Psi_V^3, \Psi_E^3, \Psi_F^3, \Psi_I^3\big]$ be the basis of all shape functions.

The interior block of the mass and stiffness matrix on the triangle $\triangle_s$ are denoted by

$$M_{II,s,\nabla_3} = \int_{\triangle_s} [\Psi_I^3]^\top [\Psi_I^3] := \Big[\mu_{ijk;lmn}^{s,3}\Big]_{i,l=2;j,m,l,n=1}^{i+j+k\le p; l+m+n\le p} \quad \text{and} \tag{21}$$

$$K_{II,s,\nabla_3} = \int_{\triangle_s} [\nabla\Psi_I^3]^\top \cdot [\nabla\Psi_I^3] := \Big[a_{ijk;lmn}^{s,3}\Big]_{i,l=2;j,k,m,n=1}^{i+j+k\le p; l+m+n\le p}, \tag{22}$$

respectively.

Theorem 5.3. *The inner block of the mass matrix M_{II,s,∇_3} has in total $\mathcal{O}(p^3)$ nonzero matrix entries. More precisely, $\mu_{ijk;mln} = 0$ if $|i-l| > 2$, $|i-l+j-m| > 4$ or $|i-l+j-m+k-n| > 6$.*

The inner block of the stiffness matrix K_{II,s,∇_3} has in total $\mathcal{O}(p^3)$ nonzero matrix entries. More precisely, $\mu_{ijk;mln} = 0$ if $|i-l| > 2$, $|i-l+j-m| > 3$ or $|i-l+j-m+k-n| > 4$.

Proof. Evaluation of the matrix entries using the algorithm described in Sect. 8, see also [14, 15]. □

Remark 5.4. In [15], the auxiliary functions are defined in the more general form

$$
\begin{aligned}
u_i(x,y,z) &:= \hat{p}_i^0\left(\frac{\lambda_2-\lambda_1}{\lambda_2+\lambda_1}\right)(\lambda_2+\lambda_1)^i,\\
v_{ij}(x,y,z) &:= \hat{p}_j^{2i-a}\left(\frac{2\lambda_3-(1-\lambda_4)}{1-\lambda_4}\right)(1-\lambda_4)^j,
\end{aligned}
\tag{23}
$$

and $w_{ijk}(x,y,z) := \hat{p}_k^{2i+2j-b}(2\lambda_4-1)$,
where the integers a and b satisfy $0 \le a \le 4$, $a \le b \le 6$. The interior bubbles coincide with the functions given in [57], see also [42], if $a = b = 0$. To make this equivalence obvious use the identities (7) and (8). This choice corresponds to the sparsity optimal case for H^1 for both mass and stiffness matrix. In this case the results of Theorem 5.3 reduce to $|i-l| > 2$, $|i-l+j-m| > 3$ or $|i-l+j-m+k-n| > 4$ for the mass matrix and $|i-l| > 2$, $|i-l+j-m| > 3$ or $|i-l+j-m+k-n| > 2$ for the stiffness matrix. The auxiliary polynomials used in this paper correspond to setting $a = 1$ and $b = 2$.

Again a stencil like structure for mass and stiffness matrix is obtained. However, the elimination of the interior bubbles by static condensation with nested dissection is much more expensive in the 3D case than in the 2D case. The computational complexity is now $\mathcal{O}(p^6)$ flops in comparison to $\mathcal{O}(p^3)$ flops in the two-dimensional case.

Besides the sparsity, also the condition numbers of the local matrices are important. Figure 4 displays the condition numbers of the stiffness matrix $\hat{K}_{II,\nabla_3}$

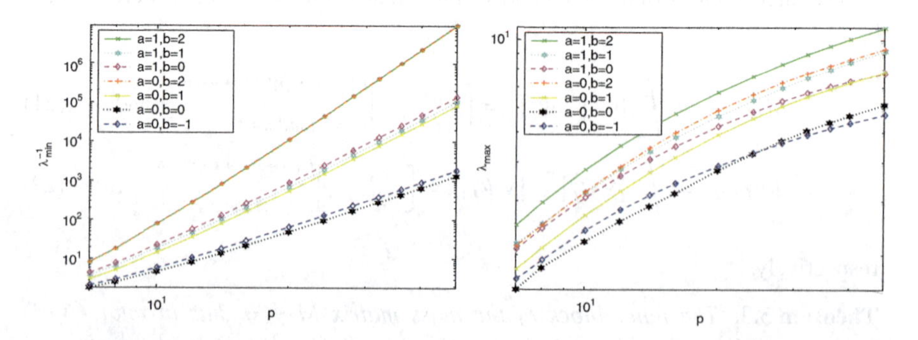

Fig. 4 Maximal (*right*) and minimal (*left*) eigenvalues for the diagonally preconditioned stiffness matrix $\hat{K}_{II,\nabla_3}$ (22) on the reference element $\hat{T}$ for different values of a and b in (23)

(22) on the reference element $\hat{T}$ for several polynomial degrees and several choices of auxiliary functions (14) and (23). Numerically, the condition number grows as least with $\mathcal{O}(p^4)$.

6 Sparsity Optimization of $H(\mathrm{div})$-Conforming Basis Functions

The following construction of $H(\mathrm{div})$-conforming finite elements applies the ideas on sparsity optimization on simplices of [14, 15, 19] to the general construction principles of $H(\mathrm{div})$-conforming high-order fe bases developed in [62] and [55]. A detailed description of both the two and three dimensional case can be found in [16]. In the sequel, we only report the results for tetrahedra.

Let Δ_s denote an arbitrary non-degenerated simplex $\Delta_s \subset \mathbb{R}^3$, its set of four vertices by $\mathcal{V} = \{V_1, V_2, V_3, V_4\}$, $V_i \in \mathbb{R}^3$, and $\lambda_1, \lambda_2, \lambda_3, \lambda_4 \in P^1(\Delta_s)$ its barycentric coordinates. Global $H(\mathrm{div})$ conformity requires normal continuity over element interfaces, which can be easily achieved by using a face-interior-based high-order finite element basis. The general construction follows [55, 62]: The set of face-based shape functions consists of low-order Raviart-Thomas shape functions and divergence-free shape functions. The set of interior based shape functions are split into a set of divergence-free fields (rotations) and a set of non-divergence-free completion functions. Using the appropriately weighted Jacobi-type polynomials of Sect. 3 the $H(\mathrm{div})$-conforming shape functions on the tetrahedron are defined as follows.

- For each face $F = [f_1, f_2, f_3]$, characterized by the vertices V_{f_1}, V_{f_2} and V_{f_3} we construct the face based basis functions as follows. First, we choose the classical Raviart-Thomas function of order zero ψ_0^F (16) and add the divergence-free higher-order face based shape functions

$$\begin{aligned}
\psi_{1j}^F &:= \nabla \times \left(\varphi_1^{[f_1,f_2]} v_{1j}^F\right), && 1 \le j \le p, \\
\psi_{ij}^F &:= \nabla \times \left(\nabla u_i^F v_{ij}^F\right) = -\nabla u_i^F \times \nabla v_{ij}^F, && 2 \le i; 1 \le j; i + j \le p + 1
\end{aligned} \tag{24}$$

where we use the face-based Jacobi-type polynomials (13) and the lowest-order Nédélec function (15) corresponding to the edge $[f_1, f_2]$. Let

$$[\Psi_0] := \left[\psi_0^{F_1}, \psi_0^{F_2}, \psi_0^{F_3}, \psi_0^{F_4}\right] \tag{25}$$

denote the row vector of low-order shape functions,

$$[\Psi^F] := \left[\left[\psi_{1j}^F\right]_{j=1}^{p}, \left[\psi_{ij}^F\right]_{i=2,j=1}^{i+j\le p+1}\right]$$

denote the row vector of the faced-based high-order shape functions of one fixed face F, and

$$[\Psi_F] := \left[[\Psi^{F1}] \, [\Psi^{F2}] \, [\Psi^{F3}] \, [\Psi^{F4}] \right] \tag{26}$$

be the row vector of all face-based high-order shape functions.

- The cell-based basis functions are constructed in two types. First we define the divergence-free shape functions by the rotations

$$\begin{aligned}
\psi^{(a)}_{1jk}(x,y,z) &:= \nabla \times \left(\varphi_1^{[1,2]}(x,y,z)\, v_{2j}(x,y,z)\, w_{2jk}(x,y,z)\right), \\
& \qquad j,k \geq 1;\ j+k \leq p, \\
\psi^{(b)}_{ijk}(x,y,z) &:= \nabla \times \left(\nabla u_i(x,y,z)\, v_{ij}(x,y,z)\, w_{ijk}(x,y,z)\right), \\
& \qquad i \geq 2;\ j,k \geq 1;\ i+j+k \leq p+2, \\
\psi^{(c)}_{ijk}(x,y,z) &:= \nabla \times \left(\nabla (u_i(x,y,z)\, v_{ij}(x,y,z))\, w_{ijk}(x,y,z)\right), \\
& \qquad i \geq 2;\ j,k \geq 1;\ i+j+k \leq p+2,
\end{aligned}$$

and complete the basis with the non-divergence free cell-based shape functions

$$\begin{aligned}
\widetilde{\psi}^{(a)}_{10k}(x,y,z) &:= \psi_0^{[1,2,3]}(x,y,z)\, w_{21k}(x,y,z), \\
& \qquad 1 \leq k \leq p-1, \\
\widetilde{\psi}^{(b)}_{1jk}(x,y,z) &:= \varphi_0^{[1,2]}(x,y,z) \times \nabla w_{2jk}(x,y,z)\, v_{2j}(x,y,z), \\
& \qquad j,k \geq 1;\ j+k \leq p, \\
\widetilde{\psi}^{(c)}_{ijk}(x,y,z) &:= w_{ijk}(x,y,z)\, \nabla u_i(x,y,z) \times \nabla v_{ij}(x,y,z), \\
& \qquad i \geq 2;\ j,k \geq 1;\ i+j+k \leq p+2,
\end{aligned}$$

where $\psi_0^{[1,2,3]}(x,y,z)$ denotes the Raviart-Thomas function (16) associated to the bottom face $[1,2,3]$ and $\varphi_0^{[1,2]}$ is the Nédélec function (15) associated to the edge $[1,2]$. The auxiliary functions u_i, v_{ij} and w_{ijk} have been defined in (14).

Finally, we denote the row vectors of the corresponding basis functions as

- $[\Psi_a] = \left[\psi^{(a)}_{1jk}(x,y,z)\right]_{j,k,\geq 1}^{j+k\leq p}$,
- $[\Psi_b] = \left[\psi^{(b)}_{ijk}(x,y,z)\right]_{i\geq 2,j,k,\geq 1}^{i+j+k\leq p+2}$,
- $[\Psi_c] = \left[\psi^{(c)}_{ijk}(x,y,z)\right]_{i\geq 2,j,k,\geq 1}^{i+j+k\leq p+2}$, for the divergence-free parts, and
- $[\widetilde{\Psi}_a] = \left[\widetilde{\psi}^{(a)}_{10k}(x,y,z)\right]_{k=1}^{p-1}$,

– $[\widetilde{\Psi}_b] = \left[\widetilde{\psi}^{(b)}_{1jk}(x,y,z)\right]_{j,k,\geq 1}^{j+k\leq p}$, and
– $[\widetilde{\Psi}_c] = \left[\widetilde{\psi}^{(c)}_{ijk}(x,y,z)\right]_{i\geq 2,j,k,\geq 1}^{i+j+k\leq p+2}$ for the non divergence-free polynomials.

The set of the interior shape functions is denoted by

$$[\Psi_I^{\cdot}] := \left[[\Psi_1]\ [\Psi_2] \right] \text{ with } \quad [\Psi_1] := \left[[\Psi_a]\ [\Psi_b]\ [\Psi_c] \right],$$
$$[\Psi_2] := \left[[\widetilde{\Psi}_a]\ [\widetilde{\Psi}_b]\ [\widetilde{\Psi}_c] \right]. \tag{27}$$

Using (25)–(27), the complete set of low-order-face-cell-based shape functions on the tetrahedron is written as

$$[\Psi_{\nabla\cdot}] := \left[[\Psi_0]\ [\Psi_F^{\cdot}]\ [\Psi_I^{\cdot}] \right]. \tag{28}$$

Let

$$K_{s,\cdot} = \int_{\Delta_s} [\nabla\cdot\Psi_{\nabla\cdot}]^\top [\nabla\cdot\Psi_{\nabla\cdot}] \tag{29}$$

be the element stiffness matrix with respect to the basis (28) and

$$M_{II,s,\cdot} = \int_{\Delta_s} [\Psi_I^{\cdot}]^\top \cdot [\Psi_I^{\cdot}] \tag{30}$$

be the block of the interior bubbles of the mass matrix. The following orthogonality results can be shown.

Theorem 6.1. *Let the set* $[\Psi_{\nabla\cdot}]$ *of basis functions be defined in* (28)*. Then, the fluxes* $[\nabla\cdot\Psi_I^{\cdot}]$ *are* L_2*-orthogonal to* $[\nabla\cdot\Psi_{\nabla\cdot}]$*. Moreover, the stiffness matrix* $K_{s,\cdot}$ (29) *is diagonal up to the* 4×4 *low-order block* $a_{\mathrm{div}}([\Psi_0],[\Psi_0])$*.*

The number of nonzero matrix entries per row in the matrix $M_{II,s,\cdot}$ (30) *is bounded by a constant independent of the polynomial degree* p*.*

Proof. The first result can be proved by straightforward computation. For the mass matrix, the assertion follows by evaluation of the matrix entries using the algorithm described in Sect. 8, see also [16]. □

Due to a construction based on the Jacobi type polynomials (14), the nonzero pattern of the matrix $M_{II,s,\cdot}$ in (30) has again a stencil like structure as the matrices M_{II,s,∇_3} and K_{II,s,∇_3} in (21) and (22) for the H^1 case. Also the growth of the condition number is as $\mathcal{O}(p^4)$. However, the absolute numbers for a fixed polynomial degree p are higher than for the H^1 case.

The divergence of the inner basis functions vanishes for the first part and coincides with the higher-order L_2-optimal Dubiner basis functions for the second part. Hence, the results for the element stiffness matrix $K_{s,\cdot}$ are strongly related to the L_2 results of Sect. 4. Namely, $K_{s,\cdot}$ is diagonal up to the low-order block. The nonzero pattern for mass and stiffness matrix is displayed in Fig. 5 for $p = 15$.

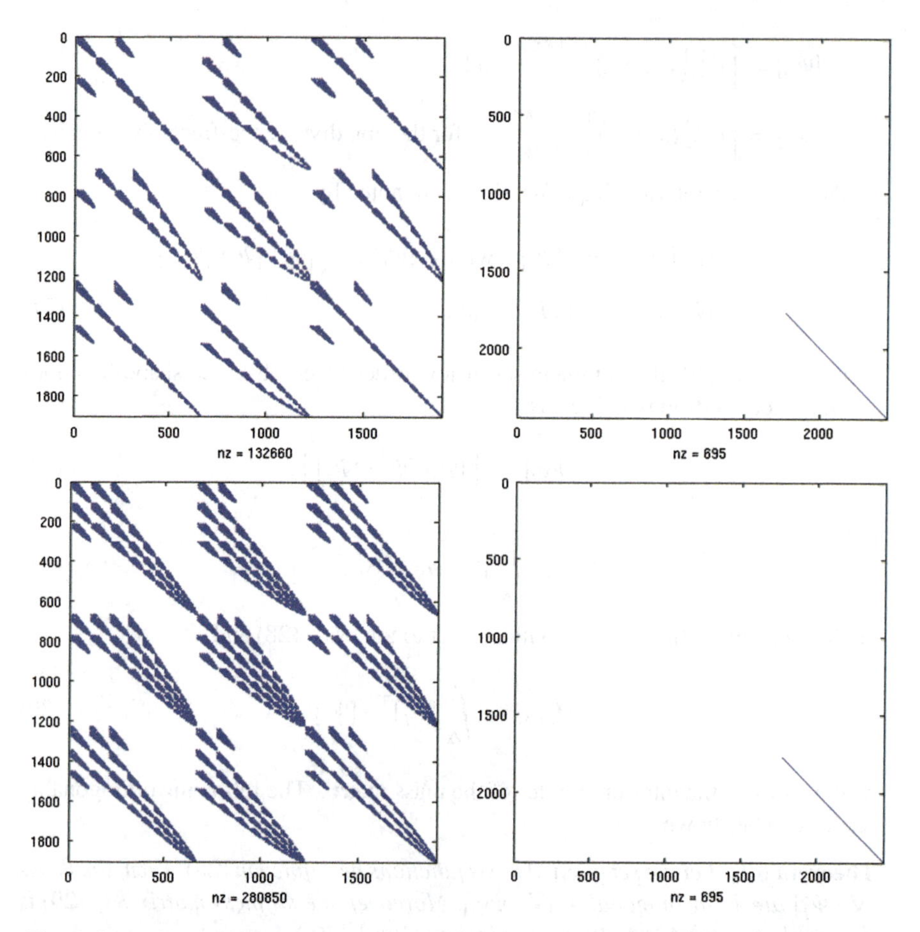

Fig. 5 Optimally weighted Jacobi-type basis $[\Psi_{\nabla\cdot}]$ for $p = 15$: *Above*: Sparsity pattern of inner block $\widehat{M}_{II,\cdot}$ of element mass (*left, above*) and element stiffness matrix $\widehat{K}_{\cdot}$ (*right, above*) on reference tetrahedron $\hat{\Delta}$. *Below*: Sparsity pattern of inner block $M_{II,s,\cdot}$ of mass matrix (*left, below*) and stiffness matrix $K_{s,\cdot}$ on a general affine tetrahedron Δ_s (*right, below*)

Besides sparsity the appropriately chosen weights imply a tremendous improvement in condition numbers of the system matrices (even for curved element geometries) as reported in [16].

7 Sparsity Optimized $H(\mathrm{curl})$-Conforming Basis Functions

The sparsity results for $H(\mathrm{curl})$-conforming basis functions included in this section will be presented in a forthcoming paper [17]. Again, the general construction principle follows [54] and [62]. The sparsity optimization will be performed only for the interior basis functions. Hence, in the sequel we restrict ourselves only to the definition of interior functions, while the edge and face based functions can be taken from [54].

The interior (cell-based) basis functions are constructed in two types. First we define the curl-free shape functions by the gradients

$$\varphi_{ijk}^{(b)}(x,y,z) := \nabla\big(u_i(x,y,z)\, v_{ij}(x,y,z)\, w_{ijk}(x,y,z)\big), \\ i \geq 2;\, j,k \geq 1;\; i+j+k \leq p+1 \tag{31}$$

and complete the basis with the non-curl free cell-based shape functions

$$\begin{aligned} \tilde{\varphi}_{1jk}^{(a)}(x,y,z) &:= \varphi_1^{[1,2]}(x,y,z) v_{1j}(x,y,z) w_{1jk}(x,y,z), \\ & \qquad j,k \geq 1;\; j+k \leq p-1, \\ \tilde{\varphi}_{ijk}^{(b)}(x,y,z) &:= \nabla u_i(x,y,z) v_{ij}(x,y,z) w_{ijk}(x,y,z), \\ & \qquad i \geq 2;\, j,k \geq 1;\; i+j+k \leq p+1, \\ \tilde{\varphi}_{ijk}^{(c)}(x,y,z) &:= \nabla\left(u_i(x,y,z) v_{ij}(x,y,z)\right) w_{ijk}(x,y,z), \\ & \qquad i \geq 2;\, j,k \geq 1;\; i+j+k \leq p+1, \end{aligned} \tag{32}$$

where $\varphi_1^{[1,2]}$ is the Nédélec function (15), and u_i, v_{ij} and w_{ijk} are defined in (14). Finally, we denote the row vectors of the corresponding basis functions as

- $[\Phi_b] = \left[\phi_{ijk}^{(b)}(x,y,z)\right]_{i\geq 2, j,k,\geq 1}^{i+j+k\leq p+1}$ as the gradient fields, and
- $[\widetilde{\Phi}_a] = \left[\widetilde{\phi}_{1jk}^{(a)}(z)\right]_{j,k=1}^{j+k\leq p-1}$,
- $[\widetilde{\Phi}_b] = \left[\widetilde{\phi}_{ijk}^{(b)}(x,y,z)\right]_{i\geq 2, j,k,\geq 1}^{i+j+k\leq p+1}$ and
- $[\widetilde{\Phi}_c] = \left[\widetilde{\phi}_{ijk}^{(c)}(x,y,z)\right]_{i\geq 2, j,k,\geq 1}^{i+j+k\leq p+1}$

as the non curl free functions. The set of interior basis functions is denoted by

$$[\Psi_I^{\times}] := \big[[\Phi_b]\; [\Phi_2]\big] \quad \text{with} \quad [\Phi_2] := \big[[\widetilde{\Phi}_a]\; [\widetilde{\Phi}_b]\; [\widetilde{\Phi}_c]\big]. \tag{33}$$

Finally, we introduce

$$K_{s,II,\times} = \int_{\Delta_s} [\nabla \times \Psi_{\nabla\times}]^\top \cdot [\nabla \times \Psi_{\nabla\times}] \text{ and } M_{s,II,\times} = \int_{\Delta_s} [\Psi_{\nabla\times}]^\top \cdot [\Psi_{\nabla\times}] \tag{34}$$

as the stiffness and mass matrix with respect to the interior bubbles (33), respectively.

Theorem 7.1. *The matrices $K_{s,II,\times}$ and $M_{s,II,\times}$* (34) *are sparse matrices having a bounded number of nonzero entries per row. The total number of nonzero entries grows as $\mathcal{O}(p^3)$.*

Proof. The result follows from the construction principle of the basis functions in (31), (32) and Theorems 5.3 and 6.1. We refer the reader for a more detailed discussion to [17]. □

8 Integration by Rewriting

In this section we present the algorithm that is used to evaluate the matrix entries for different spaces and choices of basis functions. As indicated earlier, the basic idea is to apply a rewriting procedure to the given integrands that yields a reformulation of the integrand as a linear combination of products of the form

$$\left(\frac{1-x}{2}\right)^{\alpha} p_i^{\alpha}(x) p_j^{\alpha}(x).$$

These terms then can be evaluated directly by the Jacobi orthogonality relation (4). Below we use the short-hand notation $w_\alpha(x) = \left(\frac{1-x}{2}\right)^{\alpha}$ for the weight function.

For the necessary rewriting steps several relations between Jacobi polynomials and integrated polynomials are needed that have been proven in [14, 15, 19] and are summarized in the next lemma.

Lemma 8.1 *Let $p_n^{\alpha}(x)$ and $\hat{p}_n^{\alpha}(x)$ be the polynomials defined in* (3) *and* (5). *Then for all $n \geq 1$ we have the relations*

$$\begin{aligned}
\hat{p}_n^{\alpha}(x) &= \frac{2(n+\alpha)}{(2n+\alpha-1)(2n+\alpha)} p_n^{\alpha}(x) + \frac{2\alpha}{(2n+\alpha-2)(2n+\alpha)} p_{n-1}^{\alpha}(x) \\
&\quad - \frac{2(n-1)}{(2n+\alpha-1)(2n+\alpha-2)} p_{n-2}^{\alpha}(x), \qquad \alpha \geq -1, \quad (35)
\end{aligned}$$

$$\hat{p}_n^{\alpha}(x) = \frac{2}{2n+\alpha-1}[p_n^{\alpha-1}(x) + p_{n-1}^{\alpha-1}(x)], \qquad \alpha > -1, \quad (36)$$

$$(\alpha-1)\hat{p}_n^{\alpha}(x) = (1-x)\, p_{n-1}^{\alpha}(x) + 2\, p_n^{\alpha-2}(x), \qquad \alpha > 1. \quad (37)$$

$$p_n^{\alpha-1} = \frac{1}{2n+\alpha}[(n+\alpha) p_n^{\alpha}(x) - n\, p_{n-1}^{\alpha}(x)], \qquad \alpha > -1, \quad (38)$$

After decoupling the integrands by means of the Duffy transformation, the integrals are evaluated in the order given by the dependencies of the parameters α. For each of these univariate integrals the following algorithm is executed:

1. Collect integrands depending on the current integration variable.
2. For each integrand: Rewrite integrated Jacobi polynomials in terms of Jacobi polynomials using (35), (36), or (37).
3. Collect integrands depending on the current integration variable.
4. For each integrand: Adjust Jacobi polynomials to appearing weight functions.
5. Collect integrands depending on the current integration variable.
6. For each integrand: Evaluate integrals using orthogonality relation (4).

The two steps of the algorithm that need further explanations are steps 2 and 4. Indeed, let us consider steps 2 and 4 in detail: which of the identities relating integrated Jacobi polynomials and Jacobi polynomials (35)–(37) have to be used in step 2 depends on the difference $\gamma - \alpha$ of the parameters of $\hat{p}_n^\alpha(\zeta)$ and of the weight function $w_\gamma(\zeta)$.

2. Rewrite $w_\gamma(\zeta)\hat{p}_n^\alpha(\zeta)$ in terms of Jacobi polynomials

   (a) $\gamma - \alpha \geq 0$: transform integrated Jacobi polynomials to Jacobi polynomials with same parameter using (35).

   (b) $\gamma - \alpha = -1$: transform integrated Jacobi polynomials to Jacobi polynomials with parameter $\alpha - 1$ using (36).

   (c) $\gamma - \alpha = -2$: use the mixed relation (37) to obtain

$$w_\gamma(\zeta)\hat{p}_n^{\gamma+2}(\zeta) = \frac{2}{\gamma+1}\left(w_\gamma(\zeta)p_n^\gamma(\zeta) + w_{\gamma+1}(\zeta)p_{n-1}^{\gamma+2}(\zeta)\right).$$

If none of the cases 2(a)–2(c) applies, the algorithms interrupts and returns the unevaluated integrand for further examination. Such an output can lead either to a readjustment of the parameters of the basis functions, or to the discovery of a new relation between Jacobi polynomials that needs to be added to the given rewrite rules. This finding of new, necessary identities can again be achieved with the assistance of symbolic computation, e.g., by means of Koutschan's package HolonomicFunctions [45] or Kauers' package SumCracker [43].

Rewriting the Jacobi polynomials $p_n^\alpha(\zeta)$ in terms of $p_n^\gamma(\zeta)$ fitting to the appearing weights $w_\gamma(\zeta)$ in step 4, means lifting the polynomial parameter α using (38) $(\gamma - \alpha)$ times. This transformation is performed recursively for each appearing Jacobi polynomial.

4. Rewrite the Jacobi polynomials $p_n^\alpha(\zeta)$ in terms of Jacobi polynomials fitting to the appearing weights $w_\gamma(\zeta)$ $(\gamma - \alpha > 0)$ by lifting the polynomial parameter α using (38) $(\gamma - \alpha)$-times, i.e., written in explicit form we have

$$p_n^\alpha(\zeta) = \sum_{m=0}^{\gamma-\alpha}(-1)^k\binom{\gamma-\alpha}{m}\frac{(n+\gamma-m)^{\underline{\gamma-\alpha-m}}\,n^{\underline{m}}}{(2n+\gamma-m+1)^{\underline{\gamma-\alpha+1}}}(2n-2m+\gamma+1) \times p_{n-m}^\gamma(\zeta),$$

where $a^{\underline{k}} = a(a-1)\cdot\ldots\cdot(a-k+1)$ denotes the falling factorial.

If $\gamma - \alpha < 0$ the algorithm interrupts. In this step of the algorithm polynomials down to degree $n - \gamma + \alpha$ are introduced. Hence this transformation is a costly one as it increases the number of terms significantly.

Acknowledgements This work has been supported by the FWF-projects P20121-N12 and P20162-N18, the Austrian Academy of Sciences, the Spezialforschungsbereich "Numerical and Symbolic Scientific Computing" (SFB F013) , the doctoral program "Computational Mathematics" (W1214) and the FWF Start Project Y-192 on "3D hp-Finite Elements: Fast Solvers and Adaptivity".

References

1. Abramowitz, M., Stegun, I. (eds.): Handbook of Mathematical Functions. Dover-Publications, New York (1965)
2. Ainsworth, M.: A preconditioner based on domain decomposition for h-p finite element approximation on quasi-uniform meshes. SIAM J. Numer. Anal. **33**(4), 1358–1376 (1996)
3. Ainsworth, M., Coyle, J.: Hierarchic finite element bases on unstructured tetrahedral meshes. Int. J. Num. Meth. Eng. **58**(14), 2103–2130 (2003)
4. Ainsworth, M., Demkowicz, L.: Explicit polynomial preserving trace liftings on a triangle. Math. Nachr. **282**(5), 640–658 (2009)
5. Ainsworth, M., Guo, B.: An additive Schwarz preconditioner for p-version boundary element approximation of the hypersingular operator in three dimensions. Numer. Math. **85**(3), 343–366 (2000)
6. Appell, P.: Sur des polynômes de deux variables analogues aux polynômes de jacobi. Arch. Math. Phys. **66**, 238–245 (1881)
7. Babuška, I., Craig, A., Mandel, J., Pitkäranta, J.: Efficent preconditioning for the p-version finite element method in two dimensions. SIAM J. Numer. Anal. **28**(3), 624–661 (1991)
8. Babuška, I., Guo, B.Q.: The h-p version of the finite element method for domains with curved boundaries. SIAM J. Numer. Anal. **25**(4), 837–861 (1988)
9. Babuška, I., Griebel, M., Pitkäranta, J.: The problem of selecting the shape functions for a p-type finite element. Int. Journ. Num. Meth. Eng. **28**, 1891–1908 (1989)
10. Bernardi, C., Dauge, M., Maday, Y.: Polynomials in weighted Sobolev spaces: Basics and trace liftings. Technical Report R 92039, Universite Pierre et Marie Curie, Paris (1993)
11. Bernardi, Ch., Dauge, M., Maday, Y.: The lifting of polynomial traces revisited. Math. Comp. **79**(269), 47–69 (2010)
12. Beuchler, S.: Multi-grid solver for the inner problem in domain decomposition methods for p-FEM. SIAM J. Numer. Anal. **40**(3), 928–944 (2002)
13. Beuchler, S.: Wavelet solvers for hp-FEM discretizations in 3D using hexahedral elements. Comput. Methods Appl. Mech. Engrg. **198**(13-14), 1138–1148, (2009)
14. Beuchler, S., Pillwein, V.: Shape functions for tetrahedral p-fem using integrated Jacobi polynomials. Computing **80**, 345–375 (2007)
15. Beuchler, S., Pillwein, V.: Completions to sparse shape functions for triangular and tetrahedral p-fem. In Langer, U., Discacciati, M., Keyes, D.E., Widlund, O.B., Zulehner, W. (eds.) Domain Decomposition Methods in Science and Engineering XVII, volume 60 of Lecture Notes in Computational Science and Engineering, pp. 435–442, Springer, Heidelberg (2008). Proceedings of the 17th International Conference on Domain Decomposition Methods held at St. Wolfgang / Strobl, Austria, July 3–7, 2006
16. Beuchler, S., Pillwein, V., Zaglmayr, S.: Sparsity optimized high order finite element functions for H(div) on simplices. Technical Report 2010-04, DK Computational Mathematics, JKU Linz (2010)

17. Beuchler, S., Pillwein, V., Zaglmayr, S.: Sparsity optimized high order finite element functions for $H(curl)$ on tetrahedral meshes. Technical Report Report RICAM, Johann Radon Institute for Computational and Applied Mathematics, Linz (2011) in preparation.
18. Beuchler, S., Schneider, R., Schwab, C.: Multiresolution weighted norm equivalences and applications. Numer. Math. **98**(1), 67–97 (2004)
19. Beuchler, S., Schöberl, J.: New shape functions for triangular p-fem using integrated jacobi polynomials. Numer. Math. **103**, 339–366 (2006)
20. Bossavit, A.: Computational Electromagnetism: Variational formulation, complementary, edge elements. Academic Press Series in Electromagnetism. Academic, San Diego (1989)
21. Braess, D.: Finite Elemente. Springer, Berlin (1991)
22. Braess, D.: Approximation on simplices and orthogonal polynomials. In Trends and applications in constructive approximation, volume 151 of Internat. Ser. Numer. Math., pp. 53–60. Birkhäuser, Basel (2005)
23. Chyzak, F.: Gröbner bases, symbolic summation and symbolic integration. In Gröbner bases and applications (Linz, 1998), volume 251 of London Math. Soc. Lecture Note Ser., pp. 32–60. Cambridge University Press, Cambridge (1998)
24. Chyzak, F.: An extension of Zeilberger's fast algorithm to general holonomic functions. Discrete Math. **217**(1-3), 115–134 (2000). Formal power series and algebraic combinatorics (Vienna, 1997)
25. Chyzak, F., Salvy, B.: Non-commutative elimination in Ore algebras proves multivariate identities. J. Symbolic Comput. **26**(2), 187–227 (1998)
26. Ciarlet, P.: The Finite Element Method for Elliptic Problems. North–Holland, Amsterdam (1978)
27. Costabel, M., Dauge, M., Demkowicz, L.: Polynomial extension operators for H^1, H(curl) and H(div)-spaces on a cube. Math. Comp. **77**(264), 1967–1999 (2008)
28. Demkowicz, L.: Computing with hp Finite Elements. CRC Press, Taylor and Francis (2006)
29. Demkowicz, L., Gopalakrishnan, J., Schöberl, J.: Polynomial extension operators. I. SIAM J. Numer. Anal. **46**(6), 3006–3031 (2008)
30. Demkowicz, L., Gopalakrishnan, J., Schöberl, J.: Polynomial extension operators. II. SIAM J. Numer. Anal. **47**(5), 3293–3324 (2009)
31. Demkowicz, L., Kurtz, J., Pardo, D., Paszyński, M., Rachowicz, W., Zdunek, A.: Computing with hp-adaptive finite elements. Vol. 2. Chapman & Hall/CRC Applied Mathematics and Nonlinear Science Series. Chapman & Hall/CRC, Boca Raton, FL, (2008). Frontiers: three dimensional elliptic and Maxwell problems with applications
32. Demkowicz, L., Monk, P., Vardapetyan, L., Rachowicz, W.: De Rham diagram for hp finite element spaces. Comput. Math. Apl. **39**(7-8), 29–38, (2000)
33. Deville, M.O., Mund, E.H.: Finite element preconditioning for pseudospectral solutions of elliptic problems. SIAM J. Sci. Stat. Comp. **18**(2), 311–342 (1990)
34. Dubiner, M.: Spectral methods on triangles and other domains. J. Sci. Computing **6**, 345 (1991)
35. Dunkl, C.F., Xu, Y.: Orthogonal polynomials of several variables, volume 81 of Encyclopedia of Mathematics and its Applications. Cambridge University Press, Cambridge (2001)
36. Eibner, T., Melenk, J.M.: An adaptive strategy for hp-FEM based on testing for analyticity. Comput. Mech. **39**(5), 575–595 (2007)
37. George, A.: Nested dissection of a regular finite element mesh. SIAM J. Numer. Anal. **10**, 345–363 (1973)
38. Guo, B., Cao, W.: An iterative and parallel solver based on domain decomposition for the hp-version of the finite element method. J. Comput. Appl. Math. **83**, 71–85 (1997)
39. Ivanov, S.A., Korneev, V.G.: On the preconditioning in the domain decomposition technique for the p-version finite element method. Part I. Technical Report SPC 95-35, Technische Universität Chemnitz-Zwickau, December 1995
40. Ivanov, S.A., Korneev, V.G.: On the preconditioning in the domain decomposition technique for the p-version finite element method. Part II. Technical Report SPC 95-36, Technische Universität Chemnitz-Zwickau, December 1995

41. Jensen, S., Korneev, V.G.: On domain decomposition preconditioning in the hierarchical $p-$version of the finite element method. Comput. Methods. Appl. Mech. Eng. **150**(1–4), 215–238 (1997)
42. Karniadakis, G.M., Sherwin, S.J.: Spectral/HP Element Methods for CFD. Oxford University Press, Oxford (1999)
43. Kauers, M.: SumCracker – A Package for Manipulating Symbolic Sums and Related Objects. J. Symbolic Comput. **41**(9), 1039–1057 (2006)
44. Korneev, V., Langer, U., Xanthis, L.: On fast domain decomposition methods solving procedures for hp-discretizations of 3d elliptic problems. Comput. Meth. Appl. Math. **3**(4), 536–559, (2003)
45. Koutschan, C.: HolonomicFunctions (User's Guide). Technical Report 10-01, RISC Report Series, University of Linz, Austria, January 2010
46. Melenk, J.M., Pechstein, C., Schöberl, J., Zaglmayr, S.: Additive Schwarz preconditioning for p-version triangular and tetrahedral finite elements. IMA J. Num. Anal. **28**, 1–24 (2008)
47. Melenk, J.M., Gerdes, K., Schwab, C.: Fully discrete hp-finite elements: Fast quadrature. Comp. Meth. Appl. Mech. Eng. **190**, 4339–4364 (1999)
48. Monk, P.: Finite Element Methods for Maxwell's Equations. Numerical Mathematics and Scientific Computation. The Clarendon Press Oxford University Press, New York (2003)
49. Munoz-Sola, R.: Polynomial liftings on a tetrahedron and applications to the h-p version of the finite element method in three dimensions. SIAM J. Numer. Anal. **34**(1), 282–314 (1996)
50. Nédélec, J.C.: Mixed finite elements in $\mathbb{R}^3$. Numerische Mathematik **35**(35), 315–341 (1980)
51. Pavarino, L.F.: Additive schwarz methods for the p-version finite element method. Numer. Math. **66**(4), 493–515 (1994)
52. Petkovšek, M., Wilf, H.S., Zeilberger, D.: $A = B$. A K Peters Ltd., Wellesley, MA (1996)
53. Quateroni, A., Valli, A.: Numerical Approximation of partial differential equations. Number 23 in Springer Series in Computational Mathematics. Springer, Berlin (1997)
54. Schöberl, J., Zaglmayr, S.: High order Nédélec elements with local complete sequence properties. International Journal for Computation and Mathematics in Electrical and Electronic Engineering (COMPEL) **24**, 374–384 (2005)
55. Schöberl, J., Zaglmayr, S.: hp finite element De Rham sequences on hybrid meshes, in preparation
56. Schwab, C.: $p-$ and $hp-$finite element methods. Theory and applications in solid and fluid mechanics. Clarendon Press, Oxford (1998)
57. Sherwin, S.J., Karniadakis, G.E.: A new triangular and tetrahedral basis for high-order finite element methods. Int. J. Num. Meth. Eng. **38**, 3775–3802 (1995)
58. Solin, P., Segeth, K., Dolezel, I.: Higher-Order Finite Element Methods. Chapman and Hall, CRC Press (2003)
59. Szegö, G.: Orthogonal Polynomials. AMS Colloquium Publications, Vol. XXIII. 3rd edn. (1974)
60. Tricomi, F.G.: Vorlesungen über Orthogonalreihen. Springer, Berlin (1955)
61. Wilf, H.S., Zeilberger, D.: An algorithmic proof theory for hypergeometric (ordinary and "q") multisum/integral identities. Invent. Math. **108**(3), 575–633 (1992)
62. Zaglmayr, S.: High Order Finite Elements for Electromagnetic Field Computation. PhD thesis, Johannes Kepler University, Linz, Austria (2006)
63. Zeilberger, D.: A fast algorithm for proving terminating hypergeometric identities. Discrete. Math. **80**, 207–211 (1990)
64. Zeilberger, D.: A holonomic systems approach to special functions identities. J. Comput. Appl. Math. **32**(3), 321–368 (1990)
65. Zeilberger, D.: The method of creative telescoping. J. Symbolic Comput. **11**, 195–204 (1991)

Fast Solvers and A Posteriori Error Estimates in Elastoplasticity

Peter G. Gruber, Johanna Kienesberger, Ulrich Langer, Joachim Schöberl, and Jan Valdman

Abstract The paper reports some results on computational plasticity obtained within the Special Research Program "Numerical and Symbolic Scientific Computing" and within the Doctoral Program "Computational Mathematics" both supported by the Austrian Science Fund FWF under the grants SFB F013 and DK W1214, respectively. Adaptivity and fast solvers are the ingredients of efficient numerical methods. The paper presents fast and robust solvers for both 2D and 3D plastic flow theory problems as well as different approaches to the derivations of a posteriori error estimates. In the last part of the paper higher-order finite elements are used within a new plastic-zone concentrated setup according to the regularity of the solution. The theoretical results obtained are well supported by the results of our numerical experiments.

1 Introduction

The theory of plasticity has a long tradition in the engineering literature. These classical results on plasticity together with the introduction of the Finite Element Method (FEM) into engineering computations provides the basis for the modern computational plasticity (see [59] and the references therein). The rigorous mathematical analysis of plastic flow theory problems and of the numerical methods for their solution started in the late 70ies and in the early 80ies by the work of C. Johnson [33, 34], H. Matthies [43, 44], V.G. Korneev and U. Langer [42], and others. Since then many mathematical contributions to Computational Plasticity have been made. We here only refer to the monographs by J.C. Simo and T.J.R. Hughes [53] and W. Han and B.D. Reddy[31], to the habilitation theses by

U. Langer (✉) · P.G. Gruber · J. Kienesberger · J. Schöberl · J. Valdman
Institute of Computational Mathematics, Johannes Kepler University Linz, 4040 Linz, Austria
e-mail: ulanger@numa.uni-linz.ac.at

U. Langer and P. Paule (eds.), *Numerical and Symbolic Scientific Computing*,
Texts and Monographs in Symbolic Computation, DOI 10.1007/978-3-7091-0794-2_3,

C. Carstensen [12] and C. Wieners [57], to the collection [54], and the references given therein.

The incremental elastoplasticity problem can be reformulated as a minimization problem for a convex but not-smooth functional, where the unknowns are the displacements u and the plastic strains p. One method to deal with this non-smoothness relies on regularization techniques which were initially studied in [37]. However, eliminating the plastic strains p and using Moreau's theorem, we see that the reduced functional, that is now only a functional in the displacements u, is actually continuously Fréchet differentiable. The elimination of the plastic strains can be done locally and with the help of symbolic techniques. Unfortunately, the second derivative of the reduced functional does not exist. As a remedy, the concept of slanting functions, introduced by X. Chen, Z. Nashed, and L. Qi in [17], allows us to construct and analyze generalized Newton methods which show fast convergence in all our numerical experiments. More precisely, we can prove super-linear convergence of these generalized Newton methods at least in the finite element setting.

The second part of this paper is devoted to the a posteriori error analysis of elastoplastic problems. Two different techniques were developed: the first one is exploring a residual-type estimator respecting certain oscillations, and the second one is based on functional a posteriori estimates introduced by S. Repin [48].

Finally, we consider spatial discretizations of the incremental plasticity problems based on hp finite element techniques. A straightforward application of the classical h-FEM yields algebraic convergence. However, the regularity results presented in [6, 41], namely H^2_{loc} regularity of the displacements in the whole domain, and C^∞ regularity apart from plastic zones and the boundary of the computational domain, justify the application of high order finite element methods in the elastic part, but not necessarily in the plastic part. A few hp-adaptive strategies, as well as a related technique, the so-called Boundary Concentrated Finite Element Method (BC-FEM) introduced by B.N. Khoromskij and J.M. Melenk [35], are discussed in this paper.

The rest of the paper is organized as follows: In Sect. 2, we describe the initial-boundary value problem of elastoplasticity which is studied in this paper. Section 3 is devoted to the incremental elastoplasticity problems and strategies for their solution. In Sect. 4 we derive a posteriori error estimates which can be used in the adaptive h-FEM providing an effective spatial discretization in every incremental step. Section 5 deals with the use of the hp-FEM in elastoplasticity. Finally, we draw some conclusions.

2 Modeling of Elastoplasticity

There are many mathematical models describing the elastoplastic behavior of materials under loading. In this paper we follow the description given by C. Carstensen in [12–15]. The classical equations of elastoplasticity can be found in the standard literature on plasticity, see, e.g., [31,53]. Let us first recall these describing relations.

Let $\Theta := [0, T]$ be a (pseudo) time interval, and let Ω be a bounded domain in $\mathbb{R}^3$ with a Lipschitz continuous boundary $\Gamma := \partial\Omega$. In the quasi-static case which is considered throughout this paper, the equilibrium of forces reads

$$-\operatorname{div}(\sigma(x,t)) = f(x,t) \qquad \forall\ (x,t) \in \Omega \times \Theta\,, \tag{1}$$

where $\sigma(x,t) \in \mathbb{R}^{3\times 3}$ is called Cauchy's stress tensor and $f(x,t) \in \mathbb{R}^3$ represents the volume force acting at the material point $x \in \Omega$ at the time $t \in \Theta$. Let $u(x,t) \in \mathbb{R}^3$ denote the displacements of the body, and let

$$\varepsilon(u) := \frac{1}{2}\left(\nabla u + (\nabla u)^T\right) \tag{2}$$

be the linearized Green-St. Venant strain tensor. In elastoplasticity, the total strain ε is additively split into an elastic part e and a plastic part p, that is,

$$\varepsilon = e + p. \tag{3}$$

We assume a linear dependence of the stress on the elastic strain by Hooke's law

$$\sigma = \mathbb{C}e. \tag{4}$$

Since we assume the material to be isotropic, the single components of the elastic stiffness tensor $\mathbb{C} \in \mathbb{R}^{3\times3\times3\times3}$ are defined by $\mathbb{C}_{ijkl} := \lambda\delta_{ij}\delta_{kl} + \mu(\delta_{ik}\delta_{jl} + \delta_{il}\delta_{jk})$. Here, $\lambda > 0$ and $\mu > 0$ denote the Lamé constants, and δ_{ij} the Kronecker symbol.

Let the boundary Γ be split into a Dirichlet part Γ_D and a Neumann part Γ_N such that $\Gamma = \overline{\Gamma_D \cup \Gamma_N}$. We assume the boundary conditions

$$u = u_D \ \text{ on } \Gamma_D \quad \text{ and } \quad \sigma \cdot n = g \ \text{ on } \Gamma_N\,, \tag{5}$$

where $n(x,t)$ denotes the exterior unit normal, $u_D(x,t) \in \mathbb{R}^3$ denotes a prescribed displacement and $g(x,t) \in \mathbb{R}^3$ denotes a prescribed traction. If $p = 0$ in (3), the system (1)–(5) describes the linear elastic behavior of the continuum Ω.

Two more properties, incorporating the admissibility of the stress σ with respect to a certain hardening law and the time evolution of the plastic strain p, are required to describe the plastic behavior of some body Ω. Therefore, we introduce the hardening parameter α and define the generalized stress (σ, α), which we call admissible if for a given convex yield functional ϕ the inequality

$$\phi(\sigma, \alpha) \leq 0. \tag{6}$$

holds. The explicit form of ϕ depends on the choice of the hardening law, see, e.g., formula (9) for isotropic hardening. The second property addresses the time development of the generalized plastic strain $(p, -\alpha)$, described by the normality rule

$$\langle(\dot{p}, -\dot{\alpha}), (\tau, \beta) - (\sigma, \alpha)\rangle_F \leq 0 \qquad \forall\, (\tau, \beta) \text{ which satisfy } \phi(\tau, \beta) \leq 0, \tag{7}$$

where $\dot{p}$ and $\dot{\alpha}$ denote the first time derivatives of p and α, respectively. Finally, let

$$p(x, 0) = p_0(x) \quad \text{and} \quad \alpha(x, 0) = \alpha_0(x) \qquad \forall\, x \in \Omega, \tag{8}$$

for given initial values $p_0 : \Omega \to \mathbb{R}^{3\times 3}_{\text{sym}}$ and $\alpha_0 : \Omega \to [0, \infty[$.

Problem 1 (classical formulation). Find (u, p, α), which satisfies (1)–(8).

In this paper we concentrate on the isotropic hardening law, where the hardening parameter α is a scalar function $\alpha : \Omega \to \mathbb{R}$ and the yield functional ϕ is defined by

$$\phi(\sigma, \alpha) := \begin{cases} \|\operatorname{dev} \sigma\|_F - \sigma_y(1 + H\alpha) & \text{if } \alpha \geq 0, \\ +\infty & \text{if } \alpha < 0. \end{cases} \tag{9}$$

Here, the Frobenius norm $\|A\|_F := \langle A, A\rangle_F^{1/2}$ is defined by the matrix scalar product $\langle A, B\rangle_F := \sum_{ij} a_{ij} b_{ij}$ for $A = (a_{ij}) \in \mathbb{R}^{3\times 3}$ and $B = (b_{ij}) \in \mathbb{R}^{3\times 3}$. The deviator is defined for square matrices by $\operatorname{dev} A = A - \frac{\operatorname{tr} A}{\operatorname{tr} I} I$, where the trace of a matrix is defined by $\operatorname{tr} A = \langle A, I\rangle_F$ and I denotes the identity matrix. The real material constants $\sigma_y > 0$ and $H > 0$ are called yield stress and modulus of hardening, respectively.

3 The Incremental Elastoplasticity Problems and Solvers

We turn to the specification of proper function spaces. For a fixed time $t \in \Theta$, let

$$u \in V := \left[H^1(\Omega)\right]^3, \quad p \in Q := [L_2(\Omega)]^{3\times 3}_{\text{sym}}, \quad \alpha \in L_2(\Omega).$$

Further, let $V_D := \{v \in V \mid v_{|\Gamma_D} = u_D\}$, and $V_0 := \{v \in V \mid v_{|\Gamma_D} = 0\}$, with

$$\langle u, v\rangle_V := \int_\Omega \left(u^T v + \langle\nabla u, \nabla v\rangle_F\right) \mathrm{d}x, \qquad \|v\|_V := \langle v, v\rangle_V^{1/2},$$

$$\langle p, q\rangle_Q := \int_\Omega \langle p, q\rangle_F \,\mathrm{d}x, \qquad \|q\|_Q := \langle q, q\rangle_Q^{1/2}.$$

Starting from Problem 1, one can derive a uniquely solvable time dependent variational inequality for unknown displacement $u \in \{v \in H^1(\Theta; V) \mid v_{|\Gamma_D} = u_D\}$ and plastic strain $p \in H^1(\Theta; Q)$ (see [31, Theorem 7.3] for details). However, the numerical treatment requires a time discretization. Therefore, we pick a fixed number of time ticks $0 = t_0 < t_1 < \ldots < t_{N_\Theta} = T$ out of Θ, and define

$$u_k := u(t_k),\quad p_k := p(t_k),\quad \alpha_k := \alpha(t_k),\quad f_k := f(t_k),\quad g_k := g(t_k),\quad \ldots,$$

and approximate time derivatives by the backward difference quotients

$$\dot{p}_k \approx (p_k - p_{k-1}) \,/\, (t_k - t_{k-1}) \quad \text{and} \quad \dot{\alpha}_k \approx (\alpha_k - \alpha_{k-1}) \,/\, (t_k - t_{k-1}) \,.$$

Consequently, the time dependent problem is approximated by a sequence of time independent variational inequalities of the second kind. Each of these variational inequalities can be equivalently expressed by a minimization problem, which by definition of the set of extended real numbers, $\overline{\mathbb{R}} := \mathbb{R} \cup \{\pm\infty\}$, reads [14, Example 4.5]:

Problem 2. Find $(u_k, p_k) \in V_D \times Q$ such that $J_k(u_k, p_k) = \inf_{(v,q)\in V_D\times Q} J_k(v,q)$, where $J_k : V_D \times Q \to \overline{\mathbb{R}}$ is defined by

$$J_k(v,q) := \frac{1}{2}\|\varepsilon(v) - q\|_{\mathbb{C}}^2 + \psi_k(q) - l_k(v), \tag{10}$$

with

$$\langle q_1, q_2\rangle_{\mathbb{C}} := \int_\Omega \langle \mathbb{C}\, q_1(x), q_2(x)\rangle_F \,\mathrm{d}x, \qquad \|q\|_{\mathbb{C}} := \langle q, q\rangle_{\mathbb{C}}^{\frac{1}{2}}, \tag{11}$$

$$\tilde{\alpha}_k(q) := \alpha_{k-1} + \sigma_y H \|q - p_{k-1}\|_F, \tag{12}$$

$$\psi_k(q) := \begin{cases} \int_\Omega \left(\frac{1}{2}\tilde{\alpha}_k(q)^2 + \sigma_y \|q - p_{k-1}\|_F\right) \mathrm{d}x & \text{if } \operatorname{tr}(q - p_{k-1}) = 0, \\ +\infty & \text{else,} \end{cases} \tag{13}$$

$$l_k(v) := \int_\Omega f_k \cdot v \,\mathrm{d}x + \int_{\Gamma_N} g_k \cdot v \,\mathrm{d}s. \tag{14}$$

The convex functional J_k expresses the mechanical energy of the deformed system at the kth time step. It is smooth with respect to the displacements v, but not with respect to the plastic strains q. Notice, that no minimization with respect to the hardening parameter α_k is necessary. It is computed in the post-processing by $\alpha_k = \tilde{\alpha}_k(p_k)$, with $\tilde{\alpha}_k$ defined as in (12). A short summary on the modeling of Problem 2 starting from the classical formulation can be found in [40]. The problem is uniquely solvable due to [22, Proposition 1.2 in Chap. II].

J. Valdman together with M. Brokate and C. Carstensen published results on the analysis [10] and numerical treatment [11] of multi-yield elastoplastic models based on the PhD-thesis of J. Valdman [55] and its extension. The main feature of the multi-yield models is a higher number of plastic strains $p_1, \ldots, p_N$ used for more realistic modeling of the elastoplastic-plastic transition. Since the structure of the minimization functional in the multi-yield plasticity model remains the same as for the single-yield model, it was possible to prove the existence and uniqueness of the solution of the corresponding variational inequalities and design a FEM based solution algorithm. In terms of a software development, an existing

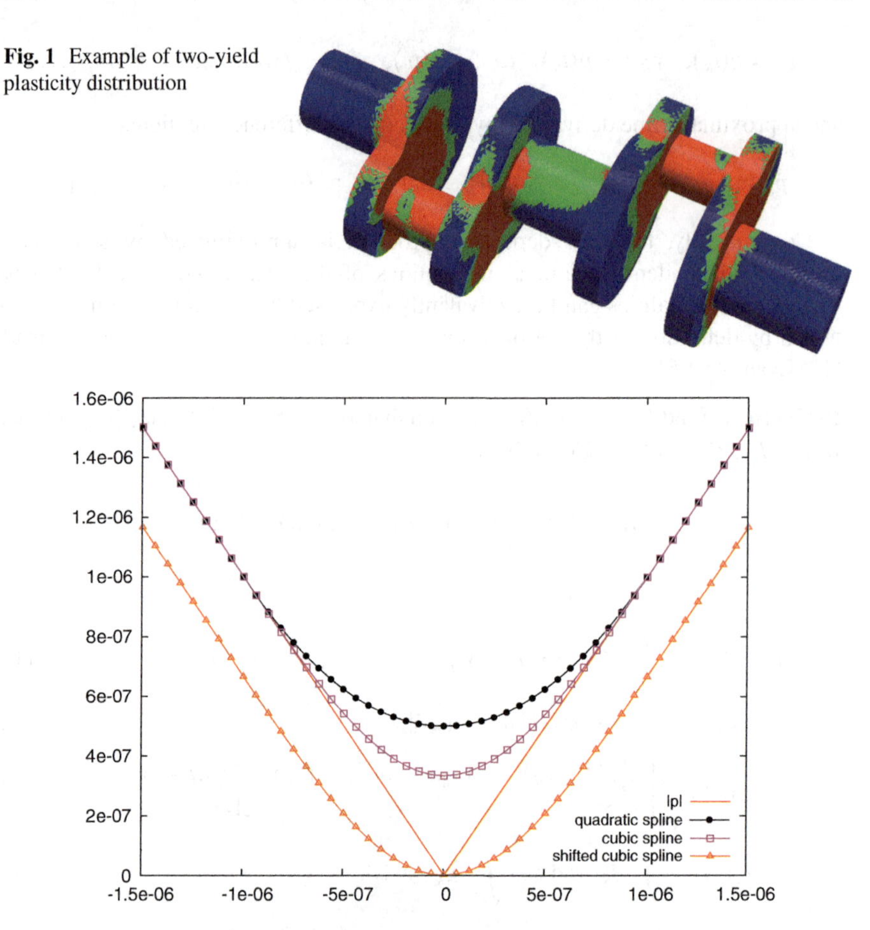

Fig. 1 Example of two-yield plasticity distribution

Fig. 2 Plot of $|p|$ and its regularizations

elastoplasticity package [36], written as a part of the NETGEN/NGSolve software of J. Schöberl, was modified to make the computations of a two-yield elastoplastic problem feasible [39]. Figure 1 displays elastic (blue), first (red) and second (green) plastic deformational zones of the shaft model. The numerical treatment of the two-yield problem requires to resolve the plastic-strain increment matrices P_1 and P_2 from a local minimization problem with a convex but non-smooth functional. Since there are typically millions of such minimizations, iterative techniques such as alternating minimizations, Newton based methods or even partially exact analytical solutions were studied in [32].

The first class of algorithms is based on a regularization of the objective, where the modulus is smoothed for making the objective $J_k^{(\delta)}$ twice differentiable. Figure 2 shows the modulus $|p| := \|p_k - p_{k-1}\|_F$ and possible regularizations $|p|^{(\delta)}$ depending on the regularization parameter δ, where δ is here chosen as 10^{-6}. The quadratic regularization has a smooth first derivative within the interval

$(-\delta, \delta)$, but the second derivative is piecewise constant and discontinuous. Thus, the local quadratic convergence of Newton type methods cannot be guaranteed. The piecewise cubic spline has a piecewise linear continuous second derivative. Thus, Newton type methods can be applied. As a final choice of regularization, the cubic spline function is shifted to the origin, so that $|p|^{(\delta)} = 0$ holds for $p = 0$.

For instance, in case of a quadratic regularization (green), we have

$$|p|_\delta := \begin{cases} |p| & \text{if } |p| \geq \delta, \\ \frac{1}{2\delta}|p|^2 + \frac{\delta}{2} & \text{if } |p| < \delta, \end{cases}$$

with a small regularization parameter $\delta > 0$.

The algorithm is based on alternating minimization with respect to the two variables, and on the reduction of the objective to a quadratic functional with respect to the plastic strains. This can be interpreted as a linearization of the nonlinear elastoplastic problem.

The minimization problem with respect to the plastic part of the strain is separable and the analytical solution $p^{(\delta)}(u)$ can be calculated in explicit form. Problem 2 formally reduces to

$$J_k^{(\delta)}(u) = \min_v \; J_k^{(\delta)}(v, p^{(\delta)}(v)). \tag{15}$$

After the finite element (FE) discretization and the elimination of plastic strains, the FE displacement field results from the solution of a linear Schur complement system. The solution of this linear system can efficiently be computed by a multigrid preconditioned conjugate gradient solver, see [38, 39].

Using Moreau's theorem, that is well known in the scope of convex analysis [46], we can avoid the regularization of the original functional J_k. The formula for minimizing $J_k(u, p)$ with respect to the plastic strain p for a given displacement u is explicitly known [2], i. e., we know a function $\tilde{p}_k(\varepsilon(u))$, such that there holds

$$J_k(u) := J_k(u, \tilde{p}_k(\varepsilon(u))) = \inf_q J_k(u, q).$$

In detail, the plastic strain minimizer reads as follows

$$\tilde{p}_k(\varepsilon(v)) = \xi \, \max\{0, \|\operatorname{dev} \sigma_k(\varepsilon(v))\|_F - \sigma_y\} \frac{\operatorname{dev} \sigma_k(\varepsilon(v))}{\|\operatorname{dev} \sigma_k(\varepsilon(v))\|_F} + p_{k-1}, \tag{16}$$

with the constant $\xi := \left(1 + \sigma_y^2 H^2\right)^{-1}$, the trial stress $\sigma_k(\varepsilon(v)) := \mathbb{C}\,(\varepsilon(v) - p_{k-1})$, and the deviatoric part $\operatorname{dev} \sigma := \sigma - (\operatorname{tr} \sigma/3)\, I$. Thus, it remains to solve a minimization problem with respect to one variable only, i.e. $J_k(u) \to \min$. The theorem of Moreau says, that, due to the specific structure of $J_k(u, p)$, the functional $J_k(u)$ is continuously Fréchet differentiable and strictly convex. Moreover, the explicit form of the derivative is also provided. The Gâteaux differential is given by the relation

$$\mathrm{D}\, J_k(v; w) = \langle \varepsilon(v) - \tilde{p}_k(\varepsilon(v)),\, \varepsilon(w)\rangle_{\mathbb{C}} - l_k(w). \tag{17}$$

Hence, it suffices to find u such that the first derivative of F vanishes. This approach was first discussed in the master thesis [25] by P.G. Gruber. Several numerical examples can also be found in [24, 29, 30].

The second derivative of J_k does not exist. As a remedy, the concept of slanting functions, introduced by X. Chen, Z. Nashed, and L. Qi in [17], allows the application of the following Newton-like method: Let $v^0 \in V_D$ be a given initial guess for the displacement field. Then, for $j = 0, 1, 2, \ldots$ and given v^j, find $v^{j+1} \in V_D$ such that

$$(\mathrm{D}\, J_k)^o\, (v^j; v^{j+1} - v^j, w) = -\mathrm{D}\, J_k(v^j; w) \tag{18}$$

holds for all $w \in V_0$, where the slanting function of $\mathrm{D}\, J_k$ is defined by the identity

$$(\mathrm{D}\, J_k)^o\, (v\,;\, w_1, w_2) = \langle \varepsilon(w_1) - \tilde{p}_k{}^o(\varepsilon(v)\,;\, \varepsilon(w_1)),\, \varepsilon(w_2)\rangle_{\mathbb{C}} \quad \forall\, w_1, w_2 \in V_0.$$

Here, $(\mathrm{D}\, J_k)^o$ is a slanting function for $\mathrm{D}\, J_k$ in (17) if and only if $\tilde{p}_k{}^o$ serves as a slanting function for $\tilde{p}_k$ in (16). By using the definition $\beta_k(\varepsilon(v)) := 1 - \sigma_y\, \|\mathrm{dev}\, \sigma_k(\varepsilon(v))\|_F^{-1}$, and the abbreviations $\beta_k(\varepsilon(v)) := \beta_k$ and $\sigma_k(\varepsilon(v)) := \sigma_k$, a candidate for the slanting function reads

$$\tilde{p}_k{}^o(\varepsilon(v)\,;\, q) = \begin{cases} 0 & \text{if } \beta_k \leq 0, \\ \xi \left(\beta_k\ \mathrm{dev}\, q + (1 - \beta_k)\ \frac{\langle \mathrm{dev}\, \sigma_k, \mathrm{dev}\, q\rangle_F}{\|\mathrm{dev}\, \sigma_k\|_F^2}\ \mathrm{dev}\, \sigma_k \right) & \text{else.} \end{cases}$$

Utilizing this concept, P. G. Gruber and J. Valdman [29, 30] were able to prove the local super-linear convergence of the resulting Newton-like solver in the spatial discretized case (see Table 1). It is still an open problem, if $\tilde{p}_k^o$ serves as a slanting function for $\tilde{p}_k$ in the continuous setting, i. e., when $\tilde{p}_k$ is a Nemytskii operator mapping from $L_p(\Omega)$ into $L_2(\Omega)$ with $p \geq 2$. The assumption, that the Newton-like iterates v^j of (18) are in the Sobolev space $W^{1,2+\epsilon_j}(\Omega)$, where $(\epsilon_j)_{j\in\mathbb{N}}$ is a strictly decreasing positive sequence, may be helpful in this respect, see also [23]. At least the convergence of a damped Newton-like method could be shown in the dissertation of P. Gruber [27] by using a concept of E. Zeidler [58]. An extension of the numerical solver to other kinds of time-dependent models with internal variables, as discussed in [28], is possible and left for future investigation.

The slant Newton method is tested on a benchmark problem in computational plasticity [54]. The left plot of Fig. 3 shows the mesh for the right upper quarter of a plate with geometry $(-10, 10) \times (-10, 10) \times (0, 2)$ and a circular hole of the radius $r = 1$ in the middle. One elastoplastic time step is performed, where a surface load g with the intensity $|g| = 450$ is applied to the plate's upper and lower edge in outer normal direction. Due to the symmetry of the domain, the solution is calculated on

Table 1 Convergence behavior of the slant Newton method for different refinement levels

DOF:	717	5736	45888	367104
step 1:	1.000e+00	1.000e+00	1.000e+00	1.000e+00
step 2:	1.013e−01	1.254e−01	1.367e−01	1.419e−01
step 3:	7.024e−03	6.919e−03	7.159e−03	6.993e−03
step 4:	1.076e−04	9.359e−05	1.263e−04	1.176e−04
step 5:	2.451e−08	6.768e−07	1.744e−06	1.849e−06
step 6:	7.149e−15	6.887e−12	4.874e−09	1.001e−08
step 7:			4.298e−13	2.368e−14

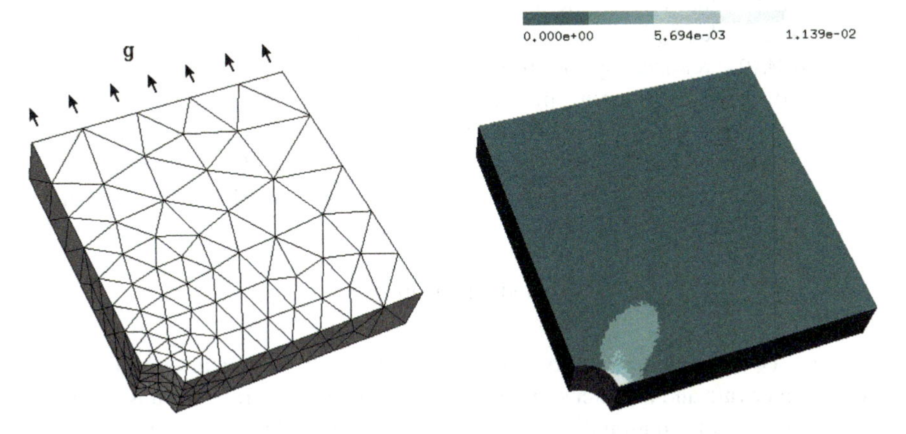

Fig. 3 Coarsest triangulation (*left*) and the Frobenius norm of the plastic strain field p (*right*)

one quarter of the domain only. Thus, homogeneous Dirichlet boundary conditions in the normal direction (gliding conditions) are considered for both symmetry axes. The material parameters are set to

$$\lambda = 1.1074 * 10^5, \quad \mu = 8.0194 * 10^4, \quad \sigma_y = 450\,\sqrt{2/3}, \quad H = 0.5.$$

Differently to the original problem in [54], the modulus of hardening H is nonzero, i. e. hardening effects are considered. The numerical results for the original problem (H = 0) can be found in [24]. The two plots in Fig. 3 show the coarsest tetrahedral FE-mesh with the applied traction g (left), and the Frobenius norm of the plastic strain field p (right). Table 1 outlines the convergence of the slant Newton method, where the initial values for the displacement are chosen to be zero at each level of refinement. The number of degrees of freedom (DOF) is given in the first row. The following rows show the super-linear convergence of the Newton iterates with respect to the Cauchy test $\|u_i - u_{i-1}\|/(\|u_i\| + \|u_{i-1}\|)$. The implementation was done in the NETGEN/NGSolve software package developed by J. Schöberl [51].

4 Adaptive h-FEM and A Posteriori Error Estimates for Elastoplasticity

The efficient numerical treatment of problems with poor regularity of the solution can be realized with adaptive mesh refinement techniques based on a posteriori error estimators. An $h-$finite element adaptive algorithm consists of successive loops of the form

$$\text{SOLVE} \rightarrow \text{ESTIMATE} \rightarrow \text{MARK} \rightarrow \text{REFINE} \tag{19}$$

designed to produce more efficient meshes by targeted local refinements with less computational effort. The a posteriori error analysis of (19) started with the pioneering work of [4] for a two-point elliptic boundary value problem and with the step MARK realized by the max refinement rule. This marking rule currently employed in the engineering literature consists in looking at the elements with the largest error and refining these in order to achieve a better accuracy. Let $\eta^2 := \sum_M \eta_M^2$ denote a typical reliable error estimator with local contributions η_M associated with an edge, face, or element M in the current mesh, the max refinement rule marks a subset $\mathcal{M}$ according to

$$L \in \mathcal{M} \text{ if and only if } \eta_L \geq \Theta \max_M \eta_M \tag{20}$$

with $0 \leq \Theta \leq 1$. The analysis of [4], however, does not provide information on the convergence rate and its extension to higher dimensions still remains unsolved. It is only after the contribution of Dörfler [20] with the introduction of a new marking strategy for error reduction (hereafter referred to as bulk criterion or fixed fraction criterion) that the convergence analysis of AFEMs has experienced significant development. With such criterion, one defines the set $\mathcal{M}$ of the marked objects using the rule

$$\sum_{M \in \mathcal{M}} \eta_M^2 \geq \Theta \, \eta^2 \tag{21}$$

with $0 \leq \Theta \leq 1$. The condition (21) together with local discrete efficiency estimates, and the Galerkin orthogonality yields a linear error reduction rate for the energy norm towards a preassigned tolerance TOL in finite steps for the Poisson problem.

In [16], a proof of convergence of AFEM with indication of the rate of convergence for the primal formulation of plasticity is provided under the application of the bulk criterion (21). Applications include several plasticity models: linear isotropic-kinematic hardening, linear kinematic hardening, multi-surface plasticity as model for nonlinear hardening laws, and perfect plasticity. Exploiting properties of a non-differentiable energy functional J, and the reliability of a new edge-based residual error estimate, we obtain the following results:

(i) *Energy reduction*: for some data oscillations $\text{osc}_\ell^2 \geq 0$ and positive constants ρ_E, C with $\rho_E < 1$ there holds

$$J(w_{\ell+1}) - J(w) \leq \rho_E(J(w_\ell) - J(w)) + C\ \mathrm{osc}_\ell^2.$$

Here, $J(w)$ denotes a minimal energy and $J(w_\ell)$ and $J(w_{\ell+1})$ are energies on refined triangulations $\mathscr{T}_\ell$ and $\mathscr{T}_{\ell+1}$.

(ii) *R–linear convergence for the stresses*: up to oscillation terms there holds

$$|||\sigma - \sigma_\ell|||_{\mathbb{C}^{-1};\Omega} \leq \alpha_\ell \quad \text{for } \ell = 0, 1, 2, \ldots$$

with $\alpha_\ell \to 0$ and linear convergent, and $|||\cdot|||_{\mathbb{C}^{-1};\Omega}$ the energy norm induced by the Hooke tensor $\mathbb{C}$. Here, σ denotes the stress at the exact solution and σ_ℓ its approximation on the triangulation $\mathscr{T}_\ell$.

In [50], the framework introduced in the book [47] is applied to elastoplasticity, where the estimates are derived by the analysis of the variational problem and its dual counterpart. A computable upper bound of the error is obtained on a purely functional level without exploitation of specific properties of the approximation or the method used for its computation. Estimates of such a type are often called "functional a posteriori estimates". Application to linear isotropic hardening allows us to express another reliability estimate

$$\frac{1}{2}|||w - v|||^2 \leq \mathscr{M}(v, \tau, \lambda) \tag{22}$$

which bounds an error of a discrete solution v, i.e. its distance from the exact solution w by an expression on the right-hands side called a functional majorant $\mathscr{M}(v, \tau, \lambda)$. The functional majorant can be generally minimized with respect to free parameters τ, λ to keep the estimate (22) as sharp as possible. Numerical verification of this estimate will be the topic of the forthcoming paper, where it should be profited from the experience in problems with nonlinear boundary conditions [49] and an application of a multigrid preconditioned solver to a majorant computation [56].

5 High Order FEM for Elastoplasticity: *hp*-FEM and BC-FEM

In nowadays computer simulations of elastoplasticity, adaptive h-FEM (as presented in Sect. 4) is probably the most propagated and well known discretization technique. However, as computers become faster, and parallelization is no longer just a scientific topic, the mixture of low and high order finite element methods (hp-FEM) becomes more and more attractive in daily practice. Applying a high order method means to increase the polynomial degree of the shape functions on an element instead of refining it. The major drawback of a high order method is the expensive assembling of the system matrix. As long as this handicap can be settled

Algorithm 1 The hp-adaptive Algorithm:

Require: A mesh $\mathcal{T}$, a polynomial degree vector $(p_K)_{K\in\mathcal{T}}$, a Finite Element Solution u_{FE}.
Ensure: A refined mesh $\mathcal{T}_{\mathrm{ref}}$, a new polynomial degree vector $(p_K)_{K\in\mathcal{T}_{\mathrm{ref}}}$.

1: Determine which elements to refine $\rightarrow \mathcal{T}_h$.
2: Determine where the polynomial degree should be increased $\rightarrow \mathcal{T}_p$.
3: Obtain a preliminary refined mesh $\rightarrow \mathcal{T}'_{\mathrm{ref}}$.
4: Elimination of hanging nodes $\rightarrow \mathcal{T}_{\mathrm{ref}}$.
5: Increase the polynomial degree $p_K = p_K + 1$ for all elements $K \in \mathcal{T}_{\mathrm{ref}} \cap \mathcal{T}_p$. In particular: Elements to which an h-refinement is applied inherit the polynomial degree from their father.

(e.g., by finding recurrences via symbolic computation [5, 8, 9]), the application of such methods are definitely worth their price. The idea of hp-FEM [3, 52] is to increase the polynomial degree locally on elements, where the solution has high regularity. In such areas of the domain we can expect local exponential convergence of the approximate towards the solution. On other elements, i. e. where the regularity is low, mesh refinement is applied, which locally yields algebraic convergence. Moreover, by choosing proper hp-adaptive refinement strategies, an exponential convergence rate can be achieved globally [3].

In elastoplasticity, the solution in each time step is known to be in $H^2_{\mathrm{loc}}(\Omega)$, and, moreover, analytic in balls where the plastic strain p vanishes [6, 41]. Thus, the application of an hp-FEM is a natural choice. In those parts of the interior domain, where the material reacts purely elastic, the polynomial degree is increased, whereas the mesh is refined in plastic areas and towards rough boundary data or geometry.

The basic hp-adaptive algorithm reads as follows:

Note, that Items 3–5 are straight forward, whereas, one still has to decide on the exact realization of Items 1 and 2. In general, the set of all adaptive strategies divides into two classes: strategies which are problem dependent, and those which are not. In problem dependent strategies, the decision whether to refine in h, or in p, or not at all, relies on the evaluation of problem dependent quantities, typically the error estimator. Algorithms of this class can be found, e.g., in [1, 21]. Problem independent algorithms, such as discussed in [18, 19], estimate the regularity of the solution without using problem dependent quantities.

Due to the lack of a reliable and efficient error estimator for elastoplasticity, the use of problem independent algorithms is a natural choice. The application of an algorithm presented in [21] to elastoplastic problems in two dimensions is discussed in [26]. This adaptive algorithm is based on the following idea:

Expressing the solution u to the (elastoplastic) problem as an expansion with respect to orthogonal Legendre polynomials

$$u = \sum_{p,q\in\mathbb{N}_0} u_{pq}\psi_{pq} \tag{23}$$

results in a sequence of coefficients u_{pq}, which decays exponentially if and only if the solution u is analytic:

Proposition 1. *Define on the reference triangle* $\hat{K}$ *the* $L_2(\hat{K})$*-orthogonal basis* ψ_{pq}, $p,q \in \mathbb{N}_0$ *by*

$$\psi_{pq} = \tilde{\psi}_{pq} \circ D^{-1}, \quad \tilde{\psi}_{pq} = P_p^{(0,0)}(\eta_1)\left(\frac{1-\eta_2}{2}\right)^p P_q^{(2p+1,0)}(\eta_2),$$

where $P_p^{(\alpha,\beta)}$ *is the (well known) p-th Jacobi polynomial with respect to the weight* $\eta \mapsto (1-\eta)^\alpha(1+\eta)^\beta$ *and D the Duffy transformation, defined as in [21]. Let* $u \in L_2(\hat{K})$ *be written as in* (23)*. Then u is analytic on* $\overline{\hat{K}}$ *if and only if there exist constants* $C, b > 0$ *such that* $|u_{pq}| \leq C\, e^{-b(p+q)}$ *for all* $p,q \in \mathbb{N}_0$.

Proof. See [45]. □

Since the true solution u is not available, the idea for the hp-adaptive algorithm is to estimate the decay of the coefficients u_{pq} of the expansion of the finite element solution $u_{FE|K} \circ F_K = \sum_{p,q} u_{pq}\psi_{pq}$ instead. If the decay is exponentially, then the polynomial degree p will be increased, otherwise, the mesh will be refined:

Additionally to the presented adaptive strategy in Algorithm 1, a different discretization approach applied to elastoplasticity is investigated in [26]. This approach is still of an hp-adaptive Finite Element type, but with a slightly different aim: Considering a general boundary value problem, where the regularity of the solution is known to be low at the boundary and high in the interior of the domain, the parameters h and p are chosen to be small in a neighborhood of the boundary and to be growing towards the interior of the domain. This growth is done in a manner, such that

- The convergence rate is of the same order as in h-FEM.
- The number of total unknowns is proportional to the number of unknowns on the boundary (such as in BEM).

Algorithm 2 Realization of Items 1 and 2 in Algorithm 1:

Require: A mesh $\mathcal{T}$, a polynomial degree vector $(p_K)_{K\in\mathcal{T}}$, a parameter $b > 0$, a Finite Element Solution u_{FE}.

Ensure: The marked elements $\mathcal{T}_p$ and $\mathcal{T}_h$.

1: For all elements $K \in \mathcal{T}$ compute the expansion coefficients

$$u_{ij,K} = \|\psi_{ij}\|^{-2}_{L_2(\hat{K})} \langle u_{\mathrm{FE}|K} \circ F_K, \psi_{ij}\rangle_{L_2(\hat{K})}$$

for $0 \leq i+j \leq p_K$.

2: Estimate the decay coefficient b_K by a least squares fit of

$$\ln|u_{ij,K}| \approx C_K - b_K(i+j).$$

3: Determine $\mathcal{T}_p = \{K \in \mathcal{T} \mid b_K \geq b\}$ and $\mathcal{T}_h = \{K \in \mathcal{T} \mid b_K < b\}$.

Table 2 Comparison of the degrees of freedom at each numerical example

DOFs at Level	1	2	3	4	5	6
Plate with Hole (h-FEM)	2018	7810	30722	121858	485378	1937410
Plate with Hole (BC-FEM)	2018	5010	14658	37874	103050	307330
Screw Wrench (h-FEM)	474	1778	6882	27074	107394	427778
Screw Wrench (BC-FEM)	474	1618	4266	10290	24490	58474

Due to the second property, the method is called a Boundary Concentrated Finite Element Method (BC-FEM) [35]. The method exploits the knowledge about the regularity of the solution in a way, that it searches for the smallest (and sparse) system which allows for the same convergence rate as is obtained in a classical h-FEM.

In elastoplasticity, BC-FEM can be applied for the purely elastic region, where the solution is known to be analytic [6], whereas the plastic region, where the solution is known to be in $H^2_{\text{loc}}(\Omega)$ [41], is discretized by using h-FEM. However, the interface between plastic ($\|p\| > 0$) and elastic ($\|p\| = 0$) parts of the domain is not known in advance, since the calculation of the plastic strain field p depends on the displacement field, as it is pointed out in (16). Thus, one has to estimate, which parts of the domain will be plastic at the next step of refinement. This task can be again handled by Algorithm 2. The resulting method has the same accuracy as a classical h-FEM, i.e., the error satisfies $\|u - u_h\|_{H^1(\Omega)} = O(h)$, but the number of degrees of freedom is significantly smaller: Considering h-FEM in two dimensions ($d = 2$), the number degrees of freedom is roughly $O(N^2)$, with $N = h^{-1}$ denoting the number of nodes on the boundary of the domain, whereas in BC-FEM it is $O(N_E) + O(N_P^2)$, where N_E is the number of nodes on the boundary of the purely elastic sub-domain, and N_P the number of nodes on the boundary of the plastic sub-domain (compare Table 2). It is possible to generalize the primal and dual domain decomposition solvers proposed in [7] for solving interface-concentrated finite element equations to the plastic-zone concentrated finite element equations which we have to solve at each incremental step.

Finally, we present the results of the following two numerical experiments. More examples regarding some adaptive strategies in hp-FEM for elastoplastic problems can be found in [26].

- A *plate with a hole* $\{x \in [-10, 10]^2 : \|x\| \geq 1\}$ is torn on the top and the bottom edge in normal direction with a traction of intensity $|g| = 450$. Due to the symmetry of the problem, only the top right quarter is considered in the numerical simulation. The material parameters are chosen as follows: Young's modulus $E = 20690$, Poisson ratio $\nu = 0.29$, yield stress $\sigma_y = 450\sqrt{2/3}$, and modulus of hardening $H = 0.1$. On the left of Fig. 4 one can see the mesh after 5 steps of BC-refinement. The elements are colored from blue to red, indicating increasing polynomial degree. On the right of Fig. 4, the elastic (blue) and plastic (red) zones are plotted. Figure 5 shows the adaptive mesh (left), and a zoom

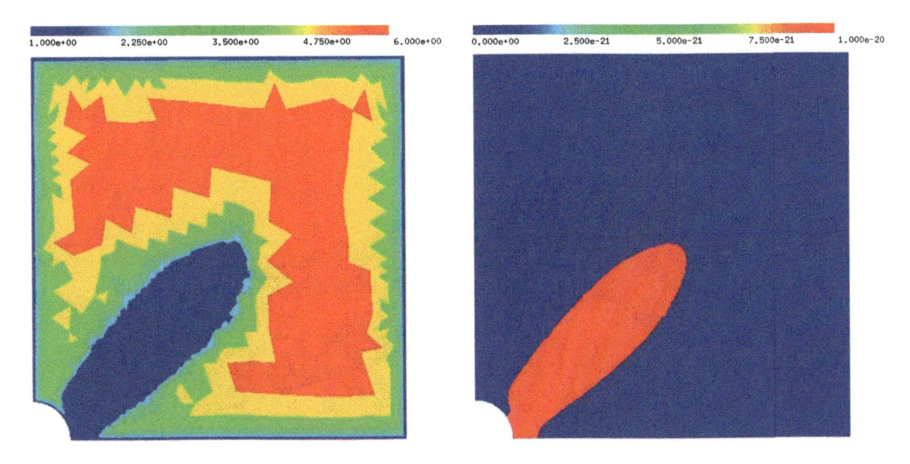

Fig. 4 Plate with a hole: polynomial order (*left*) and plastic zones (*right*)

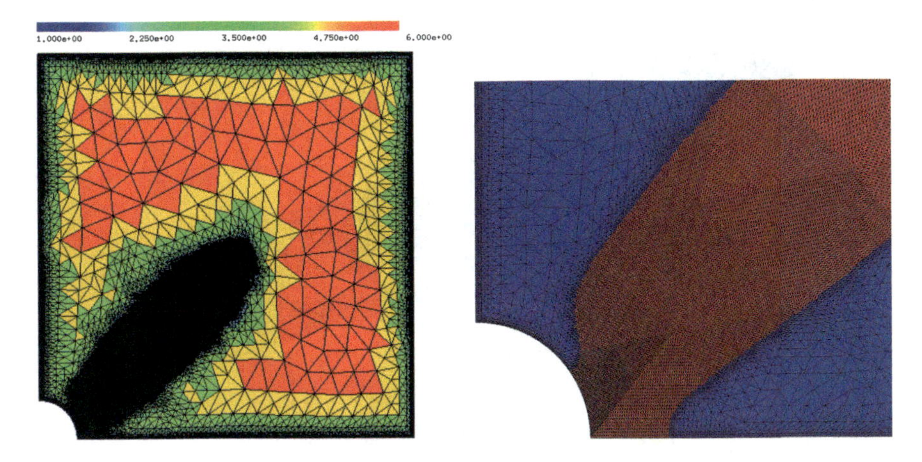

Fig. 5 Plate with a hole: the adaptive mesh

towards the boundary and elastoplastic interface (right). Plastic zones are red, elastic zones are blue.

- A *screw wrench* sticks on a screw (homogeneous Dirichlet condition) and is pressed down at its handhold in normal direction with an intensity $|g| = 1e6$. The material parameters are chosen as follows: Young's modulus $E = 2e8$, Poisson ratio $\nu = 0.3$, yield stress $\sigma_y = 1e6$, modulus of hardening $H = 0.01$. On top of Fig. 6, one can see the mesh after 5 steps of BC-refinement. The elements are colored from blue to red, indicating increasing polynomial degree. On bottom of Fig. 6, the elastic (blue) and plastic (red) zones are plotted.

Table 2 shows the number of degrees of freedom for both examples in case of an h-FEM and a BC-FEM discretization.

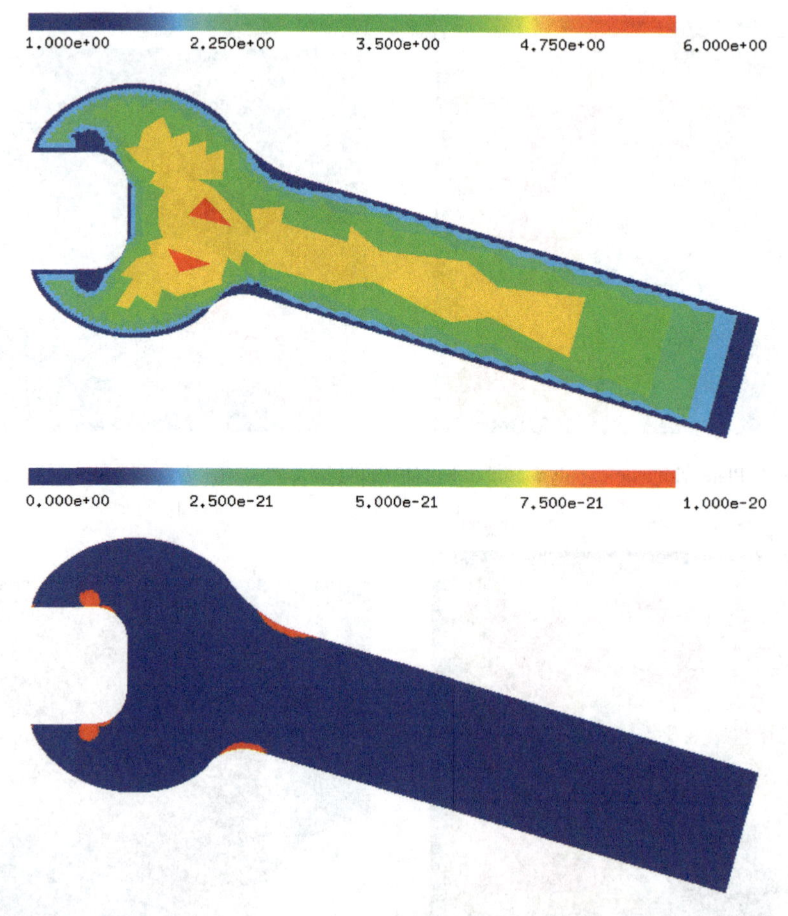

Fig. 6 Screw Wrench: polynomial order (*top*) and plastic zones (*bottom*)

6 Conclusion

We presented two strategies to deal with the non-smoothness of the functional arising at each incremental step in elastoplasticity. The first one uses traditional regularization techniques whereas the second one makes use of Moreau's theorem for the reduced functional. Generalized Newton-methods are derived and analyzed on the basis of the concept of slanting functions. Furthermore, we proposed residual-based and functional-based a posteriori error estimates for elastoplastic problems which can be used in an AFEM. In some cases the convergence of the AFEM can be shown. Finally, we studied the use of higher-order finite elements in elastoplasticity. The approximation quality of higher-order elements strongly depends on the local regularity of the solution. The new plastic-zone concentrated finite element approximation used low-order elements in the plastic zones and boundary or, more precisely, interface concentrated finite element approximations in the elastic zone where higher and higher order finite elements are used in dependence on the distance to the elastic-plastic interface and the boundary. Regularity detectors can be used to predict the elastic-plastic interface at each incremental step.

References

1. Ainsworth, M., Senior, B.: An adaptive refinement strategy for hp-finite element computations. In: Proceedings of international centre for mathematical sciences on Grid adaptation in computational PDES: theory and applications, pp. 165–178. Elsevier Science Publishers B. V., Amsterdam, The Netherlands (1998)
2. Alberty, J., Carstensen, C., Zarrabi, D.: Adaptive numerical analysis in primal elastoplasticity with hardening. Comput. Methods Appl. Mech. Eng. **171**(3–4), 175–204 (1999)
3. Babuska, I., Guo, B.: The h-p version of the finite element method – Part 1: The basic approximation results. Comput. Mech. **1**, 21–41 (1986)
4. Babuška, I., Vogelius, M.: Feedback and adaptive finite element solution of one-dimensional boundary value problems. Numer. Math. **44**(1), 75–102 (1984)
5. Becirovic, A., Paule, P., Pillwein, V., Riese, A., Schneider, C., Schöberl, J.: Hypergeometric summation algorithms for high order finite elements. Computing **78**(3), 235–249 (2006)
6. Bensoussan, A., Frehse, J.: Regularity results for nonlinear elliptic systems and applications. Applied Mathematical Sciences, vol. 151. Springer, Berlin (2002)
7. Beuchler, S., Eibner, T., Langer, U.: Primal and dual interface concentrated iterative substructuring methods. SIAM J. Numer. Anal. **46**(6), 2818–2842 (2008)
8. Beuchler, S., Pillwein, V.: Sparse shape functions for tetrahedral p-FEM using integrated Jacobi polynomials. Computing **80**(4), 345–375 (2007)
9. Beuchler, S., Pillwein, V.: Completions to sparse shape functions for triangular and tetrahedral p-FEM. In: Langer, U., Discacciati, M., Keyes, D., Widlund, O., Zulehner, W. (eds.) Domain Decomposition Methods in Science and Engineering XVII, *Lecture Notes in Computational Science and Engineering*, vol. 60, pp. 435–442. Springer, Heidelberg (2008)
10. Brokate, M., Carstensen, C., Valdman, J.: A quasi-static boundary value problem in multi-surface elastoplasticity: Part 1 – analysis. Math. Meth. Appl. Sci. **27**(14), 1697–1710 (2004)
11. Brokate, M., Carstensen, C., Valdman, J.: A quasi-static boundary value problem in multi-surface elastoplasticity: Part 2 – numerical solution. Math. Meth. Appl. Sci. **28**(8), 881–901 (2005)
12. Carstensen, C.: Nonlinear interface problems in solid mechanics: Finite element and boundary element couplings. Habilitationsschrift, Universität Hannover (1993)
13. Carstensen, C.: Coupling of fem and bem for interface problems in viscoplasticity and plasticity with hardening. SIAM J. Numer. Anal. **33**(1), 171–207 (1996)
14. Carstensen, C.: Domain decomposition for a non-smooth convex minimization problem and its application to plasticity. Numer. Linear Algebra Appl. **4**(3), 177–190 (1997)
15. Carstensen, C.: Numerical analysis of the primal elastoplasticity with hardening. Numer. Math. (82), 577–597 (2000)
16. Carstensen, C., Orlando, A., Valdman, J.: A convergent adaptive finite element method for the primal problem of elastoplasticity. Internat. J. Numer. Methods Engrg. **67**(13), 1851–1887 (2006)
17. Chen, X., Nashed, Z., Qi, L.: Smoothing methods and semismooth methods for nondifferentiable operator equations. SIAM J. Numer. Anal. **38**(4), 1200–1216 (2001)
18. Demkowicz, L., Rachowicz, W., Devloo, P.: A fully automatic hp-adaptivity. In: Proceedings of the Fifth International Conference on Spectral and High Order Methods (ICOSAHOM-01) (Uppsala), vol. 17, pp. 117–142 (2002)

19. Demkowicz, L., Šolín, P.: Goal-oriented hp-adaptivity for elliptic problems. Comput. Methods Appl. Mech. Engrg. **193**(6–8), 449–468 (2004)
20. Dörfler, W.: A convergent adaptive algorithm for Poisson's equation. SIAM J. Numer. Anal. **33**(3), 1106–1124 (1996)
21. Eibner, T., Melenk, J.M.: An adaptive strategy for hp-FEM based on testing for analyticity. Comput. Mech. **39**(5), 575–595 (2007)
22. Ekeland, I., Témam, R.: Convex Analysis and Variational Problems. SIAM, New York (1999)
23. Griesse, R., Meyer, C.: Optimal control of static plasticity with linear kinematic hardening. Tech. rep. (2009). Weierstrass Institute for Applied Analysis and Stochastics, WIAS Preprint 1370
24. Gruber, P., Valdman, J.: Implementation of an elastoplastic solver based on the Moreau-Yosida theorem. Math. Comput. Simul. **76**(1–3), 73–81 (2007)
25. Gruber, P.G.: Solution of elastoplastic problems based on the Moreau-Yosida theorem. Master's thesis, Institute of Computational Mathematics, Johannes Kepler University Linz, Austria (2006)
26. Gruber, P.G.: Adaptive strategies for hp-FEM in elastoplasticity. DK-Report 2010-02, Johannes Kepler University Linz, DK W1214 "Doctoral Program on Computational Mathematics" (2010)
27. Gruber, P.G.: Fast solvers and adaptive high-order FEM in elastoplasticity. Ph.D. thesis, Institute of Computational Mathematics, Johannes Kepler University Linz, Austria (2011)
28. Gruber, P.G., Knees, D., Nesenenko, S., Thomas, M.: Analytical and numerical aspects of time-dependent models with internal variables. Z. angew. Math. Mech. **90**, 861–902 (2010)
29. Gruber, P.G., Valdman, J.: Newton-like solver for elastoplastic problems with hardening and its local super-linear convergence. In: Kunisch, K., Of, G., Steinbach, O. (eds.) Numerical Mathematics and Advanced Applications. Proceedings of ENUMATH 2007, pp. 795–803. Springer, Berlin (2008)
30. Gruber, P.G., Valdman, J.: Solution of one-time-step problems in elastoplasticity by a slant Newton method. SIAM J. Sci. Comput. **31**(2), 1558–1580 (2009)
31. Han, W., Reddy, B.D.: Plasticity, *Interdisciplinary Applied Mathematics*, vol. 9. Springer, New York (1999)
32. Hofinger, A., Valdman, J.: Numerical solution of the two-yield elastoplastic minimization problem. Computing **81**(1), 35–52 (2007)
33. Johnson, C.: Existence theorems for plasticity problems. J. Math. Pures Appl. **55**, 431–444 (1976)
34. Johnson, C.: On plasticity with hardening. J. Math. Anal. Appls. **62**, 325–336 (1978)
35. Khoromskij, B.N., Melenk, J.M.: Boundary concentrated finite element methods. SIAM J. Numer. Anal. **41**(1), 1–36 (2003)
36. Kienesberger, J.: Multigrid preconditioned solvers for some elastoplastic problems. In: Lirkov, I., Margenov, S., Waśniewski, J., Yalamov, P. (eds.) Proceedings of LSSC 2003, Lecture Notes in Computer Science, vol. 2907, pp. 379–386. Springer, Berlin (2004)
37. Kienesberger, J.: Efficient solution algorithms for elastoplastic problems. Ph.D. thesis, Institute of Computational Mathematics, Johannes Kepler University Linz (2006)
38. Kienesberger, J., Langer, U., Valdman, J.: On a robust multigrid-preconditioned solver for incremental plasticity problems. In: Blaheta, R., Starý, J. (eds.) Proceedings of IMET 2004 – Iterative Methods, Preconditioning & Numerical PDEs, pp. 84–87. Institute of Geonics AS CR Ostrava (2004)
39. Kienesberger, J., Valdman, J.: Multi-yield elastoplastic continuum-modeling and computations. In: Dolejsi, V., Feistauer, M., Felcman, J., Knobloch, P., Najzar, K. (eds.) Numerical mathematics and advanced applications. Proceedings of ENUMATH 2003, pp. 539–548. Springer, Berlin (2004)
40. Kienesberger, J., Valdman, J.: An efficient solution algorithm for elastoplasticity and its first implementation towards uniform h- and p- mesh refinements. In: Castro, A.B., Gomez, D., Quintela, P., Salgado, P. (eds.) Numerical mathematics and advanced applications: Proceedings of ENUMATH 2005, pp. 1117–1125. Springer, Berlin (2006)

41. Knees, D., Neff, P.: Regularity up to the boundary for nonlinear elliptic systems arising in time-incremental infinitesimal elasto-plasticity. SIAM J. Math. Anal. **40**, 21–43 (2008)
42. Korneev, V.G., Langer, U.: Approximate solution of plastic flow theory problems. Teubner-Texte zur Mathematik, vol. 69. Teubner-Verlag, Leipzig (1984)
43. Matthies, H.: Existence theorems in thermoplasticity. J. Mécanique **18**(4), 695–712 (1979)
44. Matthies, H.: Finite element approximations in thermo-plasticity. Numer. Funct. Anal. Optim. **1**(2), 145–160 (1979)
45. Melenk, J.M.: hp-finite element methods for singular perturbations, *Lecture Notes in Mathematics*, vol. 1796. Springer, Berlin (2002)
46. Moreau, J.J.: Proximité et dualité dans un espace hilbertien. Bull. Soc. Math. France **93**, 273–299 (1965)
47. Neittaanmäki, P., Repin, S.: Reliable methods for computer simulation, *Studies in Mathematics and its Applications*, vol. 33. Elsevier Science B.V., Amsterdam (2004)
48. Repin, S.: A Posteriori Estimates for Partial Differential Equations, *Radon Series on Computational and Applied Matehmatics*, vol. 4. Walter de Gruyter, Berlin, New York (2008)
49. Repin, S., Valdman, J.: Functional a posteriori error estimates for problems with nonlinear boundary conditions. J. Numer. Math. **16**(1), 51–81 (2008)
50. Repin, S., Valdman, J.: Functional a posteriori error estimates for incremental models in elasto-plasticity. Cent. Eur. J. Math. **7**(3), 506–519 (2009)
51. Schöberl, J.: Netgen – an advancing front 2d/3d-mesh generator based on abstract rules. Comput. Visual. Sci. **1**, 41–52 (1997)
52. Schwab, C.: P- and hp- finite element methods: theory and applications in solid and fluid mechanics. Oxford University Press, Oxford (1998)
53. Simo, J.C., Hughes, T.J.R.: Computational inelasticity, Interdisciplinary Applied Mathematics, vol. 7. Springer, New York (1998)
54. Stein, E.: Error-controlled Adaptive Finite Elements in Solid Mechanics. Wiley, Chichester (2003)
55. Valdman, J.: Mathematical and numerical analysis of elastoplastic material with multi-surface stress-strain relation. Ph.D. thesis, Christian-Albrechts-Universität zu Kiel (2002)
56. Valdman, J.: Minimization of functional majorant in a posteriori error analysis based on H(div) multigrid-preconditioned CG method. Advances in Numerical Analysis **2009**(Article ID 164519), 15 pages (2009). DOI 10.1155/2009/164519
57. Wieners, C.: Multigrid methods for finite elements and the application to solid mechanics. Theorie und Numerik der Prandtl-Reuß Plastizität (2000). Habilitationsschrift, Universität Heidelberg
58. Zeidler, E.: Nonlinear functional analysis and its applications. III. Springer-Verlag, New York (1985). Variational methods and optimization, Translated from the German by Leo F. Boron
59. Zienkiewicz, O.: The Finite Element Method, 3rd Expanded & rev. ed. edn. McGraw-Hill, New York (1977)

A Symbolic-Numeric Algorithm for Genus Computation

Mădălina Hodorog and Josef Schicho

Abstract We report on a symbolic-numeric algorithm for computing the genus of a plane complex algebraic curve defined by a squarefree polynomial with exact and inexact coefficients. For the inexact data we are given a positive real number, which measures the error (noise) level in the coefficients. The symbolic numeric algorithm proceeds as follows: (i) firstly, we compute the numerical singularities of the plane complex algebraic curve by using subdivision methods; (ii) secondly, we compute the link of each singularity by intersecting the plane complex algebraic curve with a small sphere centered in the singularity; (iii) we then compute the Alexander polynomial of each link of the singularity by using combinatorial methods from knot theory; (iv) from the computed Alexander polynomial we compute the delta-invariant of each singularity; (iv) finally, from the delta-invariants of all the singularities we derive a formula for the genus of the plane complex algebraic curve. The computed results are approximate and thus we interpret them using regularization principles. We perform several test experiments, which indicate that the computed results are valid in the presence of noisy coefficients.

1 Introduction

To raise new questions, new possibilities, to regard old problems from a new angle, requires creative imagination and marks real advance in science

Albert Einstein

M. Hodorog (✉) · J. Schicho
Johann Radon Institute for Computational and Applied Mathematics, Altenbergerstr. 69, 4040 Linz, Austria
e-mail: madalina.hodorog@oeaw.ac.at; josef.schicho@oeaw.ac.at

U. Langer and P. Paule (eds.), *Numerical and Symbolic Scientific Computing*, Texts and Monographs in Symbolic Computation, DOI 10.1007/978-3-7091-0794-2_4,

The genus computation problem is a classical subject in computer algebra. Presently, several symbolic algorithms are available for computing the genus of plane algebraic curves over an algebraically closed field, see [17, 19, 32]. There exist also good implementations for these algorithms in several packages of some well-known computer algebra systems such as: Maple, Magma, Singular, Axiom. We shortly recall these packages, for further details see [16, 18, 20, 22, 28]:

1. Algcurves package developed at the Florida University, by Mark van Hoeij, written in Maple.
2. CASA (Computer algebra system for algebraic geometry) package developed at the Research Institute for Symbolic Computation, Hagenberg Austria, written in Maple.
3. GHS (Gaundry, Hess, Smart) attack package developed at Berlin University, written in Magma.
4. normal.lib package developed at Kaiserslautern University, written in Singular.
5. PAFF (Package for algebraic function fields in one variable) package developed at INRIA-Roquencort, by Gaétan Haché, written in Axiom.

On the other hand, there are situations when computing with numerical coefficients is preferable, for instance when the coefficients are obtained by measurements. At present numerical algorithms for genus computation are reported in the literature. In [4] the genus of any one-dimensional irreducible component of an algebraic set is computed using the homotopy continuation method, whereas in [30] the numerical genus of a plane algebraic curve with all of its singularities ordinary and affine is computed. In this paper, we propose such an algorithm for the computation of the genus of plane complex algebraic curves which makes use of the advantages of both symbolic and numeric algorithms. The method is based on combinatorial techniques from knot theory (see [10,25]), that allow us to successfully analyse the singularities of the plane complex algebraic curve by computing their topology. Previous research and results have successfully shown that the topology of the singularities of a plane complex algebraic curve is mainly determined by their algebraic link, see [27]. The algebraic link can be uniquely identified by its corresponding Alexander polynomial, see [40]. From the Alexander polynomial we derive a general formula for the delta-invariant of each singularity of the plane complex algebraic curve, which allows us to compute the value for the genus of the plane complex algebraic curve.

We have to pay further attention while formulating the genus computation problem due to the existence of numeric errors. As expected, we deduce that the value of the computed genus is highly sensitive to tiny perturbations of the coefficients of the defining polynomial of the input plane complex algebraic curve. Therefore we have to take a decision regarding the interpretation of the results or the reformulation of the input problem. We intend to use the same approach proposed by Z. Zeng for solving other numerical problems from algebraic geometry such as: the computation of the greatest common divisors of polynomials, the computation of the rank of matrices or the computation of the solutions of systems of polynomial equations, see [12, 41]. Other approaches are also taken into considerations, see

[11,34]. Further tests in these different directions of research will guide us in making the best decision.

In this paper we present a symbolic-numeric algorithm for computing the genus of plane complex algebraic curves. In Sect. 2 we introduce the genus computation problem and we propose a strategy for solving it. In Sect. 3 we describe the steps that solve the genus computation problem. In Sect. 4 we present the numerical difficulties for the genus computation problem, which arise when numerical data are used; we test several approaches before proposing the final remedy for these difficulties. We include several tests and experiments in Sect. 5, while in Sect. 6 we outline the conclusions and the future directions of research.

2 The Genus Computation Problem

There are no big problems, there are just a lot of little problems.
Henry Ford

2.1 *Genus of Plane Algebraic Curves*

The objects of our study are the plane algebraic curves over the field of complex numbers. We define the plane algebraic curves over an algebraically closed field in the following way:

Definition 1. Let K be an algebraically closed field, and $f(x, y) \in K[x, y]$ a nonconstant squarefree polynomial in x and y with coefficients in K. A plane algebraic curve over K is defined as the set of all solutions in K^2 of the equation $f(x, y) = 0$, i.e. the set $C = \{(x, y) \in K^2 | f(x, y) = 0\}$. For the curve C, f is called the defining polynomial of C. The degree of the polynomial $f(x, y)$ is called the degree of the curve C.

For a plane complex algebraic curve we are interested in a special type of points and that is its singularities (or singular points or multiple points). Informally, the singularities of an algebraic curve are the points where the curve has nasty behaviour such as a cusp or a point of self-intersection. A cusp is a point at which two branches of the curve meet such that the tangents at each branch are equal, while a point of self-intersection (or double point) is a point at which two branches of the curve meet such that the tangents at each branch are distinct, see Fig. 1.

Formally, we can give the following definition for the singularities of a plane algebraic curve over an algebraically closed field:

Definition 2. Let K be a field (i.e. the complex numbers) and $f(x, y) \in K[x, y]$ be a polynomial in x and y with coefficients over the field K. Let $C = \{(x, y) \in K^2 | f(x, y) = 0\}$ be a plane algebraic curve defined over K and $(a, b) \in C$ be a point on the curve, i.e. $f(a, b) = 0$. The point (a, b) is a singularity of C if the x and y partial derivatives of f are both zero at the point (a, b), i.e.

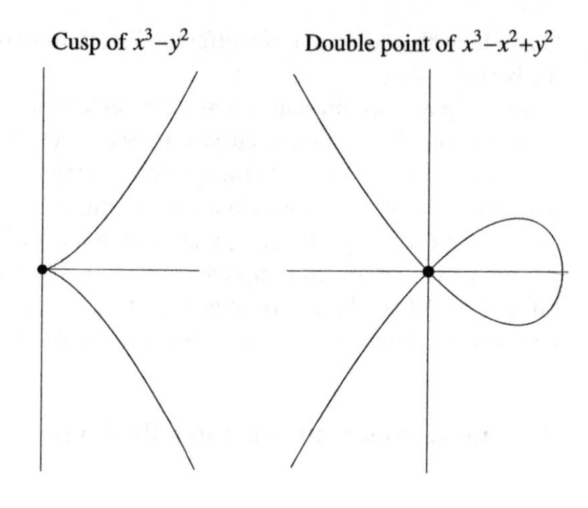

Fig. 1 Cusp in the origin of the curve $x^3 - y^2$ and double point in the origin of the curve $x^3 - x^2 + y^2$

$$\left(\frac{\mathrm{d}f}{\mathrm{d}x}(a,b), \frac{\mathrm{d}f}{\mathrm{d}y}(a,b)\right) = (0,0).$$

In the theory of plane algebraic curves, one is interested in computing their genus, which is a birational invariant that plays an important role in the rational parametrization property of plane algebraic curves. From the theory we know that an irreducible plane algebraic curve is rational parametrizable if and only if its genus is 0. The main purpose of this paper is to compute the genus of plane complex algebraic curves.

For algebraic curves with only ordinary singularities we have a method for computing the genus, based on the multiplicities of the ordinary singularities. We will not focus on the details of this method, as this is not the purpose of our paper. We advise the reader to consult [7, 15, 33, 38] for more information on this method.

We compute the genus of a plane algebraic curve over an algebraically closed field using the following definition:

Definition 3. Let C be a plane algebraic curve defined over an algebraically closed field K, $Sing(C)$ the set of singularities of C, and d the degree of C. Then the genus of the plane algebraic curve, denoted with $genus(C)$, is computed using the following formula:

$$genus(C) = \frac{1}{2}(d-1)(d-2) - \sum_{P \in Sing(C)} \delta\text{-invariant}(P),$$

where $genus(C) \in \mathbb{Z}$.

We notice that the computation of the genus reduces to the computation of the *δ-invariant* of each singularity of the curve. We present the method for computing the δ-invariant of each singularity in detail in Sect. 3.

Thus the problem that we want to solve is the following: given a plane complex algebraic curve whose defining polynomial contains numeric coefficients, the degree of the curve and the set of its singularities, we want to compute the value for the genus of the given plane complex algebraic curve.

2.2 Strategy for Solving the Problem

In order to solve the genus computation problem, we first divide it into several subproblems (some of which are interdependent), we solve each of these subproblems and then we combine the solutions to these subproblems to get the solution to our original problem. We divide the genus computation problem into the following subproblems:

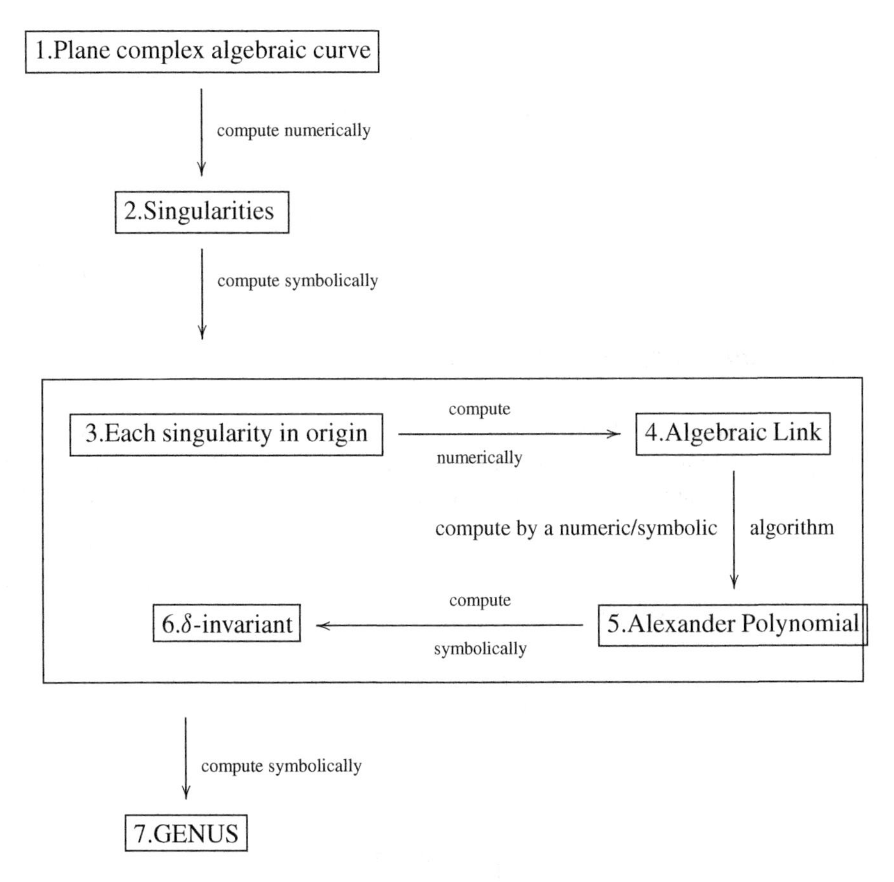

and that is:

1. We compute the singularities of the plane complex algebraic curve.
2. We translate each computed singularity in the origin.

3. We compute the algebraic link for each translated singularity.
4. We compute the Alexander polynomial for each singularity from the algebraic link.
5. We derive a formula for the δ-invariant for each singularity from the Alexander polynomial.
6. We compute the genus from the δ-invariants of all the singularities.

We use for our implementation the AXEL (see [39]) algebraic geometric modeler developed in the Galaad research team at INRIA, Sophia-Antipolis, which provides algebraic tools for the manipulation and the computation with implicit algebraic curves and implicit algebraic surfaces.

3 Why Knot? Alternative Solution to the Genus Computation Problem

Mathematician 1: Okay, so there are three steps to your algorithm. Step one is the input and step three is the output. What is step two?
Mathematician 2: Step two is when a miracle occurs.
Mathematician 1: Oh, I see. Uh, perhaps you could explain that second step a bit more?

K.O.Geddes, S.R.Czapor, G.Labahn

3.1 Computing the Singularities of the Algebraic Curve

The first subproblem that we solve is to compute the singularities of the plane complex algebraic curve C. Given the defining polynomial $F(z,w) \in \mathbb{C}[z,w]$ of C with numerical coefficients, we compute the set of all its singularities, that is the set $Sing(C) = \{(z_0, w_0) \in \mathbb{C}^2 | F(z_0, w_0) = 0, \frac{dF}{dz}(z_0, w_0) = 0, \frac{dF}{dw}(z_0, w_0) = 0\}$. In order to compute the set $Sing(C)$ we need to solve an overdetermined system of polynomial equations in $\mathbb{C}^2$:

$$\begin{cases} F(z_0, w_0) = 0 \\ \dfrac{dF}{dz}(z_0, w_0) = 0 \\ \dfrac{dF}{dw}(z_0, w_0) = 0 \end{cases} \tag{1}$$

or equivalently, by replacing $F(z,w) = F(x+iy, u+iv) = s(x,y,u,v) + it(x,y,u,v)$, an overdeterminate system of polynomial equations in $\mathbb{R}^4$:

$$\begin{cases} s(x_0, y_0, u_0, v_0) = t(x_0, y_0, u_0, v_0) = 0 \\ \frac{\mathrm{d}s}{\mathrm{d}x}(x_0, y_0, u_0, v_0) = \frac{\mathrm{d}t}{\mathrm{d}x}(x_0, y_0, u_0, v_0) = 0 \\ \frac{\mathrm{d}s}{\mathrm{d}u}(x_0, y_0, u_0, v_0) = \frac{\mathrm{d}t}{\mathrm{d}u}(x_0, y_0, u_0, v_0) = 0 \end{cases} . \tag{2}$$

For solving the systems (1), (2), we use subdivision methods introduced by [1,2,24,29] and implemented in Mathemagix [23] and Axel [39]. The subdivision methods compute the real solutions for $Sing(C)$. In the future, we intend to use algebraic methods to compute the complex solutions for $Sing(C)$ as proposed in [8]. We also mention the work in progress in the same direction of research done in parallel by [5], and by [26]. Further work in this direction is still required so that we can use these methods for our systems. Using Axel's subdivision algorithm we compute a list of points in $\mathbb{R}^2$ denoted with S with two properties:

1. Each real singularity of the algebraic curve C is close to one of the points in S.
2. The first derivatives of the defining polynomial of the algebraic curve in each point from S is small.

For $Sing(C)$ we compute all distinct singularities in the projective space. To do this we homogenize the equation of C and dehomogenize it with respect to different variables. We get three affine open subsets of the projective curve, and we have to be careful not to return singularities in the overlaps twice. We give a schematic summary of this algorithm:

Algorithm 3 Singularities of the algebraic curve C SING(F, d)

Input: $C = \{(z, w \in \mathbb{C}^2 | F(z, w) = 0)\}$, $F \in \mathbb{C}[z, w]$ has numeric coefficients
d the degree of C
Output: $Sing(C)$
where $Sing(C)$ is the set of all singularities of C.

1. Homogenize $F(z, w)$ w.r.t. u obtaining $F_1(z, w, u)$
   a. Dehomogenize $F_2(z, w) := F_1(z, w, 1)$
   b. Get S_1 by solving the system of equations $F_2 = \mathrm{d}_z F_2 = \mathrm{d}_w F_2 = 0$
   c. Dehomogenize $F_3(w, u) := F_1(1, w, u)$
   d. Get S_2 by solving the system $F_3 = \mathrm{d}_w F_3 = \mathrm{d}_u F_3 = u = 0$
   e. Dehomogenize $F_4(z, u) := F_1(z, 1, u)$
   f. Get S_3 by solving the system $F_4 = \mathrm{d}_z F_4 = \mathrm{d}_u F_4 = z = u = 0$
2. Return $Sing(C) = S_1 \cup S_2 \cup S_3$

3.2 *Computing the Algebraic Link of an Isolated Singularity*

The second subproblem that we need to solve is to compute the algebraic link for each isolated singularity of the plane complex algebraic curve. In this subsection we first present the main reasons for studying the algebraic link of an isolated singularity of a plane complex algebraic curve, we then define the algebraic link of an isolated singularity, and we conclude with giving a method for computing the algebraic link of an isolated singularity.

We consider a plane complex algebraic curve as a real two-dimensional subset in $\mathbb{C}^2 \cong \mathbb{R}^4$. We need to study and to understand the topology of these subsets near their singularities, which can be determined by the corresponding algebraic link associated to each singularity.

Milnor proved the following important result concerning the topology of complex hypersurfaces:

Theorem 1 (Milnor[27]). *Let $V \subset \mathbb{C}^{n+1}$ be a hypersurface in $\mathbb{C}^{n+1}$, i.e. an algebraic variety defined by a single polynomial. Assume $\mathbf{0} \in V$ and $\mathbf{0}$ is an isolated singularity, i.e. there is no other singularity on a sufficiently small neighborhood of $\mathbf{0}$; S_ϵ is the sphere centered in $\mathbf{0}$ and of radius ϵ; and D_ϵ is the disk centered in $\mathbf{0}$ of radius ϵ. Then, for sufficiently small ϵ, $K = S_\epsilon \cap V$ is a $(2n-1)$-dimensional nonsingular set and $D_\epsilon \cap V$ is homeomorphic to the cone over K.*

In the curve case, $n = 1$ and all singularities are isolated. Next we describe how one can compute the algebraic link associated to an isolated singularity of a plane complex algebraic curve. What we also want is to move the computed algebraic link from $\mathbb{R}^4$ to $\mathbb{R}^3$, and the stereographic projection allows us to accomplish this goal.

We compute the algebraic link of an isolated singularity of the plane complex algebraic curve C in the following way: we consider the curve C which has an isolated singularity in the origin; we take the sphere centered in the origin and of a small radius ϵ; we intersect the curve C with this sphere obtaining a set in the 4-dimensional space, which based on Theorem 1 is an algebraic link for sufficiently small radius. Next, we follow Brauner and Heergaard technique [6] to move the algebraic link from the 4-dimensional space to the 3-dimensional space using the stereographic projection. The stereographic projection allows us not only to project the 4-dimensional link into the 3-dimensional space, but it actually preserves all the topological properties of the link from the the 4-dimensional space into the 3-dimensional space.

In 3-dimensions, the stereographic projection is a certain mapping that projects a sphere onto a plane. It is constructed in the following way: we take a sphere; we draw a line from the north pole of the sphere to a point P in the equator plane to intersect the sphere at a point Q. The stereographic projection of P is Q. The map is defined at the sphere minus the north pole. In fact, the stereographic projection gives an explicit homeomorphism from the unit sphere minus the north pole to the Euclidean plane.

More generally, the stereographic projection may be applied to a n-sphere S^n in the $(n+1)$-dimensional Euclidean space R^{n+1} in the following way:

Definition 4. Consider an n-sphere $S^n = \{(x_1, x_2, \ldots, x_{n+1}) \subset \mathbb{R}^{n+1} | x_1^2 + x_2^2 + \ldots + x_{n+1}^2 = 1\} \in \mathbb{R}^{n+1}$ in the $(n+1)$ dimensional Euclidean space $\mathbb{R}^{n+1}$, and $Q(0, 0, 0, \ldots, 1) \in S^n$ the north point of the n-sphere. If H is a hyperplane in $\mathbb{R}^{n+1}$

not containing Q, then the stereographic projection of the point $P \in S^n \setminus Q$ is the point $P^{'}$ of the intersection of the line QP with H. The stereographic projection is a homeomorphism from $S^n \setminus Q \to \mathbb{R}^n$.

We now describe the method used for computing the algebraic link of an isolated singularity. For the given algebraic curve C considered as a real two-dimensional subset in $\mathbb{C}^2 \cong \mathbb{R}^4$:

$$C = \{(x, y, u, v) \in \mathbb{R}^4 | F(x, y, u, v) = 0\} \subset \mathbb{C}^2 \cong \mathbb{R}^4$$

for which the origin $(0, 0, 0, 0)$ is a singularity, that is:

$$\left(F(0,0,0,0), \frac{\delta F}{\delta x}(0,0,0,0), \frac{\delta F}{\delta y}(0,0,0,0), \frac{\delta F}{\delta u}(0,0,0,0), \frac{\delta F}{\delta v}(0,0,0,0)\right)$$
$$= (0,0,0,0,0),$$

we consider the 3-sphere centered in the origin and of small radius ϵ:

$$\begin{aligned} S^3 &= \{(z, w) \in \mathbb{C}^2 | |z|^2 + |w|^2 = \epsilon^2\} \\ &= \{(x, y, u, v) \in \mathbb{R}^4 | x^2 + y^2 + u^2 + w^2 = \epsilon^2\} \subset \mathbb{R}^4. \end{aligned}$$

We intersect the given curve with this sphere:

$$X = C \bigcap S^3 = \{(x, y, u, v) \in \mathbb{R}^4 | F(x, y, u, v) = 0,\ x^2 + y^2 + u^2 + v^2 = \epsilon^2\} \subset \mathbb{R}^4,$$

obtaining X, a real 1-dimensional set in the 4-dimensional space.
We take a point on the sphere which is not on the curve, that is:

$$P(0,0,0,\epsilon) \in S^3 \setminus C(F(0,0,0,\epsilon) \neq 0),$$

and we apply the stereographic projection for projecting the set X from the 4-dimensional space into the 3-dimensional space. We define the stereographic projection using the following homeomorphism:

$$f : S^3 \setminus \{P\} \subset \mathbb{R}^4 \to \mathbb{R}^3$$

$$(x, y, u, v) \to (a, b, c)^T = \left(\frac{x}{\epsilon - v}, \frac{y}{\epsilon - v}, \frac{u}{\epsilon - v}\right).$$

Based on Milnor's results we know that for sufficiently small ϵ the image of X under the stereographic projection f is a link, i.e. $f(X)$ is a link. Next we compute this set $f(X)$:

$$f(X) = \{(a, b, c) \in \mathbb{R}^3 | \exists (x, y, u, v) \in C \cap S^3 : (a, b, c) = f(x, y, u, v)\}.$$

We notice that we can rewrite $f(X)$ in the following way:

$$f(X) = \{(a,b,c) \in \mathbb{R}^3 | \exists (x,y,u,v) = f^{-1}(a,b,c) \in C \cap S^3\},$$

since f is an homeomorphism, and so it is a bijection, and therefore f is invertible and it admits an inverse. We compute the inverse f^{-1}:

$$f^{-1} : \mathbb{R}^3 \to S^3 \setminus \{P\}$$

$$(a,b,c) \to (x,y,u,v) = \left(\frac{2a\epsilon}{n}, \frac{2b\epsilon}{n}, \frac{2c\epsilon}{n}, \frac{-\epsilon + a^2\epsilon + b^2\epsilon + c^2\epsilon}{n}\right),$$

where $n = 1 + a^2 + b^2 + c^2$.
Now we can finally compute the set $f(X)$:

$$f(X) = \{(a,b,c) \in \mathbb{R}^3 | f^{-1}(a,b,c) \in V(F)\}$$

$$= \left\{(a,b,c) \in \mathbb{R}^3 | \left(\frac{2a\epsilon}{n}, \frac{2b\epsilon}{n}, \frac{2c\epsilon}{n}, \frac{-\epsilon + a^2\epsilon + b^2\epsilon + c^2\epsilon}{n}\right) \in V(F)\right\}$$

$$= \left\{(a,b,c) \in \mathbb{R}^3 | F\left(\frac{2a\epsilon}{n}, \frac{2b\epsilon}{n}, \frac{2c\epsilon}{n}, \frac{-\epsilon + a^2\epsilon + b^2\epsilon + c^2\epsilon}{n}\right) = 0\right\}$$

$$= \left\{(a,b,c) \in \mathbb{R}^3 | \begin{array}{l} G := Re\left(F\left(\frac{2a\epsilon}{n}, \frac{2b\epsilon}{n}, \frac{2c\epsilon}{n}, \frac{-\epsilon + a^2\epsilon + b^2\epsilon + c^2\epsilon}{n}\right)\right) = 0, \\ H := Im\left(F\left(\frac{2a\epsilon}{n}, \frac{2b\epsilon}{n}, \frac{2c\epsilon}{n}, \frac{-\epsilon + a^2\epsilon + b^2\epsilon + c^2\epsilon}{n}\right)\right) = 0 \end{array}\right\}.$$

We notice that G and H are two polynomials in (a,b,c) with real coefficients. Their common zero set in $\mathbb{R}^3$ is equal to the algebraic link.

We give a schematic summary of the algorithm used to compute the algebraic link of an isolated singularity of a plane complex algebraic curve.

3.3 Computing the Alexander Polynomial of an Algebraic Link

3.3.1 Knot Theory and the Alexander Polynomial

For our purpose, we distinguish between the following types of knots:

Definition 5. 1. A knot is a piecewise linear or a differentiable simple closed curve in the 3-dimensional space $\mathbb{R}^3$.

2. A link is a finite union of disjoint knots. The individual knots which make up a link are called the components of the link. A knot will be considered a link with one component (Fig. 2 produced with Mathematica).

Algorithm 4 Algebraic link of the isolated singularity $(0, 0)$ ALGLINK(F, ϵ)

Input: $F \in \mathbb{C}[z, w]$ with (0,0) an isolated singularity , $\epsilon \in \mathbb{R}^*$ with $\epsilon > 0$
Output: $G, H \in \mathbb{R}[a, b, c]$
where the common zero set of G, H is the algebraic link of F in $(0, 0)$.

1. Substitute $z = x + iy$ and $w = u + iv$ in the defining polynomial $F(z, w)$ of the plane curve C:

$$C = \{(z, w) \in \mathbb{C}^2 | F(z, w) = 0\} \Leftrightarrow C = \{(x, y, u, v) \in \mathbb{R}^4 | F(x, y, u, v) = 0\},$$

that has an isolated singularity in the origin $(0, 0, 0, 0)$.
2. Rewrite $F(x, y, u, v) = R(x, y, u, v) + iI(x, y, u, v)$, with $R(x, y, u, v), I(x, y, u, v) \in \mathbb{R}[x, y, u, v]$ and obtain:

$$C = \{(x, y, u, v) \in \mathbb{R}^4 | R(x, y, u, v) = I(x, y, u, v) = 0\}.$$

3. Consider the sphere centrated in the origin and of small radius ϵ:

$$S^3 = \{(x, y, u, v) \in \mathbb{R}^4 | x^2 + y^2 + u^2 + v^2 - \epsilon^2 = 0\}.$$

4. Obtain $X = C \cap S^3 \subset \mathbb{R}^4$ and P a point on the sphere but not on the curve;
5. Introduce the inverse:

$$f^{-1} : \mathbb{R}^3 \to S^3 \setminus \{P\}$$

$$(a, b, c) \to (x, y, u, v) = (\tfrac{2a\epsilon}{n}, \tfrac{2b\epsilon}{n}, \tfrac{2c\epsilon}{n}, \tfrac{-\epsilon+a^2\epsilon+b^2\epsilon+c^2\epsilon}{n}),$$

where $n = 1 + a^2 + b^2 + c^2$.
6. Compute $f(X)$ using the inverse f^{-1} finding G, H:

$$f(X) = \left\{ (a, b, c) | \begin{array}{l} G := R(\frac{2a\epsilon}{n}, \frac{2b\epsilon}{n}, \frac{2c\epsilon}{n}, \frac{-\epsilon+a^2\epsilon+b^2\epsilon+c^2\epsilon}{n}) = 0, \\ H := I(\frac{2a\epsilon}{n}, \frac{2b\epsilon}{n}, \frac{2c\epsilon}{n}, \frac{-\epsilon+a^2\epsilon+b^2\epsilon+c^2\epsilon}{n}) = 0 \end{array} \right\}.$$

7. Return $G, H \in \mathbb{R}[a, b, c]$ as computed in step 6.

3. A link is called algebraic if it arises as the intersection of an algebraic curve with a sufficiently small sphere, as described in Subsection 3.2.

In our approach, we approximate a differentiable algebraic link, namely the intersection of G and H computed in Subsection 3.2, by a piecewise linear algebraic link, as we will explain in Subsubsection 3.3.2. From now on, we only consider piecewise linear links. When we work with knots, we work with their projection in the 2-dimensional space.

Definition 6. A regular projection is a linear projection for which no three points on the knot project to the same point, and no vertex projects to the same point as any other point on the knot. A crossing point is an image of two knot points of such a regular projection from $\mathbb{R}^3$ to $\mathbb{R}^2$. Then:

Fig. 2 Trefoil knot. The figure is produced with Mathematica

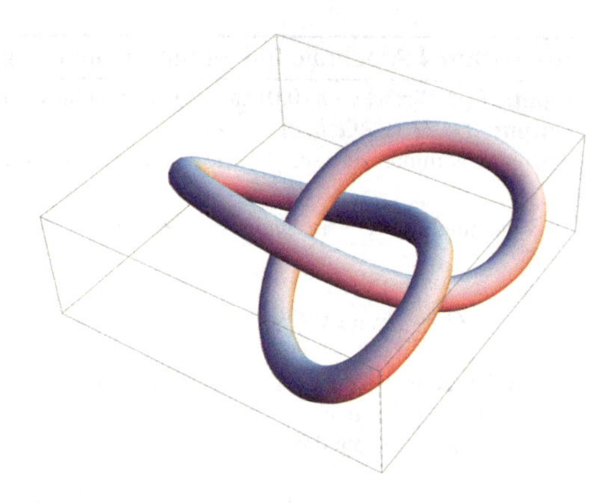

1. A link diagram (or simply diagram) is the image under regular projection, together with the information on each crossing point telling which branch goes over and which goes under. Thus we speak about overcrossings and undercrossings.
2. A diagram together with an arbitrary orientation of each knot in the link is called an oriented diagram.

We are interested in the following elements of a diagram:

Definition 7. 1. A crossing is lefthanded if the underpass traffic goes from left to right or it is righthanded if the underpass traffic goes from right to left. We denote a lefthanded crossing with -1 and a righthanded crossing with $+1$.
2. An arc is the part of a diagram between two undercrossings (Fig. 3). Whether lefthanded or righthanded, each crossing is determined by three arcs and we denote the overgoing arc with i, and the undergoing arcs with j and k (Fig. 4). We notice that the number of arcs in a link diagram is equal to the number of crossings in the same link diagram.

The main problem in knot theory is to distinguish between different links and to establish whether two links are equivalent or not. We define the equivalence of links by the following definition called (ambient) isotopy:

Definition 8. We define a homeomorphism as a continuous bijective function with a continuous inverse. Then we say that two links are equivalent if there exists an orientation-preserving homeomorphism on $\mathbb{R}^3$ that maps one link onto the other.

To prove that two links are not equivalent we use the notion of link invariants:

Definition 9. A link invariant is a function from link diagrams to some discrete set ($\mathbb{Z}$ or $\mathbb{Z}[t]$) which is unchanged under the Reidemeister moves of type I, II or III (see Fig. 5).

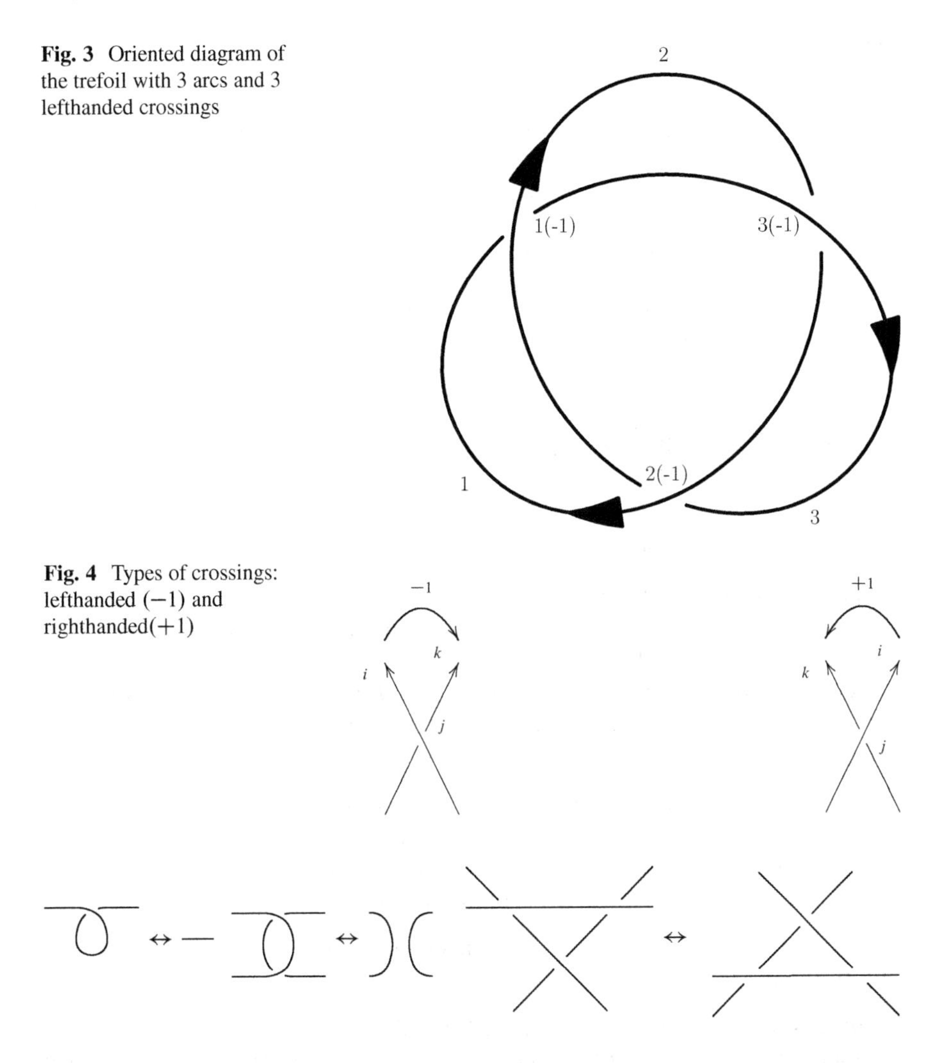

Fig. 3 Oriented diagram of the trefoil with 3 arcs and 3 lefthanded crossings

Fig. 4 Types of crossings: lefthanded (-1) and righthanded$(+1)$

Fig. 5 Reidemeister moves of type I, II, III

Some link invariants are: the tricolorability, the unknotting number, the Jones polynomial, the Alexander polynomial. For further details the reader can consult [10, 25]. At present, there exists no complete invariant for links.

Still for our purpose we are interested only in the invariants of the algebraic links. An important result in this direction of research was proved by Yamamoto in 1984 (see [40]), who showed that the Alexander polynomial is a complete invariant for the algebraic links, that is the Alexander polynomial uniquely defines all the algebraic links up to an (ambient) isotopy.

We now focus our attention on the definition and on the computation of the Alexander polynomial of a link. The Alexander polynomial was introduced by

Alexander in 1928 (see [3]). It depends on the fundamental group of the complement of the link in $\mathbb{R}^3$ and we define it as follows:

Definition 10. Let L be a link with m components. The multivariate Alexander polynomial is a Laurent polynomial $\Delta_L \in \mathbb{Z}[t_1^{\pm 1}, \ldots, t_m^{\pm 1}]$, which is uniquely defined for each link up to a factor of $\pm t_1^{k_1} \ldots t_m^{k_m}$, with $k_i \in \mathbb{Z}$ and up to a substitution $t_i := \frac{1}{t_i}$, for all $i \in \{1, \ldots, m\}$ ([9]).

We follow [25] in our approach to compute the Alexander polynomial. We distinguish three steps when computing the Alexander polynomial Δ_L of an oriented link diagram $D(L)$:

$$\boxed{D(L)} \longrightarrow \boxed{\text{Labelling matrix}(L)} \longrightarrow \boxed{\text{Prealexander matrix}(L)} \longrightarrow \boxed{\Delta_L}$$

First of all, we compute the labelling matrix of $D(L)$ defined as follows:

Definition 11. Let $D(L)$ be an oriented link diagram with m components and n crossings $x_q : q \in \{1, \ldots, n\}$. We denote the arcs of $D(L)$ with the labels $\{1, \ldots, n\}$ and separately the crossings of $D(L)$ with the labels $\{1, \ldots, n\}$. We denote the labelling matrix of $D(L)$ with $LM(L) \in \mathcal{M}(n, 4, \mathbb{Z})$. We define $LM(L) = (b_{ql})_{q,l}$ with $q \in \{1, \ldots, n\}, l \in \{1, \ldots, 4\}$ row by row for each crossing x_q as follows:

- On position b_{q1} we store the type of the crossing x_q ($+1$ or -1).
- On position b_{q2} we store the label of the arc i of the crossing x_q in $D(L)$.
- On position b_{q3} we store the label of the arc j of the crossing x_q in $D(L)$.
- On position b_{q4} we store the label of the arc k of the crossing x_q in $D(L)$.

Secondly, we compute the prealexander matrix of $D(L)$ defined using the labelling matrix $LM(L)$ as follows:

Definition 12. Let $D(L)$ be an oriented link diagram with m components and n crossings $x_q : q \in \{1, \ldots, n\}$. We denote the arcs and the crossings of $D(L)$ as in Definition 11. We consider $LM(L)$ the labelling matrix of $D(L)$ as in Definition 11. We denote the prealexander matrix of L with $PM(L) \in \mathcal{M}(n, n, \mathbb{Z}[t_1, t_2, \ldots, t_m])$. We define $PM(L)$ row by row for each crossing x_q depending on $LM(L)$. For x_q we consider the variable t_s, where $s \in \{1, \ldots, m\}$ is the s-th knot component of $D(L)$, which contains the overgoing arc that determines the crossing x_q. Then:

- If the crossing x_q is righthanded, i.e. $b_{q1} = +1$ in $LM(L)$ then at position b_{q2} of $PM(L)$ we store the label $1 - t_s$, at position b_{q3} we store the label -1 and at position b_{q4} we store the label t_s.

- If the crossing x_q is lefthanded, i.e. $b_{q1} = -1$ in $LM(L)$ then at position b_{q2} of $PM(L)$ we store the label $1 - t_s$, at position b_{q3} we store the label t_s and at position b_{q4} we store the label -1.
- If two or all of the positions b_{q2}, b_{q3}, b_{q4} have the same value, then we store the sum of the corresponding labels at the corresponding position.
- All other entries of the matrix are 0.

Finally, we define the Alexander polynomial of $D(L)$ depending on the number of components in L:

Definition 13. Let $D(L)$ be an oriented link diagram with m components and n crossings, $LM(L)$ be its labelling matrix as in Definition 11 and $PM(L)$ be its prealexander matrix as in Definition 12.

1. Univariate case, (L has one component, $m = 1$, see [25]). The univariate Alexander polynomial $\Delta_L(t_1) \in \mathbb{Z}[t_1^{\pm 1}]$ is the normalized polynomial computed as the determinant of any $(n-1) \times (n-1)$ minor of the prealexander matrix of $D(L)$.
2. Multivariate case, (L has more than one component, $m \geq 2$, see [9]). The multivariate Alexander polynomial $\Delta_L(t_1, \ldots, t_m) \in \mathbb{Z}[t_1^{\pm 1}, \ldots, t_m^{\pm 1}]$ is the normalized polynomial computed as the greatest common divisor of all the $(n-1) \times (n-1)$ minor determinants of the prealexander matrix of $D(L)$.

A normalized polynomial is a polynomial in which the term of the lowest degree is a positive constant.

Example 1. We compute the Alexander polynomial of the oriented diagram of the trefoil knot L from Fig. 3. We denote the arcs and separately the crossings of the diagram with the labels $\{1, 2, 3\}$. We compute the labelling matrix of L with Definition 11:

$$LM(L) = \left(\begin{array}{c|cccc} & \text{type} & \text{label}_i & \text{label}_j & \text{label}_k \\ \hline c_1 & -1 & 2 & 1 & 3 \\ c_2 & -1 & 1 & 3 & 2 \\ c_3 & -1 & 3 & 2 & 1 \end{array}\right)$$

From $LM(L)$ we compute the prealexander matrix of $D(L)$ with Definition 7 and Definition 12. We notice that L has only one knot component so $s = 1$ in Definition 12:

$$PM(L) = \left(\begin{array}{c|ccc} & \text{label}_i & \text{label}_j & \text{label}_k \\ \hline c_1 & 2 & 1 & 3 \\ -1 & 1-t_1 & t_1 & -1 \\ \hline c_2 & 1 & 3 & 2 \\ -1 & 1-t_1 & t_1 & -1 \\ \hline c_3 & 3 & 2 & 1 \\ -1 & 1-t_1 & t_1 & -1 \end{array}\right) = \left(\begin{array}{c|ccc} & 1 & 2 & 3 \\ \hline c_1 & t_1 & 1-t_1 & -1 \\ c_2 & 1-t_1 & -1 & t_1 \\ c_3 & -1 & t_1 & 1-t_1 \end{array}\right)$$

From $PM(L)$ we compute the Alexander polynomial with Definition 13:

$$\det\left(\text{Minor}_{33}(PM(L))\right) = \det\begin{pmatrix} t_1 & 1-t_1 \\ 1-t_1 & -1 \end{pmatrix} = -t_1^2 + t_1 - 1$$

$$\Delta_L(t_1) = \text{Normalize}(-t_1^2 + t_1 - 1) = t_1^2 - t_1 + 1$$

We give a schematic summary of the algorithm used to compute the Alexander polynomial of an algebraic link diagram $D(L)$.

Algorithm 5 Alexander polynomial for $D(L)$ ALEXPOLY$(D(L), m, n)$

Input: $D(L)$ oriented algebraic link diagram with m components, n crossings
Output: $\Delta_L(t_1, \ldots, t_m) \in \mathbb{Z}[t_1^{\pm 1}, \ldots, t_m^{\pm 1}]$
where $\Delta_L(t_1, \ldots, t_m)$ is the Alexander polynomial of $D(L)$.

1. Denote the arcs and separately the crossings of $D(L)$ with $\{1, \ldots, n\}$.
2. Compute $LM(L)$ the labelling matrix of $D(L)$.
3. Compute $PM(L)$ the prealexander matrix of $D(L)$.
4. If $m = 1$ then:

   a. Compute M any $(n-1) \times (n-1)$ minor of $PM(L)$.
   b. Compute D the determinant of the minor M.
   c. Return $\Delta_L(t_1) = \mathit{Normalize}(D)$.

5. If $m \geq 2$ then:

   a. Compute all the $(n-1) \times (n-1)$ minors of $PM(L)$.
   b. Compute G the greatest common divisor of all the computed minors in 5.(a).
   c. Return $\Delta_L(t_1, \ldots t_m) = \mathit{Normalize}(G)$.

3.3.2 Alexander Polynomial and Computational Geometry

In Subsection 3.3 we noticed that in order to compute the Alexander polynomial of an algebraic link L we need to compute the diagram of L denoted with $D(L)$. We compute L using the stereographic projection method described in Subsection 3.2 and using the Axel system [39] for the actual implementation. We remember that for the plane complex algebraic curve C with $F(z,w) \in \mathbb{C}[z,w], \epsilon \in \mathbb{R}^*, \epsilon > 0$ and $(0,0)$ an isolated singularity, we compute two polynomials $G, H \in \mathbb{R}[a,b,c]$. We have shown that the algebraic link L of $(0,0)$ is the zero common set of G, H, that is L is a smooth and closed implicit algebraic curve in $\mathbb{R}^3$ given as the intersection of two implicit algebraic surfaces in $\mathbb{R}^3$ whose defining polynomials are G, H. Axel uses certified algorithms to compute a piecewise linear approximation L' for L, which is isotopic to L [24]. L' is computed as a graph $G = \langle \mathcal{P}, \mathcal{E} \rangle$, where $\mathcal{P}$ is a set of points (or vertices) together with their Euclidean coordinates and $\mathcal{E}$ is a set of edges connecting them. From now on we denote $L' := \mathit{Graph}(L)$.

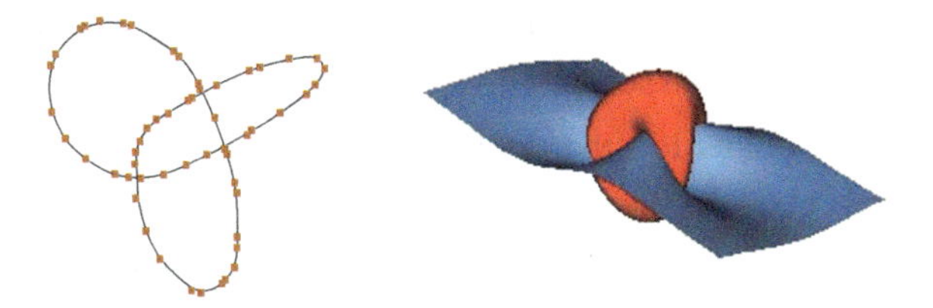

Fig. 6 Algebraic link as intersection of surfaces in Axel

Example 2. For $C^4 = \{(z,w) \in \mathbb{C}^2 | z^3 - w^2 = 0\} \subset \mathbb{R}^4$ with $(0,0)$ isolated singularity and $\epsilon = 1$ we get L, the algebraic link of $(0,0)$ by the stereographic projection method proposed in Subsection 3.2:

$$f(C^4 \cap S) := L = \{(a,b,c) \in \mathbb{R}^3 | G := ReF(\ldots) = 0, H := ImF(\ldots) = 0\}$$

and with Axel we get *Graph*(L) as the intersection of the two surfaces G, H:

$$Graph(L) = \langle \mathcal{P}, \mathcal{E} \rangle, \ \mathcal{P} = \{p = (m,n,q) \in \mathbb{R}^3\}, \ \mathcal{E} = \{(i,j) | i,j \in \mathcal{P}\}$$

such that *Graph*$(L) \cong_{isotopic} L$, see Fig. 6.

Next, from the output computed by Axel, *Graph*(L), we need to compute $D(L)$, the diagram of the algebraic link L. We then use $D(L)$ to compute the Alexander polynomial of L with the algorithm proposed in Subsection 3.3.

We compute the elements of $D(L)$, i.e. the arcs of the diagram and the number of knot components in the diagram, plus for each knot component its crossings with their types. We develop new computational algorithms for computing $D(L)$ given *Graph*(L). The main idea is that all these algorithms operate on the data structure *Graph*(L) returned by Axel. Each point in the graph is given as a 4-tuple $p(index, x, y, z)$, where *index* is an integer that uniquely identifies each point, and $(x,y,z) \in \mathbb{R}^3$ are the Euclidean coordinates of p. We use *xyzcoord(index)* for denoting the x, y, z coordinates of *index*, *xycoord*$(index)$ for denoting the x, y coordinates of *index*, *xcoord* for denoting the x coordinate of *index*, and *ycoord* for denoting the y coordinate of *index*. Each edge in the graph is given by a pair $e(source, destination)$, where *source* is the index of the source point of e, and *destination* is the index of the destination point of e. For simplicity reasons, we denote the pair $e(source, destination) := e(s,d)$. We consider the edges of *Graph*(L) to be "small" edges, i.e. the projection of any edge of *Graph*(L) has at most one crossing point. Here we shortly describe these computational algorithms, for more information the reader can consult [21].

The first algorithm is an adapted version of the Bentley–Ottman algorithm [13]. For $Graph(L) \in \mathbb{R}^3$ with the set of points $p_i = (x_i, y_i, z_i) \in \mathbb{R}^3$ we consider its projection in $\mathbb{R}^2$ with the set of points $p_i = (x_i, y_i) \in \mathbb{R}^2$. We also consider no vertical edges in the projection. This algorithm computes the intersection points of all the edges of the projection of *Graph*(L) and some extra information:

1. For each intersection point p the pair of edges (e_i, e_j) that contains p.
2. And each pair of edges (e_i, e_j) is ordered, i.e. e_i is under e_j in $\mathbb{R}^3$.

These intersection points together with the extra information coincide with the crossings of $D(L)$. Our adapted Bentley-Ottman algorithm operates as follows:

- The edges of the projection of $Graph(L)$ are oriented from left to right and they are ordered in a list $E = \{e_0, \ldots, e_N\}$ as follows: (1) by the x-coordinates of their source points; (2) if the x-coordinates of the source points of two edges coincide, then the two edges are ordered by the two slopes of their supporting lines; (3) if the x-coordinates of the source points and the slopes of two edges coincide, then the two edges are ordered by the y-coordinates of their destination points. The ordering criteria is necessary for the correctness of the algorithm.
- We consider a vertical sweep line l that sweeps the plane from left to right. While l moves, it intersects several edges from E. The list of edges that intersect l at one point during the sweeping procees, denoted SW, is called the sweep list. SW changes while l sweeps the plane. The algorithm is based on the key observation that SW is updated only at certain points of the edges from E called event points. The sweep list SW is ordered in this algorithm by the y-coordinates of the intersections of the edges of E with the sweep line l.
- We notice that in E each $index$ appears two times in E. Due to this property, we can manage SW in a simpler way in our adapted Benttley-Ottman algorithm than in the original version.
- While we traverse E, we insert the current edge $e_m(s_m, d_m)$ from E in SW in the right position and that is: (1) we search for an edge $e_n(s_n, d_n)$ in SW such that its destination coincide with the source of $e_m \in E$, i.e. $d_n = s_m$; if we find such an $e_n \in SW$ we replace it with $e_m \in E$; (2) if such an edge $e_n \in SW$ does not exist, we insert e_m in SW depending on its position against the current edges from SW. We assume $SW = \{e_{i_1}, e_{i_2}, e_{i_3}, \ldots, e_{i_k}\}$, with $e_{i_q} \in E$ for all $q \in \{1, \ldots, k\}$. There exists a unique index j with $0 \leq j \leq k$ such that $ycoord(s_m)$ is larger than the y-coordinates of all the intersections of $e_{i_1}, \ldots, e_{i_j}$ with l and smaller than the y-coordinates of all the intersections of $e_{i_{j+1}}, \ldots, e_{i_k}$ with l. This index j can be found by checking all the signs of the determinants $det\big[\big(xycoord(s_m), 1\big), \big(xycoord(s_{i_j}), 1\big), \big(xycoord(d_{i_j}), 1\big)\big]$. Then we insert e_m in SW between the two edges e_{i_j} and $e_{i_{j+1}}$ and we obtain $SW = \{e_{i_1}, e_{i_2}, \ldots, e_{i_j}, e_m, e_{i_{j+1}}, \ldots, e_{i_k}\}$. When we insert an edge from E into SW on the right position we have to additionally update SW depending on the encountered event points:

  – We test each inserted edge in SW against its two neighbours for intersection. If an intersection point p is found we report it together with the ordered pair of edges that contains it. In addition we swap the edges that intersect in SW. As opposed to the original Bentley-Ottman algorithm after swaping the edges in SW, we do not test the edges against their new neighbours for intersections because we consider only "small" edges.

- We test each inserted edge in SW against its two neighbours for common destination. In addition, when two edges are swapped in SW after reporting their intersection point, we test them against their new neighbours for common destination. Whenever we find two consecutive edges with common destinations we erase them from SW. As opposed to the original Bentley-Ottman algorithm after deleting edges from SW, we do not test the new neighbours for intersection because we consider only "small" edges.

The second algorithm constructs the knot components of the diagram from the projection of *Graph*(L). It also returns the total number of knot components. We consider E as in the previous algorithm. We denote a positive edge in $\mathbb{R}^2$ with $e(s,d)$, and its corresponding negative edge with $-e(d,s)$. The positive edges are oriented from left to right, while the negative ones are oriented from right to left. We denote the knots with $K_i, i \in \mathbb{N}$. All K_i have the properties:

1. For each edge $e_k(s_k, d_k) \in K_i$ there exists $e_{k+1}(s_{k+1}, d_{k+1}) \in K_i$ with $d_k = s_{k+1}$.
2. For $K_i = \{e_0(s_0, d_0), \ldots, e_n(s_n, d_n)\}$: $d_n = s_0$.

As opposed to the list E, which contains only positive edges oriented from left to right, each list K_i contains both positive and negative edges. We initialize the first knot K_0 with the first edge $e_0(s_0, d_0)$ from E. Next we look for the edge e_n in E which has a common index, either source or destination, with d_0. If we find $e_n(d_0, d_n) \in E$ then we insert $e_n(d_0, d_n)$ in K_0 as a positive edge. If we find $e_n(s_n, d_0) \in E$ then we insert $-e_n(d_0, s_n)$ in K_0 as a negative edge. After we insert e_n in K_0 we erase it from E. We will always find such an edge e_n in E, because each *index* such as d_0 appears two times in E. We continue with inserting edges in K_0 from E until the *destination* of an inserted edge coincide with s_0 the source of the first edge from K_0. We apply the same strategy to constructs all the knots K_i of $D(L)$ until E is empty, increasing i each time a new knot starts being constructed. At the end of the algorithm, the index i returns the total number of knot components of $D(L)$.

The third algorithm constructs the arcs for each knot component of the link. It also decides the type of crossings (righthanded or lefthanded) for each knot component. For constructing the arcs, we consider E as in the previous algorithms. This algorithm operates on the outputs of the previous two algorithms, i.e. the list of intersection points I together with the list of ordered pairs of edges E_I, and the lists of edges for all the knot components $K_i, i \in \mathbb{N}$. The key point of the algorithm is to search in K_i all the undergoing edges from E_I and to splitt them in two parts. For instance, we assume that for $E = \{e_0, \ldots, e_n, e_m, \ldots, e_l, e_k \ldots, e_t, e_s, \ldots, e_{last}\}$, we compute the following outputs with the previous two algorithms:

$$I = \{(x_1, y_1), (x_2, y_2), (x_3, y_3)\}, E_I = \{(-e_n, e_m), (e_l, e_k), (e_s, -e_t)\}$$

$$K_0 = \{e_0, \ldots, e_k, \ldots, e_s, \ldots, e_m, \ldots, e_l, \ldots, -e_t, \ldots, -e_n, \ldots, -e_1\}$$

We search the three undergoing edges $-e_n, e_l, e_s$ one by one in K_0 and we replace them with $-e_n \to (-e_n^d, -e_n^u), e_l \to (e_l^d, e_l^u), e_s \to (e_s^d, e_s^u)$ obtaining:

$$K_0' = \{e_0, ..., e_k, ..., e_s^d, e_s^u, ..., e_m, .., e_l^d, e_l^u, ..., -e_t, ..., -e_n^d, -e_n^u, ..., -e_1\}.$$

From Definition 7, we conjecture that an arc contains the list of edges from a modified knot component $K_i', i \in \mathbb{N}$ starting with an edge of type $e_j^u, j \in \mathbb{N}$ from K_i' and ending with the next consecutive edge of type $e_k^d, k \in \mathbb{N}$ from K_i'. While we insert the edges from K_i' into the list of edges representing the arcs we erase them from K_i'. Thus from the modified loop K_0' we compute the following three arcs until K_0' is empty:

$$K_0' = \{e_0, ..., e_k, ..., e_s^d, [e_s^u, ..., e_m, .., e_l^d], e_l^u, ..., -e_t, ..., -e_n^d, -e_n^u, ..., -e_1\}$$
$$arc_0 = \{e_s^u, \ldots, e_m, \ldots, e_l^d\}$$
$$K_0' = \{e_0, ..., e_k, ..., e_s^d, [e_l^u, ..., -e_t, .., -e_n^d], -e_n^u, ..., -e_1\}$$
$$arc_1 = \{e_l^u, ..., -e_t, ..., -e_n^d\}$$
$$K_0' = \{[e_0, ..., e_k, ..., e_s^d], [-e_n^u, ..., -e_1]\}$$
$$arc_2 = \{e_n^u, ..., -e_1, e_0, ..., e_k, ..., e_s^d\}$$

For deciding the type of crossings, we observe that in each knot component for a positive edge $e_i(s_i, d_i) : xcoord(s_i) < xcoord(d_i)$ and for a negative edge $-e_j(s_j, d_j) : xcoord(s_j) > xcoord(d_j)$. Each type of crossing depends on the pair of edges (e_{under}, e_{over}) that contains the corresponding intersection point, and that is:

1. On the orientation of e_{under}, and e_{over}, i.e. whether they are oriented from left to right (positive) or from right to left (negative).
2. On the comparison relation between the slope of e_{under} and the slope of e_{over}.

Depending on these three parameters, we have 2^3 possible cases for deciding the type of crossings. For instance, we consider a crossing c determined by the pair of ordered edges $\big(-e_l(s_l, d_l), e_k(s_k, d_k)\big)$, for which $-e_l$ is the undergoing edge and e_k is the overgoing edge in $\mathbb{R}^3$. We have $xcoord(s_l) > xcoord(d_l)$ for the negative undergoing edge e_l, and $xcoord(s_k) < xcoord(d_k)$ for the positive overgoing edge e_k. If additionally we suppose that $slope(e_l) < slope(e_k)$, then c is a lefthanded crossing.

We give the schematic algorithm for the computation of the diagram $D(L)$ of a differentiable algebraic link L computed as in Subsection 3.2 and approximated by a piecewise linear algebraic link $Graph(L)$.

Algorithm 6 Diagram of piecewise linear links DIAGRAM($Graph(L)$)

Input: $Graph(L) = \langle \mathscr{P}, \mathscr{E} \rangle$ piecewise linear algebraic link which approximates
L a differentiable algebraic link as computed in Subsection3.2
$\mathscr{P}$ set of points with their euclidean coordinates
$\mathscr{E}$ set of edges connecting them

Output: $D(L)$
where $D(L)$ is the diagram of $Graph(L) \cong_{isotopic} L$.

1. Compute the crossings of $D(L)$.
   a. Compute I the intersections of the edges of E.
   b. Compute E_I the pairs of ordered edges containing each intersection.
2. Compute $K_i, i \in \mathbb{N}$ the lists of edges from E for all the knots of $D(L)$.
3. Compute the arcs of $D(L)$ and the type of crossings in $D(L)$.
4. Return $D(L)$.

3.4 Computing the Delta-Invariant of an Isolated Singularity

We use Milnor results [27] for computing the δ-invariant of the isolated singularity $(0,0)$ of a plane complex algebraic curve:

- We consider μ a positive integer that measures the amount of degeneracy at the critical point $(0,0)$ of the complex polynomial $F(z,w)$. In fact, μ is the Milnor number. It is shown that μ is the degree of the characteristic polynomial Δ of the link $L = V \cap S_\epsilon$ determined by $V = F^{-1}(0)$. The characteristic polynomial Δ coincides with the Alexander polynomial $\Delta_L(t)$ if L has one knot component, and $\Delta = \dfrac{(t-1)}{\pm t^i}\Delta_L(t,\ldots,t)$ if L has more than one knot components. We observe that μ is the degree of the characteristic polynomial Δ. Based on this observation we deduce that μ is the degree of the Alexander polynomial if L has one knot component, and μ is the degree of the Alexander polynomial $+1$ if L has more than one knot components.
- We consider r the number of local analytic branches of $V = F^{-1}(0)$ with $L = V \cap S_\epsilon$ passing through origin. That is r is the number of knot components in the link L determined by V, i.e. r is the number of variables in the Alexander polynomial of the link L.

We base our algorithm for the computation of the δ-invariant on the following theorem proved by Milnor:

Theorem 2 ([27]). *Suppose that r branches of the curve $V = F^{-1}(0)$ pass through the origin $s = (0,0)$, which is an isolated singularity for V. Then the delta-invariant of the isolated singularity $s = (0,0)$ denoted with δ_s is related to the Milnor number μ by the equation $2\delta_s = \mu + r - 1$ and it is always an integer.*

We give the schematic algorithm for the computation of the δ-invariant of the isolated singularity $(0,0)$.

Algorithm 7 Delta-invariant of the isolated singularity $(0, 0)$ DELTA(Δ_L, μ, r)

Input: $\Delta_L(t_1, \ldots, t_m)$ the Alexander polynomial of L
L the algebraic link of the isolated singularity $s = (0, 0)$,
d the degree of Δ_L, m the number of variables in Δ_L
Output: $\delta_s \in \mathbb{Z}^*_+$
where δ_s is the delta-invariant of $s = (0, 0)$.

1. If $m = 1$ then return $\delta_s = \dfrac{d}{2}$
2. If $m \geq 2$ then return $\delta_s = \dfrac{d + m}{2}$

Algorithm 8 Genus of a plane complex algebraic curve GENUS(F, d, ϵ)

Input: $C = \{(z, w) \in \mathbb{C}^2 | F(z, w) = 0\}$,
$F(z, w) \in \mathbb{C}[z, w]$ with numeric coefficients,
d the degree of C, $\epsilon \in \mathbb{R}^*, \epsilon > 0$
Output: $genus(C) \in \mathbb{Z}$
where $genus(C)$ is the approximate genus of C.

1. *sumDeltaInv* $= 0$.
2. Compute $Sing(C) =$ SING(F, d).
3. For each $s_i = (z_i, w_i) \in Sing(C)$ do:
   a. Move s_i in $(0, 0) : C = \{(z + z_i, w + w_i) \in \mathbb{C}^2 | F(z + z_i, w + w_i) = 0\}$.
   b. Compute $L =$ ALGLINK(F, ϵ) (L is approximated by $Graph(L)$).
   c. Compute $D(L) =$ DIAGRAMLINK$(Graph(L))$.
   d. Compute $\Delta_L(t_1, \ldots, t_m) =$ ALEXPOLY$(D(L), m, n)$.
   e. Compute $\delta_{s_i} =$ DELTA(Δ_L, μ, r).
   f. *sumDeltaInv* $=$ *sumDeltaInv* $+ \delta_{s_i}$.
4. Return $genus(C) = \dfrac{(d-1)(d-2)}{2} -$ *sumDeltaInv*.

3.5 Computing the Genus of the Algebraic Curve

We now give the schematic algorithm for computing the genus of a plane complex algebraic curve whose defining polynomial has numeric coefficients. The computed genus is the approximate genus, which is defined as the lowest possible genus of a curve defined by a nearby polynomial. We discuss the notion of approximate genus in detail in Sect. 4.

4 What Precisely Means "Approximate Computation"?

It is the mark of an instructed mind to rest satisfied with the degree of precision to which the nature of the subject admits and not to seek exactness when only an approximation of the truth is possible.

Aristotle

The fact that the genus computation problem is ill-posed and our desire to compute the genus for approximate inputs seems to be incompatible at first glance. On the other hand, similar situations have been considered before: for ill-posed problems in the field of partial differential and integral equations, regularization theory (see [14]) has been introduced. In approximate algebraic computation, some principles have been established that aim at computing discontinuous properties of approximately given data (see [35, 41]). A probabilistic interpretation of a similar situation can be found in [31]. In this section, we discuss the regularization theory approach as much as it is relevant here, and relate it to our specific situation.

Let A and B be metric spaces. Let $F : A \to B$ be a function that is not continuous. Then the problem of computing $F(y)$ for given $y \in A$ is ill-posed, as the desired output does not continuously depend on the given input. Instead of computing F, we approximate F by a set of functions that converge towards F. More precisely, let $R : A \times \mathbb{R}_+ \to B$ be a function that is continuous. The additional input parameter $\epsilon \in \mathbb{R}_+$ is called "regularization parameter". For $y \in A$, a perturbation of y is a function from $\mathbb{R}_+$ to A, $\epsilon \to y_\epsilon$, such that $d(y, y_\epsilon) \leq \epsilon$ for all $\epsilon \in \mathbb{R}_+$.

We say that R is a *regularization* of F if there exists a bijective monotonic function $\alpha : \mathbb{R}_+ \to \mathbb{R}_+$, also known as "parameter choice rule", such that for any $y \in A$ and any perturbation $\epsilon \to y_\epsilon$ of y, we have

$$\lim_{\epsilon \to 0} R(y_\epsilon, \alpha(\epsilon)) = F(y). \tag{3}$$

One consequence of this is "convergence for exact data":

$$\lim_{\epsilon \to 0} R(y, \epsilon) = F(y)$$

(just set $y_\epsilon = y$, the constant perturbation). More important, however, is that if we know the "error level" δ of the input, then we may assume that the input we have is of the form y_δ for some perturbation. Then we calculate the value $R(y_\delta, \alpha(\delta))$. This value is an estimate for $F(y)$. If we could decrease the error level, then the limit of these estimates would be $F(y)$.

In our situation, A is the set of coefficient vectors of polynomials of fixed degree, B is discrete, and F is the function that assigns to the given polynomial the Alexander polynomial of the singular point at the origin of the defined curve. (If the origin is not a singular point, then the Alexander polynomial is 1.) The function R assigns to $\epsilon > 0$ the Alexander polynomial of the ϵ-link, arising as the intersection of the curve with the ϵ-sphere (see Algorithm 2). Note that this function is not everywhere defined, because it may happen that the intersection has singularities and is therefore not a link. But it is easy to see that the function R is continuous in its domain of definition. Since B is discrete, this amounts to saying that R is constant in every connected component of its domain of definition. The Alexander polynomial can only change if the intersection has singularities, otherwise we have an isotopy, which leaves the Alexander polynomial fixed.

Note that the convergence for exact data is a straightforward consequence of Theorem 1. On the other hand, we admit that we cannot (yet) say that computing the ϵ-link is a regularization for the problem of computing the link of a singularity, because we do not yet know the parameter choice rule α satisfying (3). An observation that may help is that the Alexander polynomial is constant in each family with constant Milnor number [37] and that the Milnor number function is an upper semicontinuous function of the coefficients of the algebraic curve [36]. Unfortunaly, regularization theory for discontinuous algebraic algorithms is not as well developed as the theory for differential equations.

Discontinuities arise not only in Algorithm 2, but also in Algorithm 1 where we compute the singularities of a curve given by coefficients. (It is clear that singularities disappear under generic pertubations, for instance.) Using subdivision, we can compute "numerical singularities", which are points in the plane where the value of the polynomial and its derivatives are small. Moreover, it is guaranteed that every singularity lies in the vicinity of such a computed numerical singularity.

5 Numerical Experiments

There is no such thing as a failed experiment, only experiments with unexpected outcomes.

R. B. Fuller

In this paper, we will give some experimental evidence for the statement that our algorithm is a regularization as explained in Sect. 4. All the experiments, numerical and symbolical, are done with the software, *GENOM3CK*-Symbolic numeric techniques for *GEN*us c*OM*putation of *C*omplex algebrai*C* *C*urves using *K*not theory. GENOM3CK is implemented and included as a library in the free system Axel [39], written in C++ with Qt Script for Applications (QSA).

As evidences for the **convergence for exact data property** we consider an input polynomial $F(x, y) \in \mathbb{C}[x, y]$ with both exact and inexact coefficients and we compute $A_\epsilon(F(x, y))$ with the approximate algorithm A_ϵ. We compute $A_\epsilon(F(x, y))$ with the approximate algorithm for different values of the parameter ϵ. We obtain several outputs such as: the singularities of the input curve defined by $F(x, y)$, the algebraic link of each singularity (i.e. the topology of the singularity), the Alexander polynomial of each algebraic link, the delta-invariant of each singularity, and the genus of the curve. The computation of the Alexander polynomial, delta-invariant and the genus depends on the computation of the algebraic link of each singularity. From the experiments, we observe that the approximate solution computed with A_ϵ converges to the exact solution as ϵ tends to 0.

Example 3. We consider $F(x, y) = x^2 - xy - y^3$. We notice that $x^2 - xy = x(x - y)$ thus $F(x, y)$ has a vertical tangent $x = 0$ in $\mathbb{C}^2$. In order to assure a valid stereographic projection in $\mathbb{R}^3$ we make the substitution $\{x \to y, y \to x\}$ in $F(x, y)$ obtaining $F(x, y) = -x^3 - xy + y^2$, and thus we consider this polynomial as the input of the problem. We use Arnold's results concerning the analysis of curve singularities and we deduce that the algebraic link of the singularity $(0, 0)$ of the polynomial $-x^3 - xy + y^2$ is the same as the algebraic link of the singularity $(0, 0)$ of the polynomial $-xy + y^2$ which is the Hopf link, and which represents the exact solution for the algebraic link of the singularity $(0, 0)$ of $F(x, y)$. We notice in Table 1 that the approximate solution converges to the exact solution as ϵ tends to 0.

We can consider the input polynomial with both exact and inexact coefficients, such as $F(x, y) = -x^3 - xy + y^2 - 0.01$. We observe in Table 2 that the approximate solution converges to the exact solution when ϵ tends to 0. This is an evidence for property (3) from Sect. 4, which is also called the **convergence for noisy data property**.

Example 4. We consider $F(x, y) = x^2 - y^2 - y^3$. We use Arnold's results concerning the analysis of curve singularities and we deduce that the algebraic link of the singularity $(0, 0)$ of $F(x, y)$ is the same as the algebraic link of the singularity $(0, 0)$ of the polynomial $x^2 - y^2$ which is the Hopf link, and which represent the exact solution for the algebraic link of the singularity $(0, 0)$ of $F(x, y)$. We notice in Table 3 that the approximate solution converges to the exact solution as ϵ tends to 0.

As evidences for the **continuity property** we consider an input curve defined by the polynomial $F(x, y) \in \mathbb{C}[x, y]$ with exact and inexact coefficients and we compute $A_\epsilon(F(x, y))$ with the approximate algorithm A_ϵ. The continuity property of A_ϵ states that small changes in the input polynomial $F(x, y)$ produce constant output for the computed approximate solution. To observe this we proceed in the following way:

Table 1 Convergence of $-x^3 - xy + y^2$ with exact coefficients

Equation and ϵ		Link	Alexander, δ invariants, genus		
$-x^3 - xy + y^2$	1.00	Trefoil knot	$\Delta(t_1) = t_1^2 - t_1 + 1$	$\delta = 1$	$g = 0$
$-x^3 - xy + y^2$	0.5	Trefoil knot	$\Delta(t_1) = t_1^2 - t_1 + 1$	$\delta = 1$	$g = 0$
$-x^3 - xy + y^2$	0.25	Hopf link	$\Delta(t_1, t_2) = 1$	$\delta = 1$	$g = 0$
$-x^3 - xy + y^2$	0.14	Hopf link	$\Delta(t_1, t_2) = 1$	$\delta = 1$	$g = 0$

Table 2 Convergence of $-x^3 - xy + y^2 - 0.01$ with inexact coefficients

Equation and ϵ		Link	Alexander, δ invariants, genus		
$-x^3 - xy + y^2 - 0.01$	1.00	Trefoil knot	$\Delta(t_1) = t_1^2 - t_1 + 1$	$\delta = 1$	$g = 0$
$-x^3 - xy + y^2 - 0.01$	0.5	Hopf link	$\Delta(t_1, t_2) = 1$	$\delta = 1$	$g = 0$
$-x^3 - xy + y^2 - 0.01$	0.25	Hopf link	$\Delta(t_1, t_2) = 1$	$\delta = 1$	$g = 0$
$-x^3 - xy + y^2 - 0.01$	0.22	Hopf link	$\Delta(t_1, t_2) = 1$	$\delta = 1$	$g = 0$

Table 3 Convergence of $x^2 - y^2 - y^3$ with exact coefficients

Equation and ϵ		Link	Alexander, δ invariants, genus		
$x^2 - y^2 - y^3$	1.00	1 singularity curve	–	–	–
$x^2 - y^2 - y^3$	0.7	Hopf link	$\Delta(t_1, t_2) = 1$	$\delta = 1$	$g = 0$
$x^2 - y^2 - y^3$	0.5	Hopf link	$\Delta(t_1, t_2) = 1$	$\delta = 1$	$g = 0$
$x^2 - y^2 - y^3$	0.19	Hopf link	$\Delta(t_1, t_2) = 1$	$\delta = 1$	$g = 0$

Table 4 Continuity for perturbations of type I of $-x^3 - xy + y^2$

Perturbations I and ϵ		$\sigma = 10^{-e}, e \in \mathbb{N}^*$	Link	Invariants
$-x^3 - xy + y^2 - 10^{-e}$	0.5	$\{10^{-2}, \ldots, 10^{-10}\}$	Trefoil knot	$\Delta(t_1) = t_1^2 - t_1 + 1$ $\delta = 1$ $g = 0$
$-x^3 - xy + y^2 - 10^{-e}$	0.25	$\{10^{-2}, \ldots, 10^{-10}\}$	Hopf link	$\Delta(t_1, t_2) = 1$ $\delta = 1$ $g = 0$

- We consider a polynomial $p(x, y) \in \mathbb{C}[x, y]$ which contains only exact coefficients.
- For $\sigma \in \mathbb{R}^*$, we slightly perturbed the coefficients of the polynomial $p(x, y)$ obtaining some new polynomials denoted with $p_\sigma(x, y)$ that we call perturbations of the polynomial $p(x, y)$. We call σ the perturbation of the exact polynomial $p(x, y)$.
- We consider several values for the parameter ϵ. For each of these values, we execute the approximate algorithm A_ϵ on the perturbed polynomials $p_\sigma(x, y)$ for different values of $\sigma \in \mathbb{R}^*$. The perturbed polynomials $p_\sigma(x, y)$ represent the input polynomials $F(x, y)$ with exact and inexact coefficients, i.e. $F(x, y) = p_\sigma(x, y)$, for $\sigma \in \mathbb{R}^*$.

We distinguish between two types of perturbations:

1. Perturbations of type I: For these types of perturbations, $p_\sigma(x, y)$ is of the following form: $p_\sigma(x, y) = p(x, y) + \sigma$, where $p(x, y)$ is the exact polynomial and $\sigma \in \mathbb{R}^*$ is a real number different from 0.
2. Perturbations of type II: For these types of perturbations, $p_\sigma(x, y)$ is of the following form: $p_\sigma(x, y) = p(x, y) + \sigma q(x, y)$, where $p(x, y)$ is the exact polynomial, $\sigma \in \mathbb{R}^*$ and $q(x, y) \in \mathbb{C}[x, y]$ is an arbitrary exact polynomial.

From the experiments, we observe that for the perturbed polynomials the approximate computed solution is preserved, that is for small changes of the input data we obtain constant output for the computed approximate solution.

Example 5. For the exact polynomial $p(x, y) = -x^3 - xy + y^2$, we consider perturbations of type I of the form $p_\sigma(x, y) = -x^3 - xy + y^2 - \sigma$, with $\sigma \in \{10^{-2}, \ldots, 10^{-10}\}$. We notice in Table 4 that for perturbations of type I of $-x^3 - xy + y^2$ we obtain constant approximate solution.

For the perturbations of type II we consider the exact polynomial $p(x, y) = -x^3 - xy + y^2$, the arbitrary exact polynomial $q(x, y) = -x^3 - 2xy + y^2$ and $\sigma \in \{10^{-1}, \ldots, 10^{-10}\}$, obtaining the perturbed polynomials $p_\sigma(x, y) = p(x, y) + \sigma q(x, y) = -x^3 - xy + y^2 + \sigma(-x^3 - 2xy + y^2) = -(1+\sigma)x^3 - (1+2\sigma)xy + (1+$

Table 5 Continuity for perturbations of type II of $-x^3 - xy + y^2$

Perturbations II and ϵ		$\sigma = 10^{-e}, e \in \mathbb{N}^*$	Link	Invariants
$-(1+10^{-e})x^3 - (1+2\cdot 10^{-e})$ $xy + (1+10^{-e})y^2$	0.15	$\{10^{-1}, \ldots 10^{-10}\}$	Hopf link	$\Delta(t_1, t_2) = 1$ $\delta = 1$ $g = 0$
$-(1+10^{-e})x^3 - (1+2\cdot 10^{-e})$ $xy + +(1+10^{-e})y^2$	0.14	$\{10^{-1}, \ldots 10^{-10}\}$	Hopf link	$\Delta(t_1, t_2) = 1$ $\delta = 1$ $g = 0$

Table 6 Continuity for perturbations of type I of $x^2 - y^2 - y^3$

Perturbations I and ϵ		$\sigma = 10^{-e}, e \in \mathbb{N}^*$	Link	Invariants
$x^2 - y^2 - y^3 - 10^{-e}$	0.5	$\{10^{-1}, \ldots, 10^{-10}\}$	Hopf link	$\Delta(t_1, t_2) = 1$ $\delta = 1$ $g = 0$
$x^2 - y^2 - y^3 - 10^{-e}$	0.14	$\{10^{-1}, \ldots, 10^{-10}\}$	Hopf link	$\Delta(t_1, t_2) = 1$ $\delta = 1$ $g = 0$

Table 7 Continuity for perturbations of type II of $x^2 - y^2 - y^3$

Perturbations II and ϵ		$\sigma = 10^{-e}, e \in \mathbb{N}^*$	Link	Invariants
$(1+10^{-e})x^2 - (1+3\cdot 10^{-e})y^2$ $-(1+4\cdot 10^{-e})y^3$	0.25	$\{10^{-1}, \ldots 10^{-10}\}$	Hopf link	$\Delta(t_1, t_2) = 1$ $\delta = 1$ $g = 0$
$(1+10^{-e})x^2 - (1+3\cdot 10^{-e})y^2$ $-(1+4\cdot 10^{-e})y^3$	0.14	$\{10^{-1}, \ldots 10^{-10}\}$	Hopf link	$\Delta(t_1, t_2) = 1$ $\delta = 1$ $g = 0$

$\sigma)y^2$. For $\sigma = 0.1$ we obtain the perturbed polynomial $p_{\sigma \leftarrow 0.1} = -1.1x^3 - 1.2xy + 1.1y^2$; for $\sigma = 0.01$ we obtain the perturbed polynomial $p_{\sigma \leftarrow 0.01} = -1.01x^3 - 1.02xy + 1.01y^2$; for $\sigma = 0.001$ we obtain the perturbed polynomial $p_{\sigma \leftarrow 0.001} = -1.001x^3 - 1.002xy + 1.001y^2$, etc. We notice in Table 5 that for perturbations of type II of $-x^3 - xy + y^2$ we obtain constant approximate solution.

Example 6. For the exact polynomial $p(x, y) = x^2 - y^2 - y^3$, we consider perturbations on type I of the form $p_\sigma(x, y) = x^2 - y^2 - y^3 - \sigma$, with $\sigma \in \{10^{-1}, \ldots, 10^{-10}\}$. We notice in Table 6 that for perturbations of type I of $x^2 - y^2 - y^3$ we obtain constant approximate solution.

For the perturbations of type II we consider the exact polynomial $p(x, y) = x^2 - y^2 - y^3$, the arbitrary exact polynomial $q(x, y) = x^2 - 3y^2 - 4y^3$ and $\sigma \in \{10^{-1}, \ldots, 10^{-10}\}$, obtaining the perturbed polynomials $p_\sigma(x, y) = p(x, y) + \sigma q(x, y) = x^2 - y^2 - y^3 + \sigma(x^2 - 3y^2 - 4y^3) = (1+\sigma)x^2 - (1+3\sigma)y^2 - (1+4\sigma)y^3$. For $\sigma = 0.1$ we obtain the perturbed polynomial $p_{\sigma \leftarrow 0.1} = 1.1x^2 - 1.3y^2 - 1.4y^3$; for $\sigma = 0.01$ we obtain the perturbed polynomial $p_{\sigma \leftarrow 0.01} = 1.01x^2 - 1.03y^2 - 1.04y^3$; for $\sigma = 0.001$ we obtain the perturbed polynomial $p_{\sigma \leftarrow 0.001} = 1.001x^2 - 1.003y^2 - 1.004y^3$, etc. We notice in Table 7 that for perturbations of type II of $x^2 - y^2 - y^3$ we obtain constant approximate solution.

6 Conclusion and Future Work

If I have seen further than others, it is by standing upon the shoulders of giants.

Isaac Newton

For each input plane complex algebraic curve C defined by the polynomial $F(z,w)$ with numeric coefficients, GENOM3CK performs the following computational operations:

1. It computes the set of all distinct real singularities in the projective real plane of C.
2. It computes and visualizes the algebraic link L of each singularity of the input curve C in the three-dimensional space; for each algebraic link L, which is a smooth, implicitly defined closed algebraic curve in $\mathbb{R}^3$, it computes and visualizes the two implicit algebraic surfaces that define the algebraic link L. In fact these surfaces represent the Milnor fibration.
3. It computes the diagram of each algebraic link L.
4. It computes the Alexander polynomial of each algebraic link L.
5. It computes the δ-invariant of each singularity.
6. It computes the genus of the curve C.
7. It also computes the time needed for performing each of these operations.

We have reported on a symbolic-numeric algorithm for genus computation of plane complex algebraic curves whose defining polynomials have coefficients of limited accuracy, i.e the coefficients of the polynomial are both exact and inexact data. We have successfully realized a complete automatization for the steps of the proposed symbolic-numeric algorithm in the GENOM3CK library using Axel, an algebraic geometric modeler. The library allows us to compute several invariants of an input plane complex algebraic curve, such as: the algebraic link, the Alexander polynomial and the delta-invariant of each singularity of the curve. In addition, the library allows us to analyse the performance of the proposed symbolic-numeric algorithm. As expected, the test experiments indicate the efficiency of the proposed symbolic-numeric algorithm. Moreover, we use the library in order to offer practical evidences for the convergence and the continuity properties of the proposed symbolic-numeric algorithm. These tests also indicate that the proposed symbolic numeric algorithm can be described using principles from regularization theory and approximate algebraic computation. Using these principles, we intend to give a precise meaning to the notion of approximate genus of the input plane complex algebraic curve computed using the proposed symbolic-numeric algorithm.

Acknowledgements Many thanks to Bernard Mourrain who also contributed to the implementation of GENOM3CK and offered important computational and mathematical support and guidance whenever required. Many thanks to Julien Wintz, who contributed to the implementation of the library in its starting phase. We would like to especially thank Esther Klann and Ronny Ramlau, and the other colleagues from the "Doctoral Program-Computational Mathematics" for their helpful discussions and comments, which contributed with many useful insights to handling the numerical part of our problem.

References

1. Alberti, L., Mourrain, B.: Regularity criteria for the topology of algebraic curves and surfaces. In Proceeding IMA Conference of the Mathematics of Surfaces, pp. 1–28 (2007)
2. Alberti, L., Mourrain, B.: Visualization of implicit algebraic curves. In Proceeding 15th Pacific Conference on Computer Graphics and Applications, pp. 303–312 (2007)
3. Alexander, J.W.: Topological invariant of knots and links. Trans. Am. Math. Soc. **30**, 275–306 (1928)
4. Bates, D.J., Peterson, C., Sommese, A.J., Wampler, C.W.: Numerical computation of the genus of an irreducible curve within an algebraic set. J. Pure. Appl. Algebra **215**(8), 1844–1851 (2011)
5. Béla, S., Jüttler, B.: Fat arcs for implicitly defined curves. Mathematical Methods for Curves and Surfaces. Lecture Notes in Computer Science, vol. 5862, pp. 26–40. Springer, New York (2010)
6. Brauner, K.: Zur Geometrie der Funktionen zweier komplexer Veränderlichen Abh. Math. Sem. Hamburg **6**, 1–54 (1928)
7. Brieskorn, E., Knorrer, H.: Plane Algebraic Curves. Birkhäuser, Berlin (1986)
8. Busé, L., Khalil, H., Mourrain, B.: Resultant-based methods for plane curves intersection problems. In Proceeding CASC 2005, vol. 3718, pp. 75–92 (2005)
9. Cimasoni, D.: Studying the multivariable Alexander polynomial by means of Seifert surfaces. Bol. Soc. Mat. Mexicana (3), **10**, 107–115 (2004)
10. Colin, C.A.: The knot book. An elementary introduction to the mathematical theory of knots. W.H. Freeman and Company, USA (2004)
11. Corless, R.M., Watt, S.M., Zhi, L.: QR factoring to compute the GCD of univariate approximate polynomials. IEEE Trans. Signal Process **52**, 3394–3402 (2004)
12. Dayton, B.H., Zeng, Z.:1 The approximate GCD of inexact polynomials. part ii: A multivariate algorithm. In Proceeding 2004 Internat. Symp. Symbolic Algebraic Comput, pp. 320–327 (2004)
13. de Berg, M., Krefeld, M., Overmars, M., Schwarzkopf, O.: Computational geometry: algorithms and applications. Second edition. Springer, Berlin (2008)
14. Engl, H.W., Hanke, M., Neubauer, A.: Regularization of inverse problems. Kluwer Academic Publishers Group, Dordrecht (1996)
15. Fulton, W.: Algebraic curves-An introduction to algebraic geometry. Addison-Wesley, Redwood City California (1989)
16. Greuel, G.M., Pfister, G.: A Singular introduction to commutative algebra. Springer, Berlin (2002)
17. Gutierrez, J., Rubio, R., Schicho, J.: Polynomial parametrization of curves without affine singularities. Comput. Aided Geomet. Des. **19**, 223–234 (2002)
18. Haché, G.: Computation in algebraic function fields for effective construction of algebraic-geometric codes. In: Cohen, G., Giusti, M., T. Mora (eds.) Applied Algebra, Algebraic Algorithms and Error-Correcting Codes. Lecture Notes in Computer Science, vol. 948, pp. 262–278. Springer, Berlin (1995)
19. Hess, F.: Computing Riemann-Roch spaces in algebraic function fields and related topics. Symbolic Comput. **33**, 425–445 (2002)
20. Hess, F.: Generalising the GHS attack on the elliptic curve discrete logarithm. LMS Comput. Math. **7**, 167–192 (2004)
21. Hodorog, M., Schicho, J.: Computational geometry and combinatorial algorithms for the genus computation problem. Doctoral Program "Computational Mathematics", Linz, Austria, 7 (2010)
22. Deconinck, B., Patterson, M.: Computing with plane algebraic curves and riemann surfaces: The algorithms of the maple package "Algcurves". In: Bobenko, A.I., Klein, C. (eds.) Computational Approach to Riemann Surfaces. Lecture Notes in Mathematics, vol. 2013, pp. 67–123. Springer, Berlin (2011)

23. van der Hoeven, J., Lecerf, G., Mourrain, B., Trebuchet, P., Berthomieu, J., Diatta, D.N., Mantzaflaris, A.: The quest of modularity and efficiency for symbolic and certified numeric computation. ACM SIGSAM Communications in Computer Algebra (2011)
24. Liang, C., Mourrain, B., Pavone, J.P.: Subdivision methods for 2d and 3d implicit curves, chapter 11, pp. 199–214. Springer, Geometric Modeling and Algebraic Geometry (eds. Jüttler B. Piene R.) edition, August (2008)
25. Livingston, C.: Knot theory. Mathematical Association of America, Washington, DC, USA (1993)
26. Mantzaflaris, A., Mourrain, B., Tsigaridas, E.: Continued fraction expansion of real roots of polynomial systems. In Proceeding 2009 SNC Conference on Symbolic-Numeric Computation, pp. 85–94 (2009)
27. Milnor, J.: Singular points of complex hypersurfaces. Princeton University Press and the University of Tokyo Press, New Jersey (1968)
28. Mnuk, M., Winkler, F.: CASA – A system for computer aided constructive algebraic geometry. In Proceeding International Symposium on Design and Implementation of Symbolic Computation Systems, pp. 297–307 (1996)
29. Mourrain, B., Pavone, J.P.: Subdivision methods for solving polynomial equations. J. Symbolic Comput. **44**(3), 292–306 (2009)
30. Pérez-Díaz, S., Sendra, J.R., Rueda, S.L., Sendra, J.: Approximate parametrization of plane algebraic curves by linear systems of curves. Comput. Aided Geomet. Des. **27**(2), 212–231 (2010)
31. Pikkarainen, H.K., Schicho, J.: A Bayesian model for root computation. Math. in Comp. Sci. **2**, 567–586 (2009)
32. Sendra, J.R., Winkler, F.: Parametrization of algebraic curves over optimal field extensions. Symbolic Comput. **23**, 191–208 (1997)
33. Sendra, J.R., Winkler, F., Diaz, S.P.: Rational algebraic curves. A computer algebra approach. Springer, Berlin (2008)
34. Shuhong, G., Kaltofen, E., May, J., Yang, Z., Zhi, L.: Approximate factorization of multivariate polynomials via differential equations. Symbolic Comput. **43**, 359–376 (2008)
35. Stetter, H.J.: Numerical polynomial algebra. SIAM, Philadelphia (2004)
36. Tougeron, J.C.: Ideaux de fonctions differentiables. Springer, Berlin (1972)
37. Tráng, L.D., Ramanujam, C.P.: The invariance of Milnor's number implies the invariance of the topological type. Amer. J. Math. **98**, 67–78 (1976)
38. Walker, R.J.: Algebraic curves. Springer, New York (1978)
39. Wintz, J.: Algebraic methods for geometric modelling. PhD thesis, University of Nice, Sophia-Antipolis (2008)
40. Yamamoto, M.: Classification of isolated algebraic singularities by their Alexander polynomials. Topology, **23**, 277–287 (1984)
41. Zeng, Z.: Computing multiple roots of inexact polynomials. Math. Comp. **74**, 869–903 (2005)

The "Seven Dwarfs" of Symbolic Computation

Erich L. Kaltofen

Abstract We present the Seven Dwarfs of Symbolic Computation, which are sequential and parallel algorithmic methods that today carry a great majority of all exact and hybrid symbolic compute cycles.

SymDwf 1. Exact linear algebra, integer lattices

SymDwf 2. Exact polynomial and differential algebra, Gröbner bases

SymDwf 3. Inverse symbolic problems, e.g., interpolation and parameterization

SymDwf 4. Tarski's algebraic theory of real geometry

SymDwf 5. Hybrid symbolic-numeric computation

SymDwf 6. Computation of closed form solutions

SymDwf 7. Rewrite rule systems and computational group theory

We will elaborate on each dwarf and compare with Colella's seven and the Berkeley team's thirteen dwarfs of scientific computing.

1 Introduction

Phillip Colella [7] in his 2004 presentation "Defining Software Requirements for Scientific Computing" about DARPA's High Productivity Computing Systems (HPCS) program gave his list of the now-famous *"Seven Dwarfs"* of algorithms for high-end simulation in the physical sciences.

HPCS 1. Structured Grids
HPCS 2. Unstructured Grids
HPCS 3. Fast Fourier Transform
HPCS 4. Dense Linear Algebra
HPCS 5. Sparse Linear Algebra
HPCS 6. Particles
HPCS 7. Monte Carlo

E.L. Kaltofen (✉)
North Carolina State University, Department of Mathematics, Raleigh, NC 27695-8205, USA
http://www.math.ncsu.edu/~kaltofen

U. Langer and P. Paule (eds.), *Numerical and Symbolic Scientific Computing*,
Texts and Monographs in Symbolic Computation, DOI 10.1007/978-3-7091-0794-2_5,

The dwarfs in allusion to the fairy tale mine compute cycles for golden results. Recently, the term "killer kernels" has been used to replace the notion of dwarf, but the dwarfs seem more like library procedures than operating system kernels. Following Colella, researches in parallel computation at the University of California at Berkeley, who include David Patterson and Katherine Yelick, have modified and upgraded to 13 dwarfs, where "A dwarf is an algorithmic method that captures a pattern of computation and communication [http://view.eecs.berkeley.edu/wiki/Dwarf_Mine]:"

Berkeley 1. Dense Linear Algebra
Berkeley 2. Sparse Linear Algebra
Berkeley 3. Spectral Methods
Berkeley 4. N-Body Methods
Berkeley 5. Structured Grids
Berkeley 6. Unstructured Grids
Berkeley 7. MapReduce
Berkeley 8. Combinational Logic
Berkeley 9. Graph Traversal
Berkeley 10. Dynamic Programming
Berkeley 11. Backtrack and Branch-and-Bound
Berkeley 12. Graphical Models
Berkeley 13. Finite State Machines

Both lists are notably numerical computing oriented. They exclude symbolic computation, i.e., methods with exact arithmetic, or logic programming, say rewriting via rules, altogether. However, they inspire to make a corresponding list, and here we will do so for symbolic computation. Bruno Buchberger [5] in his 1985 editorial in the first issue of *the Journal of Symbolic Computation* makes an attempt to define the discipline of symbolic computation. We adopt his breadth and view symbolic computation to include all of computer algebra [18, 25] and also algebraic methods for analysis, statistics and combinatorics, logic programming, computational geometry and program synthesis. The report [3] offers a then glimpse into the future of symbolic computation and has made several accurate predictions (see, e.g., Sect. 6 below).

Here we add to this taxonomy via our seven dwarfs of symbolic computation. Our methods are oriented to mid-level and high performance computation tasks, and should not be considered comprehensive. A subject on the boundary not included is computational number theory. The important application of symbolic computation to mathematics education is not discussed. Education tasks can be compute intensive. For example, the automatic grading of the Maple homework worksheets of our calculus classes by NCSU's egrader software consumes an entire night. On the low performance side, micro symbolic computation systems for compact devices such as cell phones constitute an important educational application of the discipline: vastly more people world-wide own cell phones than computers.

We presented the list in the talk "The Seven Dwarfs of Symbolic Computation and the Discovery of Reduced Symbolic Models" [http://www.math.ncsu.edu/~kaltofen/bibliography/07/SNSC07.pdf] at 4th International Conference on Symbolic and Numerical Scientific Computing SNSC '08 at RISC Linz, Hagenberg, Austria, on July 24, 2008. In the following, we briefly discuss each dwarf and give selected references, which are meant to highlight some past and current results and not as a complete survey as other important work could not be included.

2 Exact Linear Algebra, Integer Lattices

Important breakthroughs in exact linear algebra actually happened later than those in polynomial algebra, notably after Buchberger's Gröbner basis algorithm. One is the discovery of exact sparse iterative algorithms based on the numeric Krylov and Lanczos algorithms [29, 51] and their block versions [8, 26, 48] whose probabilistic analysis for small coefficient fields is being completed today [11]. The algorithms are available in the open source LinBox library [www.linalg.org], callable from the SAGE and Maple platforms, and put to important use. A second breakthrough are the lattice basis reduction algorithms [12, 37] that today have greatly improved implementations [40] and are used extensively for discovery of exact identities from numeric approximations ([20], "the inverse symbolic calculator" [http://oldweb.cecm.sfu.ca/projects/ISC/ISCmain.html]).

We observe additional trends today: Strassen's fast matrix multiplication algorithm and cache-efficient BLAS libraries improve performance of exact linear algebra [9]; characteristic polynomials and integer Smith normal forms of sparse integer matrices [10, 15] are important invariants, for instance in computing the so-called bar code of a persistent topology of data; and structured exact linear problem solvers such as the matrix Berlekamp/Massey algorithm [32] form a fundamental ingredient in sparse solvers.

Exact linear algebra algorithms are easily underestimated. Great progress has been made in the past ten years, and the software has a wide range of applications. Exact solutions are not only needed for finite field entries, but also for diophantine problems and when the exact input forms an ill-conditioned matrix.

3 Exact Polynomial and Differential Algebra, Gröbner Bases

Polynomial arithmetic including the computation of multivariate polynomial greatest common divisors, factorizations, and triangular and other canonical forms for polynomial systems constitute the heart of computer algebra. Classical tools include resultant computation and Hensel lifting and modern tools Buchberger's Gröbner basis algorithm. Truncated power series are represented by polynomials and thus included in this dwarf.

The calculus of differentiation and differential ideals allows manipulation of differential equations as polynomials with a derivative operator. In addition, one can interpret the derivative (or difference) operator as a new symbol and construct composed operators as polynomials with variables and derivative (difference) symbols. Those operator rings are generalized to Ore extensions and have an additional, special, non-commutative multiplication. Two references are [13, 43]. See also Sect. 7.

Efficient implementations of polynomial factoring and Gröbner basis algorithms, for instance Jean-Charles Faugere's FGb which is also callable from within Maple, make a serious use of the methods as easy as, say, Matlab gives access to numerical linear algebra. Today's applications are abundant, e.g., cryptosystems have be broken with them.

Basic polynomial arithmetic of multivariate polynomials forms the core infrastructure of any symbolic manipulation system, and efficiency improvements can still being made: any speedup will speed many application algorithms. This is the more true with the arrival of multicore and multiprocessor workstations.

4 Inverse Symbolic Problems, e.g., Interpolation and Parameterization

Interpolation and curve fitting are basic and important operations to build mathematical models from data. Zippel's [53] and Ben-Or and Tiwari's [1] sparse multivariate polynomial algorithms are a fundamental contribution from symbolic computation to the task of function/model recovery. The paradigm of early termination via randomization has successfully been exploited [27]. In Sect. 6 we point to new numerical methods that were derived from the exact symbolic algorithms. More recently polynomial and rational function recovery with very high degree terms have been achieved [14, 17, 34]. There the values are determined at roots of unity to prevent size explosion. Beyond polynomial and rational function recovery is, for instance, recovery of algebraic functions and differential equations from series solutions.

The circle as an implicitly represented curve $x^2 + y^2 = 1$ can be rationally parameterized as $x = \cos(\alpha) = (1 - t^2)/(t^2 + 1)$, $y = \sin(\alpha) = 2t/(t^2 + 1)$ with $-\infty \leq t = \tan(\alpha/2) \leq \infty$. Not all real curves can be so parameterized, for instance elliptic curves. A reference is the book [45]. Parametric curves form basic objects in geometric rendering.

Interpolation and Chinese remaindering forms the recovery step in computing with homomorphic images, where a computation is split by first computing the solution for various values of a symbolic parameter and then the symbolic solution is interpolated from those values. Because each value can be processed separately and no intermediate degree/size growth occurs, the paradigm constitutes a powerful and parallel/distributed approach.

5 Tarski's Algebraic Theory of Real Geometry

Tarksi's algorithm for eliminating quantifiers in sentences formed on semi-algebraic sets makes most of Euclidean geometry and real polynomial optimization decidable. Unfortunately, the general method solves problems in a high complexity class (super-exponential). Nonetheless, George Collins's cylindrical algebraic decomposition algorithm is implemented and has solved non-trivial problems. References are the collection [6] and [4], which has references to newer work.

A fundamental quantifier elimination problem is to determine whether a multivariate polynomial $f(x_1, \ldots, x_n)$ has a real root, which we shall call Seidenberg's problem. For instance, deciding if a polynomial can attain negative values, i.e., is not positive semidefinite, is equivalent to deciding if $f(x_1, \ldots, x_n)x_{n+1}^2 + 1$ has a real root. Thus all (unconstrained) polynomial inequalities are reduced to Seidenberg's problem. A more general fundamental problem is to compute a sample point in each connected component of the real solution set of a system of polynomial equations.

Modern software, such as RAGlib [44], analyzes the real critical values via Gröbner basis computation. A variant of Tarski's quantifier elimination problem that weakens the pre- and post conditions and thus lowers the intrinsic complexity can be based on such real polynomial software [21].

Hilbert's Problem on polynomial sums-of-squares and Artin's Theorem offers an additional approach to real polynomial optimization, which is made possible by numerical non-linear optimization and discussed in Sect. 6.

6 Hybrid Symbolic-Numeric Computation

The use of approximate, floating point, arithmetic and approximations of irrational functions by polynomials and rational functions is as old as logarithm tables and Taylor series and Padé fractions. Section 2.12.3 in [18] describes what constitutes hybrid symbolic-numeric computation. Our description already contains the fundamental concept of computing a nearest polynomial, measured in some distance norm, that satisfies a property which the input polynomial does not. Classical properties are having non-trivial polynomial greatest common divisors and factors, or common solutions (the nearest consistent system) or solutions that have real components (the nearest polynomial with a real root) or higher multiplicities (contracting clusters of zeros to a single common point). The inputs are not exact, because of physical measurement or because the scalars come from a floating point computation, and therefore lack the needed property. The sought property may have to be avoided, and a lower bound on the distance yields a condition number. New work and references are found in the proceedings [24, 47, 49].

Because there is a gradual transition to mostly numerical solution of, say, algebraic geometry problems, e.g., via programs like Bertini [http://www.nd.edu/~sommese/bertini/] and PHCpack [http://www.math.uic.edu/~jan/PHCpack/phcpack.html], the symbolic computation component in the hybrid approach is sometimes dismissed. Clearly, the algorithms for sparse approximate interpolation [16, 30] are based on the exact sparse polynomial interpolation algorithms by Zippel and by Ben-Or and Tiwari. Those hybrid algorithms have applications to sparse signal processing and compressive sensing. The approximate Buchberger-Möller algorithm has found an application in analyzing data from oil wells [http://www.algebraic-oil.uni-passau.de/].

Any positive semidefinite polynomial f with real (rational) coefficients (see Sect. 5) can be written as a finite sum

$$f(x_1,\dots,x_n) = \frac{1}{g_0(x_1,\dots,x_n)^2} \sum_{i=1}^{k} g_i(x_1,\dots,x_n)^2, \tag{1}$$

where g_i are polynomials with real (rational) coefficients. If there exist g_i with $g_0 = 1$, f is said to be SOS, but not all f are, e.g., Motzkin's polynomial. Any polynomial inequality $f \geq h$ is equivalent to $f - h$ being positive semidefinite; h in global optimization is the real infimum (or a rational lower bound) of all values of f. Therefore, any g_i satisfying $f - h = 1/g_0^2 \sum_i g_i^2$ constitute a proof (exact certificate) for the inequality/optimum. Two recent developments have made it possible to compute such certificates. The first are the numerical optimization algorithms for semidefinite programming. The second is a symbolic technique for converting an imprecise SOS with floating point coefficients to an exact identity over the rational numbers [28, 33, 41]. Among the recent successes are the proof of the Monotone Column Permanent Conjecture for $n = 4$ [31], which was completed shortly before the general conjecture could be established, the Bessis-Moussa-Villani (BMV) conjecture for $m \leq 13$ [35], new SOS proofs for many known inequalities, and a deformation analysis approach to Seidenberg's problem of Sect. 5 [22]. Optimization with additional polynomial inequality constraints are handled by various so-called Positivstellensätze [38].

7 Computation of Closed Form Solutions

Robert Risch's 1970 solution of Hardy's problem to determine if an indefinite integral can be expressed in closed form as an expression in elementary functions is a hallmark of early symbolic computation. Closed form solutions to differential equations and the inclusion of special functions, possibly defined by lower order differential equations constitutes an active area of research. References are the book [46] and [52], which has references to newer work. A connection to differential elimination theory of Sect. 3 should be noted.

Algorithms for closed form solutions for discrete summations, difference equations, and combinatorial counts form an active subarea of symbolic computation which could be named "symbolic combinatorics" (Michael F. Singer). The members of Peter Paule's research group, some of who are part of the Austrian DK research grant "Numerical and Symbolic Scientific Computing," have made significant recent contributions to the area of symbolic combinatorics: http://www.risc.uni-linz.ac.at/research/combinat/risc/publications/. An example is the closed form solution for the generating function for counting so-called Gessel walks, which turned out to be an algebraic function in three variables [2], which was discovered in collaboration with the Algorithms Project at INRIA [http://algo.inria.fr/index.html].

8 Rewrite Rule Systems and Computational Group Theory

Computational group and representation theory is a traditional subject lying in the intersection of symbolic computation and combinatorics. Famous popular examples are to compute the minimum number of moves necessary for solving Rubik's cube puzzle from any configuration [36], which was recently completed on a Google data center http://cube20.org. Group decomposition plays a major role in the synthesis of high performance FFT library [42].

Bruno Buchberger included rewrite rule systems as a subject of symbolic computation, motivated perhaps by the interpretation of his Gröbner basis algorithm as a critical-pair/completion method (Knuth-Bendix completion). Rewrite techniques are often deployed for expression simplification in symbolic computation. The RTA conference series [http://rewriting.loria.fr/rta/] covers the many applications beyond symbolic computation (see also the Coq proof assistant http://www.lix.polytechnique.fr/coq/). Algebraic techniques are also be applied to algorithm synthesis, such as automatic differentiation [19] and the transposition principle for matrix-times-vector products or elimination of divisions from algebraic algorithms.

Acknowledgements I thank Bruno Salvy for his thoughtful comments.

This material is based on work supported in part by the National Science Foundation under Grants CCF-0830347, CCF-0514585 and DMS-0532140.

References

1. Ben-Or, M., Tiwari, P.: A deterministic algorithm for sparse multivariate polynomial interpolation. In Proceeding of the Twentieth Annual ACM Symposium on Theory of Computing, pp. 301–309, ACM Press, New York (1988)
2. Bostan, A., Kauers, M.: The complete generating function for Gessel walks is algebraic. In Proceedings of the AMS, (2010); with an Appendix by Mark van Hoeij. To appear. http://www.risc.uni-linz.ac.at/people/mkauers/publications/bostan10.pdf
3. Boyle, A., Caviness, B.F. (ed.): Future Directions for Research in Symbolic Computation. SIAM, Philadelphia (1989); Report of a Workshop on Symbolic and Algebraic Computation April 29–30, 1988 Washington DC. Anthony C. Hearn Workshop Chairperson. http://www.cis.udel.edu/~caviness/wsreport.pdf
4. Brown, C.W.: Fast simplification of Tarski formulas. In ISSAC'09 Proceedings of the 2009 International Symposium on Symbolic and Algebraic Computation, pp. 63–70, New York, NY, USA (2009)
5. Buchberger, B.: Symbolic computation (an editorial). J. Symbolic Comput. **1**(1), 1–6 (1985)
6. Caviness, B.F., Johnson, J.R. (ed.): Quantifier Elimination and Cylindrical Algebraic Decomposition. Springer, Berlin (1998)
7. Colella, P.: Defining software requirements for scientific computing. Slide of 2004 presentation included in David Patterson's 2005 talk, (2004); http://www.lanl.gov/orgs/hpc/salishan/salishan2005/davidpatterson.pdf
8. Coppersmith, D.: Solving homogeneous linear equations over GF(2) via block Wiedemann algorithm. Math. Comput. **62**(205), 333–350 (1994)

9. Dumas, J.-G., Giorgi, P., Pernet, C.: Dense linear algebra over finite fields: the FFLAS and FFPACK packages. ACM Trans. Math. Software **35**(3), 1–42 (2008)
10. Dumas, J.-G., Saunders, B.D., Villard, G.: On efficient sparse integer matrix Smith normal form computation. J. Symbolic Comput. **32**(1/2), 71–99 (2001); Special issue on Computer Algebra and Mechanized Reasoning: Selected St. Andrews' ISSAC/Calculemus Contributions. Guest editors: T. Recio and M. Kerber
11. Eberly, W.: Yet another block Lanczos algorithm: How to simplify the computation and reduce reliance on preconditioners in the small field case. In [Watt 2010], page to appear, July 2010 International Symposium on Symbolic and Algebraic Computation, ACM, New York, NY, USA (2010)
12. Ferguson, H.R.P., Forcade, R.W.: Multidimensional Euclidean algorithms. J. Reine Angew. Math. **334**, 171–181 (1982)
13. Gao, X.-S., der Hoeven, J.V., Yuan, C.M., Zhang, G.-L.: Characteristic set method for differential-difference polynomial systems. J. Symb. Comput. **44**(9), 1137–1163 (2009)
14. Garg, S., Schost, É.: Interpolation of polynomials given by straight-line programs. Theor. Comput. Sci. **410**(27–29), 2659–2662 (2009). ISSN 0304-3975. http://www.csd.uwo.ca/~eschost/publications/interp.pdf
15. Giesbrecht, M.: Fast computation of the Smith form of a sparse integer matrix. Comput. Complex. **10**, 41–69 (2001)
16. Giesbrecht, M., Labahn, G., Shin Lee, W.: Symbolic-numeric sparse interpolation of multivariate polynomials. J. Symbolic Comput. **44**, 943–959 (2009)
17. Giesbrecht, M, Roche, D.S.: Interpolation of shifted-lacunary polynomials. Comput. Complex. **19**(3), 333–354 (2010)
18. Grabmeier, J., Kaltofen, E., Weispfenning, V. (ed.): Computer Algebra Handbook. Springer, Heidelberg, Germany (2003). ISBN 3-540-65466-6. 637 + xx pages + CD-ROM.
19. Griewank, A.: Evaluating Derivatives: Principles and Techniques of Algorithmic Differentiation. SIAM Publications, Philadephia (2008)
20. Håstad, J., Just, B., Lagarias, J.C., Schnorr, C.P.: Polynomial time algorithms for finding integer relations among real numbers. SIAM J. Comput. **18**(5), 859–881 (1989)
21. Hong, H., Safey El Din, M.: Variant real quantifier elimination: Algorithm and application. In ISSAC'09, ACM, pp. 183–190 (2009)
22. Hutton, S.E., Kaltofen, E.L., Zhi, L.: Computing the radius of positive semidefiniteness of a multivariate real polynomial via a dual of Seidenberg's method. In [Watt 2010], page to appear, July 2010 International Symposium on Symbolic and Algebraic Computation, pp. 227–234, New York, NY, USA (2010); http://www.math.ncsu.edu/~kaltofen/bibliography/10/HKZ10.pdf
23. Jeffrey, D. (ed.): ISSAC 2008. ACM Press. ISBN 978-1-59593-904-3
24. Kai, H., Sekigawa, H. (ed.): SNC'09 Proceeding 2009 International Workshop on Symbolic-Numeric Computation, pp. 28–31, ACM Press, New York, NY, USA (2009). ISBN 978-1-60558-664-9
25. Kaltofen, E.: Computer algebra algorithms. In: Traub, J.F. (eds.) Annual Review in Computer Science, vol. 2, pp. 91–118. Annual Reviews Inc., Palo Alto, California (1987); http://www.math.ncsu.edu/~kaltofen/bibliography/87/Ka87_annrev.pdf
26. Kaltofen, E.: Analysis of Coppersmith's block Wiedemann algorithm for the parallel solution of sparse linear systems. Math. Comput. **64**(210), 777–806 (1995); http://www.math.ncsu.edu/~kaltofen/bibliography/95/Ka95_mathcomp.pdf
27. Kaltofen, E., Lee, W.-S.: Early termination in sparse interpolation algorithms. J. Symbolic Comput. **36**(3–4), 365–400 (2003); Special issue Internat. Symp. Symbolic Algebraic Comput. (ISSAC 2002). Guest editors: M. Giusti & L. M. Pardo. http://www.math.ncsu.edu/~kaltofen/bibliography/03/KL03.pdf
28. Kaltofen, E., Li, B., Yang, Z., Zhi, L.: Exact certification of global optimality of approximate factorizations via rationalizing sums-of-squares with floating point scalars. In ISSAC'08, pp. 155–163 ACM Press, New York, NY, USA (2008); http://www.math.ncsu.edu/~kaltofen/bibliography/08/KLYZ08.pdf

29. Kaltofen, E., Saunders, B.D.: On Wiedemann's method of solving sparse linear systems. In: Mattson, H.F., Mora, T., Rao, T.R.N. (eds.) Proceeding AAECC-9, Lect. Notes Comput. Sci., vol. 539, pp. 29–38, Springer, Heidelberg, Germany (1991); http://www.math.ncsu.edu/~kaltofen/bibliography/91/KaSa91.pdf
30. Kaltofen, E., Yang, Z., Zhi, L.: On probabilistic analysis of randomization in hybrid symbolic-numeric algorithms. In ISSAC '07 Proceedings of the 2007 international symposium on Symbolic and algebraic computation, pp. 11–17 ACM Press, New York, NY, USA (2007); http://www.math.ncsu.edu/~kaltofen/bibliography/07/KYZ07.pdf
31. Kaltofen, E., Yang, Z., Zhi, L.: A proof of the Monotone Column Permanent (MCP) Conjecture for dimension 4 via sums-of-squares of rational functions. In SNC'09, pp. 65–69 (2009a); http://www.math.ncsu.edu/~kaltofen/bibliography/09/KYZ09.pdf
32. Kaltofen, E., Yuhasz, G.: On the matrix Berlekamp-Massey algorithm, (2006); Manuscript, 29 pages. Submitted
33. Kaltofen, E.L., Li, B., Yang, Z., Zhi, L.: Exact certification in global polynomial optimization via sums-of-squares of rational functions with rational coefficients, January (2009b); Accepted for publication in J. Symbolic Comput. http://www.math.ncsu.edu/~kaltofen/bibliography/09/KLYZ09.pdf
34. Kaltofen, E.L., Nehring, M.: Supersparse black box rational function interpolation. In: Leykin, A. (ed.) Proc. 2011 Internat. Symp. Symbolic Algebraic Comput, ISSAC 2011, pp. 177–185. Association for Computing Machinery, New York, (2011). ISBN 978-1-4503-0675-1
35. Klep, I., Schweighofer, M.: Sums of Hermitian squares and the BMV conjecture. J. Stat. Phys. **133**, 739–760 (2008)
36. Kunkle, D., Cooperman, G.: Harnessing parallel disks to solve Rubik's cube. J. Symbolic Comput. **44**(7), 872–890 (2009); http://www.ccs.neu.edu/home/gene/papers/jsc09.pdf
37. Lenstra, A.K., Lenstra, Jr., H. W., Lovász, L.: Factoring polynomials with rational coefficients. Math. Ann. **261**, 515–534 (1982)
38. Marshall, M.: Positive Polynomials and Sums of Squares. Amer. Math. Soc. **146**, 187 (2008)
39. May, J.P. (ed.): ISSAC 2009 Proceeding 2009 International Symposium Symbolic Algebraic Computation, ACM, (2009). ISBN 978-1-60558-609-0
40. Novocin, A., Stehlé, D., Villard, G.: An LLL-reduction algorithm with quasi-linear time complexity: extended abstract. In: Proc. 43rd Annual ACM Symp. Theory Comput., pp. 403–412. ACM, New York (2011)
41. Peyrl, H., Parrilo, P.A.: Computing sum of squares decompositions with rational coefficients. Theor. Comput. Sci. **409**, 269–281 (2008)
42. Püschel, M., Moura, J.M.F., Johnson, J., Padua, D., Veloso, M., Singer, B., Xiong, J., Franchetti, F., Gacic, A., Voronenko, Y., Chen, K., Johnson, R.W., Rizzolo, N.: SPIRAL: Code generation for DSP transforms. Proc. IEEE **93**(2), 232–275 (2005); special issue on "Program Generation, Optimization, and Adaptation", http://spiral.ece.cmu.edu:8080/pub-spiral/pubfile/paper_1.pdf
43. Rosenkranz, M., Regensburger, G.: Solving and factoring boundary problems for linear ordinary differential equations in differential algebras. J. Symbolic Comput. **43**(8), 515–544 (2008). ISSN 0747-7171
44. Safey El Din, M.: Computing the global optimum of a multivariate polynomial over the reals. In ISSAC'08 Proceedings of the twenty-first international symposium on Symbolic and algebraic computation, ACM Press, New York, NY (2008)
45. Sendra, J.R., Winkler, F., Pérez-Díaz, S.: Rational Algebraic Curves A Computer Algebra Approach, *Algorithms and Computation in Mathematics*. vol. 22, Springer, Heidelberg, Germany (2007). ISBN ISSN 1431-1550, ISBN 978-3-540-73724-7
46. van der Put, M., Singer, M.F.: Galois Theory of Linear Differential Equations, *Grundlehren der mathematischen Wissenschaften*. vol. 328 Springer, Heidelberg, Berlin (2003); http://www4.ncsu.edu/~singer/papers/dbook.ps

47. Verschelde, J., Watt, S.M. (ed.): SNC'07 Proceeding 2007 International Workshop on Symbolic-Numeric Computation, ACM Press, New York, NY, USA (2007). ISBN 978-1-59593-744-5
48. Villard, G.: Further analysis of Coppersmith's block Wiedemann algorithm for the solution of sparse linear systems. In: Küchlin, W. (eds) ISSAC 97 Proceeding 1997 International Symposium Symbolic Algebraic Computation, pp. 32–39, ACM Press, New York, NY, USA (1997). ISBN 0-89791-875-4
49. Wang, D., Zhi, L. (ed.): Symbolic-Numeric Computation. Trends in Mathematics. Birkhäuser Verlag, Basel, Switzerland (2007). ISBN 978-3-7643-7983-4
50. Watt, S.M. (ed.): Proceeding 2010 International Symposium Symbolic Algebraic Computation ISSAC 2010, Association for Computing Machinery (2010). ISBN 978-1-4503-0150-3
51. Wiedemann, D.: Solving sparse linear equations over finite fields. IEEE Trans. Inf. Theory **32**(1), 54–62 (1986)
52. Yuan, Q., van Hoeij, M.: Finding all Bessel type solutions for linear differential equations with rational function coefficients. In [Watt 2010], page to appear, July 2010, pp.37–44 (2010)
53. Zippel, R.: Interpolating polynomials from their values. J. Symbolic Comput. **9**(3), 375–403 (1990)

Computer Algebra Meets Finite Elements: An Efficient Implementation for Maxwell's Equations

Christoph Koutschan, Christoph Lehrenfeld, and Joachim Schöberl

Abstract We consider the numerical discretization of the time-domain Maxwell's equations with an energy-conserving discontinuous Galerkin finite element formulation. This particular formulation allows for higher order approximations of the electric and magnetic field. Special emphasis is placed on an efficient implementation which is achieved by taking advantage of recurrence properties and the tensor-product structure of the chosen shape functions. These recurrences have been derived symbolically with computer algebra methods reminiscent of the holonomic systems approach.

1 Introduction

This paper is dedicated to a successful cooperation between symbolic computation and numerical analysis. The goal is to simulate the propagation of electromagnetic waves using finite element methods (FEM). Such simulations play an important role for constructing antennas, electric circuit boards, bodyworks, and many other devices where electromagnetic radiation is involved. The numerical simulation of such physical phenomena helps to optimize the shape of components and saves the engineer from doing a long and expensive series of experiments.

Finite element methods serve to approximate the solution of partial differential equations on a given domain $\Omega \subseteq \mathbb{R}^d$ subject to certain constraints (e.g., boundary conditions). The domain Ω is partitioned into small elements (typically triangles or

C.Koutschan (✉)
Research Institute for Symbolic Computation, Johannes Kepler University, Linz, Austria

C. Lehrenfeld
Institut für Geometrie und Praktische Mathematik, RWTH Aachen, Germany

J. Schöberl
Center for Computational Engineering Science, RWTH Aachen, Germany

U. Langer and P. Paule (eds.), *Numerical and Symbolic Scientific Computing*, Texts and Monographs in Symbolic Computation, DOI 10.1007/978-3-7091-0794-2_6,

tetrahedra) and the solution is approximated on each element by means of certain shape functions. In our application we deal with Maxwell's equations which relate the magnetic and the electric field. In Sect. 2 we describe how the problem can be discretized using FEM and in Sect. 3 we give the details concerning an efficient implementation.

An important ingredient for the fast execution of some operations in the FEM are certain difference-differential relations that were derived with computer algebra methods. The methods that we employ, originate in Zeilberger's holonomic systems approach [3,10,13] whose basic idea is to define functions and sequences in terms of differential equations and recurrence equations plus initial values (these equations have to be linear with polynomial coefficients). Luckily the shape functions used in the chosen FEM discretization fit into the holonomic framework since they are defined in terms of orthogonal polynomials. Section 4 explains how the desired relations have been computed.

2 FEM Formulation of Maxwell's Equations

In order to describe electromagnetic wave propagation problems, we consider the loss-free time-domain Maxwell's equations

$$\varepsilon \frac{\partial E}{\partial t} = \operatorname{curl} H,$$

$$\mu \frac{\partial H}{\partial t} = -\operatorname{curl} E,$$

subject to appropriate initial and boundary conditions. Here $E = E(x,t)$ denotes the electric and $H = H(x,t)$ the magnetic field strength (with $x = (x_1, x_2, x_3)$ the space variables and t the time), and ε and $\mu > 0$ are the permittivity and the permeability, respectively. When discretizing these equations with the finite element method, we go over to a weak formulation by multiplying both equations with test functions $e(x)$ and $h(x)$ and integrating over the whole domain $\Omega \subset \mathbb{R}^3$. The solution of the Maxwell's equations then has to fulfill the conditions

$$\begin{aligned} \frac{\partial}{\partial t}(\varepsilon E, e)_\Omega &= (\operatorname{curl} H, e)_\Omega, \\ \frac{\partial}{\partial t}(\mu H, h)_\Omega &= -(\operatorname{curl} E, h)_\Omega \end{aligned} \tag{1}$$

for all test functions e and h, where $(\cdot,\cdot)_\Omega$ is the short notation for the $L^2(\Omega)$ inner product $(a,b)_\Omega = \int_\Omega ab \,\mathrm{d}x$. Then we replace both the magnetic and electric field as well as the test functions by finite-dimensional approximations on a triangulation $\mathcal{T}_h$ of the domain Ω. Herein h denotes some characteristic length of the elements in $\mathcal{T}_h$ (not to be confused with the test function h).

Conforming finite elements ensure that the finite-dimensional approximations are within a space which is appropriate for the partial differential equations under consideration. For Maxwell's equations this space is $H(\mathrm{curl}, \Omega)$ which demands tangential components to be continuous across element interfaces. The discontinuous Galerkin finite element method (DG) neglects this conformity condition when building up a discrete basis for the approximation, but instead has to incorporate stabilization terms to achieve a consistent and stable formulation. This is normally done by applying integration by parts and replacing fluxes at element boundaries with *numerical fluxes* [1, 7, 8, 11]. The latter approach has the major advantage that the mass matrices M_ε and M_μ, i.e., the matrices that arise when discretizing $(\varepsilon E, e)_\Omega$ and $(\mu H, h)_\Omega$, respectively, are block-diagonal which makes the application of their inverses computationally more efficient.

We consider the approximation space

$$V_h^k = \left\{ v \in \left(L^2(\Omega)\right)^3 : v|_T \in \left(\mathcal{P}^k(T)\right)^3 \;\; \forall T \in \mathcal{T}_h \right\}$$

that consists of functions which are piecewise polynomial up to degree k. By integration by parts of (1) on each element $T \in \mathcal{T}_h$, and by adding a consistent stabilization term on all element boundaries we get (again for all test functions e and h)

$$\frac{\partial}{\partial t} \sum_{T \in \mathcal{T}_h} (\varepsilon E, e)_T = \sum_{T \in \mathcal{T}_h} \big((H, \mathrm{curl}\, e)_T + (H^* \times \nu, e)_{\partial T} \big),$$

$$\frac{\partial}{\partial t} \sum_{T \in \mathcal{T}_h} (\mu H, h)_T = \sum_{T \in \mathcal{T}_h} \big(-(\mathrm{curl}\, E, h)_T + (E^* - E, h \times \nu)_{\partial T} \big),$$

where ν denotes the outer normal on each element boundary and H^*, E^* are the numerical fluxes. The properties of different DG formulations mainly depend on the choice of the numerical fluxes. As all derivatives are now shifted to the electric field E and the according test functions e, it is reasonable to approximate the electric field of one degree higher than the magnetic field. So we choose the approximation spaces V_h^{k+1} for E and e and V_h^k for H and h.

2.1 Numerical Flux

Several choices for the numerical flux are used in practice. Our goal here is to derive a numerical flux which ensures that the numerical approximation fulfills the following two important properties which are already fulfilled on the continuous level:

1. Conservation of the energy $\frac{1}{2}(\varepsilon E, E)_\Omega + \frac{1}{2}(\mu H, H)_\Omega$
2. Non-existence of *spurious modes*

On the one hand using dissipative fluxes avoids spurious modes and is often used, but as it introduces dissipation, the energy of the system is not conserved. On the other hand the standard approach for energy conserving methods is the so called *central flux*. Its mayor disadvantage is, that it introduces non-physical modes, spurious modes.

Nevertheless we start with this approach to derive the *stabilized central flux* formulation which gets rid of both problems. A more extensive discussion of numerical fluxes (including the stabilized central flux) for Maxwell's equations can be found in [6, 8.2].

The central flux takes the averaged values of neighboring elements for the numerical flux, i.e., $H^* = \{\{H\}\}$ and $E^* = \{\{E\}\}$ with $\{\{\cdot\}\}$ denoting the averaging operator, and ends up with a semi-discrete system of the form

$$\frac{\partial}{\partial t}\begin{pmatrix} M_\varepsilon & \\ & M_\mu \end{pmatrix}\begin{pmatrix} E \\ H \end{pmatrix} = \begin{pmatrix} & -C_h^T \\ C_h & \end{pmatrix}\begin{pmatrix} E \\ H \end{pmatrix} \tag{2}$$

where C_h denotes the discrete curl operator stemming from the central flux formulation. The matrix on the left side is symmetric and positive definite whereas the matrix on the right side is antisymmetric. Then the evolution matrix for the modified unknowns $(M_\varepsilon^{\frac{1}{2}} E, M_\mu^{\frac{1}{2}} H)^T$ is also antisymmetric and thus the proposed energy is conserved. Nevertheless this matrix has a lot of eigenvalues close to zero which correspond to the discretization, but not to the physical behavior of the system. To motivate the modification which will stabilize the formulation, let us have a brief look at the problem in frequency domain, i.e., for time-harmonic electric and magnetic fields. Then the discrete problem in frequency domain reads (with frequency ω):

$$0 = (i\omega)^2(M_\varepsilon E, e) + (M_\mu^{-1} C_h E, C_h e). \tag{3}$$

The problem with non-physical zero eigenvalues now manifests in $(C_h E, C_h e)$ being only positive semidefinite. We overcome this issue by adding a stabilization bilinear form $S(E, e)$ to (3) as proposed in [6].

$$S(E,e) := \sum_{F \in \mathcal{F}_h} \frac{\alpha}{h}([\![E]\!] \times \nu, [\![e]\!] \times \nu)_F$$

with $\alpha > 0$, where $\mathcal{F}_h$ is the union of all element boundaries and $[\![\cdot]\!]$ denotes the jump operator, i.e., the difference between values of adjacent elements. This stabilization bilinearform eliminates the nontrivial kernel of C_h and is consistent as $[\![E]\!] \times \nu$ is zero for the exact solution. Before we can translate the formulation back to the time domain, we introduce a new variable which is defined as

$$H^F := \frac{([\![E]\!] \times \nu)\alpha}{i\omega h}$$

The new unknown H^F is also piecewise polynomial on each face.

If we go back to the time-domain formulation we end up with the following formulation (note that relations between $[\![\cdot]\!]$ and $\{\!\!\{\cdot\}\!\!\}$ were used):

$$\begin{aligned}
\frac{\partial}{\partial t}\sum_{T\in\mathcal{T}_h}(\varepsilon E,e)_T &= \sum_{T\in\mathcal{T}_h}\big((H,\operatorname{curl} e)_T + (\{\!\!\{H\}\!\!\}\times\nu,e)_{\partial T}\big)\\
&\quad + \sum_{F\in\mathcal{F}_h}(H^F\times\nu,[\![e]\!])_F,\\
\frac{\partial}{\partial t}\sum_{T\in\mathcal{T}_h}(\mu H,h)_T &= \sum_{T\in\mathcal{T}_h}\left(-(h,\operatorname{curl} E)_T + \left(\frac{1}{2}[\![E]\!]\times\nu,h\right)_{\partial T}\right),\\
\frac{\partial}{\partial t}\sum_{F\in\mathcal{F}_h}\frac{\alpha}{h}(H^F,h^F)_F &= \sum_{F\in\mathcal{F}_h}([\![E]\!]\times\nu,h^F)_F.
\end{aligned}$$

For p-robust behavior α should scale with p^2, where p is the polynomial degree. This is motivated by the symmetric interior penalty method for elliptic equations (see e.g. [1]) where a scaling of α with p^2 in the bilinearform S is necessary for stability to dominate over some terms stemming from inverse inequalities which scale with p^2 (see also [7]).

We again achieve a system of the form (2) where the vector H now consists of element and face unknowns and the matrix representing the discrete curl operator is the stabilized central flux curl operator now. Thus we conclude that the method now conserves energy, and spurious modes, introduced by the central flux, vanish.

2.2 *Numerical Examples (Spherical Vacuum Resonator)*

We consider a spherical domain $\Omega := \{x \in \mathbb{R}^3 : \|x\|_2 \leq 1\}$ and the frequency domain formulation of the Maxwell's equations subject to perfect electrical boundary conditions

$$\begin{aligned}
\left.\begin{aligned} i\omega\varepsilon E &= \operatorname{curl} H,\\ i\omega\mu H &= -\operatorname{curl} E,\end{aligned}\right\} &\quad \text{on} \quad \Omega,\\
E\times\nu = 0 \qquad &\quad \text{on} \quad \partial\Omega,
\end{aligned}$$

To demonstrate the opportunities of higher order discretizations we consider a coarse mesh consisting of 30 elements and increase the polynomial degree to increase the spatial resolution. We are interested in the error of the eight smallest resonance frequencies. Therefore we compare the eigenvalues of the numerical discretization with those of a reference solution. In Fig. 1 we observe the expected exponential convergence of the method.

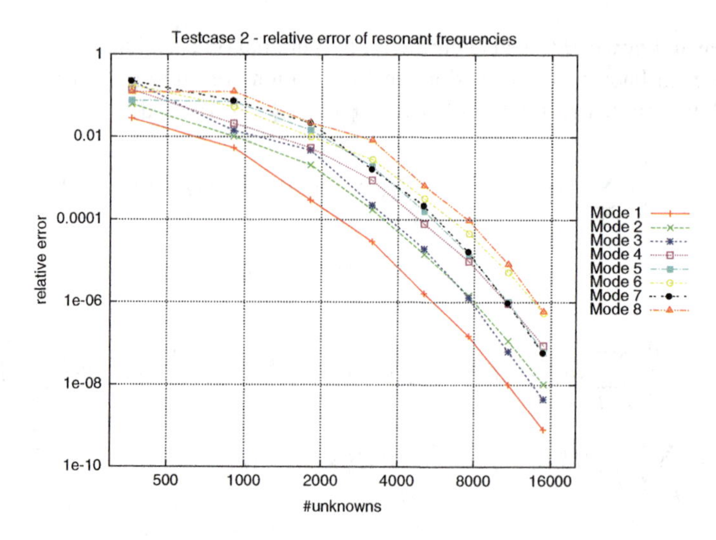

Fig. 1 Convergence of the resonance frequencies after p-refinement

3 Computational Aspects

As the spatial discretization conserves energy, we consider symplectic time integration methods which conserve the energy on a time-discrete level. The simplest one is the *symplectic Euler* method which discretizes the semi-discrete system (2) in the following way:

$$H^{n+1} = H^n + \Delta t \; M_\mu^{-1} C_h E^n$$

$$E^{n+1} = E^n - \Delta t \; M_\varepsilon^{-1} C_h^T H^{n+1}$$

with the stability condition

$$\Delta t \leq 2 \left(\rho (M_\mu^{-\frac{1}{2}} C_h M_e^{-1} C_h^T M_\mu^{-\frac{1}{2}}) \right)^{-1}$$

The matrix $M_\mu^{-\frac{1}{2}} C_h M_e^{-1} C_h^T M_\mu^{-\frac{1}{2}}$ is symmetric and the spectral radius ρ can be estimated once by an iterative method like the power iteration. When shifting the electric or the magnetic field by a half time-step we can reconstruct the well-known *leap frog* method. Nevertheless for our considerations it is less important which time integration scheme is used as long as it is explicit. The matrix multiplications with C_h and C_h^T (see Sect. 3.2) as well as with M_μ^{-1} and M_ε^{-1} (see Sect. 3.3) decide about the computational efficiency of an implementation.

The advantage of discontinuous Galerkin methods becomes evident now. The mass matrices can be inverted in an element by element fashion and also the discrete curl operations only need information of (element-)local and adjacent degrees of freedom, which allows for straightforward parallelization. Element matrices such as mass matrices and the discrete curl operation can be stored once and applied at each time step. This is how far one comes just because of the formulation itself.

With appropriate choices for the local shape functions we can use advanced techniques to execute those operations with a lower complexity than local matrix-vector multiplications. Furthermore we don't even have to store the element matrices, s.t. the techniques presented below are also much more memory-efficient.

The following ingredients are essential for the techniques proposed below, which enhance the implementation of the DG method:

1. Definition of an L^2-orthogonal basis of polynomial shape functions in tensor-product form[1] on a reference element $\hat{T}$
2. Use of curl-conforming (*covariant*) transformation for evaluations on the physical element T
3. Use of recurrences for the polynomial shape functions to evaluate gradients and curls
4. Use of tensor-product structure to evaluate traces[2]

3.1 *Local Shape Functions*

For stability and fast computability we choose the L_2-orthogonal Dubiner basis [5,9]. Here, the basis functions on the reference element are constructed in a tensor-product form of Jacobi polynomials $P_i^{(\alpha,\beta)}$ for each spatial component (note that the Legendre polynomials $P_i = P_i^{(0,0)}$ are just a special case). For example, on the reference triangle spanned by the points $(0,0)$, $(1,0)$ and $(0,1)$ the shape functions take the form

$$\varphi_{i,j}(x,y) = P_i\left(\frac{2y}{1-x} - 1\right) \cdot (1-x)^i \cdot P_j^{(2i+1,0)}(2x-1). \tag{4}$$

They are orthogonal on the reference triangle, and gradients can be evaluated by means of recurrence relations as demonstrated in Sect. 3.2.2. Due to the tensor-product form traces can be evaluated very fast, see Sect. 3.2.3.

3.2 *Discrete* curl *Operations*

At each time step we have to evaluate terms like $(H, \operatorname{curl} e)_T$ on each element T and $(\{H\} \times \nu, [\![e]\!])_F$ on each face F. Similar expressions have to be evaluated for the electric field E.

[1]These are polynomials which are products of univariate polynomials.

[2]Values at a boundary.

3.2.1 Covariant Transformation

Let $\Phi : \hat{T} \to T$ be a diffeomorphic mapping from the reference element to some physical element T. Then the covariant transformation of a function $\hat{u}$ defined on the reference element $\hat{T}$ is

$$u := (F^{-1})^T \hat{u} \circ \Phi^{-1} \qquad \text{with} \qquad F = \nabla\Phi.$$

If we define the shape functions on the mapped elements as the covariant transformed shape functions on the reference element, then the tangential component on the mapped element depends only on the tangential component of the reference element. The transformation is called curl-conforming as it ensures that for any function $\hat{u} \in H(\text{curl}, \hat{\Omega})$ the covariant transformed function u lies in $H(\text{curl}, \Omega)$. Furthermore it preserves certain integrals, s.t. the following relations hold for the covariant transformations $H, e \in H(\text{curl}, T)$ of $\hat{H}, \hat{e} \in H(\text{curl}, \hat{T})$:

$$\left| \int_T H \,\text{curl}\, e \,\text{d}x \right| = \left| \int_{\hat{T}} \hat{H} \,\text{curl}\, \hat{e} \,\text{d}x \right|,$$

$$\left| \int_{\partial T} (H \times \nu) e \,\text{d}s \right| = \left| \int_{\partial \hat{T}} (\hat{H} \times \nu) \hat{e} \,\text{d}s \right|.$$

This means that the integrals of these forms appearing in the formulation are independent of the geometry of the particular elements. The matrices can be computed once on the reference element. This trick was published in [4].

3.2.2 Evaluating Gradients

For computing curls it is sufficient to evaluate gradients, since the curl is a certain linear combination of derivatives. We write the corresponding function $\hat{E}$ in modal representation, i.e.,

$$\hat{E} = \sum_{\alpha} a_\alpha \varphi_\alpha, \; a_\alpha \in \mathbb{R}^3,$$

where the sum ranges over the finite collection of (scalar) shape functions defined on the reference element (in 2D the multi-index α is (i, j) and in 3D $\alpha = (i, j, k)$). With the use of the covariant transformation, we just have to consider the integral on the reference element $\hat{T}$:

$$\int_{\hat{T}} \hat{h} \,\text{curl}\, \hat{E} \,\text{d}x.$$

The idea is now to take advantage of recurrence relations between derivatives of Jacobi polynomials and Jacobi polynomials itself. We aim for an operation which gives the coefficients $b_\alpha \in \mathbb{R}^3$ representing the gradient

$$\nabla \hat{E} = \sum_{\alpha} b_\alpha \varphi_\alpha.$$

Then L^2-orthogonality can be used to evaluate the complete integral very fast.

For ease of presentation let's consider the far more easy case of evaluating the derivative of a scalar one-dimensional function $v(x) = \sum_{i=0}^{n} v_i P_i(x),\ v_i \in \mathbb{R}$ given in a modal basis of Legendre polynomials P_i, which fulfill the relation

$$P'_{i+1}(x) = P'_{i-1}(x) + (2i + 1)P_i(x). \tag{5}$$

Then the problem is to find the modal representation of

$$v'(x) = \sum_{i=0}^{n} v_i P'_i(x) = \sum_{i=0}^{n-1} w_i P_i(x).$$

Let's show the first step, i.e., how we get the highest order coefficient w_{n-1}:

$$\begin{aligned} v'(x) &= \sum_{i=0}^{n} v_i P'_i(x) = \sum_{i=0}^{n-1} v_i P'_i(x) + v_n P'_n(x) \\ &= \sum_{i=0}^{n-1} v_i P'_i(x) + v_n P'_{n-2}(x) + v_n(2n-1)P_{n-1} \\ &= \sum_{i=0}^{n-1} \tilde{v}_i P'_i(x) + w_{n-1}P_{n-1}(x) \end{aligned}$$

where we used the recurrence relation (5) for $P'_n(x)$ and thus get $w_{n-1} = v_n(2n-1)$. For the remaining polynomial $\sum_{i=0}^{n-1} \tilde{v}_i P'_i(x)$ of degree $n-1$ we can apply the same procedure to get w_{n-2}. This can be continued until also w_0 and thereby the complete polynomial representation $\sum_{i=0}^{n-1} w_i P_i(x)$ of $v'(x)$ is determined.

An efficient C++ implementation of this procedure was achieved by *template meta-programming*, where the compiler can generate optimized code for all elements up to an a priori chosen maximal polynomial order.

The same basically also works in three dimensions with Jacobi polynomials, but the relations are far more complicated, see Sect. 4, and need three nested loops.

The overall costs for the evaluation of the element curl integral scales *linearly* with the number of unknowns N on one element which is much better than the matrix-vector multiplication which already has complexity $\mathcal{O}(N^2)$.

3.2.3 Evaluating Traces

The boundary integrals that have to be evaluated can make use of the tensor-product form to evaluate traces. Again we don't want those traces to be evaluated pointwise but in a modal sense and recurrences for the Jacobi polynomials make the transformation from volume element shape functions to face shape functions with $\mathcal{O}(N)$ operations possible. The procedure therefore is similar to the evaluation of the gradient in the previous section.

3.3 Mass Matrix Operations

So far we dealt only with the discrete curl operations. So the only thing that is left to talk about is the application of the inverse mass matrices. Due to the covariant transformation we have

$$
\begin{aligned}
((M_\varepsilon)_{\alpha,\beta})_{l,m} &= \int_T \varepsilon\,(\varphi_\alpha e_l^T)\,(\varphi_\beta e_m)\,\mathrm{d}x \\
&= \int_{\hat{T}} |\det(F)|\,\varepsilon\,(\hat{\varphi}_\alpha e_l^T) F^{-1}(F^{-1})^T (\hat{\varphi}_\beta e_m)\,\mathrm{d}x
\end{aligned} \tag{6}
$$

with φ_α denoting the scalar-valued shape functions and e_n the n-th unit vector. Note also the block structure of M_ε that is indicated by the above notation. In some FEM applications, symbolic methods related to those described in Sect. 4, can be used to prove the sparseness of the corresponding system matrix, see [12].

3.3.1 Flat Elements

Let's assume the material parameters ε and μ are piecewise constant and the elements are flat, i.e., $\nabla\Phi = F = const$ on each element. Then the integral (6) simplifies to

$$
\int_T \varepsilon\,(\varphi_\alpha e_l^T)\,(\varphi_\beta e_m)\,\mathrm{d}x = |\det(F)|\,\varepsilon\,(F^{-1}(F^{-1})^T)_{l,m} \int_{\hat{T}} \hat{\varphi}_\alpha \hat{\varphi}_\beta\,\mathrm{d}x
$$

and as $\int_{\hat{T}} \hat{\varphi}_\alpha \hat{\varphi}_\beta\,\mathrm{d}x = \delta_{\alpha,\beta}$ the matrix is (3×3)-block-diagonal and the inversion is trivial. The computational effort is obviously of order $\mathcal{O}(N)$ where N is the number of unknowns.

3.3.2 Curved Elements

If we consider curved elements or non-constant material parameters ε and μ, the approach has to be modified as the mass matrix arising from (6) may be fully occupied. Let's go a step back and consider a similar scalar problem[3] with a non-constant coefficient ε:

$$
\begin{aligned}
\text{Given:}\quad & f(v) = \int_T f v\,\mathrm{d}x \\
\text{Find:}\quad & u,\ \text{s.t.}\ \int_T \varepsilon u v\,\mathrm{d}x = \int_T f v\,\mathrm{d}x
\end{aligned}
$$

[3] Extensions to 3D are straightforward.

We now transform back to the reference element $\hat{T}$ and get

$$\int_T \varepsilon uv \, \mathrm{d}x = \int_{\hat{T}} |\det(F)| \, \varepsilon uv \, \mathrm{d}x = \int_{\hat{T}} \hat{u}\tilde{v} \, \mathrm{d}x$$

where $\tilde{v} = |\det(F)| \, \varepsilon \hat{v}$. If we now approximate $\tilde{v}$ with the same basis we used for v before, the mass matrix is diagonal again. Nevertheless the evaluation of the functional $f(v)$ has to be transformed as well:

$$\int_T f v \, \mathrm{d}x = \int_T \frac{1}{|\det(F)| \, \varepsilon} f \tilde{v} \, \mathrm{d}x = \int_{\hat{T}} \frac{1}{\varepsilon} f \tilde{v} \, \mathrm{d}x$$

To evaluate the last term we will use numerical integration. But as (in our application) f is not given pointwise, but in a modal sense, we have to calculate a pointwise representation for the numerical integration of $\int_T f v \, \mathrm{d}x$ first:

$$\text{Given:} \quad f(v) = \int_T f v \, \mathrm{d}x = \int_{\hat{T}} |\det(F)| \, f \hat{v} \, \mathrm{d}x$$

$$\text{Find:} \quad f_i, \text{ s.t.} \int_T f v \, \mathrm{d}x = \sum_i |\det(F)|(x_i) \, f_i \omega_i v(x_i)$$

Then we can divide (on each integration point) by ε and with those new coefficients we can, by numerical integration, get a good approximation to $\int_{\hat{T}} \frac{1}{\varepsilon} f \tilde{v} \, \mathrm{d}x$. The "reverse numerical integration" and the numerical integration used here can be accelerated by the use of the sum factorization technique. Doing so the complexity of both "reverse numerical integration" and the numerical integration is $\mathcal{O}(p^4)$, where p is the polynomial degree. Note that the approximate inverse $\tilde{M}_\varepsilon^{-1}$ obtained by this method is still symmetric and positive definite.

3.4 Overall Computational Effort

In the previous sections we saw that the overall computational effort scales linearly with the degrees of freedom N as long as the elements are flat and coefficients are piecewise constant. Even for curved elements (and variable coefficients) the computational effort is only of order $\mathcal{O}(N^{\frac{4}{3}})$. Furthermore no element matrices have to be stored. Only the geometric transformations and the local topology have to be kept in the memory.

3.5 Timings

Let's also state some exemplary numbers that were achieved for this method and its implementation on an Intel Xeon CPU 5160 at 3.00 GHz (64 bit) (single core) for a

Table 1 Timings for flat elements (left), using $\mathcal{O}(1)$ floating point operations per dof and curved elements (right) using $\mathcal{O}(p)$) floating point operations per dof

Order p	Time (μs)	Order p	Time (μs)
1	0.61	1	4.89
2	0.58	2	2.54
3	0.71	3	1.93
4	0.79	4	1.79
5	1.16	5	2.06
6	1.24	6	2.17
7	1.32	7	2.33
8	1.53	8	2.67
9	1.66	9	2.88
10	1.74	10	3.04

tetrahedral mesh with 2078 elements. The costs for one step of the symplectic Euler method per 6 scalar degrees of freedom are listed in Table 1.

4 Symbolic Derivation of Relations

In this section we want to describe the symbolic methods that were employed for finding the desired relations for the polynomial shape functions. These relations allow for efficient computation of the discrete curl operations and traces as described in Sect. 3.2. They have been computed by following the holonomic systems approach [3, 10, 13], which works for all functions that satisfy sufficiently many linear differential equations or recurrences or mixed ones; these relations have to have polynomial coefficients. A large class of functions (like rational or algebraic functions, exponentials, logarithms, and some of the trigonometric functions) as well as a multitude of special functions is covered by this framework. Part of it are algorithms for the "basic arithmetic" (that we will refer to as "closure properties"), i.e., given two implicit descriptions for functions f and g, respectively, we can compute such descriptions for $f + g$, fg, and for functions obtained by certain substitutions into f or g. All computations in this section have been performed in Mathematica using our package `HolonomicFunctions` (it is freely available from the website http://www.risc.uni-linz.ac.at/research/combinat/software/).

4.1 Introductory Example

For demonstration purposes we show how to derive automatically the rewriting formula (5) for Legendre polynomials $P_n(x)$. It is well known that these orthogonal polynomials satisfy some linear relations, e.g., the second order differential equation

$$(x^2 - 1)P_n''(x) + 2xP_n'(x) - n(n + 1)P_n(x) = 0$$

or the three term recurrence

$$(n+2)P_{n+2}(x) - (2n+3)xP_{n+1}(x) + (n+1)P_n(x) = 0.$$

We will represent such linear relations in the convenient operator notation, using the symbols D_x for the partial derivative with respect to x, and S_n for denoting the shift operator with respect to n. Then the two relations above are written as

$$(x^2-1)D_x^2 + 2xD_x + (-n^2-n)$$

and

$$(n+2)S_n^2 + (-2nx-3x)S_n + (n+1),$$

respectively, and we identify operators and relations with each other. The operators can be regarded as elements of a (noncommutative) polynomial ring in S_n and D_x with coefficients being rational functions in $\mathbb{Q}(n,x)$. We can obtain additional relations for $P_n(x)$ by combining the given relations linearly, or by shifting and differentiating them. In the operator setting these operations correspond to addition and multiplication (from the left) and we can refer to the set of all operators obtained in this way as the annihilating left ideal generated by the initially given operators. In the following we will represent annihilating ideals by means of their Gröbner bases; these are special sets of generators that allow for deciding the ideal membership problem (i.e., the question whether some relation is indeed valid for the function under consideration) and for obtaining unique representatives of the residue classes modulo the ideal (see [2]). All algorithms mentioned below will require Gröbner bases as input. A Gröbner basis of the annihilating ideal of the Legendre polynomials is given by

$$G = \left\{(n+1)S_n + (1-x^2)D_x + (-nx-x), (x^2-1)D_x^2 + 2xD_x + (-n^2-n)\right\}.$$

Our main task will be to find elements with certain properties in an annihilating ideal; this can be done via an ansatz as we demonstrate now. The relation (5) that we are going to recover connects $P'_{n+2}(x)$, $P'_n(x)$, and $P_{n+1}(x)$, and its coefficients are free of x. These facts translate to an ansatz operator of the form

$$A = c_1(n)D_xS_n^2 + c_2(n)D_x + c_3(n)S_n$$

where the coefficients c_i are rational functions in $\mathbb{Q}(n)$, and hence free of x as required. We have to determine the c_i such that the operator A is an element of the left ideal I generated by G, so that $A(P_n(x)) = 0$. For this purpose we use the Gröbner basis G to compute the unique representation of the residue class of A modulo I (it is achieved by reduction). We have $A \in I$ if and only if the residue class is represented by the zero operator and hence we can equate all its coefficients to zero, obtaining the following two equations

$$c_1(2nx^2+3x^2-n-2)+c_2(n+1)+c_3(x^2-1)=0,$$
$$c_1(n+1)(2n+3)x+c_3(n+1)x=0.$$

Note that in these equations the variable x occurs, since it is contained in the coefficients of G. We get a solution that is free of x by performing a coefficient comparison with respect to this variable. This yields in the end the linear system

$$\begin{pmatrix} -n-2 & n+1 & -1 \\ 2n+3 & 0 & 1 \\ (n+1)(2n+3) & 0 & n+1 \end{pmatrix} \begin{pmatrix} c_1 \\ c_2 \\ c_3 \end{pmatrix} = 0$$

whose solution is

$$c_1=-1, \quad c_2=1, \quad c_3=2n+3,$$

and this gives rise to the desired relation.

Now what do we do if we don't know the exact shape of the ansatz as given here by A? Then we have to include all possible monomials $D_x^i S_n^j$ up to some total degree into our ansatz. Looping over the degree, we will finally find the relation, but the effort can be tremendous. Therefore, as a preprocessing step, we determine the shape of the ansatz by modular computations. This means plugging in concrete values for some of the variables and reducing all integers in the coefficients modulo some prime. These techniques have been described in detail in [10] and they are crucial for getting results in a reasonable time.

All these steps have been implemented in the package `HolonomicFunctions` and it computes the relation (5) immediately:

```
In[1]:= << HolonomicFunctions.m

HolonomicFunctions package by Christoph Koutschan, RISC-Linz,
Version 1.3 (25.01.2010)
—→ Type ?HolonomicFunctions for help

In[2]:= FindRelation[Annihilator[LegendreP[n, x]], Eliminate → x]
```

Out[2]= $\{S_n^2 D_x + (-2n-3)S_n - D_x\}$

4.2 *Relations for the Shape Functions*

A core functionality of our package `HolonomicFunctions` [10] is to execute closure property algorithms (e.g., for addition, multiplication, and substitution) on functions represented by their annihilating ideals. We can now use these algorithms to obtain annihilating ideals for the shape functions φ, since their definition in terms of Jacobi and Legendre polynomials involves just the above mentioned operations.

4.2.1 The 2D Case

We first consider triangular finite elements in two dimensions. For these, the shape functions are defined as in (4). Analogously to the one-dimensional example in Sect. 3.2.2 we want to express the partial derivatives (with respect to x and y, respectively) in terms of the original shape functions. So the goal is to find relations (free of x and y) that connect the partial derivatives with the original function. More concretely, we are looking for a relation that allows to express some linear combination of shifts of $\frac{\mathrm{d}}{\mathrm{d}x}\varphi_{i,j}(x,y)$ as a linear combination of shifts of $\varphi_{i,j}(x,y)$ (and similarly for y). This corresponds to an operator of the form

$$\sum_{(m,n)\in\mathbb{N}^2} c_{1,m,n}(i,j)D_x S_i^m S_j^n + \sum_{(m,n)\in\mathbb{N}^2} c_{0,m,n}(i,j)S_i^m S_j^n \tag{7}$$

where the yet unknown coefficients $c_{d,m,n} \in \mathbb{Q}(i,j)$ do not depend on x and y, and the sums have finite support.

Since we have to find such a relation in the annihilating ideal for $\varphi_{i,j}(x,y)$, it is natural to start by computing a Gröbner basis for this ideal. The package `HolonomicFunctions` provides a command **Annihilator** that analyzes a given mathematical expression and performs the necessary closure properties for obtaining its annihilating ideal. So in our example we can just type

```
In[3]:= ann = Annihilator[(1 − x)^i * LegendreP[i, 2y/(1 − x) − 1] *
            JacobiP[j, 2i + 1, 0, 2x − 1], {S[i], S[j], Der[x], Der[y]}];
```

and after a second we have the result (which is already respectable in size, namely 340kB, corresponding to about 10 pages of output).

Having implemented noncommutative Gröbner bases, our first attempt was to use them for eliminating the variables x and y. But it soon turned out that this attempt did not produce optimal results, and in addition the computations were very time-consuming. Therefore we came up with the ansatz described in Sect. 4.1. We use it now to compute the desired relations (both computations take less than a minute):

```
In[4]:= FindRelation[ann, Eliminate → {x, y}, Pattern → {_, _, 0 | 1, 0}] // Factor
```

Out[4]=
$$\{(2i+j+5)(2i+2j+5)S_iS_j^2D_x + (j+3)(2i+2j+5)S_j^3D_x$$
$$+\ 2(2i+3)(i+j+3)S_iS_jD_x - 2(2i+1)(i+j+3)S_j^2D_x$$
$$-\ 2(i+j+3)(2i+2j+5)(2i+2j+7)S_iS_j - (j+1)(2i+2j+7)S_iD_x$$
$$-\ 2(i+j+3)(2i+2j+5)(2i+2j+7)S_j^2 - (2i+j+3)(2i+2j+7)S_jD_x\}$$

```
In[5]:= FindRelation[ann, Eliminate → {x, y}, Pattern → {_, _, 0, 0 | 1}] // Factor
```

Out[5]=
$$\{(2i+j+6)(2i+j+7)(2i+2j+7)S_i^2S_j^2D_y - (j+3)(j+4)(2i+2j+7)S_j^4D_y$$
$$-\ 4(j+2)(i+j+4)(2i+j+6)S_i^2S_jD_y + 4(j+3)(i+j+4)(2i+j+5)S_j^3D_y$$
$$+\ (j+1)(j+2)(2i+2j+9)S_i^2D_y - 4(2i+3)(i+j+4)(2i+2j+7)(2i+2j+9)S_iS_j^2$$
$$-\ (2i+j+4)(2i+j+5)(2i+2j+9)S_j^2D_y\}$$

Here the option **Pattern** specifies the admissible exponents for the operators, e.g., in the first case we allow any exponent for the shift operators, whereas D_x may occur with power at most 1 only, and D_y must not appear at all in the result.

4.2.2 The 3D Case

When dealing with tetrahedra in three dimensions, the shape functions are denoted by $\varphi_{i,j,k}(x,y,z)$ and are defined by

$$(1-x-y)^i(1-x)^j P_i\left(\tfrac{2z}{1-x-y}-1\right) P_j^{(2i+1,0)}\left(\tfrac{2y}{1-x}-1\right) P_k^{(2i+2j+2,0)}(2x-1).$$

Again they have the nice property of being L^2-orthogonal on the reference tetrahedron

$$T = \{(x,y,z) \in \mathbb{R}^3 \mid x \geq 0 \wedge y \geq 0 \wedge z \geq 0 \wedge x+y+z \leq 1\}.$$

Computing an annihilating ideal for $\varphi_{i,j,k}(x,y,z)$ is already much more involved than in the 2D case:

```
In[6]:= phi = (1 − x − y)^i (1 − x)^j LegendreP[i, 2z/(1 − x − y) − 1]
            JacobiP[j, 2i + 1, 0, 2y/(1 − x) − 1] JacobiP[k, 2i + 2j + 2, 0, 2x − 1];
In[7]:= Timing[ann = Annihilator[phi, {Der[x], S[i], S[j], S[k]}];]
Out[7]= {359.686, Null}
```

The Gröbner basis for this annihilating ideal is about 117MB in size (corresponding to several thousand of printed pages). Note also that it is more efficient to consider only one derivation operator, and compute annihilating ideals for each of the cases $\frac{\mathrm{d}}{\mathrm{d}x}$, $\frac{\mathrm{d}}{\mathrm{d}y}$, and $\frac{\mathrm{d}}{\mathrm{d}z}$ separately (this applies to the 2D case, too).

In principle, the desired relations for the 3D case can be found in the same way as for two dimensions. As described in Sect. 4.1 we find by means of modular computations that the ansatz (for the case $\frac{\mathrm{d}}{\mathrm{d}x}$) contains the 16 monomials

$$S_i S_j S_k^2 D_x,\ S_i S_k^3 D_x,\ S_j^2 S_k^2 D_x,\ S_j S_k^3 D_x,\ S_i S_j S_k D_x,\ S_i S_k^2 D_x,\ S_j^2 S_k D_x,\ S_j S_k^2 D_x,$$
$$S_i S_j S_k,\ S_i S_j D_x,\ S_i S_k^2,\ S_i S_k D_x,\ S_j^2 S_k,\ S_j^2 D_x,\ S_j S_k^2,\ S_j S_k D_x.$$

However, in order to compute the corresponding coefficients, we did not succeed with the standard approach used in Sect. 4.2.1. Instead, we had to employ modular techniques again for many interpolation points, and then interpolate and reconstruct the solution.

5 Conclusion

We have presented an efficient implementation for solving the time-domain Maxwell's equations with a finite element method that uses discontinuous Galerkin elements. Besides many other optimizations that speed up the whole simulation, the usage of certain recurrence relations for the shape functions allows for a fast evaluation of gradients and traces. These relations have been derived symbolically with computer algebra methods.

It is widely believed that the mathematical subjects "numerical analysis" and "symbolic computation" do not have much in common, or even that they are kind of orthogonal. Experts from both areas can barely communicate with each other unless they don't talk about work. It was the great merit of the project SFB F013 "Numerical and Symbolic Scientific Computing" that had been established in 1998 at the Johannes Kepler University of Linz, Austria, to bring together these two communities to identify potential collaborations. We consider our results as a perfect example for such a fruitful cooperation.

Acknowledgements We would like to thank Veronika Pillwein for making contact between the first- and the last-named author and for kindly supporting our work by interpreting between the languages of symbolics and numerics. Christoph Koutschan was supported by the Austrian Science Fund (FWF): SFB F013 and P20162-N18, and partially by NFS-DMS 0070567 as a postdoctoral fellow.

References

1. Arnold, D.N., Brezzi, F., Cockburn, B., Marini, D.: Unified analysis of discontinuous Galerkin methods for elliptic problems. SIAM J. Numer. Anal. **39**(5), 1749–1779 (2002)
2. Buchberger, B.: Ein Algorithmus zum Auffinden der Basiselemente des Restklassenrings nach einem nulldimensionalen Polynomideal. Ph.D. thesis, University of Innsbruck, Innsbruck, Austria (1965)
3. Chyzak, F.: An extension of Zeilberger's fast algorithm to general holonomic functions. Discrete Math. **217**(1–3), 115–134 (2000)
4. Cohen, G., Ferries, X., Pernet, S.: A spatial high-order hexahedral discontinuous Galerkin method to solve Maxwell's equations in time domain. J. Comput. Phys. **217**, 340–363 (2006)
5. Dubiner, M.: Spectral methods on triangles and other domains. J. Sci. Comput. **6**(4), 345–390 (1991)
6. Hesthaven, J.S., Warburton, T.: Nodal Discontinuous Galerkin Methods–Algorithms, Analysis and Applications. Text in Applied Mathematics. Springer, Berlin (2007)
7. Hesthaven, J.S., Warburton, T.: On the constants in hp-finite element trace inverse inequalities. Comput. Methods Appl. Mech. Eng. **192**, 2765–2773 (2003)
8. Houston, P., Perugia, I., Schötzau, D.: Mixed discontinuous Galerkin approximation of the Maxwell operator. SIAM J. Numer. Anal. **42**(1), 434–459 (2004)
9. Karniadakis, G.E., Sherwin, S.J.: Spectral/hp Element Methods for Computational Fluid Dynamics. Oxford Science Publications, Oxford (2005)
10. Koutschan, C. Advanced Applications of the Holonomic Systems Approach. Ph.D. thesis, RISC, Johannes Kepler University, Linz, Austria (2009)
11. Perugia, I., Schötzau, D., Monk, P.: Stabilized interior penalty methods for the time-harmonic Maxwell equations. Comput. Methods Appl. Mech. Eng. **191**, 4675–4697 (2002)
12. Pillwein, V.: Computer Algebra Tools for Special Functions in High Order Finite Element Methods. Ph.D. thesis, Johannes Kepler University, Linz, Austria (2008)
13. Zeilberger, D.: A holonomic systems approach to special functions identities. J. Comput. Appl. Math. **32**(3), 321–368 (1990)

A Symbolic Approach to Generation and Analysis of Finite Difference Schemes of Partial Differential Equations

Viktor Levandovskyy and Bernd Martin

Abstract We discuss three symbolic approaches for the generation of a finite difference scheme of a general single partial differential equation (PDE). We concentrate on the case of a linear PDE with constant coefficients and prove, that these three approaches are equivalent. We systematically use another symbolic technique, namely the cylindrical algebraic decomposition, in order to derive conditions for the von Neumann stability of a given difference scheme. We demonstrate algorithmic symbolic approaches for the computation of both continuous resp. discrete dispersion relations of a linear PDE with constant coefficients resp. a finite difference scheme. We present an implementation of tools for the generation of schemes in the computer algebra system SINGULAR. Numerous examples are computed with our implementation and presented in details. Some of the methods we propose can be generalized to nonlinear PDEs as well as to the case of variable coefficients and to the case of systems of equations.

1 Introduction

The finite difference method for linear PDEs belongs to a very classical topics in mathematics. Its exposition in the classical books like [26] often relies on huge experience, gathered in last centuries. In particular, some important steps are based on *a posteriori* analysis. A pure algebraist is often confused with such exposition and asks, whether there is a way to split the whole picture into a purely analytic

V. Levandovskyy (✉)
Lehrstuhl D für Mathematik, RWTH Aachen, Templergraben 64, 52062 Aachen, Germany
e-mail: viktor.levandovskyy@math.rwth-aachen.de

B. Martin
Lehrstuhl Algebra und Geometrie, BTU Cottbus, 03013 Cottbus, Germany
e-mail: martin@math.tu-cottbus.de

U. Langer and P. Paule (eds.), *Numerical and Symbolic Scientific Computing*, Texts and Monographs in Symbolic Computation, DOI 10.1007/978-3-7091-0794-2_7,

and an algebraic part and how is it possible to automatize the process of scheme generation and further analysis of its properties. The ideas to generate a finite difference scheme in an algebraic (or a symbolic) way are folklore, see for instance [9,21] for approaches and older implementations.

Terminologically, we address a *difference scheme polynomial* as symbolic polynomial expression involving unknown function and partial shift (or difference) operators. A fully described difference scheme also includes initial and/or boundary conditions in addition to the difference scheme polynomial. However, since the generation of a difference scheme polynomial is independent on initial and/or boundary conditions, through this paper we call a difference scheme polynomial also a *difference scheme*, if no confusion arises.

In the article [13], Gerdt et al. used for the first time several new ideas like the application of integral relations in discrete form (especially useful if one deals with conservation laws), the formulation of the scheme generation problem as a task for difference elimination and the systematic use of involutive and Gröbner bases. Inspired by these ideas, we present our approaches, which will make the overall picture of scheme generation and analysis (meaning investigation of von Neumann stability and dispersion) more complete for the special situation.

Namely, in this article our primary target is a single linear PDE with constant coefficients. We will comment on cases, when some methods can be applied to more general setting.

As we will see, von Neumann stability can be regarded as global result, always being a necessary condition for stability of a problem with initial and/or boundary conditions (and sometimes sufficient condition as well). Of course, one uses initial and/or boundary conditions for numerical solving, but the splitting of the whole problem into purely symbolic pre-processing and numerical post-processing seems to be the way to address such problems in the future.

The ideas of algebraic analysis suggest a separation of a problem into analytic and algebraic parts. This allows, in the case of linear PDEs, to treat systems of equations via modules over algebras (D-module theory, homological algebra etc.). There exist many algorithms and several powerful implementations. Gröbner bases and involutive bases play a fundamental role in such algorithms, see e. g. [3,22,24].

On the other side, the theories of differential algebra (e. g. [23]) and of difference algebra ([5, 20]) allow to tackle nonlinear equations as well, though the algorithms in these realms are very complicated. In particular, we do not know any implementation of a basis construction algorithm for the difference algebra. Notably, a new algorithmic approach to nonlinear difference equations might arise from the *letterplace* approach [18]. However, one needs to elaborate all details of this promising direction.

The technique of cylindrical algebraic decomposition (CAD) has its origins in real algebraic geometry. It has been applied to von Neumann stability problems already in [17, 21]. Since that time more implementations of the CAD have been evolved and their performance has been greatly enhanced.

This paper is organized as follows. We start with minimal prerequisites and revisit the basic concepts of scheme generation, paying attention to the algebraic background including Gröbner bases and elimination tools in Sect. 2.

We discuss three symbolic methods, used in applications in Sect. 3 and prove their equivalence in the case of a linear PDE with constant coefficients in Theorem 1. In cases of more general PDEs and systems of such PDEs only one method works in general, see Remark 3. For a linear PDE with constant coefficients we propose a novel way to generate finite difference scheme by using Gröbner basis for eliminating module components in a submodule of a free module. The latter can be seen as a natural generalization of the Gaussian elimination to matrices over rings. We show the merits of this method, applied to the classical equations of mathematical physics (heat, wave, advection equations) for various approximations. Note, that the method we propose can handle high order approximations, which are seldom used in the theory of PDE, but quite often arise in the theory of ODE as high order Runge-Kutta methods.

In Sect. 5 we present an algebraic and constructive formulation of von Neumann stability via ring homomorphism. We shortly revisit the concepts of cylindrical algebraic decomposition and connect its use to questions, arising from difference schemes.

In Sect. 6 we consider the λ-wave equation $u_{tt} = \lambda u_{xx}$ for a nonzero parameter λ and perform generation as well as stability analysis of several difference schemes, obtained with different approximations. We demonstrate the merits of the semi-factorized form of a difference scheme, it turns out to be especially useful for higher dimensional situations.

In Sect. 7 we show, that the determination of continuous dispersion relation for a linear PDE with constant coefficients as well as discrete dispersion relation with respect to a finite difference scheme can be algebraized to a large extent as well.

All examples have been computed with our implementation of tools for difference schemes in the freely available computer algebra system SINGULAR [6]. The corresponding library `findifs.lib` is distributed with SINGULAR starting with version 3-1-2. For the cylindrical algebraic decomposition we use the commercial system MATHEMATICA; indeed there are freely available systems like QEPCAD [2] and REDLOG [7], which can compute the decomposition as well.

In Appendix we present a detailed example of the use of our toolbox together with short introduction to the system SINGULAR.

Often a general skepticism is met about the use of symbolic methods connected with numerical analysis. We want to stimulate a discussion between scientists of both fields based on a realistic viewpoint.

2 Algebraization of Differential and Difference Equations

2.1 Types of Operator Algebras

At first we fix a computable field k, the base field, it is mostly the field of rational numbers $\mathbb{Q}$ or complex rational numbers $\mathbb{Q}[i]$. (Computing with real or complex numbers is in principle possible, but only with a fixed precision, i.e. with a rational approximation, or one can compute with algebraic extensions in roots of polynomials - this being not so interesting for our purpose.) We can extend the base field by indeterminate constants, i.e. rational functions in the constants: $K = k(a, b, c, \ldots)$. K is called the field of constants.

We fix a set of variables $x := (x_1, \ldots, x_n)$ and an 'algebra' of functions $C = C(x)$, for instance differentiable functions or functions in discrete (shiftable) arguments. C is not our object of computation. Instead we consider various operator algebras, consisting of operators, which act on C. There are many operators, which one can handle in this framework, for example:

1. Multiplication with a variable: $x_i : u(x) \in C \mapsto x_i u(x) \in C$.
2. Multiplication with a function $f \in C$: $m_f : u(x) \in C \mapsto f(x) \cdot u(x) \in C$.
3. Partial differentiation: $\partial_i : u(x) \in C \mapsto \frac{\partial u(x)}{\partial x_i} \in C$.
4. Partial shift operators: $T_i : u(x) \in C \mapsto u(x_1, \ldots, x_i + 1, \ldots, x_n) \in C$.
5. Partial λ-shift operators: $T_i^\lambda : u(x) \in C \mapsto u(x_1, \ldots, x_i + \lambda, \ldots, x_n) \in C$, clearly $T_i = T_i^1$.
6. Partial difference operators: $\Delta_i = T_i^\lambda - 1;\ u(x) \in C \mapsto T_i^\lambda(u(x)) - u(x) \in C$;
7. q-dilation operators: $D_d : u(x) \in C \mapsto u(x_1, \ldots, qx_i, \ldots, x_n) \in C$ for $q \in K$.

Fix a set S of operators, we consider the operator algebra $\mathcal{O} := K\langle S\rangle$ being the subalgebra of all (linear) operators $Hom_K(C, C)$ generated by S, i.e. the smallest linear subspace closed under composition of operators. As long as S consists of a finite number of pairwise commuting and independent operators the resulting algebra is isomorphic to a polynomial ring: $K[t_1, \ldots, t_m]$. Otherwise we get (non-commutative) quotient algebra of the free algebra $K\langle S\rangle$ by the two-sided ideal of all relations of S.

*Example 1 (**Algebras with constant coefficients**).* The algebras of linear partial differential and shift (or difference) operators with constant coefficients are commutative K-algebras, isomorphic to $K[x_1, \ldots, x_n]$. We denote them by $K[\partial_1, \ldots, \partial_n]$ and $K[T_1, \ldots, T_n]$ respectively.

*Example 2 (**Algebra with polynomial coefficients**).* The algebra of linear partial differential operators with **polynomial** coefficients is the Weyl algebra. It is non-commutative but is has simple commuting relations. We denote this algebra by $K\langle x, \partial \mid \partial x = x\partial + 1\rangle$, what means that this algebra is generated by the variables x and ∂ over the field K. Moreover, multiplication is defined on its generators: $x \cdot \partial = x\partial, \partial \cdot x = x\partial + 1$ and extended to arbitrary products inductively. The generalization to the multivariate case is easy, the variable x_i commutes with any x_k and also with variable ∂_j except for the case $j = i$, here the relation as above applies.

Why do the variables satisfy this relation? Consider the multiplication of two operators, $x := x_i$ and $\partial := \frac{\partial}{\partial x_i}$. Take some differentiable function $f(x)$ and apply the Leibnitz rule to the product: $(\partial x)(f) = \partial(xf) = x\partial(f) + f = (x\partial + 1)(f)$. Hence, in the operator form $(\partial x - x\partial - 1)(f) = 0$.

Consider the algebra of linear λ-shift operators with **polynomial** coefficients, having in mind $\lambda = \Delta x$. As above, we can derive the relation between operators $T := T^{\lambda}$ and x. For any function f of discrete arguments, $(Tx)(f) = T(xf) = T(x)T(f) = (x + \Delta x)T(f) = (xT)(f) + \Delta x \cdot T(f)$. This relation can be expressed in the operator form as $Tx = xT + \Delta x \cdot T$. The algebra, corresponding to the difference operator $\Delta = T^{\Delta x} - 1$ has the relation $\Delta x = x\Delta + \Delta + 1$. These algebras are so called G-algebras, in which Gröbner basis algorithms exist and are implemented in the system SINGULAR:PLURAL ([15]), see e. g. [19].

*Example 3 (**Algebra with coefficients in rational functions**).* Algorithmic computations are possible in the algebras whose coefficient fields are rational functions in x: $K(x)\langle \partial \mid \partial x = x\partial + 1\rangle$ and $K(\Delta x, x)\langle T \mid Tx = xT + \Delta x \cdot T\rangle$, which are called **rational** Weyl algebras resp. **rational** shift algebras. Algebraically speaking, a passage from a polynomial algebra to a rational algebra may be achieved by means of so-called Ore localization.

*Example 4 (**Differential and Difference Algebra**).* In order to handle non-linear differential resp. difference equations with polynomial nonlinearities, one can consider a full differential resp. difference algebra $K[\{O_1^{\beta_1} \cdots O_n^{\beta_n} u \mid \beta_i \in \mathbb{N}\}]$, where O_i stands for a partial differential resp. difference operator. Note, that $O^{\beta}u := O_1^{\beta_1} \cdots O_n^{\beta_n} u$ is a variable, representing $O^{\beta}(u)$, where $u = u(x_1, \ldots, x_m)$ symbolizes an unknown function in variables $x_1, \ldots, x_m$. Note, that such an algebra is commutative and its infinitely many generators are algebraically independent. In particular, such an algebra is not Noetherian.

The given nonlinear equations can be taken as generators of the differential resp. difference ideal. Such ideal is defined to be a minimal ideal, containing given equations, which is closed under the action of corresponding differential resp. difference operators in the corresponding algebra.

Since such algebras are not Noetherian, Gröbner basis-like algorithms are not terminating in general. Nevertheless, some parts of the theory in the linear situation can be extended to this general situation.

In this paper we work algorithmically with linear partial differential operators with constant coefficients. However, in some parts we address and discuss more general situations as well.

2.2 *Presentation of a System of Differential Equations*

Any linear partial differential equation (depending of its kind) defines an element of an algebra of corresponding linear differential operators. A solution of a system of equations, if it exists, must be a solution to any equation from the left ideal generated by the equations of the system in the algebra. Hence, the solution does not depend on the choice of a basis (that is, a generating set) of the ideal. A first possible application of symbolic algebra is to compute a better basis of the ideal, such as *Gröbner basis* or an *involutive basis* like *Janet basis* (see [11, 22, 25]) – as far as it is possible. The advantage of such a pre-processing could be: check the consistency of the system of equations, find hidden constraints or integrability conditions of the system, determine the dimension of the solution space.

These data are well known for standard equations from mathematical physics, but the methods we propose are methodologically applicable to any system of equations. Let us recall a small example (by W. Seiler [25]) as an illustration.

Example 5.

$$\begin{cases} u_z + y\,u_x = 0 \\ u_y = 0 \end{cases} \Longrightarrow \begin{cases} u_{yz} + y\,u_{xy} + u_x = 0 \\ u_{xy} = u_{yz} = 0 \end{cases} \Longrightarrow u_x = 0$$

Hence, the initial system is equivalent to $\{u_x = u_y = u_z = 0\}$.

Let $\mathscr{F}$ be a space of functions and $\mathscr{O}$ be an algebra of operators, acting on $\mathscr{F}$. We denote by $a \bullet f$ the action of an operator $a \in \mathscr{O}$ on a function $f \in \mathscr{F}$. In the case of a linear system (S)

$$S_i = \sum_{j=1}^{n} D_{ij} \bullet u_j, \; i = 1, \ldots m, \;\; D_{ij} \in \mathscr{O}$$

we associate to (S) the submodule $P = P(S) \subset \mathscr{O}^n$ generated by the columns of the presentation matrix $D \in Mat(m,n;\mathscr{O})$, and, finally, a factor-module $M(S) := \mathscr{O}^n/P(S)$. We can simplify the system finding a special presentation matrix, or we can read properties of the system from computable invariants of the module $M(S)$.

In the example above, the system can be written as $\begin{pmatrix} \partial_z + y\partial_x \\ \partial_y \end{pmatrix} \bullet (u) = \begin{pmatrix} 0 \\ 0 \end{pmatrix}$. Hence, the system algebra is $\mathscr{O} = K(x,y,z)\langle \partial_x, \partial_y, \partial_z \mid \partial_x x = x\partial_x + 1, \partial_y y = y\partial_y + 1, \partial_z z = z\partial_z + 1\rangle$ (that is the 3rd rational Weyl algebra) and the presentation matrix for the system module $M(S)$, written in columns of the original presentation (that is, transposed to the usual row presentation) is $P(S) = (\partial_z + y\partial_x, \partial_y) \in \mathscr{O}^{1\times 2}$. As a submodule of $\mathscr{O}$, $P(S)$ is an ideal and it has two polynomial generators $\{\partial_z + y\partial_x, \partial_y\}$. The Gröbner basis of $P(S)$ is equal to $Q(S) = \langle \partial_x, \partial_y, \partial_z\rangle$ and hence, $M(S) = \mathscr{O}/Q(S) \cong K(x,y,z)$ as $\mathscr{O}$-module. Thus, $\dim_{K(x,y,z)} M(S) = 1$ is the dimension of the space of holomorphic solutions of S.

2.3 Gröbner Basis Algorithm and Elimination Tools

The notion of Gröbner basis can be given in a common way for different classes of algebras. Recall the basic notation for monomials and monomial ordering. We shall use the short notation $\partial^\alpha := \partial_1^{\alpha_1}\partial_2^{\alpha_2}\ldots\partial_n^{\alpha_n}$, $\alpha \in \mathbb{N}^n$. Finitely generated operator algebras, which we are dealing with, have infinite dimension as K-vector spaces. The infinite set of monomials constitutes a K-basis.

- For operator algebras in operators $\{\partial_1, \ldots, \partial_n\}$ with constant coefficients, the monomials are $\{\partial^\alpha \mid \alpha \in \mathbb{N}^n\}$, they form the basis of the algebra over the field K.
- In the case where the coefficients are polynomials in $\{x_1, \ldots, x_m\}$, the monomials are $\{x^\alpha \cdot \partial^\beta \mid \alpha \in \mathbb{N}^m, \beta \in \mathbb{N}^n\}$ and they form the basis of the algebra over K.
- When the coefficients are rational functions, the monomials $\{\partial^\alpha \mid \alpha \in \mathbb{N}^n\}$ constitute the basis of the algebra over $K(x_1, \ldots, x_m)$.

We are dealing not only with ideals of an algebra $\mathcal{O}$, but also with submodules of the free module $\mathcal{O}^r = \oplus_{i=1}^r \mathcal{O} e_i$, where e_i stands for the canonical i-th basis vector. We extend the notion of a monomial to A^r by supplying a monomial with one of the unit vectors. Clearly, if $\mathrm{Mon}(\mathcal{O}) := \{m_\alpha\}$ is the set of monomials of $\mathcal{O}$, bijective to $\mathbb{N}^r$, then a monomial of $\mathcal{O}^r$ is $m_\alpha e_i$ with $\alpha \in \mathbb{N}^n, 1 \leq i \leq r$.

Definition 1. A (global) monomial ordering on an algebra $\mathcal{O}$ is a total ordering $\prec$ on the set of monomials $\mathrm{Mon}(\mathcal{O})$ bounded from below and compatible with the multiplication, i.e. it fulfills the following conditions for all $\alpha, \beta, \gamma \in \mathbb{N}^n$:

- $1 \prec m_\alpha$.
- $m_\alpha \prec m_\beta \Rightarrow m_\alpha m_\gamma \prec m_\beta m_\gamma$.

Since $\prec$ is total, any nonzero polynomial $f \in \mathcal{O}$ can be uniquely sorted according to its monomials. The highest term (that is a monomial times a nonzero coefficient) is called the leading term of f. We say, that $m_\alpha \mid m_\beta$ (m_α divides m_β), if $\forall 1 \leq i \leq n$ $\alpha_i \leq \beta_i$. Note, that divisibility induces a partial ordering.

Any monomial ordering can be extended to a module monomial ordering in several ways. The most common ways are: either sorting module monomials first by the monomial ordering and then by the number of the component, or first by the component and then by the monomial ordering.

Definition 2. Given a monomial ordering $\prec$ of $\mathcal{O}$, then monomial orderings $\prec_{top}$ (term-over-position) and $\prec_{pot}$ (position-over-term) on the set of monomial of $\mathcal{O}^r$ are defined by:

$$(m_\alpha, e_i) \prec_{top} (m_\beta, e_j) \text{ iff } (m_\alpha \prec m_\beta \text{ or if } m_\alpha = m_\beta, \text{then } i < j),$$

respectively

$$(m_\alpha, e_i) \prec_{pot} (m_\beta, e_j) \text{ iff } (i < j \text{ or if } i = j, \text{then } m_\alpha \prec m_\beta).$$

Definition 3. A Gröbner basis of a submodule $M \subset \mathcal{O}^r$ is a finite subset $G \subset M$ that satisfies the following property. For any $f \in M \setminus \{0\}$, there exists a element of the basis $g \in G$, such that the leading monomial $lm(f) = m_\alpha e_i$ is divisible by $lm(g) = m_\beta e_i$, i. e. $m_\beta \mid m_\alpha$.

An immediate application of Gröbner basis is the *normal form* of a vector of polynomials. Namely, if $G = \{g_1, \ldots, g_m\}$ is a Gröbner basis of a submodule $M \subset \mathcal{O}^r$, then $\forall\, v \in \mathcal{O}^r\ \exists\, w, a_i \in \mathcal{O}$, such that

$$v = \sum_{i=1}^{m} a_i g_i + w, \text{ where either } a_i g_i = 0 \text{ or } \operatorname{lm}(v) \preceq \operatorname{lm}(a_i g_i) \text{ and either } w = 0 \text{ or}$$

$\operatorname{lm}(w) \preceq \operatorname{lm}(v)$. One denotes $w = \mathrm{NF}(v, G)$ and calls w a *normal form* of v with respect to G. Note, that $v \in M = \langle G \rangle$ if and only if $w = \mathrm{NF}(v, G) = 0$.

A Gröbner basis $G = \{g_1, \ldots, g_m\}$ is called *reduced*, if for any $1 \le i \le m$ and $j \ne i$, no monomial of g_j is divided by $\operatorname{lm}(g_i)$. Having a Gröbner basis, a reduced Gröbner one is computed in a finite number of steps. Recall, that a nonzero element from a ring is called *monic* or *normalized*, it its leading coefficient is 1. Any nonzero element can be made monic. Notably, monic reduced Gröbner with respect to a fixed monomial ordering is unique as well as monic reduced normal form of an element.

There are effective ways to compute a Gröbner basis, like the *Buchberger's Algorithm*, *involutive algorithm* and *Faugère's F4* or *F5 algorithm*. Gröbner bases have been implemented in all major computer algebra systems. More details for the commutative case can be found in any standard textbook on computer algebra, e. g. in [16]. See [4, 19] for the non-commutative case of operators with variable coefficients.

Note, that the result of a Gröbner basis algorithm (with respect to position-over-term ordering), applied to a module generated by the columns of a constant matrix, is the row-reduced normal form by Gaussian elimination.

The Gröbner basis algorithm with respect to certain monomial orderings can be used to eliminate some of the variables $\{u_i \mid i \in I\}$ of a given system, i.e. to compute a basis of $M_I := M \cap \mathcal{O}_I^r, \mathcal{O} = \mathcal{O}_I\langle u_i, i \in I\rangle$.

Lemma 1. (Elimination of variables). *Let* $\prec$ *be an elimination monomial ordering for* $\{u_i \mid i \in I\}$ *on* $\mathrm{Mon}(\mathcal{O})$ *(that is,* $m_\alpha \in \mathcal{O}_I$ *and* $j \notin I$ *implies* $m_\alpha \prec u_j$*). Let* G *be a Gröbner basis of* M*, then* $G \cap \mathcal{O}_I$ *is a Gröbner basis of* M_I.

Obviously, a lexicographical ordering of monomials by $u_1 > u_2 > \ldots > u_n$ induces an elimination monomial ordering for any set $\{u_i, \ldots, u_n\}$.

We can also eliminate module components. Usually it is much easier than the elimination of variables.

Lemma 2. (Elimination of components). *Let* G *be a Gröbner basis of a submodule* $M \subset \mathcal{O}^r$ *with respect to the module monomial ordering* $\prec_{pot}$*, let* $F_s := \mathcal{O}e_1 \oplus \cdots \oplus \mathcal{O}e_s \subset \mathcal{O}^r$ *be the free submodule of the first* s *components, then* $G \cap F_s$ *is a Gröbner basis of* $M \cap F_s$.

The proofs of both lemmata on elimination are easy and can be found in e. g. [16,19] for various situations.

Remark 1. We want to stress the fact, that the proposed algorithm, based on the operator formulation will be faster, than the algorithm in difference algebra, which is used by Gerdt et al. in [13], when applied to a linear PDE with constant coefficients.

The difference in complexity lies in the number of variables/components and the intrinsic differences between two similar-looking elimination concepts. Computing in the case, when the functions and discretizations of their derivatives $u, u_t, u_{xx}, \ldots$ appear as variables (difference algebra approach), one has to distinguish between the two multiplications, firstly the action of difference operators on u's (denoted by $\bullet$), and secondly the composition of difference operators (denoted by $\cdot$). The involutive basis approach with its partition of variables into multiplicative and non-multiplicative ones forbids the multiplications between u's in the Gröbner basis algorithm. In addition, one has to employ a complicated elimination ordering, which respects the special role of u's.

We do not use unknown functions u at all by passing to the submodule of a free module of finite rank over a ring of partial difference operators. The linearity of equations allows to consider them as linear operators, presented by polynomials in difference operators with constant coefficients, involving parameters. Thus, we need less variables, and we use simple and efficient module orderings, which eliminate components. The attention of Gröbner basis algorithm is shifted from single polynomials to their components, which results in easier and faster computation, not speaking on optimized memory usage.

3 Three Equivalent Approaches and the Main Theorem

Assume we are dealing with m spatial variables $x_1, \ldots, x_m$ and one temporal variable $t = x_{m+1}$. We denote $x^\alpha := x_1^{\alpha_1} \cdots x_m^{\alpha_m} t^{\alpha_{m+1}}$ for $\alpha \in \mathbb{N}^{m+1}$ and $|\alpha| = \sum \alpha_i$. Then we use notations

$$u_{x^\alpha} = u_\alpha := \frac{\partial^{|\alpha|} u}{\partial x^\alpha} = \frac{\partial^{|\alpha|} u}{\prod \partial x_i^{\alpha_i}}.$$

A single linear PDE with constant coefficients, to which we further refer as to P, can be written as follows

$$\sum_{\beta \in B} c_\beta u_{x^\beta} = 0, \tag{1}$$

where $B \subset \mathbb{N}^{m+1}$ is a finite set and for $\beta \in B$ one has $c_\beta \in K \setminus \{0\}$.

We introduce a uniform rectangular grid on $\mathbb{R}^{m+1}$ with steps $\triangle x_1, \ldots, \triangle x_m, \triangle t$. Thus, a point on the grid can be presented as $\bar{\imath} = (i_1 \triangle x_1, \ldots, i_m \triangle x_m, n \triangle t) \in \mathbb{G} := \mathbb{Z}\triangle x_1 \times \cdots \times \mathbb{Z}\triangle x_n \times \mathbb{Z}\triangle t$ (where $\mathbb{G}$ can be identified with $\mathbb{Z}^{m+1}$). Let us write the PDE P in an arbitrary *interior* point (that is, the one lying far enough from the initial point and/or boundary region) of the grid $\bar{\imath}$.

$$\sum_{\beta \in B} c_\beta u^{\bar{\imath}}_{x^\beta} = \sum_{\beta \in B} c_\beta u_{x^\beta}(i_1 \triangle x_1, \ldots, i_m \triangle x_m, n \triangle t) = 0. \tag{2}$$

Define $\Gamma := \{\gamma \in \mathbb{N}^{m+1} \mid |\gamma| < |\beta| \forall \beta \in B\}$, a finite subset of $\mathbb{N}^{m+1}$. On the grid $\mathbb{G}$, one needs to give an approximation to a function $u^{\bar{\iota}}_{x^\beta}$ by a finite linear combination of expressions $u^{\bar{\kappa}}_{x^\gamma}$ of order, lower than the order of u_{x^β}, that is $\forall \bar{\iota} \in \mathbb{G}, \forall \beta \in B \setminus \{0\}$

$$u^{\bar{\iota}}_{x^\beta} = \sum_{\gamma \in \Gamma, \bar{\kappa} \in \mathbb{G}} d_{\gamma,\bar{\kappa}} u^{\bar{\kappa}}_{x^\gamma} \text{ with } |\gamma| < |\beta|, d_{\gamma,\bar{\kappa}} \in K. \tag{3}$$

We refer to this approximation as to $A^{\bar{\iota}}_\beta$.

Assume we are given a set A of approximations to the terms $u^{\bar{\iota}}_{x^\beta}$. We call an approximation **global**, if it is defined on the whole interior region through its definition in an arbitrary point $\bar{\iota}$. In such a case, $A^{\bar{\iota}}_\beta$ depends only on β and shift operators (see below). From now on we assume, that we are dealing with global approximations only.

Remark 2. The restriction to global approximations on uniform rectangular grid is not essential for the theory. However, the restriction holds in order to simplify the exposition. Allowing different subdomains with different grids on them can be approached in a similar fashion. Namely, on each subdomain we proceed as in the global case and obtain a difference scheme polynomial. In addition, there will be equations from compatibility conditions, which arise from the specific decomposition of a domain.

Any expression of the form $E = u^{\bar{\iota}}_{x^\beta} - \sum_{\gamma \in \Gamma, \bar{\kappa} \in \mathbb{G}} d_{\gamma,\bar{\kappa}} u^{\bar{\kappa}}_{x^\gamma}$ can be brought (by sorting its terms with respect to the order of u_γ) to the form $H - L$ where $H = H(E)$ is the sum of terms of the highest order and $L = L(E) = E - H$.

We say, that a general problem of approximation of a partial differential equation, given by the set of global approximations A is **admissible**, if:

1. All $E \in A$ are written in the form $H(E) - L(E)$.
2. $H(E) = k u^{\bar{\kappa}}_{x^\gamma}$ for $k \in K \setminus \{0\}$ (that is $H(E)$ consists of precisely one term).
3. For all $\beta \in B \setminus \{0\}$ (that is for any u_{x^β}, appearing in the equation with non-zero coefficient except for u itself) a unique approximation E from A exists.

From now on we assume, that an admissible set of approximations is given.

After sorting we can reveal inconsistencies in a given set of global approximations. It might happen, that the given set is not complete (there are no approximations for some u_{x^β} after the proper sorting) or inconsistent (two nonequal approximation for some u_{x^β}). In practice one is interested in approximations, having certain order in $\Delta x_i, \Delta t$.

On the grid $\mathbb{G}$ we have natural shift operators $T_{x_i} : v^{\bar{\iota}} \mapsto v^{\bar{\iota}+\epsilon_i}$, where ϵ_i is the i-th canonical basis vector. That is $T_{x_i}(v(\ldots, \iota_i, \ldots)) = v(\ldots, \iota_i + \Delta x_i, \ldots))$. Clearly T_{x_i} is the well-known forward shift operator, which is invertible, since its inverse is the associated backward shift operator. Thus we allow exponents of monomials of T_{x_i} to be integers. For an exponent vector $\alpha \in \mathbb{Z}^{m+1}$, denote $T^\alpha = T^{\alpha_1}_{x_1} \cdot \ldots \cdot T^{\alpha_m}_{x_m} T^{\alpha_{m+1}}_t$. In what follows we will use the field $K(T) := K(T_{x_1}, \ldots, T_{x_m}, T_t)$.

The equation 3 can be rewritten in a single generic point $\bar{\iota}$ of the grid in terms of shift operators:

Lemma 3. *In the notations from above, there exist exponent vectors* $\delta(\bar{\iota}), \delta(\bar{\kappa})$, $\bar{\kappa} \in G$ *such that there are two equivalent formulas*

$$T^{\delta(\bar{\iota})}(u_{x^\beta})^{\bar{\iota}} = \sum_{\gamma \in \Gamma, \bar{\kappa} \in \mathbb{G}} d_{\gamma,\bar{\kappa}} T^{\delta(\bar{\kappa})}(u_{x^\gamma})^{\bar{\iota}}, (u_{x^\beta})^{\bar{\iota}} = \sum_{\gamma \in \Gamma, \bar{\kappa} \in \mathbb{G}} d_{\gamma,\bar{\kappa}} T^{\delta(\bar{\kappa})-\delta(\bar{\iota})}(u_{x^\gamma})^{\bar{\iota}}. \quad (4)$$

Proof. For $1 \leq k \leq m+1$ set $\tau_k := \min\{\kappa_k, \iota_k \mid \kappa \in \mathbb{G}, \gamma \in \Gamma, d_{\gamma,\bar{\kappa}} \neq 0\}$. By setting $\delta(\bar{\iota}) := \bar{\iota} - \bar{\tau}$ and respectively $\delta(\bar{\kappa}) := \bar{\kappa} - \bar{\tau}$ we obtain two exponent vectors for the monomials in shift operators. □

According to the lemma, we will derive and encode approximations for functions on the grid by shift operators. This allows us to drop the grid point notation as soon as shift operators are present. In other words, we use shift operators to formulate the problem in a generic point of the grid.

Several approaches exist for the computation of a finite difference scheme of a single partial differential equation with constant coefficients.

3.1 *Mimicking Difference Algebra Approach*

Consider the formal consequences of equalities P as in (1) and A_β as in (3) over the commutative finitely generated ring $R_B := K(T)[u_{x^\beta} \mid \beta \in B]$. Recall, that the variables $\{u_{x^\beta}\}$ are algebraically independent. In other words, we consider an ideal I of the ring R_B, generated by $P \cup \{A_\beta \mid \beta \in B\}$. Since B is finite, the ring R_B is Noetherian and contains the subring $R := K(T)[u]$. The ideal $J := I \cap K(T)[u]$ is computable (e. g. by the elimination of all but one variables as in Lemma 1). Since R is a principal ideal domain, J is generated by a single element, say $p \in K(T)[u]$. Clearing denominators, we obtain a polynomial expression $\tilde{p} \in K[T][u]$. Dividing by its leading coefficient, we obtain monic $f \in K[T][u]$, which is the unique result.

Note, that we do not work with *difference ideal*, but with an algebraic ideal in a difference ring.

3.2 *Algebraic Analysis Approach*

We order the set $\{u_{x^\beta} \mid \beta \in B\}$ according to the monomial ordering and write the resulting ordered list as a column vector $U = [u_{x^{\beta_{max}}}, \ldots, u]^T$. Since P and A_β are linear equations with coefficients in $K[T]$ in the entries of U, we put each equation as a row in a matrix M, with entries in $K[T]$, such that $M \bullet U = 0$, where $\bullet$ stands for the action of shift operators with coefficients in K on functions in discrete arguments. Then we can perform algebraic operations from the left on the matrix M, without engaging the unknown functions u_β, as it is done in algebraic analysis. We compute the intersection of $K[T]$-module M with the free submodule, generated by u, the last component of the vector U. The latter intersection is an ideal $J \subset K[T]$ of all polynomials p in shift operators, such that $p \bullet u = 0$. Define a K-linear map $\star : K[T] \to K[T][u]$, which sends T^α for $\alpha \in \mathbb{N}^n$ to $T^\alpha u$. The latter can be interpreted as an element from the difference algebra.

3.3 Term Rewriting System Approach

Consider the equations from Lemma (4) in the monic form, that is

$$u_{x^\beta} = \sum_{\gamma \in \Gamma, \bar{\kappa} \in \mathbb{G}} d_{\gamma,\bar{\kappa}} T^{\delta(\bar{\kappa})-\delta(\bar{\iota})} u_{x^\gamma}.$$

Let us treat them as rewriting rules for symbols $\{u_{x^\beta} \mid \beta \in B \setminus \{0\}\}$, which substitutes every u_{x^β} with the sum on the right hand side. We denote this system by S. Since S involves u_{x^γ} if and only if $\gamma \prec \beta$, we do the following. At first, we order occurring variables with respect to the monomial ordering, getting $\{u_{x^{\beta_{max}}}, \ldots, u\}$. Then, in the same sequence, the variable u_{x^ε} is substituted with the right hand side of the corresponding approximation A_ε. The result of the substitution does not contain variables, which are higher than u_{x^ε} with respect to the monomial ordering. In such a way we obtain an equivalent rewriting system, each right hand side of which depends only on u. Then we apply this new rewriting system to the operator P.

Theorem 1. *Consider a single linear partial differential equation with constant coefficients P as in (1). Assume that the set of given approximations A is admissible and its elements are written in a point of a grid $\mathbb{G}$ as in (4). Let us define the following polynomials:*

1. *f, a monic polynomial from $K[T][u]$ satisfying $I \cap K(T)[u] = \langle f \rangle$ in the notations of 3.1.*
2. *$g = \star(\tilde{g}) = \tilde{g}u$, a monic polynomial from $K[T][u]$; $\tilde{g} \in K[T]$ satisfies $J = \langle \tilde{g} \rangle$, such that for $r := |B|$ one has $K[T]^r \supset M \cap K[T]e_r = J \subset K[T]$ in the notations of 3.2.*
3. *h, a monic polynomial from $K[T][u]$ satisfying $P \to_S h$ in the notations of 3.3.*

Then $f = g = h$, that is the three methods are equivalent.

Proof. a) A_β is already a Gröbner basis in $K(T)[u_{x^\beta}]$ by the product criterion, because the leading monomials of its elements are coprime, since the set $\{u_{x^\beta} \mid \beta \in B\}$ is algebraically independent. Since $\mathrm{NF}(P, A_\beta) \in K(T)[u]$, we obtain that $\{\mathrm{NF}(P, A_\beta)\} \cup A_\beta$ is a Gröbner basis of $P \cup A_\beta$. The uniqueness follows from the uniqueness of monic reduced normal form [16].

b) Proceeding with the vector as above and starting with higher leading monomials, the matrix representation M of the set A_β is already upper triangular with entries in $K[T]$. Moreover, the last row has exactly two nonzero elements (say, the last two ones in that row). Thus M is already in a row-reduced form. Making complete reduction of the rows will produce a matrix M', where each row contains exactly two nonzero elements: $(0,\dots,0,f_i(T),0,\dots,0,f_r(T))$ with $f_i, f_r \in K[T]$. Hence M' simplifies the set of approximations and corresponds to the completely reduced Gröbner basis.

   Now, we append to M (or M', what is equivalent) the row P', corresponding to the equation P, written in the operator form. The computation of the Gaussian elimination of the resulting matrix amounts in reductions of P' with the rows of row-reduced M'. The result of such reductions is a row vector with the only nonzero entry $w(T) \in K[T]$ at the last position. Since it is a constant multiple of the reduced normal form $\text{NF}(P', M')$ and M' is a Gröbner basis, the ideal J from the statement is a principal ideal, generated by $w(T)$. Let us define $\tilde{g} \in K[T]$ to be the normalized monic $w(T)$ and put $g = \tilde{g}u \in K[T][u]$. Then, since $\tilde{g}$ is the single operator of smallest order, acting on u, we conclude that g is the difference scheme polynomial.

c) The application of rewriting rules in the sequence, as described above, leads to the normal form $\text{NF}(P, A_\beta)$. Since by a) A_β is a Gröbner basis with respect to any monomial well-ordering, monic normal form is unique. Fixing a monomial ordering, we can produce another rewriting system S' by applying rewriting rules to every right hand side of S, starting ascendingly from the smallest nonzero $\beta \in B$. Then S' becomes $\{(u_{x^\beta} \to f_\beta(T)u) \mid \beta \in B \setminus \{0\}\}$, where $f_\beta(T) \in K[T^{\pm 1}] \subset K(T)$. It is straightforward, that S' does not depend on the sequence of reductions like S anymore. Since the set of approximations A is admissible, it follows that S' is confluent. The reduction of P with respect to S' is the same as with respect to S, hence its monic form is unique.

   Since in each of the proofs above we have guaranteed uniqueness and showed that difference scheme polynomial has been computed, the final claim follows.

□

Remark 3. Note, that the equivalences of the previous Theorem do not hold in general. Algebraic Analysis Approach 3.2 works only for linear PDE, since it relies on the module structure, which is linear per definition. Both a) and b) do not necessarily deliver a difference scheme in the case of variable coefficients due to different concepts of discretization. As soon as one deals with algebras, where x and T_x do not commute, the left normal form of a vector (the computation uses subtractions of left multiples of an approximation and thus invokes non-commutative multiplication) is not necessarily the result of rewriting of any term u_{x^β} (which just plugs the right hand side expression into the place where the term resides and does not invoke non-commutative multiplication).

Provided all approximations from A are linear with respect to u_β as in (3), the Term Rewriting System Approach 3.3 will successfully lead to a finite difference scheme for a nonlinear PDE with variable coefficients.

4 Generation of Difference Schemes

Armed with the methods from the previous section, we proceed with the generation of schemes for a linear equation with constant coefficients. We prefer the method of algebraic analysis from 3.2, in contrast to Gerdt et al. [13], who used the method of difference algebra because of the reasons of practical complexity, which is significantly lower if one uses the approach 3.2. However, Gerdt et al. systematically follow the difference algebra approach for nonlinear equations. Notably, in [13] they have obtained an interesting nice-behaving scheme with cubic nonlinearities. The original equation contains quadratic nonlinearities, but the new non-traditional scheme does not contain switches as traditional schemes.

A large class of equations might be written in a so-called *conservation law form*, which can be obtained e.g. by applying the Green's formula. For example, the equation $\frac{\partial Q}{\partial x} - \frac{\partial P}{\partial y} = 0$ is equivalent to the equation $\oint_{\Gamma} P dx + Q dy = 0$ for arbitrary piecewise smooth closed contour Γ.

We choose some discretized integration contours and approximations rules for the integrals and proceed as above. The difference schemes, which we obtain by elimination are fully consistent by construction [13] .

4.1 Approximation Rules and Their Operator Form

A general way for approximation of a PDE consists in the application of integral relations (like $\int_{t_n}^{t_{n+1}} u_{tt}(x,t)dt = u_t(x,t_{n+1}) - u_t(x,t_n)$) together with further approximations of derivatives (like u_t) and integrals.

Contour approximations. Many possibilities exist for choosing contours and approximations. We are using rather rectangular than quadratic grids, the two most frequently used approximations on contours are node points of the rectangle and midpoints of the grid with double distance, as illustrated by the pictures below.

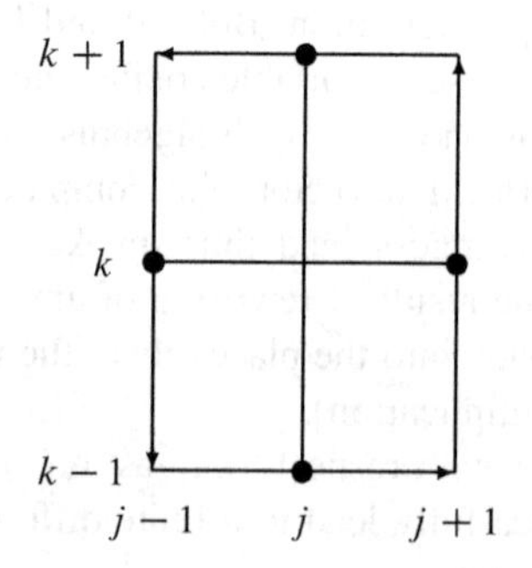

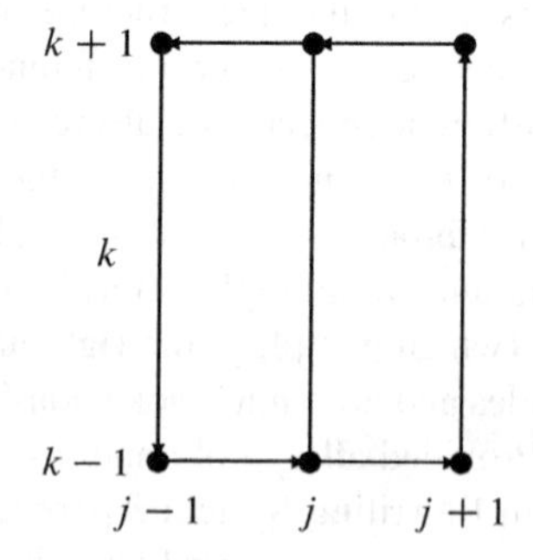

By applying the Green's formula we lower the order of an equation by 1. The approximation formulas derived from the contour are usually more complicated, than the approximations derived from the original equation and integral relations. This is not a problem for an implementation, since complicated manipulations with polynomial expressions can be performed effectively with modern computer algebra systems.

Approximation of derivatives via Taylor series. Applying the Taylor expansion up to the 2nd order, we obtain $u(x \pm \Delta x) = u(x) \pm \Delta x u_x(x) + \frac{\Delta x^2}{2} u_{xx}(x) + \mathcal{O}(\Delta x^3)$. Hence, we can approximate as follows:

$u_x(x) = \frac{u(x+\Delta x)-u(x)}{\Delta x} + \mathcal{O}(\Delta x)$ (forward difference)

or $u_x(x) = \frac{u(x)-u(x-\Delta x)}{\Delta x} + \mathcal{O}(\Delta x)$ (backward difference). Subtracting these two equalities we obtain $u(x+\Delta x)-u(x)+u(x)-u(x-\Delta x) = 2\Delta x u_x(x)+\mathcal{O}(\Delta x^3)$, hence $u_x(x) = \frac{u(x+\Delta x)-u(x-\Delta x)}{2\Delta x} + \mathcal{O}(\Delta x^2)$ (*central* 1*st order difference*).
Adding these two equalities and rewriting the result, we obtain

$$\frac{u(x+\Delta x)-2u(x)+u(x-\Delta x)}{\Delta x^2} = u_{xx}(x) + \mathcal{O}(\Delta x^2)(\textit{central 2 nd order difference}).$$

Approximation of integrals. Closed Newton-Cotes formulas give rise to so-called trapezoid an pyramid rules, whereas open Newton-Cotes formulas lead us to midpoint rule. The trapezoid rule is expressed as follows:

$$\int_{x_0}^{x_0+\Delta x} f(x)dx = \frac{1}{2}\Delta x(f(x_0)+f(x_0+\Delta x)) - \frac{1}{12\Delta x^3} f''(\xi),\ x_0 \le \xi \le x_0+\Delta x.$$

We obtain as approximation for $u_x(x)$:

$$u(x_0 + \Delta x) - u(x_0) = \int_{x_0}^{x_0+\Delta x} u_x(x)dx = \frac{1}{2}\Delta x(u_x(x_0) + u_x(x_0 + \Delta x)),$$

and hence $(T_x - 1) \bullet u = \frac{1}{2}\Delta x(T_x + 1) \bullet u_x$.
Pyramid (or Simpson's) rule looks as follows:

$$\int_{x_0}^{x_0+2\Delta x} f(x)dx = \frac{1}{3}\Delta x(f(x_0)+4f(x_0+\Delta x)+f(x_0+2\Delta x)) - \frac{1}{90\Delta x^5} f^{(4)}(\xi),$$

hence its difference form is $\frac{1}{3}\Delta x \cdot (T_x^2 + 4T_x + 1) \bullet u_x = (T_x^2 - 1) \bullet u$.
Open Newton-Cotes formula for one point

$$\int_{x_0}^{x_0+2\Delta x} f(x)dx = 2\Delta x f(x_0 + \Delta x) + \mathcal{O}(\Delta x^2 f'),$$

leads us to the midpoint formula $\Delta x \cdot T_x \bullet u_x = (T_x^2 - 1) \bullet u$.

Summary. We gather the most used approximations in difference operator form:

- **Forward difference** $(\Delta x,\ 1 - T_x) \bullet (u_x, u)^T = 0$
- **Backward difference** $(\Delta x \cdot T_x,\ 1 - T_x) \bullet (u_x, u)^T = 0$
- A 1st **order central appr.** $(2\Delta x \cdot T_x,\ 1 - T_x^2) \bullet (u_x, u)^T = 0$
- A 2nd **order central appr.** $(-\Delta x^2 \cdot T_x,\ (1 - T_x)^2) \bullet (u_{xx}, u)^T = 0$
- **Trapezoid rule** $(\frac{1}{2}\Delta x \cdot (T_x + 1), 1 - T_x) \bullet (u_x, u)^T = 0$
- **Midpoint rule** $(2\Delta x \cdot T_x, 1 - T_x^2) \bullet (u_x, u)^T = 0$.
- **Pyramid rule** $(\frac{1}{3}\Delta x \cdot (T_x^2 + 4T_x + 1), 1 - T_x^2) \bullet (u_x, u)^T = 0$
- **Lax method**[1] $(2\Delta t \cdot T_x, T_x^2 - 2T_t T_x + 1) \bullet (u_t, u)^T = 0$
- **Parametric temporal difference** for $0 \le \theta \le 1$: $(\Delta t \cdot (\theta T_t + (1 - \theta)), 1 - T_t) \cdot (u_t, u)^T = 0$. If $\theta = 0$ resp. $\theta = 1$, it becomes forward resp. backward difference.

We assume that the difference scheme involves quantities $\Delta x_1, \ldots, \Delta x_m, \Delta t$ and originates from a typical set of approximations. The difference scheme is of the smallest difference order by construction, hence the associated shift polynomial p is irreducible. In many situations we want to present p as the sum of products of operators. We propose the following notation considered in application to von Neumann stability.

Definition 4. A **semi-factorized** presentation of a linear difference scheme of order $O(\Delta x_1^{b_1}, \ldots, \Delta x_m^{b_m}, \Delta t^c)$ is the sum $p = \Delta x_1^{b_1} p_1 + \ldots + \Delta x_m^{b_m} p_m + \Delta t^c p_t$ for $p_i \in K[T]$, such that p_i does not involve Δx_i in its coefficients and most (if not all) p_j do not involve $\Delta x_1, \ldots, x_m, \Delta t$.

Unlike nodal form, a semi-factorized form allows compact descriptions of very complicated and higher dimensional schemes. Note, that in the examples it turns out, that there exists a unique (up to constant factors) semi-factorized presentation. We have a method for computing a semi-factorized form constructively.

Example 6. Consider the 1D heat equation $u_t - a^2 u_{xx} = 0$ with parameter a. We approximate u_t with backwards difference $\Delta t \cdot T_t \bullet u_t = (T_t - 1) \bullet u$, resp. in the nodes of the grid, $\Delta t \cdot (u_t)_i^{m+1} = (u)_i^{m+1} - (u)_i^m$. u_{xx} is approximated with the 2nd order weighted centered space method, that is

$$\Delta x^2 \cdot T_x \bullet u_{xx} = (\theta T_t + (1 - \theta)) \cdot (T_x - 1)^2 \bullet u, \text{ where } 0 \le \theta \le 1.$$

We obtain the following matrix formulation of the problem

$$\begin{pmatrix} 1 & -a^2 & 0 \\ -\Delta t \cdot T_t & 0 & T_t - 1 \\ 0 & -\Delta x^2 T_x T_t & (\theta T_t + (1 - \theta)) \cdot (Tx - 1)^2 \end{pmatrix} \bullet \begin{pmatrix} u_t \\ u_{xx} \\ u \end{pmatrix} = 0.$$

[1]Used in the discretization of the advection equation

By computing a Gröbner basis (with the algebraic analysis approach), we obtain a single polynomial in shift operators for the scheme $-a^2\Delta t\theta T_x^2 T_t + a^2\Delta t(\theta-1)T_x^2 + (2a^2\Delta t\theta + \Delta x^2)T_x T_t - (2a^2\Delta t(\theta-1) + \Delta x^2)T_x - a^2\Delta t\theta T_t + a^2\Delta t(\theta-1)$

Its semi-factorized form is $\Delta x^2 T_x(T_t - 1) - a^2\Delta t(T_x - 1)^2(\theta T_t + 1 - \theta) = 0$. In the following example we show SINGULAR code for obtaining these objects and for producing a nodal presentation of the scheme, which is

$$\frac{1}{a^2\Delta t}\cdot(u_{j+1}^{n+1}-u_{j+1}^{n})-\frac{\theta}{\Delta x^2}\cdot(u_{j+2}^{n+1}-2u_{j+1}^{n+1}+u_j^{n+1})-\frac{(1-\theta)}{\Delta x^2}(u_{j+2}^{n}-2u_{j+1}^{n}+u_j^{n})=0.$$

The obtained scheme is called FTCS if $\theta = 0$, BTCS if $\theta = 1$ and Crank-Nicholson, if $\theta = \frac{1}{2}$.

This scheme is consistent with the original differential equation for any $\theta \in \mathbb{R}$. Since $\frac{u_{j+1}^{n+1}-u_{j+1}^{n}}{\Delta t} = u_t + \mathcal{O}(\Delta t)$ and $\frac{u_{j+2}^{n+1}-2u_{j+1}^{n+1}+u_j^{n+1}}{\Delta x^2} = u_{xx} + \mathcal{O}(\Delta x^2)$, we have

$$\tfrac{1}{a^2}u_t - \theta u_{xx} - (1-\theta)u_{xx} = \tfrac{1}{a^2}u_t - u_{xx} = \mathcal{O}(\Delta t) + \mathcal{O}(\Delta x^2).$$

The order of the scheme is $(\Delta t, \Delta x^2)$.

Example 7. In this example, we demonstrate computations with SINGULAR and with `findifs.lib`. In the matrix formulation above the *parameters* are $\Delta t, \Delta x, a, \theta$. We introduce an additional parameter d, which will be needed later for the check of stability. The *variables* of the ring are T_t and T_x. We define the ring in SINGULAR and the matrix of equations as follows:

```
ring r = (0,a,dx,dt,theta,d),(Tx,Tt),(c,Dp);
matrix M[3][3]=
1, -a^2, 0,              // the equation itself
-dt*Tt, 0, Tt-1,         // appr. u_t with backward difference
0, -dx^2*Tt*Tx,(theta*Tt+(1-theta))*(Tx-1)^2; // appr. u_xx
```

where u_{xx} is approximated with the 2nd order weighted centered space method. We transpose the matrix and call the `std` routine for the Gröbner basis computation.

```
module R = module(transpose(M)); module S = std(R);
print(S);
=> 0,      0,             1,
   0,      (-a^2*dt)*Tt,(-a^2),
   S[3,1],Tt-1,           0
```

The first column vector of the resulting matrix is the only one with non-zero entry only in the 3rd component. The symbol `S[3,1]` is displayed since this entry, which is the difference scheme polynomial, is big in size.

```
poly p = S[3,1]; p; // assign and print the answer
=>(-a^2*dt*theta)*Tx^2*Tt+(a^2*dt*theta-a^2*dt)*Tx^2+
  (2*a^2*dt*theta+dx^2)*Tx*Tt+(-2*a^2*dt*theta+2*a^2*dt-
  dx^2)*Tx+(-a^2*dt*theta)*Tt+(a^2*dt*theta-a^2*dt)
```

We proceed with the construction of the semi-factorized form.

```
LIB "findifs.lib"; // load the library for schemes
ideal I = decoef(p,dt); // see Appendix for details
I; // the sum of elements of I gives p
=>I[1]=(dx^2)*Tx*Tt+(-dx^2)*Tx
  I[2]=(-a^2*dt*theta)*Tx^2*Tt+(a^2*dt*theta-a^2*dt)*Tx^2+
       (2*a^2*dt*theta)*Tx*Tt+(-2*a^2*dt*theta+2*a^2*dt)*Tx+
       (-a^2*dt*theta)*Tt+(a^2*dt*theta-a^2*dt)
```

Next, we can obtain the semi-factorized operator form of the scheme:

```
factorize(I[1]); // we suppress the output
factorize(I[2]); // factors with multiplicities
=> [1]:
     _[1]=(-a^2*dt)
     _[2]=Tx-1
     _[3]=(theta)*Tt+(-theta+1)
   [2]:
     1,2,1
```

The semi-factorized form is $\Delta x^2 T_x(T_t - 1) - a^2 \Delta t (T_x - 1)^2(\theta T_t + 1 - \theta) = 0$.

```
list L; L[1] = theta;
difpoly2tex(I,L); // see Appendix
=> \frac{-1}{a^{2} \tri t}\cdot (u^{n+1}_{j+1}-u^{n}_{j+1})+...
```

The string above in tex format (we showed above only a part of it) is the nodal presentation of the scheme, which was obtained already in the previous example.

5 Symbolic Methods for von Neumann Stability Analysis

5.1 Stability Rings, Morphisms and Polynomials

We refer the reader to e. g. [8, 26] for details about stability. Suppose that t is the temporal variable and $x_1, \dots, x_m$ are the spatial variables. We start with a finite difference scheme, written in the nodal form on a uniform orthogonal grid with steps $\Delta t, \Delta x_1, \dots, \Delta x_m$. We suppose to work in the interior region, which is bounded, say, by $L_1, \dots, L_m$ in spatial directions.

In the von Neumann stability analysis, one presents the functions on the grid as discrete Fourier modes, that is

$$\chi(u^n_{j_1 j_2 \dots j_m}) = g^n \prod_{k=1}^{m} e^{i j_k \ell_k \pi \Delta x_k},$$

where χ is a linear map, g is a new symbolic variable, $0 \le \ell_k \Delta x_k \le L_k$. We abbreviate $\beta_{j_k} := \pi \ell_k \Delta x_k$. We substitute this presentation of nodes into the equation, perform simplifications and obtain a polynomial G in one variable g with constant coefficients.

The **von Neumann stability criterion** (see e. g. [8,26]) states, that the difference scheme is stable if $|\xi| \leq 1$ for every root ξ of G.

The **Lax-Richtmeyer equivalence theorem** can be stated in the following form (adopted from [26]). A consistent scheme for a well-posed linear initial value problem is convergent if and only if it is stable. For a well-posed linear initial-boundary-value problem, however, stability is only a necessary condition for convergence.

We do not address algorithms for an algorithmic check of consistency of a difference scheme with its differential equation. Several methods using algebraic tools can be found in e. g. [9, 10, 14]. However, we demonstrate the usage of semi-factorized form for a positive conclusion about consistency in some examples.

Let A be the algebra of functions on a given grid. It carries a natural module structure over the algebra R of linear partial difference operators with constant coefficients $C[T_t, T_{x_1}, \ldots, T_{x_m}]$ over some field $C \supseteq \mathbb{Q}(\Delta t, \Delta x_1, \ldots, \Delta x_m)$. The action of R on discrete Fourier nodes by the map χ can be written as follows, for all j_k:

$$\chi(T_t^a \bullet u^n_{j_1 j_2 \ldots j_m}) = g^a \cdot \chi(u^n_{j_1 j_2 \ldots j_m}) \text{ and } \chi(T_{j_s}^b \bullet u^n_{j_1 j_2 \ldots j_m}) = e^{i j_s l_s \pi \Delta x_k} \cdot \chi(u^n_{j_1 j_2 \ldots j_m}).$$

The map χ and this action give rise to an homomorphism of C-algebras

$$\chi : C[T_t, T_{x_1}, \ldots, T_{x_m}] \longrightarrow C\big([i, sin_{x_1}, cos_{x_1}, \ldots, sin_{x_m}, cos_{x_m}]/J_m\big)[g],$$

where $J_m = \langle i^2 + 1, sin^2_{x_1} + cos^2_{x_1} - 1, \ldots, sin^2_{x_m} + cos^2_{x_m} - 1 \rangle$ is an ideal. We denote this **constructive stability morphism** by the same letter χ and note its C-linearity. It is defined by its values on the generators of the source algebra $\chi(T_t) = g$ and $\chi(T_{j_s}) = e^{i l_s \pi \Delta x_s} = \cos\beta_s + i \cdot \sin\beta_s$ for all $1 \leq s \leq m$.

The constructive nature of this approach and its applicability in computer algebra systems, lies in the following. We choose the complex-rational numbers $\mathbb{Q}[i]/\langle i^2 + 1\rangle$ as basic numeric field. We can do on demand further algebraic extensions. We avoid complex exponentials by passing to the sine and cosine and by including their algebraic relations in the factor ideal. A stability morphism can be defined in computer algebra systems.

Let $P = \sum_{a,\alpha} c_{a,\alpha} T_t^a T_x^\alpha$ be the operator form of the finite difference scheme $P \bullet u = 0$, where T_x^α stands for $T_{x_1}^{\alpha_1} \cdot \ldots \cdot T_{x_m}^{\alpha_m}$ for a multi-index $\alpha = (\alpha_1, \ldots, \alpha_m) \in \mathbb{N}^m$. Then,

$$\chi(P) = \sum_{a,\alpha} c_{a,\alpha} \chi(T_t)^a \chi(T_x)^\alpha = \sum_{a,\alpha} c_{a,\alpha} \prod_{k=1}^{m} (\cos\beta_k + i \cdot \sin\beta_k)^{\alpha_k} g^a = \sum_a d_a g^a$$

is the univariate polynomial in g, which we call the **stability polynomial** of a given difference scheme. Obviously, the degree of $\chi(P)$ is the same as the highest degree of T_t in P.

Example 8. Let us continue with the Example 6. In order to prepare the scheme for stability analysis, one can rewrite it as follows:

$$u_{j+1}^{n+1} - u_{j+1}^{n} = a^2 d\big(\theta \cdot (u_{j+2}^{n+1} - 2u_{j+1}^{n+1} + u_{j}^{n+1}) + (1-\theta)\cdot(u_{j+2}^{n} - 2u_{j+1}^{n} + u_{j}^{n})\big),$$

with $d := \frac{\Delta t}{\Delta x^2}$. We prefer to work with the semi-factorized operator form of the scheme

$$\Delta x^2 T_x(T_t - 1) - a^2 \Delta t (T_x - 1)^2(\theta T_t + 1 - \theta) = 0.$$

By creating the stability ring and performing simplification and factorization (see the next example for the SINGULAR code), we obtain the following linear polynomial in the variable g

$$(i\cos + \sin)\cdot(((-2a^2 d\theta)\sin + 2a^2 d\theta + 1)\cdot g + (2a^2 d\theta - 2a^2 d)\sin - 2a^2 d\theta$$
$$+ 2a^2 d - 1).$$

The first factor $i \cdot \cos(\beta) + \sin(\beta)$ is ignored in stability analysis, since it is of magnitude 1.

Example 9. We continue with the Example 7. Define the semi-factorized scheme again.

```
poly P  = Tx*(Tt-1)+(-a^2)*d*(Tx-1)^2*((theta)*Tt+(-theta+1));
ring r2 = (0,a,theta,d),(Tx,Tt),(c,Dp);
poly P  = imap(r,P);
```

Now, we create the stability ring ST, which will be $\mathbb{Q}(a,\theta,d)[g,i,sin,cos]$ and a map χ from r2, which is $\mathbb{Q}(a,\theta,d)[T_x,T_t]$) to ST.

```
ring ST    = (0,a,d,theta),(g,i,sin,cos),lp;
ideal Rels = std(ideal(i2+1,sin^2+cos^2-1));
map chi    = r2,ideal(sin+i*cos,g);
poly P = chi(P); // the mapping
P = NF(P,Rels); P; // reduction wrt ideal Rels
=>(-2*a^2*d*theta)*g*i*sin*cos+(2*a^2*d*theta+1)*g*i*cos+ ...
ideal FP = factorize(P); // factorization
```

The polynomial P together with its factorization have been presented in the previous example.

We obtained from a system of linear equations a single univariate polynomial in the stability ring. Next we face the following problem:

Given a univariate parametric polynomial P, find out, under which conditions on parameters all the roots of P lie in the complex unit circle.

As already mentioned in [17,21], this problem can be solved algorithmically with the help of CAD (Cylindrical Algebraic Decomposition).

5.2 Cylindrical Algebraic Decomposition

The algorithm for CAD goes back to G. Collins et al. It is one of the most important algorithms, for quantifier elimination not only in real algebraic geometry [1]. Its algorithmic complexity is high and can be double exponential in the number of variables. Nevertheless, the universality of the method makes it very powerful and applicable to various problems.

A finite set of polynomials $\{p_1, \ldots, p_m\} \in \mathbb{R}[x_1, \ldots, x_n]$ induces a decomposition (partition) of $\mathbb{R}^n$ into maximal sign-invariant cells. A **cell** in the algebraic decomposition of $\{p_1, \ldots, p_m\} \in \mathbb{R}[x_1, \ldots, x_n]$ is a maximal connected subset of $\mathbb{R}^n$, on which all the p_i are sign invariant.

Definition 5. For $n \in \mathbb{N}$, let $\pi_n : \mathbb{R}^n \to \mathbb{R}^{n-1}, (x_1, \ldots, x_{n-1}, x_n) \mapsto (x_1, \ldots, x_{n-1})$ denote the canonical projection. Let $\{p_1, \ldots, p_m\} \in \mathbb{Q}[x_1, \ldots, x_n]$. The algebraic decomposition of $\{p_1, \ldots, p_m\}$ is called **cylindrical**, if:

- For any two cells C, D of the decomposition, the images $\pi(C), \pi(D)$ are either identical or disjoint.
- The algebraic decomposition of $\{p_1, \ldots, p_m\} \cap \mathbb{Q}[x_1, \ldots, x_{n-1}]$ is cylindrical.

For instance, any algebraic decomposition of $\mathbb{R}^1$ is cylindrical.

There are several sophisticated implementations of the CAD algorithm. We are using the one from the system MATHEMATICA, where two commands, `Reduce` and `CylindricalDecomposition` are available in the context of CAD. There are also freely available systems QEPCAD by C. Brown [2] and REDLOG by A. Dolzmann et al. [7].

5.3 CAD and von Neumann Stability

Example 10. Let us continue with the examples 6, 8. Let us represent the root of a stability polynomial as $\frac{c}{d'}$, where $c = 2a^2d(1-\theta)\sin -2a^2d(1-\theta)+1$, $d' = 2a^2d\theta(1-\sin)+1$.

Since $d' \geq 0$, we have to solve the inequality $-d \leq c \leq d$, that is $c+d \geq 0$ and $d \geq c$. The first inequality $2a^2d(1-\sin) \geq 0$ is always satisfied, and the second is equivalent to $a^2d(2\theta-1)(1-\sin)+1 \geq 0$. In this example, we compare the functions `CylindricalDecomposition` and `Reduce` of MATHEMATICA

```
CylindricalDecomposition[{a^2*d*(2*theta-1)*(1-s) + 1 >= 0,
-1 <= s <= 1, a > 0, d > 0}, {theta, a, d, s}]
```

returns

$$\left(\theta < \frac{1}{2} \;\&\&\; a > 0 \;\&\&\; \left(0 < d \leq -\frac{1}{-2a^2+4a^2\theta}\right) \;\&\&\; -1 \leq s \leq 1\right) \;||$$

$$\left(d > -\frac{1}{-2a^2 + 4a^2\theta} \;\&\&\; \frac{1 - a^2 d + 2a^2 d\theta}{-a^2 d + 2a^2 d\theta} \leq s \leq 1\right) \;||$$

$$\left(\theta > \frac{1}{2} \;\&\&\; a > 0 \;\&\&\; d > 0 \;\&\& \; -1 \leq s \leq 1\right).$$

Executing more specialized call,

```
Reduce[a > 0 && d > 0 && 0 <= theta <= 1 && ForAll[s, -1 <=
s <= 1, a^2*d*(2*theta - 1)*(1-s) + 1 >= 0], {theta, d}]
```

we obtain a more informative and structured answer:

$$a > 0 \;\&\&\; (0 \leq \theta < \tfrac{1}{2} \;\&\&\; 0 < d \leq -\tfrac{1}{-2a^2+4a^2\theta}) \;||\; (\tfrac{1}{2} \leq \theta \leq 1 \;\&\&\; d > 0).$$

We conclude:

- If $\frac{1}{2} \leq \theta \leq 1$, the scheme is unconditionally stable.
- If $0 \leq \theta < \frac{1}{2}$, the scheme is stable under the condition $d = \frac{\Delta t}{\Delta x^2} \leq \frac{1}{2a^2(1-2\theta)}$.

The quantity $d = \frac{\Delta t}{\Delta x^2}$ if often called Courant (or Courant-Friedrichs-Lewy) number. It is classical to express conditions on the von Neumann stability in terms of the Courant number.

Example 11. Consider the 1D **advection equation** $u_t + au_x = 0$. We approximate u_t with the parametric temporal method and u_x with the trapezoid rule. As a result, we obtain the difference scheme in the semi-factorized form $\Delta x \cdot (T_x + 1) \cdot (T_t - 1) + 2a\Delta t \cdot (T_x - 1) \cdot (\theta T_t - (\theta - 1)) = 0$, which reads as follows in the nodal form:

$$\tfrac{1}{2a\Delta t} \cdot (u_{j+1}^{n+1} - u_{j+1}^{n} + u_j^{n+1} - u_j^n) + \tfrac{1}{\Delta x} \cdot (\theta(u_{j+1}^{n+1} - u_j^{n+1}) - (\theta - 1)(u_{j+1}^n - u_j^n)) = 0.$$

This scheme is consistent with its differential equation. The stability polynomial is linear with complex coefficients, so we present it as a fraction. The reformulated stability problem, which we have to solve, is

$$-2 \leq \frac{4a^2d^2(2\theta - 1)}{4a^2d^2(\theta - 1)^2 + \frac{1+\sin(\beta)}{1-\sin(\beta)}} \leq 0, \;\forall \beta \notin \frac{\pi}{2}\mathbb{Z}$$

Since $t := \frac{1+\sin(\beta)}{1-\sin(\beta)} \geq 0$, the right hand side inequality is equivalent to $\theta \leq \frac{1}{2}$. The left hand side is equivalent to $4a^2d^2(2\theta - 1) + 2(4a^2d^2(\theta - 1)^2 + t)) \geq 0$. Since $t \in [0, \infty)$, we have to show that $0 \leq 4a^2d^2(2\theta - 1) + 8a^2d^2(\theta - 1)^2 = 4a^2d^2(\theta^2 + (\theta - 1)^2)$, what is true for all d. Of course, computations with CAD confirm this answer.

Thus, this scheme is unconditionally stable if $\theta \leq \frac{1}{2}$ and unstable otherwise.

6 Examples for λ-Wave Equation

We consider a parametric equation $u_{tt} - \lambda^2 u_{xx} = 0,\ \lambda \neq 0$ and its higher dimensional versions. We construct finite difference schemes for several different approximations and analyze their stability.

6.1 *Conservative Law with Parametric Time Approximation*

The presentation via the conservation law is $\oint_\Gamma \lambda^2 u_x dt + u_t dx = 0$. We use trapezoid rule for the contour integral and spatial integral relations. For temporal integral relations we use parametric difference with $\theta \in [0, 1]$.

We obtain the following system of difference equations:

$$\begin{pmatrix} \Delta h \cdot (-T_x T_t + T_x + T_t - 1) & \lambda^2 \Delta t \cdot (T_x T_t - T_t - T_x + 1) & 0 \\ 0 & \frac{1}{2}\Delta x \cdot (T_x + 1) & 1 - T_x \\ \Delta t \cdot (\theta T_t + (1-\theta)) & 0 & 1 - T_t \end{pmatrix} \bullet \begin{pmatrix} u_t \\ u_x \\ u \end{pmatrix} = 0.$$

After the computation of Gröbner basis, we obtain the scheme

$$\begin{aligned} 0 = & \frac{\Delta h^2}{2\lambda^2 \Delta t^2}\left(u_j^{n+2} - 2u_j^{n+1} + u_j^n\right) - \frac{\Delta h^2}{2\lambda^2 \Delta t^2}\left(u_{j+2}^{n+2} - 2u_{j+2}^{n+1} + u_{j+2}^n\right) \\ & + \theta\left(u_{j+2}^{n+2} - 2u_{j+1}^{n+2} + u_j^{n+2}\right) - (2\theta - 1)\left(u_{j+2}^{n+1} - 2u_{j+1}^{n+1} + u_j^{n+1}\right) \\ & + (\theta - 1)\left(u_{j+2}^n - 2u_{j+1}^n + u_j^n\right). \end{aligned}$$

The stability polynomial of 2nd degree is rather complicated. However, factorization reveals a factor $g - 1$. The other factor is linear, but with complicated coefficients. We present it as $g - \frac{c}{d'}$. Since both c and d' are complex numbers, we compute absolute values of them. Then, $||d'|| = (4\theta^2 d^4 - 1) \cdot (\cos(\beta_x) - 1) - 2$ and $||c|| = ||d'|| - 4d^4(2\theta - 1)(\cos(\beta_x) - 1)$.

Hence, $||\frac{c}{d'}|| \leq 1 \Leftrightarrow 0 \leq 4d^4 \sin(\beta_x/2)^2 \dfrac{(2\theta - 1)}{(4\theta^2 d^4 - 1)\sin(\beta_x/2)^2 + 1} \leq 1.$

Consider the left hand side inequality

$$\begin{aligned} 0 \leq 4d^4 \sin(\beta_x/2)^2 \frac{(2\theta - 1)}{(4\theta^2 d^4 - 1)\sin(\beta_x/2)^2 + 1} \Leftrightarrow 0 \\ \leq (2\theta - 1)((4\theta^2 d^4 - 1)\sin(\beta_x/2)^2 + 1). \end{aligned}$$

Since $4\theta^2 d^4 > 0 \Leftrightarrow 4\theta^2 d^4 - 1 > -1 \geq -\frac{1}{\sin(\beta_x/2)^2}$, the second factor is always positive. Hence, the inequality is satisfied as soon as $\theta \geq \frac{1}{2}$.

The second inequality reads as $4d^4 \sin(\beta_x/2)^2 \frac{(2\theta-1)}{(4\theta^2 d^4 - 1)\sin(\beta_x/2)^2 + 1} \leq 1$. Then,

$$4\theta^2 d^4 \sin(\beta_x/2)^2 + 1 - \sin(\beta_x/2)^2 \geq (4d^4 \sin(\beta_x/2)^2)(2\theta - 1) \Leftrightarrow$$

$$\theta^2 + \frac{\cos(\beta_x/2)^2}{4d^4 \sin(\beta_x/2)^2} \geq (2\theta - 1) \Leftrightarrow (\theta - 1)^2 + \frac{\cos(\beta_x/2)^2}{4d^4 \sin(\beta_x/2)^2} \geq 0,$$

what is always the case. Summarizing, we obtain that this scheme is unconditionally stable, if $\theta \geq \frac{1}{2}$ and unstable otherwise.

6.2 Integral Relations and 2nd Order Central Approximations

Using direct 2nd order central approximations for both t and x, we obtain the following scheme:

$$\left(u_{j+1}^{n+2} - 2u_{j+1}^{n+1} + u_{j+1}^{n}\right) - \lambda^2 \frac{\Delta t^2}{\Delta h^2} \cdot \left(u_{j+2}^{n+1} - 2u_{j+1}^{n+1} + u_{j}^{n+1}\right) = 0.$$

We denote $d := \lambda \frac{\Delta t}{\Delta h}$, then the scheme is described by the polynomial

$$p = d^2 T_x^2 T_t - T_x T_t^2 + (-2d^2 + 2) T_x T_t - T x + d^2 T_t = T_x (T_t - 1)^2 - d^2 (T_x - 1)^2 T_t,$$

which is presented in a semi-factorized form. After simplifications the stability polynomial reads as $g^2 + (4d^2 \sin^2(a/2) - 2)g + 1 = 0$. Denote $b := -1 + 2d^2 \sin^2(a/2)$, i.e. $g^2 + 2bg + 1 = 0$. The roots are $b \pm \sqrt{b^2 - 1}$. If $b^2 > 1$, then one of the roots has modulus bigger, than one. If $b^2 = 1$, the roots are ± 1. If $b^2 < 1$, the absolute value of both roots equals $b^2 + 1 - b^2 = 1$. Hence, $b^2 \leq 1$, if and only if $d \leq 1$, that is $\frac{\Delta t}{\Delta h} \leq \frac{1}{\lambda}$. The same condition is produced with the help of CAD in MATHEMATICA.

This scheme is conditionally stable if the Courant number $d = \lambda \frac{\Delta t}{\Delta h} \leq 1$.

6.3 Explicit Integration for t and Trapezoid Rule for x

We use explicit integration (that is, a backward difference) for t and trapezoid rule for x and obtain the following scheme.

$$\frac{1}{4\Delta t^2} \cdot \left(u_{j+2}^{n+2} - 2u_{j+2}^{n+1} + u_{j+2}^{n} + 2\left(u_{j+1}^{n+2} - 2u_{j+1}^{n+1} + u_{j+1}^{n}\right) + u_j^{n+2} - 2u_j^{n+1} + u_j^{n}\right)$$

$$- \frac{\lambda^2}{\Delta h^2} \cdot \left(u_{j+2}^{n+2} - 2u_{j+1}^{n+2} + u_j^{n+2}\right) = 0.$$

The difference scheme polynomial is

$$T_x^2 T_t^2 \; -2T_x^2 T_t + 2T_x T_t^2 + T_x^2 - 4T_x T_t + T_t^2 + 2T_x - 2T_t + 1 \\ - \frac{4\lambda^2 \Delta t^2}{\Delta h^2}(T_x^2 T_t^2 - 2T_x T_t^2 + T_t^2).$$

Denote $d^2 := \frac{4\lambda^2 \Delta t^2}{\Delta h^2}$. After performing substitutions, we obtain $g^2 - 2bg + b = 0$, where $b = (1 + d^2 \tan^2(a))^{-1}$. Its solutions are straightforward: $g = b \pm \sqrt{b^2 - b}$. If $b^2 - b > 0$, we have $b > 1$ and hence one root is too big. If $b^2 - b \leq 0$, the absolute value of a root is just $b^2 + b - b^2 = b$, what is not bigger than 1. $b \leq 1$ is satisfied, since $b^{-1} = 1 + d^2 \tan^2(a)$. Hence, this scheme is unconditionally stable. With the help of CAD and MATHEMATICA, we arrive to the same conclusion.

6.4 *Higher Dimensional λ-Wave Equation*

One of the crucial advantages of our approach and its implementation is the scalability. We employ the algorithms in a very general setting. The algorithms can be easily modified for the case of more functions. In particular, we are able to generate schemes and test them for stability in a higher-dimensional setting.

Consider the approach from Subsection 6.2 which led us to a conditionally stable scheme. We apply the same approximations to all spatial variables.

Two spatial dimensions. We have $u_{tt} - \lambda^2(u_{xx} + u_{yy}) = 0$. The scheme is

$$0 = \frac{1}{\Delta t^2} \cdot \left(u_{j+1,k+1}^{n+2} - 2u_{j+1,k+1}^{n+1} + u_{j+1,k+1}^{n} \right) \\ - \frac{\lambda^2}{\Delta x^2} \cdot \left(u_{j+2,k+1}^{n+1} - 2u_{j+1,k+1}^{n+1} + u_{j,k+1}^{n+1} \right) \\ - \frac{\lambda^2}{\Delta y^2} \cdot \left(u_{j+1,k+2}^{n+1} - 2u_{j+1,k+1}^{n+1} + u_{j+1,k}^{n+1} \right).$$

In a semi-factorized form, the scheme looks as follows

$$T_x T_y (T_t - 1)^2 - d_x^2 \cdot (T_x - 1)^2 T_y T_t - d_y^2 \cdot T_x (T_y - 1)^2 T_t = 0.$$

The stability polynomial in a simplified form is

$$g^2 - 2(d_x^2 \cos(\beta_x) + d_y^2 \cos(\beta_y) - d_x^2 - d_y^2 - 2) \cdot g + 1 = 0.$$

Using CAD, we conclude, that this scheme is **conditionally stable** with the condition $d_x^2 + d_y^2 \leq 1$ for the Courant numbers $d_x := \lambda \frac{\Delta t}{\Delta x}$, $d_y := \lambda \frac{\Delta t}{\Delta y}$.

Three spatial dimensions. The equation is $u_{tt} - \lambda^2(u_{xx} + u_{yy} + u_{zz}) = 0$. The difference scheme is analogous to the two-dimensional one, in a semi-factorized form it has the following form (from which one easily deduces, how the scheme looks for higher dimensions):

$$T_x T_y T_z (T_t - 1)^2 - d_x^2 \cdot (T_x - 1)^2 T_y T_z T_t - d_y^2 \cdot T_x (T_y - 1)^2 T_z T_t - d_z^2 \cdot T_x T_y (T_z - 1)^2 T_t = 0.$$

Running CAD, we obtain, that this scheme, as its lower-dimensional analogues, is **conditionally stable** if $d_x^2 + d_y^2 + d_z^2 \leq 1$ holds for the Courant numbers d_x, d_y and $d_z := \lambda \frac{\Delta t}{\Delta z}$.

7 Dispersion Analysis

7.1 Continuous Dispersion

Recall, that a **Fourier node** in $n + 1$ dimensions is a function of the form

$$e^{i(\langle k,x\rangle - \omega t)}, \quad \langle k, x\rangle := \sum_{j=1}^{n} k_j x_j$$

Respectively, in $1 + 1$ dimensions it is just $e^{i(kx - \omega t)}$. One obtains continuous dispersion from the given linear PDE by substituting Fourier nodes into the PDE and by deriving an equation for ω in terms of k from the result. The latter equation $\omega = \omega(k)$ is called the *continuous dispersion relation*.

Example 12. For the equation $u_{tt} - \lambda^2 u_{xx} = 0$ we have

$$0 = \left(\frac{\partial}{\partial t^2} - \lambda^2 \frac{\partial}{\partial x^2}\right) e^{i(kx - \omega t)} = -e^{i(kx - \omega t)} \cdot (\omega^2 - \lambda^2 k^2).$$

Hence, $\omega = \pm\lambda k$ is the continuous dispersion relation for the λ-wave equation.

We can write down the action of partial derivatives on a Fourier mode. Namely,

$$\frac{\partial^a}{\partial t^a}(e^{i(\langle k,x\rangle - \omega t)}) = (-i\omega)^a e^{i(\langle k,x\rangle - \omega t)} \text{ and } \frac{\partial^{b_j}}{\partial x_j^{b_j}}(e^{i(\langle k,x\rangle - \omega t)}) = (ik_j)^{b_j} e^{i(\langle k,x\rangle - \omega t)}.$$

Hence, the monomial in partial differentiations has its eigenvalue

$$\frac{\partial^a}{\partial t^a} \prod_{j=1}^{n} \frac{\partial^{b_j}}{\partial x_j^{b_j}}(e^{i(\langle k,x\rangle - \omega t)}) = (-i\omega)^a \prod_{j=1}^{n} (ik_j)^{b_j} \cdot (e^{i(\langle k,x\rangle - \omega t)})$$

Let us denote $F = e^{i(\langle k,x\rangle - i\omega t)}$. Then $\partial^{\alpha}(F) = c(\alpha)\cdot F$, where $\alpha := (a, b_1, \ldots, b_n) \in \mathbb{N}^{n+1}$. Extending this action by linearity to the ring of partial differentiations with constant coefficients $R = K[\partial_t, \partial_{x_1}, \ldots, \partial_{x_n}]$, we are able to compute the eigenvalue of a linear PD operator $P \in R$, corresponding to the eigenfunction F:

$$P(F) = \sum_{\alpha} p_{\alpha}\partial^{\alpha}(F) = \left(\sum_{\alpha} p_{\alpha}c(\alpha)\right)\cdot F.$$

The continuous dispersion relation is obtained by solving with respect to ω the equation

$$\sum_{\alpha} p_{\alpha}c(\alpha) = 0,\ \ p_{\alpha} \in K,\ c_{\alpha} \in K(k_1, \ldots, k_n, \omega),$$

which is called the **continuous dispersion equation** (CDE) for P.

Example 13. For the $1+n$-dimensional heat equation $u_t - a^2\cdot\sum_{j=1}^{n} u_{x_j x_j} = 0$ the continuous dispersion equation and relation are

$$0 = -i\omega - a^2\sum_{j=1}^{n} i^2 k_j^2 \iff \omega = -ia^2\sum_{j=1}^{n} k_j^2.$$

Example 14. For $1+n$-dimensional modified λ_i-wave equation $u_{tt} - \sum_{j=1}^{n}\lambda_j^2\cdot u_{x_j x_j} = 0$ the continuous dispersion relation is $w = \pm\sqrt{\sum_{j=1}^{n}\lambda_j^2 k_j^2}$.

7.2 *Discrete Dispersion*

In the discrete case, we consider a discrete Fourier node, corresponding to the grid point $(t_m, (x_1)_{l_1}, \ldots, (x_n)_{l_n})$,

$$F_l^m = e^{i\langle k, x_{(l)}\rangle - \omega t_m},\ \ \langle k, x_{(l)}\rangle := \sum_{j=1}^{n} k_j (x_j)_{l_j}.$$

One substitutes a discrete Fourier node into the difference scheme and derives an expression of ω in terms of k from the result. The latter equation $\omega = \omega(k)$ is called the *discrete dispersion relation*. Let us write down the formula for the eigenvalue of a monomial:

$$T_t^a \prod_{j=1}^{n} T_{x_j}^{b_j}(e^{i\langle k,x_{(l)}\rangle - \omega t_m}) = (e^{-i\omega\Delta t})^a \prod_{j=1}^{n}(e^{ik_j\Delta x_j})^{b_j}\cdot(e^{i\langle k,x_{(l)}\rangle - \omega t_m}).$$

As in the continuous case, we extend this action by linearity to polynomials. For a polynomial $P \in K[T_t, T_{x_1}, \ldots, T_{x_n}]$ one has

$$P(F_l^m) = \sum_\alpha p_\alpha T^\alpha(F_l^m) = \left(\sum_\alpha p_\alpha c(\alpha)\right) \cdot F_l^m,$$

so we solve the **discrete dispersion equation** (DDE) for P

$$\sum_\alpha p_\alpha c(\alpha) = 0, \;\; p_\alpha \in K, \;\; c(\alpha) \in K(\{k_j\}, \omega)$$

and obtain the discrete dispersion relation. Note, that in contrast to the continuous case, this relation is not of polynomial form in general.

Presenting discrete Fourier nodes via trigonometrical functions, we are able to compute discrete dispersion relations symbolically. We prefer not to use the de Moivre's formula, but to express dispersion relations in terms of sine and cosine of a single argument. We work in the commutative ring $\mathbb{C}(\Delta t, \Delta x)[sin_t, cos_t, \{sin_j, cos_j\}]$ modulo the ideal $\langle \{sin_j^2 + cos_j^2 - 1\}, sin_t^2 + cos_t^2 - 1\rangle$, where $cos_j := \cos(k\Delta x_j)$, $cos_t := \cos(\omega \Delta t)$. Then,

$$T_t^a \prod_{j=1}^n T_{x_j}^{b_j}(F_l^m) = (\cos_t - i \sin_t)^a \prod_{j=1}^n (\cos_j + i \sin_j)^{b_j} \cdot (F_l^m).$$

Example 15. Consider the λ-wave equation $u_{tt} - \lambda^2 u_{xx} = 0$ and the difference scheme

$$d^2 T_x^2 T_t - T_x T_t^2 + (-2d^2 + 2) T_x T_t - T_x + d^2 T_t = 0,$$

obtained with the 2nd order central approximations for x and t, where $d = \lambda \frac{\Delta t}{\Delta x}$.

Performing computations, we obtain after simplification $d^2 \cos_x - cos_t + 1 - d^2 = 0$, that is $\cos(\omega \Delta t) = 1 - d^2(1 - \cos(k\Delta x))$. In the stability limit $d \to 1$, we have $\cos(\omega\Delta t) = \cos(k \Delta x)$, hence $\omega = \pm \frac{\Delta x}{\Delta t} k + 2\pi m, m \in \mathbb{Z}$. Since $d \to 1$ implies $\frac{\Delta x}{\Delta t} \to \lambda$, in the stability limit the discrete dispersion relation becomes $\omega = \pm \lambda k + 2\pi m$. By setting $m = 0$ we recover the continuous dispersion relation.

8 Conclusion and Future Work

The advantages of presented methods include, among other, their scalability and tendency towards automatization. Indeed, we do not make distinction between classical types of PDEs (hyperbolic, elliptic, parabolic). Thus these methods are very general. Symbolic methods are able to generate automatically many difference schemes of standard linear PDEs with constant coefficients, as it was demonstrated in [13] and by ourselves.

Moreover, for the same situation we presented an approach to determine conditions for von Neumann stability, using cylindrical algebraic decomposition, and a symbolic approach to the determination of continuous and discrete dispersion relations. The generalization of these methods to systems of equations, to the case of variable coefficients and nonlinear equations is very important. It is known to be hard in general and even the notion of stability might differ from one case to another. On the other hand, Lax-Richtmeyer equivalence theorem can be generalized to some more general, even nonlinear, situations. Thus the investigation about the applicability of symbolic methods for obtaining conditions on stability will continue primarily for the cases, where generalized Lax-Richtmeyer theorem holds.

We decided not to include the treatment of systems of linear PDEs in this paper. However, we want to remark, that by the rewriting system approach the number of the discretized equations is exactly the number of PDEs one started with. By using Gröbner or involutive bases, we get in general more equations, which reveal the interplay between discretized equations. Such interplay is not detected by the rewriting approach at all; it seems to us that such interplay has not been investigated before.

An important issue for future research is a partial algebraization of the consistency analysis of a generated scheme of the given PDE or a system of PDEs. Provided such a check, one could work with general multi-parametric schemes, where the conditions on parameters arise from the consistency check and the symbolic stability approach. This has been investigated in case of a system of linear PDEs with constant [9] and variable [14] coefficients.

Within his recent PhD thesis Christian Dingler (TU Kaiserslautern, Germany) presented a new package `findiff.lib` for SINGULAR with QEPCAD as an engine for cylindrical algebraic decomposition. This package, already distributed with SINGULAR, extends the tools for the generation of finite difference schemes to the cases of a single linear PDE and of a system of linear PDEs. Another problem for further research is the generalization of von Neumann stability for systems, which is clear only for some classes of equations. Thus further generalization of our methods will go into several directions: allowing variables coefficients and/or allowing nonlinearity.

A very important question concerns the role of differential and difference Gröbner bases for nonlinear equations in the scheme generation and stability analysis. The recent papers [12–14] show for some cases, that a systematic use of the interplay between equations can produce more universal, though more complicated, schemes.

Arising from the *letterplace* philosophy, see [18], the development of new theory and algorithms for infinite difference Gröbner bases will be of great interest.

Acknowledgements The authors express their deep gratitude to Vladimir P. Gerdt (JINR, Russia) for his interest, discussions and suggestions during the work on this paper. We would also like to thank M. Kauers (RISC, Linz, Austria), W. Zulehner (J. Kepler University of Linz, Austria), M. Fröhner (BTU Cottbus, Germany) and A. Klar (TU Kaiserslautern, Germany) for discussions on various topics around stability in this paper. We have learned many examples from the scripts and papers of colleagues, mentioned above. A special thanks goes to H. Engl (Vienna, Austria) for his constructive critics, which helped to improve the presentation of the results. At last, but not at least, we thank to anonymous referees for their remarks and questions.

The first author is grateful to the SFB F013 "Numerical and Symbolic Scientific Computing" of the Austrian FWF for partial financial support in 2005-2007.

9 Appendix. The Detailed SINGULAR Code of an Example

9.1 *A Quick Introduction to the System* SINGULAR

We want to describe shortly by examples how to read SINGULAR language and how to obtain and interpret the output – as far as it is used to generate a difference scheme. The very detailed documentation of SINGULAR can be found online at `www.singular.uni-kl.de`.

9.1.1 Definition of an Algebra

Nearly any computation with SINGULAR takes place inside of a ring, which has to be defined first. Consider the following input:

```
ring R = (0,dt,dh),(Tx,Tt),(c,dp);
```

This command defines commutative polynomial ring $R = \mathbb{Q}(dt, dh)[T_x, T_t]$ equipped with the position-over-term monomial module ordering $\prec$. Here, the ground field is $K = \mathbb{Q}(dt, dh)$, that is the field of rational functions over $\mathbb{Q}$ in transcendental parameters dt, dh. These constant parameters have the following meaning here: $dt = \Delta r$, $dh = \Delta h$ are step sizes of the grid. R is the ring in the variables T_x, T_t, corresponding to shift operators, over the field K. The monomial module ordering $\prec$ will be used in Gröbner basis computations. In the example, dp stands for the degree reverse lexicographical ordering on polynomials. A small c at the first place indicates, the polynomial vectors will be sorted first by components in descending order, i. e., $e_1 > e_2 > \ldots$ and then by the monomial ordering dp.

9.1.2 Creation of a Matrix

Starting with a linear system of PDEs with constant coefficients and approximation rules, one has to deal with an extended system $\tilde{A}U = 0$. We need only the matrix with entries in the ring R of shift operators:

```
ring R = (0,dt,dh),(Tx,Tt),(c,dp);
matrix A[3][3] =
(-Tx*Tt^2+Tx), (Tx^2*Tt - Tt), 0 ,
0, (dh/2)*(Tx+1), 1-Tx,
(dt/2)*(Tt+1), 0, 1-Tt;
```

One has to indicate row- and column-size in the definition of a matrix. On the right hand side follows a list of polynomials, describing the entries of a matrix.

9.1.3 Elimination of Components

We have to eliminate all but last components from the matrix A, In this example, the anonymous vector U stands for $(u_t, u_x, u)^t$. We want to produce within a row module of A a row, having entries only in the last component. This is done most efficiently by a Gröbner basis computation of a submodule with respect to the given monomial module ordering. The last nonzero component of the first column generator corresponds to the difference scheme.

```
module M  = transpose(A);
module M1 = std(M); // Groebner basis computation
print(M1); // we suppress its output
```

Note, that the command `print`, applied to a module, does not necessarily displays every entry completely. However, one can display every single element separately. In this example, the difference scheme polynomial is

```
M1[3,1];
=>(-dt)*Tx^2*Tt+(dh)*Tx*Tt^2+(2*dt-2*dh)*Tx*Tt+(dh)*Tx+(-dt)*Tt.
```

9.1.4 Evaluation of the Constants

There are several ways for the evaluation of the constants. One of them is to use the command `subst`. In the running example, suppose one wants to evaluate the scheme in $\Delta t = 10^{-1}, \Delta h = 10^{-2}$.

```
poly p = M1[3,1]; // the polynomial as above
poly pnew = p;
pnew = subst(pnew,dt,1/10);
pnew = subst(pnew,dh,1/100);
pnew;
=> -1/10*Tx^2*Tt+1/100*Tx*Tt^2+9/50*Tx*Tt+1/100*Tx-1/10*Tt
```

9.2 *Tools for Difference Schemes*

The library `findifs.lib` has been created to automate numerous processes during the generation of finite difference schemes. An important role is played by the routines, transforming the different forms of objects into some classical ones. One can generate complicated schemes and easily present them for instance in nodal form or in polynomial operator presentation including semi-factorized form, which is used in stability analysis.

`decoef(P,n)`; where P is a polynomial and n is a number. `decoef` decomposes the polynomial P into summands with respect to the presence of the number n in the coefficients and returns an ideal in usually two generators. For example,

```
ring r = (0,dh,dt),(Tx,Tt),dp;
poly P = (4*dh^2-dt)*Tx^3*Tt + dt*dh*Tt^2 + dh*Tt;
P;
=> (4*dh^2-dt)*Tx^3*Tt+(dh*dt)*Tt^2+(dh)*Tt
decoef(P,dt);
=>_[1]=(4*dh^2)*Tx^3*Tt+(dh)*Tt   // the part, not containing dt
  _[2]=(-dt)*Tx^3*Tt+(dh*dt)*Tt^2 // the part which contains dt
decoef(P,dh);
=>_[1]=(-dt)*Tx^3*Tt              // the part, not containing dh
  _[2]=(4*dh^2)*Tx^3*Tt+(dh*dt)*Tt^2+(dh)*Tt
```

`difpoly2tex(S,P[,Q])`; where S is an ideal, P is a list and Q is an optional list. `difpoly2tex` converts the difference scheme, given in the ideal S, to its the nodal form in a LaTeX string. The ideal S is assumed to be the result of `decoef`, list P contains parameters, which will be controlled in order to remain in numerators. The optional list Q contains polynomials, which will be added to the scheme (written in the function u) the part in terms of a function p. For example,

```
ring r = (0,dh,dt,V),(Tx,Tt),dp;
poly M = (2*dh*Tx+dt)^2*(Tt-1) + V*Tt*Tx;
M;
=> (4*dh^2)*Tx^2*Tt+(-4*dh^2)*Tx^2+(4*dh*dt+V)*Tx*Tt+
   (-4*dh*dt)*Tx+(dt^2)*Tt+(-dt^2)
ideal I = decoef(M,dt); // see above
I;
=> I[1]=(4*dh^2)*Tx^2*Tt+(-4*dh^2)*Tx^2+(V)*Tx*Tt
   I[2]=(4*dh*dt)*Tx*Tt+(-4*dh*dt)*Tx+(dt^2)*Tt+(-dt^2)
list L; L[1] = V; // V stands for nu
difpoly2tex(I,L);
=> \frac{1}{4 \tri t}\cdot (u^{n+1}_{j+2}-u^{n}_{j+2}+
   \frac{ \nu}{4 \tri h ^{2}} u^{n+1}_{j+1})+ \frac{1}{4\tri h}
   \cdot (u^{n+1}_{j+1}-u^{n}_{j+1}+\frac{ \tri t}{4 \tri h}
   u^{n+1}_{j}+ \frac{- \tri t}{4 \tri h} u^{n}_{j})
```

The last output, compiled with TeX, produces

$$\frac{1}{4\Delta t}\cdot\left(u^{n+1}_{j+2}-u^{n}_{j+2}+\frac{\nu}{4\Delta h^2}u^{n+1}_{j+1}\right)$$
$$+\frac{1}{4\Delta h}\cdot\left(u^{n+1}_{j+1}-u^{n}_{j+1}+\frac{\Delta t}{4\Delta h}u^{n+1}_{j}+\frac{-\Delta t}{4\Delta h}u^{n}_{j}\right).$$

Now let us illustrate the use of the optional list Q. Suppose there are two equations in the operator form, denoted by U and P. We want to treat them as corresponding to two different unknown functions u and p.

```
ring D = (0,ro,K,dt,dh),(Tx,Tt),(c,Dp);
poly U = (-K*dt)*Tx^2*Tt+(K*dt)*Tt;
```

```
poly P = (-2*ro*dh)*Tx*Tt+(2*ro*dh)*Tx;
list V; V[1] = K; V[2] = ro;
difpoly2tex(-U,V,-P);
=> \frac{K}{2 \tri h}\cdot (u^{n+1}_{j+2}-u^{n+1}_{j})+
   \frac{ \rho}{ \tri t} \cdot (p^{n+1}_{j+1}-p^{n}_{j+1})
```

Here, we have produced the nodal form of a scheme for two functions u and p:

$$\frac{K}{2\Delta h} \cdot \left(u_{j+2}^{n+1} - u_j^{n+1}\right) + \frac{\rho}{\Delta t} \cdot \left(p_{j+1}^{n+1} - p_{j+1}^{n}\right).$$

References

1. Basu, S., Pollack, R., Roy, M.F.: Algorithms in Real Algebraic Geometry. Algorithms and Computation in Mathematics. 10. Springer (2003)
2. Brown, C.W.: QEPCAD B: A program for computing with semi-algebraic sets using CADs. SIGSAM Bull. **37**(4), 97–108 (2003). DOI 10.1145/968708.968710. http://www.usna.edu/Users/cs/qepcad/B/QEPCAD.html
3. Chyzak, F., Quadrat, A., Robertz, D.: Linear control systems over Ore algebras. effective algorithms for the computation of parametrizations. Applicable Algebra in Engineering, Communication and Computing **16**(5), 938–1279 (2005). http://www.springerlink.com/content/y61643p573387258
4. Chyzak, F., Salvy, B.: Non-commutative elimination in Ore algebras proves multivariate identities. J. Symbolic Comput. **26**(2), 187–227 (1998)
5. Cohn, R.M.: Difference algebra. R.E. Krieger (1979)
6. Decker, W., Greuel, G.M., Pfister, G., Schönemann, H.: SINGULAR 3-1-2 — A Computer Algebra System for Polynomial Computations. Centre for Computer Algebra, University of Kaiserslautern (2010). http://www.singular.uni-kl.de
7. Dolzmann, A., Sturm, T.: Redlog: Computer algebra meets computer logic. ACM SIGSAM Bulletin **31**(2), 2–9 (1997). http://redlog.dolzmann.de
8. Fröhner, M.: Numerische Methoden in der Hydrodynamik. Wiss. Schriftenr. Tech. Hochsch. Karl-Marx-Stadt 12 (1984)
9. Ganzha, V., Vorozhtsov, E.: Computer-Aided Analysis of Difference Schemes for Partial Differential Equations. Wiley Interscience (1996)
10. Ganzha, V.G., Vorozhtsov, E.V.: Parallel implementation of stability analysis of difference schemes with Mathematica. J. Math. Sci. **108**, 1070–1088 (2002). DOI 10.1023/A:1013500723898. http://dx.doi.org/10.1023/A:1013500723898
11. Gerdt, V.: Involutive algorithms for computing Groebner bases. In: Pfister, G., Cojocaru, S., Ufnarovski V. (eds.) Computational Commutative and Non-Commutative Algebraic Geometry. IOS Press (2005)
12. Gerdt, V., Blinkov, Y.: Involution and difference schemes for the Navier-Stokes equations. Proceedings CASC 2009, Kobe, Japan (2009)
13. Gerdt, V., Blinkov, Y., Mozzhilkin, V.: Gröbner bases and generation of difference schemes for partial differential equations. SIGMA **2**, 051 (2006). http://arxiv.org/abs/math/0605334
14. Gerdt, V., Robertz, D.: Consistency of finite difference approximations for linear PDE systems and its algorithmic verification. In: Proceedings of the International Symposium on Symbolic and Algebraic Computation (ISSAC'10). ACM Press (2010). DOI 10.1145/1837934.1837950
15. Greuel, G.M., Levandovskyy, V., Motsak, A., Schönemann, H.: PLURAL. A SINGULAR 3.1 Subsystem for Computations with Non-commutative Polynomial Algebras. Centre for Computer Algebra, TU Kaiserslautern (2010). http://www.singular.uni-kl.de

16. Greuel, G.M., Pfister, G.: A SINGULAR Introduction to Commutative Algebra. Springer (2002)
17. Hong, H., Liska, R., Steinberg, S.: Testing stability by quantifier elimination. J. Symbolic Comput. **24**(2), 161–187 (1997). http://www.sciencedirect.com/science/journal/07477171
18. La Scala, R., Levandovskyy, V.: Letterplace ideals and non-commutative Gröbner bases. J. Symbolic Comput. **44**(10), 1374–1393 (2009). DOI doi:10.1016/j.jsc.2009.03.002
19. Levandovskyy, V.: Non-commutative computer algebra for polynomial algebras: Gröbner bases, applications and implementation. Ph.D. Thesis, Universität Kaiserslautern (2005). http://kluedo.ub.uni-kl.de/volltexte/2005/1883/
20. Levin, A.: Difference algebra. Algebra and Applications. Springer, New York (2008)
21. Liska, R., Drska, L.: FIDE: a REDUCE package for automation of FInite difference method for solving pDE. In: Watanabe, S., Nagata M. (eds.) Proceedings of the International Symposium on Symbolic and Algebraic Computation (ISSAC'90), pp. 169–176. ACM Press and Addison-Wesley (1990). http://www.acm.org:80/pubs/citations/proceedings/issac/96877/p169-liska/
22. Pommaret, J.F.: Partial differential control theory. Vol. 1: Mathematical tools. Vol. 2: Control systems. Mathematics and its Applications (Dordrecht) 530. Dordrecht: Kluwer Academic Publishers (2001)
23. Ritt, J.F.: Differential algebra. American Mathematical Society (AMS) (1950)
24. Saito, S., Sturmfels, B., Takayama, N.: Gröbner Deformations of Hypergeometric Differential Equations. Springer (2000)
25. Seiler, W.M.: Involution. The formal theory of differential equations and its applications in computer algebra. Algorithms and Computation in Mathematics 24. Springer, Berlin (2010). DOI 10.1007/978-3-642-01287-7
26. Thomas, J.: Numerical partial differential equations: Finite difference methods. Springer (1995)

White Noise Analysis for Stochastic Partial Differential Equations

Hermann G. Matthies

Abstract Stochastic partial differential equations arise when modelling uncertain phenomena. Here the emphasis is on uncertain systems where the randomness is spatial. In contrast to traditional slow computational approaches like Monte Carlo simulation, the methods described here can be orders of magnitude more efficients. These more recent methods are based on some kind stochastic Galerkin approximations, approximating the unknown quantities as functions of independent random variables, hence the name "white noise analysis". We outline the steps leading to the fully discrete equations, commenting on one possible numerical solution method. Key to many of the developments is tensor product structure of the solution, which must be exploited both theoretically and numerically. For two examples with polynomial nonlinearities the computations are shown to be quite explicit and can be performed largely analytically.

1 Introduction

Oftentimes, numerical simulations of real-world systems are required even though not all parameters are exactly known. The uncertainties inherent in the model result in uncertainties in the results of numerical simulations, a fact which is often ignored in common practise. Clearly, it is desirable to quantify the uncertainties in the solution depending on the model's uncertainties.

Stochastic models are one way to quantify uncertainties. Uncertain parameters are modelled by random variables, uncertain time-dependent functions by stochastic processes, and uncertain spatial properties by random fields [2, 3, 9, 50]. If the physical system is described by a partial differential equation (PDE), then the

H.G. Matthies (✉)
Institute of Scientific Computing, Technische Universität Braunschweig, 38092, Braunschweig, Germany
e-mail: wire@tu-bs.de; http://www.wire.tu-bs.de

U. Langer and P. Paule (eds.), *Numerical and Symbolic Scientific Computing*, Texts and Monographs in Symbolic Computation, DOI 10.1007/978-3-7091-0794-2_8,

combination with the stochastic model results in a stochastic PDE (SPDE) [11, 18, 27, 42]. The solution of the SPDE is a random field describing both the expected system-response and its quantitative uncertainty.

These are parametrised equations, and such parametrised equations naturally have solutions in tensor product spaces. Solution methods for such a problem range over a wide set of approaches, see [23, 24, 30, 32–34, 43, 44, 48] for some developments mainly in the field of stochastic mechanics.

Next to the well-known spatial and temporal discretisation of the partial differential equation, the stochastic processes and random fields have to be discretised, and for the purpose of computation be approximated by a finite number of random variables. For computational purposes it is advantageous to describe and approximate the problem in independent random variables, a technique also known as "white noise analysis" [17–19, 22, 28].

The next step is to compute the response and its stochastic description in terms of the stochastic input. To start, we need a description of the mathematical setting which allows one to see that such stochastic models are well-posed in the sense of Hadamard essentially if the underlying deterministic model is so, and if this also holds for every possible realisation [4,13,27,32,34,41]. This will be briefly sketched here in Sects. 2 and 3.

Solution methods [34] comprise direct integration, including *Monte Carlo* [7,45] and its relatives, as well as *deterministic* integration methods such as *Smolyak* sparse-grid methods [14, 38, 40], stochastic collocation [6, 36, 37], and stochastic Galerkin methods [1, 4, 5, 16, 20, 26, 31, 32, 41, 51, 52, 52, 53], to name a few of the more popular ones. Here a variational framework for stochastic Galerkin (SG) methods will be given, numerical experiments may be found in the references just cited. The usual deterministic part will be summarised in Sect. 4, and the stochastic discretisation will be given for a simple but important kind of choice of approximating subspaces in Sect. 5.

In Sect. 6 a very brief description of a possible numerical method for the solution of the fully discrete set of nonlinear equations is given, one that gives promising results [32–34] and observes the highly structured nature of the operations on tensor product spaces.

Two examples are given in Sect. 7 where the nonlinearity is polynomial, in Subsection 7.1 a nonlinear diffusion, and in Subsection 7.2 the stationary Navier-Stokes equation is considered. With the properties of the Hermite algebra, given in Appendices C and D, the computation of the nonlinearities can be quite explicit, with a large part of the computations performed analytically. We close with a conclusion and outlook on further work in Sect. 8.

2 Deterministic Model Problem

The model problem is formally one of stationary diffusion, and it is intended to serve as a motivating example on how SPDEs may arise. It may for example describe the seepage of groundwater through a porous subsurface rock/sand formation, or heat conduction in an inhomogeneous medium.

We first introduce the deterministic problem, where $\mathcal{G} \subset \mathbb{R}^d$ is the spatial domain of interest, u is the diffusing quantity, $\boldsymbol{\kappa}$ is the diffusion tensor in the non-linear diffusion law for the flow $\boldsymbol{q} = -\boldsymbol{\kappa}(u)\nabla u$. As the diffusion tensor may depend on u, the problem may be nonlinear. The quantity f represents sinks and sources in the domain. For simplicity we assume homogeneous Dirichlet boundary conditions. The stationary diffusion equation then is

$$-\nabla \cdot (\boldsymbol{\kappa}(\boldsymbol{x}, \boldsymbol{u}(\boldsymbol{x}))\nabla u(x)) = f(x), \qquad x \in \mathcal{G} \subset \mathbb{R}^d. \tag{1}$$

For the sake of simplicity also the conductivity tensor $\boldsymbol{\kappa}$ is represented by just a scalar field κ. None of these simplifications have any influence on what we want to show later.

For the possible solutions we choose a closed subspace of the Sobolev space $W_p^1(\mathcal{G})$, namely the completion of the compactly supported smooth functions in the W_p^1-norm

$$\mathcal{U} := \mathring{W}_p^1(\mathcal{G}), \tag{2}$$

so that the essential Dirichlet boundary conditions are satisfied, and allow for the right-hand-side $f \in \mathcal{U}^* \simeq W_q^{-1}(\mathcal{G})$, where as usual $1/p + 1/q = 1$. To describe the diffusive process, define the generalised Nemytskii-operator $\mathsf{K} : \mathcal{U} \to \mathcal{Q} := L_q(\mathcal{G}, \mathbb{R}^d)$ by

$$\mathsf{K} : u(x) \mapsto (\varkappa(x) + cu(x)^2)\nabla u(x) =: \kappa(x, u(x))\nabla u(x). \tag{3}$$

This is a continuous map from $\mathcal{U} = \mathring{W}_p^1(\mathcal{G})$ into $\mathcal{Q} = L_q(\mathcal{G}, \mathbb{R}^d)$ for $p = 4$ because of the type of nonlinearity. Additionally we require $c > 0$, $\varkappa(x) > 0$ a.e., $\varkappa \in L_\infty(\mathcal{G})$ and $1/\varkappa \in L_\infty(\mathcal{G})$.

This makes the semilinear (linear in v) form

$$\mathsf{a}(v, u) := \int_{\mathcal{G}} \nabla v(x) \cdot \mathsf{K}(u)(x)\, \mathrm{d}x \tag{4}$$

hemicontinuous in u and continuous in v, and defines a hemicontinuous nonlinear operator $\mathsf{A} : \mathcal{U} \to \mathcal{U}^*$ such that

$$\forall u, v \in \mathcal{U} : \quad \mathsf{a}(v, u) = \langle \mathsf{A}(u), v \rangle_{\mathcal{U}}, \tag{5}$$

where $\langle \cdot, \cdot \rangle_{\mathcal{U}}$ is the duality pairing between $\mathcal{U}$ and its dual $\mathcal{U}^*$. If there is no danger of confusion, we will omit the index on the duality pairing.

Proposition 1. The operator A is hemicontinuous, strictly monotone and coercive. Standard arguments on monotone operators e.g. [21, 39] allow us then to conclude that under the conditions just described, the problem to find $u \in \mathcal{U}$ such that

$$\forall v \in \mathcal{U} : \quad \mathsf{a}(v, u) = \langle \mathsf{A}(u), v \rangle = \langle f, v \rangle \tag{6}$$

has a unique solution. In the linear case this reduces to the Lax-Milgram lemma.

This result shall serve as a reference of how we would like to formulate the stochastic problem in the next Sect. 3, namely have a well-posed problem in the sense of Hadamard. In the deterministic case it is well-known that this property of well-posedness will be inherited by the numerical approximation if it is done right.

3 Stochastic Model Problem

In the stochastic case, we want to model $\varkappa$ as well as f as random fields defined over some probability space $(\Omega, \mathfrak{A}, \mathbb{P})$, where Ω is the basic probability set of elementary events, $\mathfrak{A}$ a σ-algebra of subsets of Ω, and $\mathbb{P}$ a probability measure. We require additionally

$$\varkappa(x,\omega) > 0 \quad \text{a.e.}, \quad \|\varkappa\|_{L_\infty(\mathcal{G}\times\Omega)} < \infty, \quad \|1/\varkappa\|_{L_\infty(\mathcal{G}\times\Omega)} < \infty. \tag{7}$$

The solution to (6) will also be a random field in that case, and we allow for that by choosing as a solution space

$$\mathcal{W} := \mathcal{U} \otimes \mathcal{S}, \tag{8}$$

where in this case we choose $\mathcal{S} = L_p(\Omega)$ because of the type of nonlinearity. The basic tensor product space is isomorphic to the space of finite rank linear maps $L_q(\Omega) \simeq \mathcal{S}^* \to \mathcal{U}$ which may be equipped with the Schatten-p-norm [46]. This is the ℓ_p-norm of the sequence of singular values, and we take $\mathcal{W}$ to actually be the completion of the so normed tensor product. This is a reflexive space, just as $\mathcal{U}$, and $\mathcal{U}$ is naturally isometrically embedded via $u \mapsto u \otimes 1$ as a deterministic subspace.

We define a semilinear form $\mathbf{a}$ on $\mathcal{W}$ via

$$\mathbf{a}(v,u) := \mathbb{E}\left(\mathsf{a}[\omega](u(x,\omega, v(x,\omega)))\right), \tag{9}$$

where $\mathbb{E}(\cdot)$ is the expectation on Ω. The parameter-dependent semilinear forms are just as for the deterministic problem (4):

$$\mathsf{a}[\omega](v,u) := \int_{\mathcal{G}} \nabla v(x,\omega) \cdot \mathbf{K}(x,\omega,u(x,\omega))\, dx, \tag{10}$$

where the generalised Nemytskii-operator on $\mathbf{K} : \mathcal{W} = \mathcal{U} \otimes \mathcal{S} \to \mathcal{Q} \otimes \mathcal{S}^*$ is given by

$$\mathbf{K} : u(x,\omega) \mapsto (\varkappa(x,\omega) + cu(x,\omega)^2).\nabla u(x,\omega). \tag{11}$$

Again, this defines a hemicontinuous nonlinear operator $\mathbf{A} : \mathcal{W} \to \mathcal{W}^*$ such that

$$\forall u, v \in \mathcal{W}: \quad \mathbf{a}(v,u) = \langle\!\langle \mathbf{A}(u), v \rangle\!\rangle, \tag{12}$$

where $\langle\!\langle \cdot,\cdot \rangle\!\rangle$ is the duality pairing between $\mathcal{W}^*$ and $\mathcal{W}$. Here $\mathcal{W}^*$ is isomorphic to the completion of $\mathcal{U}^* \otimes \mathcal{S}^*$ in the Schatten-q-norm.

A linear form $\mathbf{f}$ on $\mathcal{W}$ is similarly defined through its deterministic but parameter-dependent counterpart for all $v \in \mathcal{U}$

$$\langle f(\omega), v \rangle := \int_{\mathcal{G}} v(x) f(x,\omega) \, \mathrm{d}x, \tag{13}$$

by $\langle\langle \mathbf{f}, w \rangle\rangle := \mathbb{E}\left(\langle f(\omega), w(\omega)\rangle\right)$ for all $w \in \mathcal{W}$.

Proposition 2. The operator $\mathbf{A}$ is hemicontinuous, strictly monotone and coercive, and standard arguments on monotone operators (cf. Proposition 1) allow us then to conclude that the problem to find $u \in \mathcal{W}$ such that

$$\forall v \in \mathcal{W}: \quad \mathbf{a}(v,u) = \langle\langle \mathbf{A}(u), v \rangle\rangle = \langle\langle f, v \rangle\rangle \tag{14}$$

has a unique solution. In the linear case this reduces to the Lax-Milgram lemma again.

4 Discretisation in Space

Almost any technique may be used for the spatial discretisation, e.g. finite differences or finite elements, and we use a finite element discretisation of the region $\mathcal{G} \subset \mathbb{R}^d$ with a vector of ansatz-functions $\boldsymbol{\phi}(x) = [\phi_1(x), \dots, \phi_N(x)]$, e.g. [10,47]. We define $\mathcal{U}_N := \operatorname{span}\{\phi_n \mid 1 \le n \le N\} \subset \mathcal{U}$. An ansatz for the solution in terms of $\boldsymbol{\phi}(x)$ yields a semi-discretisation of (14). Similarly to the method of lines for instationary boundary value problems where the coefficients would be time-dependent, we obtain an expansion

$$u^{\text{semi}}(x,\omega) = \sum_{n=1}^{N} u_n(\omega)\phi_n(x) = \boldsymbol{\phi}(x)\boldsymbol{u}(\omega), \tag{15}$$

where the coefficients are random variables $\boldsymbol{u}(\omega) = [u_1(\omega), \dots, u_N(\omega)]^T$.

By inserting the ansatz into the SPDE (14) and applying Galerkin conditions, a system of N nonlinear stochastic equations in $\mathbb{R}^N$ results,

$$\boldsymbol{A}[\omega](\boldsymbol{u}(\omega)) = \boldsymbol{f}(\omega) \qquad \text{for } \mathbb{P}\text{-almost all } \omega \in \Omega. \tag{16}$$

Here the n-th equation is given by $a[\omega](u^{\text{semi}}(\cdot,\omega)), \phi_n) =: (\boldsymbol{A}[\omega](\boldsymbol{u}(\omega)))_n$ and $(\boldsymbol{f}(\omega))_n := \langle f(\cdot,\omega), \phi_n\rangle$. It is worth noting that almost surely in ω, the operator in (16) inherits the properties of Propositions 1 and 2 – in fact essentially uniformly in ω due to (7) – as it is a symmetric Bubnov-Galerkin projection onto the subspace $\mathcal{U}_N \otimes \mathcal{S}$ [10,32,34,47].

5 Discretisation of the Probability Space

In the following we will use a stochastic Galerkin (SG) method to fully discretise (16) [1,4,5,15,16,20,26,31,32,41,51,52]. To effect the full Galerkin approximation one still has to choose ansatz functions – effectively functions of known RVs – in which to express the unknown coefficients (RVs) $u_n(\omega)$. We choose as ansatz functions Wiener's polynomial chaos expansion (PCE) [16, 25, 31, 32, 34], i.e. multivariate Hermite polynomials H_α in Gaussian RVs. The multivariate Hermite polynomials are given in Appendix B. Reassuringly, the Cameron-Martin theorem [17–19, 22, 28] tells us that the algebra of Gaussian variables is dense in all $L_p(\Omega)$ with $1 \leq p < \infty$, hence in particular in $\mathcal{S} = L_4(\Omega)$.

For example, if we simply decide to have an approximation in K Gaussian RVs with a total polynomial degree of P to choose a finite basis, then one chooses a $\mathcal{A}$ as a finite subset of $\mathcal{J} := \mathbb{N}_0^{(\mathbb{N})}$, the set of all finite non-negative integer sequences, i.e. of multi-indices, see Appendix A.

$$\mathcal{A} = \{\alpha = (\alpha_1, \ldots, \alpha_K, \ldots) \in \mathcal{J} \mid \alpha_k = 0 \text{ for } k > K, \text{ and } |\alpha|_1 < P\},$$

where the cardinality of $\mathcal{A}$ is

$$A := |\mathcal{A}| = \frac{(K+P)!}{K!\,P!}.$$

Although the set $\mathcal{A}$ is finite and $\mathcal{J}$ is countable, there is no natural order on it; we therefore do not impose one at this point. The determination of $\mathcal{A}$ via K and P as above is in many cases too crude, not least because the cardinality changes very unevenly with changing K and M. More elaborate ways to define $\mathcal{A}$ have to be employed using different functionals than just the ℓ_1-norm.

As ansatz in the probabilistic or stochastic space we take

$$\boldsymbol{u}(\omega) = \sum_{\alpha \in \mathscr{A}} \boldsymbol{u}^\alpha H_\alpha(\boldsymbol{\theta}(\omega)), \tag{17}$$

with $\boldsymbol{u}^\alpha := [u_1^\alpha, \ldots, u_N^\alpha]^T$. Through the discretisation the stochastic space $\mathcal{S}$ has been replaced by a subspace $\mathcal{S}_\mathcal{A} := \text{span}\{H_\alpha \mid \alpha \in \mathcal{A}\}$.

The Bubnov-Galerkin method applied to (16) with the ansatz (17) requires that the weighted residuals vanish:

$$\forall \beta \in \mathcal{A}: \quad \mathbb{E}\left(\left[\boldsymbol{f}(\omega) - \boldsymbol{A}[\omega]\left(\sum_{\alpha \in A} \boldsymbol{u}^\alpha H_\alpha(\boldsymbol{\theta}(\omega))\right)\right] H_\beta(\omega)\right) = 0. \tag{18}$$

This may be concisely written – with quantities in the fully discrete space $\mathbb{R}^N \otimes \mathbb{R}^A$ denoted by an upright bold font – as

$$\mathbf{r}(\mathbf{u}) := \mathbf{f} - \mathbf{A}(\mathbf{u}) = 0, \text{ or } \mathbf{A}(\mathbf{u}) = \mathbf{f}, \tag{19}$$

where $(\mathbf{f})_n^\beta := \mathbb{E}\left((f(\omega))_n H_\beta(\omega)\right)$, $\mathbf{u} := (u_n^\alpha)$, and

$$(\mathbf{A}(\mathbf{u}))_n^\beta = \mathbb{E}\left(\left(A[\omega]\left(\sum_{\alpha\in\mathcal{A}} \boldsymbol{u}^\alpha H_\alpha(\omega)\right)\right)_n H_\beta(\omega)\right).$$

A quantity like $\mathbf{u}$ may be thought of as an array of numbers (u_n^α), exploiting the isomorphy $\mathbb{R}^N \otimes \mathbb{R}^A \simeq \mathbb{R}^{N\times A}$, or as an abstract tensor $\sum_\alpha \boldsymbol{u}^\alpha \otimes \boldsymbol{e}_\alpha$, where the $\boldsymbol{e}_\alpha$ are the canonical unit vectors in $\mathbb{R}^A$, or – in a purely linear algebra fashion – regard the symbol $\otimes$ consistently as a Kronecker product. It may be noted that (19) are $A \times N$ equations, and the system (19) inherits the properties of Propositions 1 and 2 as it is a symmetric Bubnov-Galerkin projection onto the finite dimensional subspace $\mathcal{W}_{N,\mathcal{A}} := \mathcal{U}_N \otimes \mathcal{S}_\mathcal{A}$.

Proposition 3. Convergence of the full Galerkin approximation [32, 34] with coefficients the solution from eq:stoch-nonlineq

$$u^\mathrm{f}(x,\omega) := \sum_\alpha (\boldsymbol{\phi}(x)\boldsymbol{u}^\alpha) H_\alpha(\boldsymbol{\theta}(\omega)) = \sum_{n,\alpha} u_n^\alpha \phi_n(x) H_\alpha(\boldsymbol{\theta}(\omega)), \tag{20}$$

to the solution of the SPDE u from (14) with increasing densly filling subspaces $\mathcal{W}_{N,\mathcal{A}} \subseteq \mathcal{W}$ may be established with Céa's lemma [10,47] as being quasi-optimal:

$$\|u - u^\mathrm{f}\|_\mathcal{W} \le C \inf_{v\in\mathcal{W}_{N,\mathcal{A}}} \|u - v\|_\mathcal{W}. \tag{21}$$

For better convergence estimates, one would need results on the regularity of the solution u to (14). For norms weaker than the Schatten-p-norm used in (21), one may take the results in [8], these show the benefit of not only increasing the polynomial degree, but also the total number K of RVs used in the approximation.

6 Solution Methods

We may solve the nonlinear system (19) by the BFGS method with line-searches, e.g. cf. [12,29]. In every iteration a correction of the current iterate $\mathbf{u}_k$ is computed as

$$\mathbf{u}_{k+1} - \mathbf{u}_k = -\mathbf{H}_k \mathbf{r}(\mathbf{u}_k), \tag{22}$$

$$\mathbf{H}_k = \mathbf{H}_0 + \sum_{j=1}^{k} (r_j \mathbf{p}_j \otimes \mathbf{p}_j + s_j \mathbf{q}_j \otimes \mathbf{q}_j). \tag{23}$$

The tensors $\mathbf{p}_j, \mathbf{q}_j$ and the scalars r_j, s_j are results of the previous iterations of the BFGS method, cf. [12, 29]. A preconditioner or initial $\mathbf{H}_0$ is necessary in order to obtain good convergence. Most preconditioners have the form $\mathbf{H}_0 = \boldsymbol{M} \otimes \boldsymbol{\Xi}$

with matrices $\boldsymbol{M} \in \mathbb{R}^{N\times N}$ and $\boldsymbol{\Xi} \in \mathbb{R}^{A\times A}$ [32, 49], and hence display a typical tensor-product structure. One may note that (22) is an iteration on tensors, and that the update to the operator is also in form of a rank-2-tensor. Needless to say that in actual computations, neither in (22) nor anywhere else are the tensor products like in (23) actually formed [29]. This would completely destroy the very sparse nature of the computations, but rather the components are always only used in the form of an operator and stored separately [54].

7 Polynomial Nonlinearities

While the development of the previous sections gives a general avenue to approach not only the formulation and discretisation of nonlinear SPDEs, but also the actual numerical solution process for the discrete solution, in many cases one can be more specific. Often the nonlinearity is just a polynomial in the solution (or may be represented by a power series in the solution), e.g. the Navier-Stokes Equation, where the nonlinearity is just quadratic. For this it is advantageous to have a direct representation of polynomials of random variables.

In Appendices C and D it is shown how to treat polynomial nonlinearities in terms of the Hermite-algebra and Hermite transform, and that will be employed here.

Computationally we will represent random variables $\mathsf{r}_1, \mathsf{r}_2, \ldots$ by the sequence of their PCE-coefficients $(\rho_1) = \mathscr{H}(\mathsf{r}_1), (\rho_2) = \mathscr{H}(\mathsf{r}_2)$ etc., see Appendix D. This then allows us to express products of two – see (65) and (66), or more random variables similarly to (67) – all with the help of the Hermite transform.

7.1 *Nonlinear Diffusion*

Let us take a look at the introductory example of a nonlinear diffusion equation (1) with the specific nonlinearity (11). After semi-discretisation the (16) may be written as

$$\boldsymbol{A}[\omega](\boldsymbol{u}(\omega) = (\boldsymbol{K}_0(\omega) + \boldsymbol{K}_c(\boldsymbol{u}(\omega)))\,\boldsymbol{u}(\omega) = \boldsymbol{f}(\omega), \tag{24}$$

where $\boldsymbol{u}(\omega)$ and $\boldsymbol{f}(\omega)$ are as before, and almost as a usual stiffness matrix

$$(\boldsymbol{K}_0(\omega))_{n,m} := \int_{\mathcal{G}} \nabla\phi_n(x)\cdot\varkappa(x,\omega)\nabla\phi_m(x)\,\mathrm{d}x, \tag{25}$$

and

$$(\boldsymbol{K}_c(\boldsymbol{u}(\omega)))_{n,m} := \int_{\mathcal{G}} \nabla\phi_n(x)\cdot c\,(u^{\mathrm{f}}(x,\omega))^2\nabla\phi_m(x)\,\mathrm{d}x, \tag{26}$$

with $u^{\mathrm{f}}(x,\omega) := \sum_\alpha (\boldsymbol{\phi}(x)\boldsymbol{u}^\alpha) H_\alpha(\boldsymbol{\theta}(\omega))$. This quantity may also be expressed with $\mathbf{u} = [u_n^\alpha]$ for later use as $u^{\mathrm{f}}(\mathbf{u}) := u^{\mathrm{f}}(x,\omega) = \sum_{n,\alpha} u_n^\alpha \phi_n(x) H_\alpha(\boldsymbol{\theta}(\omega))$. By denoting $u_\alpha^{\mathrm{f}}(x) := \boldsymbol{\phi}(x)\boldsymbol{u}^\alpha = \sum_n u_n^\alpha \phi_n(x)$, we recognise these coefficients to be the Hermite transform $\mathscr{H}(u^{\mathrm{f}}(x,\omega)) = (u_\alpha^{\mathrm{f}}(x))_{\alpha\in\mathcal{J}}$, see Appendix D. From this and with the notation $(u_i) = (\ldots, u_i^\alpha, \ldots)$ one sees that the PCE of $(u^{\mathrm{f}}(x,\omega))^2$ is

$$(u^{\mathrm{f}}(\mathbf{u}))^2 (u^{\mathrm{f}}(x,\omega))^2 = \mathscr{H}^{-1}(\mathbf{C}_2((u_\alpha^{\mathrm{f}}(x)), (u_\alpha^{\mathrm{f}}(x))))$$

$$= \sum_\gamma \left[\sum_{i,j} \phi_i(x)((u_i)\boldsymbol{C}_2^\gamma(u_j))\phi_j(x) \right] H_\gamma(\boldsymbol{\theta}(\omega)). \tag{27}$$

There are different ways of going on from here, the simplest seems to be to set in (25) $\boldsymbol{K}_0(\omega) = \sum_\gamma H_\gamma(\boldsymbol{\theta}(\omega))\boldsymbol{K}_0^\gamma$, with $\boldsymbol{K}_0^\gamma := \mathbb{E}\left(H_\gamma \boldsymbol{K}_0\right)/\gamma!$, as the H_γ are orthogonal. For (26) this looks just as simple, setting $\boldsymbol{K}_c(\boldsymbol{u}(\omega)) = \sum_\gamma H_\gamma(\boldsymbol{\theta}(\omega))\boldsymbol{K}_c^\gamma(u^{\mathrm{f}})$ with $\boldsymbol{K}_c^\gamma(u^{\mathrm{f}}) := \mathbb{E}\left(H_\gamma(\boldsymbol{\theta}(\omega))\boldsymbol{K}_c(\boldsymbol{u}(\omega))\right)/\gamma!$. The terms in the last expression may be facilitated with (27), so that

$$(\boldsymbol{K}_c^\gamma(u^{\mathrm{f}}))_{n,m} = \int_{\mathcal{G}} \nabla\phi_n(x)\, c \left[\sum_{i,j} \phi_i(x)((u_i)\boldsymbol{C}_2^\gamma(u_j))\phi_j(x) \right] \nabla\phi_m(x)\,\mathrm{d}x. \tag{28}$$

Both matrices now have a PCE.

Using these PCEs when computing the terms of (19) with the help (18), we obtain

$$\left(\mathbf{K}_0 + \mathbf{K}_c(u^{\mathrm{f}}(\mathbf{u}))\right)\mathbf{u} = \mathbf{f}, \tag{29}$$

where $\mathbf{f}$ and $\mathbf{u}$ are as before in (19). For $\mathbf{K}_0$ the Galerkin projections in (18) result in

$$(\mathbf{K}_0)_{\alpha,\beta} := \sum_\gamma \mathbb{E}\left(H_\alpha H_\gamma H_\beta\right) \boldsymbol{K}_0^\gamma =: \sum_\gamma \boldsymbol{\Delta}_{\alpha,\beta}^\gamma \boldsymbol{K}_0^\gamma, \text{ with} \tag{30}$$

$$\boldsymbol{\Delta}_{\alpha,\beta}^\gamma := \mathbb{E}\left(H_\alpha H_\gamma H_\beta\right) = c_{\alpha,\beta}^\gamma \gamma! \text{ (see Appendix C).} \tag{31}$$

This can be written as a tensor product

$$\mathbf{K}_0 = \sum_\gamma \boldsymbol{K}_0^\gamma \otimes \boldsymbol{\Delta}^\gamma. \tag{32}$$

Similarly, for $\mathbf{K}_c$ the Galerkin projections in (18) result in

$$(\mathbf{K}_c)_{\alpha,\beta}(u^{\mathrm{f}}(\mathbf{u})) := \sum_\gamma \boldsymbol{\Delta}_{\alpha,\beta}^\gamma \boldsymbol{K}_c^\gamma(u^{\mathrm{f}}). \tag{33}$$

This can again be written as a tensor product

$$\mathbf{K}_c(u^{\mathrm{f}}(\mathbf{u})) = \sum_\gamma \boldsymbol{K}_c^\gamma(u^{\mathrm{f}}) \otimes \boldsymbol{\Delta}^\gamma. \tag{34}$$

All the terms of the nonlinear (29) have now explicitly computed, most of them purely analytically. This shows the power of the Hermite algebra calculus for such polynomial nonlinearities, giving the explicit form of (29) as

$$\left(\sum_\gamma \boldsymbol{K}_0^\gamma \otimes \boldsymbol{\Delta}^\gamma + \sum_\gamma \boldsymbol{K}_c^\gamma(u^{\mathrm{f}}(\mathbf{u})) \otimes \boldsymbol{\Delta}^\gamma\right)\mathbf{u} = \mathbf{f}. \tag{35}$$

One should note that, regarding the discussion following (19), the operation $\otimes$ has to be interpreted according to the context. If $\boldsymbol{H} \in \mathbb{R}^{N\times N}$, $\boldsymbol{\Psi} \in \mathbb{R}^{A\times A}$, $\boldsymbol{h} \in \mathbb{R}^N$, and $\boldsymbol{\psi} \in \mathbb{R}^A$, then the operator $\boldsymbol{H} \otimes \boldsymbol{\Psi}$ acts on the tensor $\boldsymbol{h} \otimes \boldsymbol{\psi}$ as

$$(\boldsymbol{H} \otimes \boldsymbol{\Psi})\,\boldsymbol{h} \otimes \boldsymbol{\psi} := (\boldsymbol{H}\boldsymbol{h}) \otimes (\boldsymbol{\Psi}\boldsymbol{\psi}),$$

and is extended by continuity to the whole space. If, as already mentioned, the symbol $\otimes$ is consistently interpreted as a Kronecker product, one gets a fully linear algebra like description, whereas interpreting $\mathbf{u}$ as a matrix $\boldsymbol{U} = [u_n^\alpha]$, the operator acts as $\boldsymbol{H}\boldsymbol{U}\boldsymbol{\Psi}^T$.

7.2 *Stationary Navier-Stokes*

Let us take as another example the stationary incompressible Navier-Stokes equation (with appropriate boundary conditions), where the nonlinearity is quadratic:

$$v \cdot \nabla v - \frac{1}{\mathrm{Re}}\nabla^2 v + \nabla p = g, \text{ and } \nabla \cdot v = 0, \tag{36}$$

where $v(x)$ is the velocity vector at position x, the pressure at x is given by $p(x)$, the body force per unit mass is $g(x)$, and Re is the Reynolds number. Assuming that boundary conditions, or initial conditions, or right hand side g are uncertain, we model the response as random fields $v(x,\omega)$ and $p(x,\omega)$.

In a discretised version, the (36) will look like

$$\boldsymbol{N}(\boldsymbol{v},\boldsymbol{v}) + \boldsymbol{K}\boldsymbol{v} + \boldsymbol{B}\boldsymbol{p} = \boldsymbol{g}, \text{ and } \boldsymbol{B}^T\boldsymbol{v} = \boldsymbol{0}, \tag{37}$$

where the type of discretisation is not really important for the formulation of the stochastic response. The bilinear operator $\boldsymbol{N}(\cdot,\cdot)$ comes from the nonlinear convective acceleration term, $\boldsymbol{K}$ is the matrix corresponding to the diffusive part, and $\boldsymbol{B}$ is a discrete gradient; $\boldsymbol{v}$ and $\boldsymbol{p}$ are the vectors for the discrete representation of the velocity v and pressure p.

Remark 1. It may be injected here, that if the Reynolds number – or rather the viscosity as the density is constant for an incompressible flow – were to be regarded as random field, then the matrix $\boldsymbol{K}$ in (37) would be a random matrix like $\boldsymbol{K}_0$ in (24) with a corresponding PCE.

Expressing the quantities involved in their PCE

$$\boldsymbol{v}(\boldsymbol{\theta}(\omega)) = \sum_{\alpha\in\mathcal{A}} \boldsymbol{v}^{\alpha} H_{\alpha}(\boldsymbol{\theta}(\omega)), \tag{38}$$

$$\boldsymbol{p}(\boldsymbol{\theta}(\omega)) = \sum_{\beta\in\mathcal{A}} \boldsymbol{p}^{\beta} H_{\beta}(\boldsymbol{\theta}(\omega)), \tag{39}$$

$$\boldsymbol{g}(\boldsymbol{\theta}(\omega)) = \sum_{\gamma\in\mathcal{A}} \boldsymbol{g}^{\gamma} H_{\gamma}(\boldsymbol{\theta}(\omega)), \tag{40}$$

one obtains with the help of Appendices C and D

$$\sum_{\beta,\gamma\in\mathcal{A}} \boldsymbol{N}(\boldsymbol{v}^{\beta}, \boldsymbol{v}^{\gamma}) H_{\beta} H_{\gamma} + \sum_{\alpha\in\mathcal{A}} \boldsymbol{K}\boldsymbol{v}^{\alpha} H_{\alpha} + \sum_{\alpha\in\mathcal{A}} \boldsymbol{B}\boldsymbol{p}^{\alpha} H_{\alpha} = \sum_{\alpha\in\mathcal{A}} \boldsymbol{g}^{\alpha} H_{\alpha}, \tag{41}$$

and

$$\sum_{\alpha\in\mathcal{A}} \boldsymbol{B}^{T}\boldsymbol{v}^{\alpha} H_{\alpha} = \boldsymbol{0}. \tag{42}$$

With the help of (57), the nonlinear term in (41) can be rewritten as

$$\sum_{\beta,\gamma\in\mathcal{A}} \boldsymbol{N}(\boldsymbol{v}^{\beta}, \boldsymbol{v}^{\gamma}) H_{\beta} H_{\gamma} = \sum_{\alpha} \left(\sum_{\beta,\gamma} c^{\alpha}_{\beta\gamma} \boldsymbol{N}(\boldsymbol{v}^{\beta}, \boldsymbol{v}^{\gamma}) \right) H_{\alpha} \tag{43}$$

Inserting this into (41) and projecting onto each H_{α} gives

$$\forall \alpha \in \mathcal{A}: \quad \sum_{\beta,\gamma} c^{\alpha}_{\beta\gamma} \boldsymbol{N}(\boldsymbol{v}^{\beta}, \boldsymbol{v}^{\gamma}) + \boldsymbol{K}\boldsymbol{v}^{\alpha} + \boldsymbol{B}\boldsymbol{p}^{\alpha} = \boldsymbol{g}^{\alpha}, \tag{44}$$

$$\text{and } \boldsymbol{B}^{T}\boldsymbol{v}^{\alpha} = \boldsymbol{0}. \tag{45}$$

Using tensor products $\mathbf{v}$ and $\mathbf{p}$ as before, and defining in the matrix representation

$$\mathbf{N}(\mathbf{v}, \mathbf{v}) = \left[\ldots, \sum_{\beta,\gamma} c^{\alpha}_{\beta\gamma} \boldsymbol{N}(\boldsymbol{v}^{\beta}, \boldsymbol{v}^{\gamma}), \ldots \right], \tag{46}$$

this may be succinctly written as

$$\mathbf{N}(\mathbf{v}, \mathbf{v}) + (\boldsymbol{K} \otimes \boldsymbol{I})\mathbf{v} + (\boldsymbol{B} \otimes \boldsymbol{I})\mathbf{p} = \mathbf{g}, \tag{47}$$

$$\text{and } (\boldsymbol{B}^{T} \otimes \boldsymbol{I})\mathbf{v} = \boldsymbol{0}. \tag{48}$$

This is an explicit PCE representation of the nonlinear stationary incompressible Navier-Stokes equation, making the Hermite-algebra calculus quite explicit. Observe that all high-dimensional integrations were done analytically.

8 Conclusion

We have tried to provide a short introduction to nonlinear SPDEs and stochastic Galerkin methods based on white noise analysis. But the computational effort is often still very high even though there may be tremendous gains compared to the ubiquitous Monte Carlo method. The references mentioned in the introduction contain many interesting directions how the computational burden may be alleviated through adaptivity and model reduction or reduced order models. Some recent references to this kind of work may be found for example in [34, 35].

A Multi-Indices

In the above formulation, the need for multi-indices of arbitrary length arises. Formally they may be defined by

$$\alpha = (\alpha_1, \dots, \alpha_j, \dots) \in \mathcal{J} := \mathbb{N}_0^{(\mathbb{N})}, \tag{49}$$

which are sequences of non-negative integers, only finitely many of which are non-zero. As by definition $0! := 1$, the expressions

$$|\alpha|_1 := \sum_{j=1}^{\infty} \alpha_j \qquad \text{and} \qquad \alpha! := \prod_{j=1}^{\infty} \alpha_j!$$

are well defined for $\alpha \in \mathcal{J}$.

B Hermite Polynomials

As there are different ways to define – and to normalise – the Hermite polynomials, a specific way has to be chosen. In applications with probability theory it seems most advantageous to use the following definition [17–19, 22, 28]:

$$h_k(t) := (-1)^k e^{t^2/2} \left(\frac{\mathrm{d}}{\mathrm{d}t}\right)^k e^{-t^2/2}; \quad \forall t \in \mathbb{R},\ k \in \mathbb{N}_0, \tag{50}$$

where the coefficient of the highest power of t – which is t^k for h_k – is equal to unity.

The first five polynomials are:

$$h_0(t) = 1, \qquad h_1(t) = t, \qquad h_2(t) = t^2 - 1,$$
$$h_3(t) = t^3 - 3t, \qquad h_4(t) = t^4 - 6t^2 + 3.$$

The recursion relation for these polynomials is

$$h_{k+1}(t) = t\, h_k(t) - k\, h_{k-1}(t); \quad k \in \mathbb{N}. \tag{51}$$

These are orthogonal polynomials w.r.t standard Gaussian probability measure Γ, where $\Gamma(\mathrm{d}t) = (2\pi)^{-1/2} e^{-t^2/2}\, \mathrm{d}t$ – the set $\{h_k(t)/\sqrt{k!} \mid k \in \mathbb{N}_0\}$ forms a complete orthonormal system (CONS) in $L_2(\mathbb{R}, \Gamma)$ – as the Hermite polynomials satisfy

$$\int_{-\infty}^{\infty} h_m(t)\, h_n(t)\, \Gamma(\mathrm{d}t) = n!\, \delta_{n,m}. \tag{52}$$

Multi-variate Hermite polynomials will be defined right away for an infinite number of variables, i.e. for $\boldsymbol{t} = (t_1, t_2, \ldots, t_j, \ldots) \in \mathbb{R}^{\mathbb{N}}$, the space of all sequences. For $\alpha = (\alpha_1, \ldots, \alpha_j, \ldots) \in \mathcal{J}$ remember that except for a finite number all other α_j are zero; hence in the definition of the multi-variate Hermite polynomial

$$H_\alpha(\boldsymbol{t}) := \prod_{j=1}^{\infty} h_{\alpha_j}(t_j); \quad \forall \boldsymbol{t} \in \mathbb{R}^{\mathbb{N}},\ \alpha \in \mathcal{J}, \tag{53}$$

except for finitely many factors all others are h_0, which equals unity, and the infinite product is really a finite one and well defined.

The space $\mathbb{R}^{\mathbb{N}}$ can be equipped with a Gaussian (product) measure [17–19,22,28], again denoted by Γ. Then the set $\{H_\alpha(\boldsymbol{t})/\sqrt{\alpha!} \mid \alpha \in \mathcal{J}\}$ is a CONS in $L_2(\mathbb{R}^{\mathbb{N}}, \Gamma)$ as the multivariate Hermite polynomials satisfy

$$\int_{\mathbb{R}^{\mathbb{N}}} H_\alpha(\boldsymbol{t})\, H_\beta(\boldsymbol{t})\, \Gamma(\mathrm{d}\boldsymbol{t}) = \alpha!\, \delta_{\alpha\beta}, \tag{54}$$

where the Kronecker symbol is extended to $\delta_{\alpha\beta} = 1$ in case $\alpha = \beta$ and zero otherwise.

C The Hermite Algebra

Consider first the usual univariate Hermite polynomials $\{h_k\}$ as defined in Appendix B, (50). As the univariate Hermite polynomials are a linear basis for the polynomial algebra, i.e. every polynomial can be written as linear combination of Hermite polynomials, this is also the case for the product of two Hermite polynomials $h_k h_\ell$, which is clearly also a polynomial:

$$h_k(t)h_\ell(t) = \sum_{n=|k-\ell|}^{k+\ell} c_{k\ell}^n h_n(t), \tag{55}$$

where n is an index, not an exponent. The coefficients are only non-zero [28] for integer $g = (k + \ell + n)/2 \in \mathbb{N}$ and if $g \geq k \wedge g \geq \ell \wedge g \geq n$. They can be explicitly given

$$c_{k\ell}^n = \frac{k!\,\ell!}{(g-k)!\,(g-\ell)!\,(g-n)!}, \tag{56}$$

and are called the structure constants of the univariate Hermite algebra.

For the multivariate Hermite algebra, analogous statements hold [28]:

$$H_\alpha(\boldsymbol{t})H_\beta(\boldsymbol{t}) = \sum_\gamma c_{\alpha\beta}^\gamma H_\gamma(\boldsymbol{t}). \tag{57}$$

with the multivariate structure constants

$$c_{\alpha\beta}^\gamma = \prod_{j=1}^\infty c_{\alpha_j\beta_j}^{\gamma_j}, \tag{58}$$

defined in terms of the univariate structure constants (56).

From this it is easy to see that

$$\mathbb{E}\left(H_\alpha H_\beta H_\gamma\right) = \mathbb{E}\left(H_\gamma \sum_\varepsilon c_{\alpha\beta}^\varepsilon H_\varepsilon\right) = c_{\alpha\beta}^\gamma \gamma!. \tag{59}$$

Products of more than two Hermite polynomials may be computed recursively, we here look at triple products as an example, using (57):

$$H_\alpha H_\beta H_\delta = \left(\sum_\gamma c_{\alpha\beta}^\gamma H_\gamma\right) H_\delta = \sum_\varepsilon \left(\sum_\gamma c_{\gamma\delta}^\varepsilon c_{\alpha\beta}^\gamma\right) H_\varepsilon. \tag{60}$$

D The Hermite Transform

A variant of the Hermite transform maps a random variable onto the set of expansion coefficients of the PCE [18]. Any random variable which may be represented with a PCE

$$\mathsf{r}(\omega) = \sum_{\alpha\in\mathcal{J}} \varrho^\alpha H_\alpha(\boldsymbol{\theta}(\omega)), \tag{61}$$

is mapped onto

$$\mathscr{H}(\mathsf{r}) := (\varrho^{\alpha})_{\alpha \in \mathcal{J}} = (\varrho) \in \mathbb{R}^{\mathcal{J}}. \tag{62}$$

On the other hand, from a sequence indexed by $\mathcal{J}$, as a mapping $\rho : \mathcal{J} \to \mathbb{R} : \alpha \mapsto \rho^{\alpha}$, one may obtain the random variable

$$\mathscr{H}^{-1}((\rho)) = \mathscr{H}^{-1}((\rho^{\alpha})_{\alpha \in \mathcal{J}}) := \sum_{\alpha \in \mathcal{J}} \rho^{\alpha} H_{\alpha}, \tag{63}$$

which defines the inverse Hermite transform.

These sequences may be seen also as the coefficients of power series in infinitely many complex variables $z \in \mathbb{C}^{\mathbb{N}}$, namely by

$$\sum_{\alpha \in \mathcal{J}} \varrho^{\alpha} z^{\alpha},$$

where $z^{\alpha} := \prod_{j} z_j^{\alpha_j}$. This is the original definition of the Hermite transform [18].

It can be used to easily compute the Hermite transform of the ordinary product like in (57), as

$$\mathscr{H}(H_{\alpha} H_{\beta}) = (c_{\alpha\beta}^{\gamma})_{\gamma \in \mathcal{J}}. \tag{64}$$

With the structure constants (58) one defines the matrices $\boldsymbol{C}_2^{\gamma} := (c_{\alpha\beta}^{\gamma})$ with indices α and β. The Hermite transform of the product of two random variables $\mathsf{r}_1(\omega) = \sum_{\alpha \in \mathcal{J}} \varrho_1^{\alpha} H_{\alpha}(\theta)$ and $\mathsf{r}_2(\omega) = \sum_{\beta \in \mathcal{J}} \varrho_2^{\beta} H_{\beta}(\boldsymbol{\theta})$ is hence

$$\mathscr{H}(\mathsf{r}_1 \mathsf{r}_2) = \left((\varrho_1) \boldsymbol{C}_2^{\gamma} (\varrho_2)^T) \right)_{\gamma \in \mathcal{J}} \tag{65}$$

Each coefficient is a bilinear form in the coefficient sequences of the factors, and the collection of all those bilinear forms $\mathbf{C}_2 = (\boldsymbol{C}_2^{\gamma})_{\gamma \in \mathcal{J}}$ is a bilinear mapping that maps the coefficient sequences of r_1 and r_2 into the coefficient sequence of the product

$$\mathscr{H}(\mathsf{r}_1 \mathsf{r}_2) =: \mathbf{C}_2((\varrho_1), (\varrho_2)) = \mathbf{C}_2 \left(\mathscr{H}(\mathsf{r}_1), \mathscr{H}(\mathsf{r}_2) \right). \tag{66}$$

Products of more than two random variables may now be defined recursively through the use of associativity. e.g. $\mathsf{r}_1 \mathsf{r}_2 \mathsf{r}_3 \mathsf{r}_4 = (((\mathsf{r}_1 \mathsf{r}_2) \mathsf{r}_3) \mathsf{r}_4)$:

$$\begin{aligned} \forall k > 2: \quad \mathscr{H}\left(\prod_{j=1}^{k} \mathsf{r}_j \right) &:= \mathbf{C}_k((\varrho_1), (\varrho_2), \ldots, (\varrho_k)) \\ &:= \mathbf{C}_{k-1}(\mathbf{C}_2((\varrho_1), (\varrho_2)), (\varrho_3) \ldots, (\varrho_k)). \end{aligned} \tag{67}$$

Each $\mathbf{C}_k$ is again composed of a sequence of k-linear forms $\{\boldsymbol{C}_k^{\gamma}\}_{\gamma \in \mathcal{J}}$, which define each coefficient of the Hermite transform of the k-fold product.

References

1. Acharjee, S., Zabaras, N.: A non-intrusive stochastic Galerkin approach for modeling uncertainty propagation in deformation processes. Comput. Struct. **85**, 244–254 (2007)
2. Adler, R.J.: The Geometry of Random Fields. Wiley, Chichester (1981)
3. Adler, R.J., Taylor, J.E.: Random Fields and Geometry. Springer, Berlin (2007)
4. Babuška, I., Tempone, R., Zouraris, G.E.: Galerkin finite element approximations of stochastic elliptic partial differential equations. SIAM J. Num. Anal. **42**, 800–825 (2004)
5. Babuška, I., Tempone, R., Zouraris, G.E.: Solving elliptic boundary value problems with uncertain coefficients by the finite element method: the stochastic formulation. Comp. Meth. Appl. Mech. Engrg. **194**, 1251–1294 (2005)
6. Babuška, I., Nobile, F., Tempone, R.: A stochastic collocation method for elliptic partial differential equations with random input data. SIAM J. Num. Anal. **45**, 1005–1034 (2007)
7. Caflisch, R.E.: Monte Carlo and quasi-Monte Carlo methods. Acta Numerica **7**, 1–49 (1998)
8. Cao, Y.: On the rate of convergence of Wiener-Ito expansion for generalized random variables. Stochastics **78**, 179–187 (2006)
9. Christakos, G.: Random Field Models in Earth Sciences. Academic Press, San Diego, CA (1992)
10. Ciarlet, P.G.: The Finite Element Method for Elliptic Problems. North-Holland, Amsterdam (1978)
11. Da Prato, G., Zabczyk, J.: Stochastic Equations in Infinite Dimensions. Cambridge University Press, Cambridge (1992)
12. Dennis J.E. Jr., Schnabel, R.B.: Numerical methods for unconstrained optimization and nonlinear equations. Classics in applied mathematics. SIAM, Philadelphia, PA (1996)
13. Frauenfelder, Ph., Schwab, Chr., Todor, R.A.: Finite elements for elliptic problems with stochastic coefficients. Comp. Meth. Appl. Mech. Engrg. **194**, 205–228, (2005)
14. Gerstner, T., Griebel, M.: Numerical integration using sparse grids. Numer. Algorithms **18**, 209–232 (1998)
15. Ghanem, R., Spanos, P.D.: Stochastic Finite Elements – A Spectral Approach. Springer-Verlag, Berlin (1991)
16. Ghanem, R.: Stochastic finite elements for heterogeneous media with multiple random non-Gaussian properties. ASCE J. Engrg. Mech. **125**, 24–40 (1999)
17. Hida, T., Kuo, H.-H., Potthoff, J., Streit, L.: White Noise – An Infinite Dimensional Calculus. Kluwer, Dordrecht (1993)
18. Holden, H., Øksendal, B., Ubøe, J., Zhang, T.-S.: Stochastic Partial Differential Equations. Birkhäuser Verlag, Basel (1996)
19. Janson, S.: Gaussian Hilbert Spaces. Cambridge University Press, Cambridge (1997)
20. Jardak, M., Su, C.-H., Karniadakis, G.E.: Spectral polynomial chaos solutions of the stochastic advection equation. SIAM J. Sci. Comput. **17**, 319–338 (2002)
21. Jeggle, H.: Nichtlineare Funktionalanalysis. Teubner, Stuttgart (1979)
22. Kallianpur, G.: Stochastic Filtering Theory. Springer-Verlag, Berlin (1980)
23. Karniadakis, G.E., Sue, C.-H., Xiu, D., Lucor, D., Schwab, C., Tudor, R.A.: Generalized polynomial chaos solution for differential equations with random input. Research Report 2005-1, SAM, ETH Zürich, Zürich (2005)
24. Keese, A.: A review of recent developments in the numerical solution of stochastic PDEs (stochastic finite elements). Informatikbericht 2003-6, Institute of Scientific Computing, Department of Mathematics and Computer Science, Technische Universitt Braunschweig, Brunswick (2003) http://opus.tu-bs.de/opus/volltexte/2003/504/
25. Krée, P., Soize, C.: Mathematics of Random Phenomena. D. Reidel, Dordrecht (1986)
26. Le Maître, O.P., Najm, H.N., Ghanem, R.G., Knio, O.M.: Multi-resolution analysis of Wiener-type uncertainty propagation schemes. J. Comp. Phys. **197**, 502–531 (2004)

27. Lions, P.-L., Souganidis, P.E.: Fully nonlinear stochastic partial differential equations. C. R. Acad. Sci. Paris, Série I. **326**, 1085–1092 (1998)
28. Malliavin, P.: Stochastic Analysis. Springer, Berlin (1997)
29. Matthies, H., Strang, G.: The solution of nonlinear finite element equations. Int. J. Numer. Methods Engrg. **14**, 1613–1626 (1979)
30. Matthies, H.G., Brenner, C.E., Bucher, C.G., Guedes Soares, C.: Uncertainties in probabilistic numerical analysis of structures and solids – stochastic finite elements. Struct. Safety **19**, 283–336 (1997)
31. Matthies, H.G., Bucher, C.G.: Finite elements for stochastic media problems. Comp. Meth. Appl. Mech. Engrg. **168**, 3–17 (1999)
32. Matthies, H.G., Keese, A.: Galerkin methods for linear and nonlinear elliptic stochastic partial differential equations. Comp. Meth. Appl. Mech. Engrg. **194**, 1295–1331, (2005)
33. Matthies, H.G.: Quantifying Uncertainty: Modern Computational Representation of Probability and Applications. In: A. Ibrahimbegović and I. Kožar (eds.), Extreme Man-Made and Natural Hazards in Dynamics of Structures. NATO-ARW series. Springer, Berlin (2007)
34. Matthies, H.G.: Stochastic Finite Elements: Computational Approaches to Stochastic Partial Differential equations. Z. Angew. Math. Mech. **88**, 849–873 (2008)
35. Matthies, H.G., Zander, E.: Solving stochastic systems with low-rank tensor compression. Submitted to Linear Algebra and its Applications (2009)
36. Nobile, F., Tempone, R., Webster, C.G.: Sparse grid stochastic collocation method for elliptic partial differential equations with random input data. SIAM J. Numer. Anal. **46**, 2309–2345 (2008)
37. Nobile, F., Tempone, R., Webster, C.G.: An anisotropic sparse grid stochastic collocation method for partial differential equations with random input data. SIAM J. Numer. Anal. **46**, 2411–2442 (2008)
38. Novak, E., Ritter, K.: The curse of dimension and a universal method for numerical integration. In: Nürnberger, G., Schmidt, J.W., Walz, G. (eds.) Multivariate Approximation and Splines, ISNM. pp. 177–188. Birkhäuser Verlag, Basel (1997)
39. Oden, J.T.: Qualitative Methods in Nonlinear Mechanics. Prentice-Hall, Englewood Cliffs, NJ (1986)
40. Petras, K.: Fast calculation of coefficients in the Smolyak algorithm. Numer. Algorithms **26**, 93–109 (2001)
41. Roman, L.J., Sarkis, M.: Stochastic Galerkin Method for Elliptic SPDEs: A White Noise Approach. Discrete and Continuous Dynamical Systems – Series B **6**, 941–955 (2006)
42. Rozanov, Yu.: Random Fields and Stochastic Partial Differential Equations. Kluwer, Dordrecht (1996)
43. Schuëller, G.I.: A state-of-the-art report on computational stochastic mechanics. Prob. Engrg. Mech. **14**, 197–321 (1997)
44. Schuëller, G.I.: Recent developments in structural computational stochastic mechanics. In: Topping, B.H.V. (eds.) Computational Mechanics for the Twenty-First Century. pp. 281–310. Saxe-Coburg Publications, Edinburgh (2000)
45. Schuëller, G.I., Spanos, P.D. (ed.): Monte Carlo Simulation. Balkema, Rotterdam (2001)
46. Segal, I.E., Kunze, R.A.: Integrals and Operators. Springer, Berlin (1978)
47. Strang, G., Fix, G.J.: An Analysis of the Finite Element Method. Wellesley-Cambridge Press, Wellesley, MA (1988)
48. Sudret, B., Der Kiureghian, A.: Stochastic finite element methods and reliability. A state-of-the-art report. Report UCB/SEMM-2000/08, Department of Civil & Environmental Engineering, University of California, Berkeley, CA (2000)
49. Ullmann, E.: A Kronecker product preconditioner for stochastic Galerkin finite element discretizations. SIAM J. Sci. Comput. **32**, 923–946 (2010)
50. Vanmarcke, E.: Random Fields: Analysis and Synthesis. The MIT Press, Cambridge, MA (1988)
51. Wan, X., Karniadakis, G.E.: An adaptive multi-element generalized polynomial chaos method for stochastic differential equations. J. Comp. Phys. **209**, 617–642 (2005)

52. Xiu, D., Karniadakis, G.E.: Modeling uncertainty in steady state diffusion problems via generalized polynomial chaos. Comp. Meth. Appl. Mech. Engrg. **191**, 4927–4948 (2002)
53. Xu, X.F.: A multiscale stochastic finite element method on elliptic problems involving uncertainties. Comp. Meth. Appl. Mech. Engrg. **196**, 2723–2736 (2007)
54. Zander, E., Matthies, H.G.: Tensor product methods for stochastic problems. Proc. Appl. Math. Mech. **7**, 2040067–2040068 (2008)

Smoothing Analysis of an All-at-Once Multigrid Approach for Optimal Control Problems Using Symbolic Computation

Stefan Takacs and Veronika Pillwein

Abstract The numerical treatment of systems of partial differential equations (PDEs) is of great interest as many problems from applications, including the optimality system of optimal control problems that is discussed here, belong to this class. These problems are not elliptic and therefore both the construction of an efficient numerical solver and its analysis are hard. In this work we will use all-at-once multigrid methods as solvers. For sake of simplicity, we will only analyze the smoothing properties of a well-known smoother.

Local Fourier analysis (or local mode analysis) is a widely-used tool to analyze numerical methods for solving discretized systems of PDEs which has also been used in particular to analyze multigrid methods. The rates that can be computed with local Fourier analysis are typically the supremum of some rational function. In several publications this supremum was merely approximated numerically by interpolation. We show that it can be resolved exactly using cylindrical algebraic decomposition which is a well established method in symbolic computation.

S. Takacs (✉)
Doctoral Program Computational Mathematics, Johannes Kepler University, Altenbergerstr. 69, 4040 Linz, Austria
e-mail: stefan.takacs@dk-compmath.jku.at

V. Pillwein
Research Institute for Symbolic Computation, Johannes Kepler University, Altenbergerstr. 69, 4040 Linz, Austria
e-mail: veronika.pillwein@risc.jku.at

U. Langer and P. Paule (eds.), *Numerical and Symbolic Scientific Computing*, Texts and Monographs in Symbolic Computation, DOI 10.1007/978-3-7091-0794-2_9,

1 Introduction

Local Fourier analysis (or local mode analysis) is a commonly used approach for designing and analyzing convergence properties of multigrid methods. In the late 1970s A. Brandt proposed to use Fourier series to analyze multigrid methods, see, e.g., [2]. Local Fourier analysis provides a framework to analyze various numerical methods with a unified approach that gives quantitative statements on the methods under investigation, i.e., it leads to the determination of sharp convergence rates. Other work on multigrid theory such as [1, 6, 9, 10] – to mention only a few – typically just show convergence and do not give sharp or realistic bounds for convergence rates.

Local Fourier analysis can be justified rigorously only in special cases, e.g., on rectangular domains with uniform grids and periodic boundary conditions. However, results obtained with local Fourier analysis can be carried over to more general problems, see, e.g. [3]. In this sense it can be viewed as heuristic approach for a wide class of applications.

Understanding local Fourier analysis as a machinery for analyzing a multigrid method, we apply it in this paper to a model problem and some specific solvers. Still, we keep in mind that this analysis can be carried over to a variety of other problems and solvers. This type of generalization has been carried out, e.g., for methods to solve optimal control problems that have been discussed in [14], [1] and [9]. In [16] the method is explained as machinery and a local Fourier analysis software *LFA* is presented. This software can be configured using a graphical user interface and allows to approximate (numerically) smoothing and convergence rates based on local Fourier analysis approaches for various problems and multigrid approaches.

Neither the proposed multigrid method nor its application to the optimal control problem discussed in this paper are new and numerical results have been published in various papers. The proposed smoother belongs to the class of Vanka smoothers [15]. In [1] this smoother was used in a finite difference framework and – beside a second kind of analysis – local Fourier analysis was used to analyze the method. In that paper the condition characterizing the smoothing rate was approximated numerically.

The goal of this paper is to show that the analysis can be carried out in an *entirely symbolic way* and as such leads to *sharp estimates* on the smoothing rate for a collective Jacobi relaxation and collective Gauss-Seidel iteration scheme. For this purpose we restrict ourselves to the case of a one-dimensional model problem and to piecewise linear ansatz functions (Courant elements). Aiming at an audience from both numerical and symbolic mathematics we try to stay at an elementary level and keep this note self-contained.

The key for involving symbolic algorithms is a proper reformulation of the quantities to be analyzed in terms of logical formulas on polynomial inequalities. These real formulas can then be simplified by means of quantifier elimination using cylindrical algebraic decomposition. This tool has been applied earlier in the analysis of (systems of) ordinary and partial differential-difference equations [8] where the necessary conditions for stability, asymptotic stability and well-posedness of the given systems were transformed into statements on polynomial inequalities using Fourier or Laplace transforms.

This paper is organized as follows. In subsection 1.1 we introduce a simple model problem and in subsection 1.2 we propose a multigrid approach to solve the discretized optimality system of this model problem. The local Fourier analysis is introduced and carried out in Sects. 2 and 4, respectively. Section 3 gives a brief overview on quantifier elimination and cylindrical algebraic decomposition, i.e., on the symbolic methods applied in order to resolve smoothing rates symbolically.

1.1 Model Problem

As a *model problem* we consider the following optimal control problem of tracking type: Minimize

$$J(y,u) := \frac{1}{2}\|y - y_D\|^2_{L^2(\Omega)} + \frac{\alpha}{2}\|u\|^2_{L^2(\Omega)},$$

subject to the elliptic boundary value problem (BVP)

$$-\,\Delta y = u \text{ in } \Omega \qquad \text{and} \qquad y = 0 \text{ on } \partial\Omega, \tag{1}$$

where $y \in H_0^1(\Omega)$ is the state variable and $u \in L^2(\Omega)$ is the control variable. The function $y_D \in L^2(\Omega)$ is given, $\alpha > 0$ is some fixed regularization or cost parameter and Ω is a given domain with boundary $\partial\Omega$. Here, the Banach space $L^2(\Omega)$ is the set of square integrable functions on Ω with associated standard norm $\|f\|_{L^2(\Omega)} := (f,f)^{1/2}_{L^2(\Omega)}$, where $(f,g)_{L^2(\Omega)} := \int_\Omega f(x)\, g(x)\, \mathrm{d}x$. The Sobolev space $H_0^1(\Omega)$ is the set of L^2-functions vanishing on the boundary with weak derivatives in $L^2(\Omega)$.

Note that for this setting the boundary value problem is (in weak sense) uniquely solvable in y for every given control u. At first we rewrite the BVP (1) in variational form: Find $y \in H_0^1(\Omega)$ such that

$$(\nabla y, \nabla p)_{L^2(\Omega)} = (u, p)_{L^2(\Omega)}$$

holds for all $p \in H_0^1(\Omega)$, where ∇^T is the weak gradient.

Solving the model problem is equivalent to finding a saddle point of the Lagrange functional which leads to the first order optimality conditions (the Karush-Kuhn-Tucker system or, in short, *KKT system*), given by: Find $(y,u,p) \in H_0^1(\Omega) \times L^2(\Omega) \times H_0^1(\Omega)$ such that

$$\begin{aligned}
&(y,\widetilde{y})_{L^2(\Omega)} && && +\,(\nabla p,\nabla\widetilde{y})_{L^2(\Omega)} &&= (y_D,\widetilde{y})_{L^2(\Omega)}\\
& && \alpha\,(u,\widetilde{u})_{L^2(\Omega)} && -\,(p,\widetilde{u})_{L^2(\Omega)} &&= 0\\
&(\nabla y,\nabla\widetilde{p})_{L^2(\Omega)} - && (u,\widetilde{p})_{L^2(\Omega)} && &&= 0
\end{aligned}$$

holds for all $(\widetilde{y}, \widetilde{u}, \widetilde{p}) \in H_0^1(\Omega) \times L^2(\Omega) \times H_0^1(\Omega)$. The second equation immediately implies that $u = \alpha^{-1} p$, which allows a reduction to the following system (2 × 2 *formulation of the KKT system*): Find $(y, p) \in X := V \times V := H_0^1(\Omega) \times H_0^1(\Omega)$ such that

$$\begin{aligned}(y, \widetilde{y})_{L^2(\Omega)} \quad &+ (\nabla p, \nabla \widetilde{y})_{L^2(\Omega)} = (y_D, \widetilde{y})_{L^2(\Omega)} \\ (\nabla y, \nabla \widetilde{p})_{L^2(\Omega)} &- \alpha^{-1}(y, \widetilde{p})_{L^2(\Omega)} = 0\end{aligned}$$

holds for all $(\widetilde{y}, \widetilde{p}) \in X$.

For finding the approximate solution to this problem we use finite element methods (FEM). Therefore we assume to have a sequence of grids partitioning the given domain Ω starting from an initial (coarse) grid on grid level $k = 0$. The grids on grid levels $k = 1, 2, \dots$ are constructed by refinement, i.e., the grid points of level $k - 1$ are also grid points of level k. Using standard finite element techniques (Courant elements), we can construct finite dimensional subsets $V_k \subset V$, where the dimension N_k depends on the grid level k. By Galerkin principle, the finite element approximation $(y_k, p_k) \in X_k := V_k \times V_k$ fulfills

$$\begin{aligned}(y_k, \widetilde{y}_k)_{L^2(\Omega)} \quad &+ (\nabla p_k, \nabla \widetilde{y}_k)_{L^2(\Omega)} = (y_D, \widetilde{y}_k)_{L^2(\Omega)} \\ (\nabla y_k, \nabla \widetilde{p}_k)_{L^2(\Omega)} &- \alpha^{-1}(y_k, \widetilde{p}_k)_{L^2(\Omega)} = 0\end{aligned} \tag{2}$$

for all $(\widetilde{y}_k, \widetilde{p}_k) \in X_k$.

Assuming to have a (nodal) basis $\Phi_k := (\varphi_{k,i})_{i=1}^{N_k}$ for V_k, we can rewrite the optimality system (2) in matrix-vector notation as follows:

$$\underbrace{\begin{pmatrix} M_k & K_k \\ K_k & -\alpha^{-1} M_k \end{pmatrix}}_{\mathcal{A}_k :=} \underbrace{\begin{pmatrix} \underline{y}_k \\ \underline{p}_k \end{pmatrix}}_{\underline{x}_k :=} = \underbrace{\begin{pmatrix} \underline{g}_k \\ 0 \end{pmatrix}}_{\underline{f}_k :=}, \tag{3}$$

where the *mass matrix* M_k and the *stiffness matrix* K_k are given by

$$M_k := ((\varphi_{k,j}, \varphi_{k,i})_{L^2(\Omega)})_{i,j=1}^{N_k} \quad \text{and} \quad K_k := ((\nabla \varphi_{k,j}, \nabla \varphi_{k,i})_{L^2(\Omega)})_{i,j=1}^{N_k},$$

respectively, and the right hand side vector $\underline{g}_k$ is given by

$$\underline{g}_k := ((y_D, \varphi_{k,i})_{L^2(\Omega)})_{i=1}^{N_k}.$$

The symbols $\underline{y}_k$ and $\underline{p}_k$ denote the coordinate vectors of the corresponding functions y_k and p_k with respect to the nodal basis Φ_k.

1.2 Multigrid Methods and Collective Point Smoothers

In this section we briefly introduce the multigrid framework that we analyze in this paper. Starting from an initial approximation $\underline{x}_k^{(0)}$ one step of the multigrid method with $\nu_1 + \nu_2$ smoothing steps for solving a discretized equation $\mathscr{A}_k\ \underline{x}_k = \underline{f}_k$ on grid level k is given by:

- Apply ν_1 (pre-)smoothing steps

$$\underline{x}_k^{(0,m)} := \underline{x}_k^{(0,m-1)} + \tau \hat{\mathscr{A}}_k^{-1}(\underline{f}_k - \mathscr{A}_k\ \underline{x}_k^{(0,m-1)}) \qquad \text{for } m = 1, \dots, \nu_1 \tag{4}$$

  with $\underline{x}_k^{(0,0)} := \underline{x}_k^{(0)}$, where the choice of the damping parameter τ and the preconditioner $\hat{\mathscr{A}}_k$ is discussed below.

- Apply coarse-grid correction
  - Compute the defect and restrict to the coarser grid
  - Solve the problem on the coarser grid
  - Prolongate and add the result

  If the problem on the coarser grid is solved exactly (two-grid method), then we obtain

$$\underline{x}_k^{(1,-\nu_2)} := \underline{x}_k^{(0,\nu_1)} + I_{k-1}^k \mathscr{A}_{k-1}^{-1} I_k^{k-1}(\underline{f}_k - \mathscr{A}_k\ \underline{x}_k^{(0,\nu_1)}).$$

- Apply ν_2 (post-)smoothing steps

$$\underline{x}_k^{(1,m)} := \underline{x}_k^{(1,m-1)} + \tau \hat{\mathscr{A}}_k^{-1}(\underline{f}_k - \mathscr{A}_k\ \underline{x}_k^{(1,m-1)}) \qquad \text{for } m = -\nu_2 + 1, \dots, 0$$

  to obtain the next iterate $\underline{x}_k^{(1)} := \underline{x}_k^{(1,0)}$.

The smoothing steps are applied in order to reduce the high-frequency error, whereas the coarse-grid correction takes care of the low-frequency parts of the overall defect. In practice the problem on grid level $k - 1$ is handled by applying one step (V-cycle) or two steps (W-cycle) of the proposed method, recursively, and just on the coarse grid level $k = 0$ the problem is solved exactly. The convergence of the two-grid method implies the convergence of the W-cycle multigrid method under mild assumptions, so we restrict ourselves to the analysis of the two-grid method only.

The intergrid-transfer operators I_{k-1}^k and I_k^{k-1} are chosen in a canonical way: we use the canonical embedding for the prolongation operator I_{k-1}^k and its adjoint as restriction operator I_k^{k-1}.

Next we need to specify the smoothing procedure (4). The preconditioning matrix $\hat{\mathscr{A}}_k$ is typically some easy-to-invert approximation of the matrix $\mathscr{A}_k$. In case of positive definite matrices (which may result from discretizing elliptic scalar BVPs), the preconditioning matrix can be composed in an either additive or multiplicative Schwarz manner based on local problems which live on patches, boxes or, which is the easiest case, just on points. The two main pointwise methods are Jacobi relaxation (additive Schwarz method) and Gauss-Seidel iteration (multiplicative Schwarz method).

We extend these methods to block-systems by combining (again in an additive or multiplicative Schwarz manner) local problems which involve the complete system of BVPs. Therefore we first recall that standard Jacobi relaxation, which can be used as smoother for the linear system $A_k\, \underline{x}_k = \underline{f}_k$, where A_k is symmetric and positive definite, reads as

$$x_i^{(0,m+1)} := x_i^{(0,m)} + \tau\, a_{ii}^{-1} \left(f_i - \sum_{j=1}^{N_k} a_{ij}\, x_j^{(0,m)} \right),$$

where $x_i^{(0,m)}$, f_i and a_{ij} are the components of the vectors $\underline{x}_k^{(0,m)}$ and $\underline{f}_k$ and the matrix $\mathscr{A}_k$, respectively. This iteration scheme can be carried over to the saddle point problem (3), which leads to the collective Jacobi relaxation:

$$\mathbf{x}_i^{(0,m+1)} := \mathbf{x}_i^{(0,m)} + \tau\, \mathscr{A}_{ii}^{-1} \left(\mathbf{f}_i - \sum_{j=1}^{N_k} \mathscr{A}_{ij}\, \mathbf{x}_j^{(0,m)} \right), \tag{5}$$

where

$$\mathbf{x}_i^{(0,m)} := (y_i^{(0,m)}, p_i^{(0,m)})^T, \quad \mathbf{f}_i := (g_i, 0)^T \quad \text{and} \quad \mathscr{A}_{ij} := \begin{pmatrix} m_{ij} & k_{ij} \\ k_{ij} & -\alpha^{-1} m_{ij} \end{pmatrix}. \tag{6}$$

Here again $y_i^{(0,m)}$, $p_i^{(0,m)}$, g_i, m_{ij} and k_{ij} are the components of $\underline{y}_k^{(0,m)}$, $\underline{p}_k^{(0,m)}$, $\underline{g}_k$, M_k and K_k.

Collective iteration schemes can be represented in the compact notation (4) using the preconditioning matrix $\hat{\mathscr{A}}_k^{(jac)}$, given by

$$\hat{\mathscr{A}}_k^{(jac)} := \begin{pmatrix} \hat{M}_k^{(jac)} & \hat{K}_k^{(jac)} \\ \hat{K}_k^{(jac)} & -\alpha^{-1} \hat{M}_k^{(jac)} \end{pmatrix} := \begin{pmatrix} \text{diag } M_k & \text{diag } K_k \\ \text{diag } K_k & -\alpha^{-1} \text{diag } M_k \end{pmatrix},$$

i.e., $\hat{M}_k^{(jac)}$ and $\hat{K}_k^{(jac)}$ are defined as the diagonals of M_k and K_k, respectively. The damping parameter τ is chosen to be in (0, 1).

The preconditioning matrix of the *collective Gauss-Seidel iteration* can be constructed in the same way and is given by

$$\hat{\mathscr{A}}_k^{(gs)} := \begin{pmatrix} \hat{M}_k^{(gs)} & \hat{K}_k^{(gs)} \\ \hat{K}_k^{(gs)} & -\alpha^{-1} \hat{M}_k^{(gs)} \end{pmatrix},$$

where $\hat{M}_k^{(gs)}$ and $\hat{K}_k^{(gs)}$ are the lower left triangular parts (including the diagonals) of M_k and K_k, respectively and the damping parameter τ is chosen to be 1. We should mention that in both cases the preconditioning matrices $\hat{M}_k^{(jac)}$, $\hat{K}_k^{(jac)}$, $\hat{M}_k^{(gs)}$ and $\hat{K}_k^{(gs)}$ are the preconditioning matrices that represent classical Jacobi or Gauss-Seidel iteration for linear systems with system matrices M_k and K_k, respectively.

Numerical examples show good behavior of multigrid methods using such iterations as smoothing procedures and have been discussed in, e.g., [1] or [9].

We want to stress that in either case the application of the preconditioning matrix $\hat{\mathscr{A}}_k$ can be realized efficiently if the iteration is implemented analogously to standard Jacobi or Gauss-Seidel iteration, as done in (5) and (6) for the case of collective Jacobi iteration. Executing the algorithm then only vectors in $\mathbb{R}^2$ need to be multiplied with 2×2 matrices and 2×2 linear systems need to be solved. For more detailed information on how to implement collective iteration schemes see, e.g., [9].

2 Local Fourier Analysis

Convergence properties of multigrid methods for the model problem have been investigated in a wide range of papers [1, 6, 9, 10]. In this paper we want to concentrate on an analysis, where symbolic computation can contribute significantly. For the time being, we complete the first step by analyzing the smoothing iteration. As mentioned earlier, we restrict the smoothing analysis to the case of a one dimensional domain Ω. While the proposed numerical method can be applied also to higher dimensions, the analysis of this case as well as the analysis of the full two-grid cycle is ongoing work.

For the analysis of the smoothing procedure introduced in the last section, we define the *iteration matrix* of the smoothing step by

$$S_k := I - \tau \hat{\mathscr{A}}_k^{-1} \mathscr{A}_k,$$

which represents the modification of the error effected by the smoothing procedure, i.e.,

$$\underline{x}_k^{(0,m)} - \underline{x}_k^* = S_k(\underline{x}_k^{(0,m-1)} - \underline{x}_k^*),$$

where $\underline{x}_k^* := \mathscr{A}_k^{-1} \underline{f}_k$ denotes the exact solution to the system.

Certainly, if it can be shown that the spectral radius of the iteration matrix or, even better, its norm, are smaller than 1, then this yields convergence of the iterative scheme. At present time, we do not aim at proving convergence, but at showing that it is a *good smoother*. In other words, we want to show that it reduces the high frequency error terms which we do using local Fourier analysis that is introduced next.

2.1 Local Fourier Analysis Framework

Since for local Fourier analysis the boundary is neglected by assuming periodic boundary conditions that allow to extend a bounded domain Ω to the entire space $\mathbb{R}$, see [3], from now on we assume that $\Omega = \mathbb{R}$. Let us repeat that good convergence and smoothing rates computed using local Fourier analysis for simple cases, typically also indicate good behavior of the analyzed methods in more general cases.

On this domain $\Omega = \mathbb{R}$, we assume to have on each grid level $k = 0, 1, 2, \ldots$ a uniform grid with nodes

$$x_{k,n} := n\, h_k \qquad \text{for } n \in \mathbb{Z},$$

where the uniform grid size is given by $h_k = 2^{-k}$. The functions in V_k are assumed to be continuous on the domain and to be linear between two nodes (Courant elements). This way the discretized function can be specified by prescribing the values on the nodes only.

The first step of local Fourier analysis consists of constructing *Fourier vectors* that diagonalize both mass and stiffness matrix. For every $\theta \in \Theta := [-\pi, \pi)$ and every grid level k, we can define a Fourier vector $\underline{\varphi_k}(\theta) \in \mathbb{R}^{\mathbb{Z}}$ as follows:

$$\underline{\varphi_k(\theta)} := (\mathrm{e}^{\mathrm{i}\theta x_{k,n}/h_k})_{n\in\mathbb{Z}}.$$

It is easy to see that every vector in $\mathbb{R}^{\mathbb{Z}}$ can be expressed as linear combination of countable infinitely many Fourier vectors. In case of a bounded domain, just finitely many Fourier vectors or functions would be necessary. Nonetheless for the analysis, all $\theta \in \Theta = [-\pi, \pi)$ are considered.

2.2 Operators in Local Fourier Space

The Fourier vectors defined in the preceding subsection diagonalize the blocks of our system matrix. Since we consider grids with uniform mesh-size h_k, the (infinitely large) mass matrix M_k and the (infinitely large) stiffness matrix K_k can be computed explicitly:

$$M_k = \frac{h_k}{6}\begin{pmatrix} \ddots & \ddots & \ddots & & \\ & 1 & 4 & 1 & \\ & & 1 & 4 & 1 \\ & & & \ddots & \ddots & \ddots \end{pmatrix} \quad \text{and} \quad K_k = \frac{1}{h_k}\begin{pmatrix} \ddots & \ddots & \ddots & & \\ & -1 & 2 & -1 & \\ & & -1 & 2 & -1 \\ & & & \ddots & \ddots & \ddots \end{pmatrix}.$$

It is easy to see that the multiplication of one of these matrices with the vector $\underline{\varphi_k(\theta)}$ equals the multiplication of this vector with the symbol of the matrix, where the symbols are given by:

$$M_k\ \underline{\varphi_k(\theta)} = \underbrace{\frac{(4+e^{i\theta}+e^{-i\theta})h_k}{6}}_{\overline{M_k}(\theta):=}\ \underline{\varphi_k(\theta)} \quad \text{and} \quad K_k\ \underline{\varphi_k(\theta)} = \underbrace{\frac{2-e^{i\theta}-e^{-i\theta}}{h_k}}_{\overline{K_k}(\theta):=}\ \underline{\varphi_k(\theta)}.$$

Thus indeed the Fourier vectors are the eigenvectors with eigenvalues $\overline{M_k}(\theta)$ and $\overline{K_k}(\theta)$, respectively, and analogously for the preconditioning matrices. For the collective Jacobi relaxation, the preconditioning matrix itself is a diagonal matrix, therefore

$$\hat{M}_k^{(jac)}\ \underline{\varphi_k(\theta)} = \underbrace{\frac{2h_k}{3}}_{\overline{\hat{M}_k^{(jac)}}(\theta):=}\ \underline{\varphi_k(\theta)} \quad \text{and} \quad \hat{K}_k^{(jac)}\ \underline{\varphi_k(\theta)} = \underbrace{\frac{2}{h_k}}_{\overline{\hat{K}_k^{(jac)}}(\theta):=}\ \underline{\varphi_k(\theta)}$$

holds. As the block matrices $\mathscr{A}_k$ and $\hat{\mathscr{A}_k}$ are built from such matrices, we can conclude that for all $\theta \in \Theta$,

$$\text{span}\left\{\begin{pmatrix} \underline{\varphi_k(\theta)} \\ \mathbf{0} \end{pmatrix}, \begin{pmatrix} \mathbf{0} \\ \underline{\varphi_k(\theta)} \end{pmatrix}\right\}$$

is invariant under the action of those block-matrices. Hence it suffices to consider only the symbol of the block matrix $\mathscr{A}_k$, given by

$$\overline{\mathscr{A}_k}(\theta) := \begin{pmatrix} \overline{M_k}(\theta) & \overline{K_k}(\theta) \\ \overline{K_k}(\theta) & -\alpha^{-1}\overline{M_k}(\theta) \end{pmatrix},$$

and the symbol $\overline{\hat{\mathscr{A}}_k^{(jac)}}(\theta)$ defined analogously.

As mentioned earlier, the smoothing iteration shall reduce the high-frequency parts of the error. To measure this phenomenon, we introduce the smoothing rate

$$q(\tau) := \sup_{\theta \in \Theta^{(high)}} \sup_{h_k > 0} \sup_{\alpha > 0} \sigma(\theta, h_k, \alpha, \tau), \tag{7}$$

where $\Theta^{(high)} := [-\pi, \pi) \backslash [-\frac{\pi}{2}, \frac{\pi}{2})$ is the set of high frequencies and σ is defined as the spectral radius of the Fourier-transformed smoothing operator given by

$$\sigma(\theta, h_k, \alpha, \tau) := \rho\big(\underbrace{I - \tau\left(\overline{\hat{\mathscr{A}_k}}(\theta)\right)^{-1}\overline{\mathscr{A}}_k(\theta)}_{\overline{S}_k(\theta):=}\big). \tag{8}$$

In (7) the supremum is not only taken over all $\theta \in \Theta^{(high)}$, but also over all grid sizes $h_k > 0$ and choices of the parameter $\alpha > 0$. Therefore, we compute an upper bound for the smoothing rate which is independent of h_k (which allows to show optimal convergence) and the parameter α (which allows to show robust convergence). For obvious reasons, the supremum is not taken with respect to the damping parameter τ, but is adjusted within the method such that the smoothing rate is optimal for Jacobi relaxation.

In principle it would be necessary to analyze the norm of the iteration matrix in (8) rather than analyzing the spectral radius. The spectral radius, however, equals the infimum over all matrix norms, which implies that for every $\epsilon > 0$ there is a matrix norm such that

$$\|\overline{S}_k(\theta)\| \leq (1+\epsilon)\, \rho(\overline{S}_k(\theta)),$$

see [7]. For the model problem and both proposed smoothing procedures (collective Jacobi relaxation and collective Gauss-Seidel iteration) straight-forward computations show that the spectral radius of the symbol of the smoothing operator $\rho(\overline{S}_k(\theta))$ is *equal* to its norm $\|\overline{S}_k(\theta)\|_{\hat{X}}$, if the matrix-norm is chosen as

$$\|\mathcal{M}\|_{\hat{X}} := \left\| \begin{pmatrix} \alpha^{1/2} & \\ & 1 \end{pmatrix} \mathcal{M} \begin{pmatrix} \alpha^{1/2} & \\ & 1 \end{pmatrix}^{-1} \right\|_{\ell^2}, \tag{9}$$

where $\|\cdot\|_{\ell^2}$ denotes the spectral norm. Observe that the scaling of the state y and the adjoint state p in this norm equals to the scaling in the norm $\|\cdot\|_X$ in classical theory that can be found in [10].

An equivalent formulation for the definition of the smoothing rate (7) using quantifiers is: Determine λ such that

$$\forall\, \theta \in \Theta^{(high)}\ \forall\, h_k > 0\ \forall\, \alpha > 0 : \sigma^2(\theta, h_k, \alpha, \tau) \leq \lambda \tag{10}$$

holds. Then for every $\tau \in (0,1)$ the value of $q(\tau)$ is the smallest such λ.

The computation of $\sigma(\theta, h_k, \alpha, \tau)$ is straight forward, but the computation of $q(\tau)$ is non-trivial. This is where symbolic computation enters our analysis. In order to determine q (that is either a polynomial in τ or a constant) we invoke *quantifier elimination* using *cylindrical algebraic decomposition* (CAD) that is introduced in the next section. Note that for both preconditioners under consideration, (10) is a quantified formula on trigonometric polynomials (after clearing denominators). For the case of a collective Jacobi relaxation σ is given by

$$\sigma^2(\theta, h_k, \alpha, \tau) = \frac{h_k^4((\cos\theta + 2)\tau - 2)^2 + 36\alpha((\cos\theta - 1)\tau + 1)^2}{4\left(h_k^4 + 9\alpha\right)}.$$

CAD, as we detail in the next section, accepts as input only polynomial (or more general rational) inequalities over the reals. This is a complication that is easily resolved by replacing $\cos\theta$ by a real variable $c \in [-1, 1]$ and, if necessary, $\sin\theta$ by a real variable $s \in [-1, 1]$ together with Pythagoras' identity $s^2 + c^2 = 1$.

3 Quantifier Elimination Using Cylindrical Algebraic Decomposition

So far we have reformulated the task of determining the smoothing rate for our multigrid methods to the problem of resolving a quantified polynomial inequality. That is, the given statement is of the form

$$Q_1\, x_1 \ldots Q_n\, x_n : A(x_1, \ldots, x_n, y_1, \ldots, y_m),$$

where Q_i denote quantifiers (either $\forall$ or $\exists$) and $A(x_1, \ldots, x_n, y_1, \ldots, y_m)$ is a boolean combination of *polynomial* inequalities. The problem of finding an equivalent, quantifier free formula $B(y_1, \ldots, y_m)$ consisting of a boolean combination of polynomial inequalities depending only on the free variables is called quantifier elimination. The first algorithm to solve this problem over the reals was given by A. Tarski [13] in the early 1950s. His method, however, was practically not efficient. Nowadays modern implementations [4, 11, 12] of G. Collins' cylindrical algebraic decomposition [5] make it possible to carry out nontrivial computations in a reasonable amount of time.

A simple example is given by: Determine a bound $B = B(z)$ for $0 < z < 1$ such that

$$\forall\, 0 < x < 1\ \forall\, 0 < y < 1 : \frac{x}{y + z} + \frac{y}{x + z} \leq B,$$

or equivalently,

$$\forall\, x\ \forall\, y : 0 < x < 1 \wedge 0 < y < 1 \Rightarrow \frac{x}{y + z} + \frac{y}{x + z} \leq B.$$

A CAD-computation quickly yields that $B(z) \geq \frac{1}{z}$. In cases where no free variables appear in the input, B is one of the logical constants True or False. Applied to a quantifier free formula the result of a CAD-computation is an equivalent formula that is normalized in a certain sense.

When executing the algorithm first the quantifier free part of the formula is considered, i.e., in the example above the inequalities $0 < x < 1$, $0 < y < 1$ and $\frac{x}{y+z} + \frac{y}{x+z} \leq B$. These inequalities can be reformulated in terms of inequalities by clearing denominators in the appropriate way. The given polynomials define a natural decomposition of the real space (in the example $\mathbb{R}^4$) into maximal connected cells on which the polynomials are sign invariant. This decomposition is then further refined to obtain cells on which the polynomials are not only sign invariant, but also cylindrically arranged. The orientation of this cylindrical arrangement depends on the order of the variables that is fixed by the quantifiers for the bound variables and the user (or the implementation) for the free variables, respectively. In this sense one may consider variables as being on the bottom (or innermost) level or on higher levels of the resulting CAD. Once such a cylindrical decomposition is obtained the quantifiers can be eliminated by considering each of the cells in an order determined by the quantifiers. The result is a formula where all the bound variables have been eliminated and the description of the cells where the formula holds is given solely in terms of the free variables as shown in the example above.

This procedure may be a very costly one depending on the number of variables and the degrees of the polynomials and even though termination of the algorithm is proven, the actual computation might exceed the expected life-time of the authors. Although it might seem a high price to pay, the gain is an *optimal* bound for the given formula that is determined by a *proving* procedure that is not approximate in any way.

The formula for σ as stated above is a rational function in the given indeterminates. Adding the necessary case distinctions for the denominators that arise when multiplying both sides of an inequality (which is commonly handled internally by the implementations), this is still a valid input for a CAD-computation.

A major issue is the runtime complexity of CAD that depends heavily on the input parameters such as number of polynomial inequalities, polynomial degrees and number of variables. In the worst case it is doubly exponential in the number of variables and this worst case bound is not only met in theory, but often experienced in practice. As we will see below, already for the one dimensional analysis suitable substitutions of the variables are applied in order to speed up the computations. These substitutions aim at reducing the number of variables on the one hand and lowering the polynomial degrees on the other hand.

For the forthcoming analysis of the two (or even three) dimensional case, further simplifications will be necessary because of the increase in both, the number of unknowns as well as the polynomial degrees of the given formulas.

4 Computing the Smoothing Rate

Now we are in the position to state the main results of this paper, the smoothing rates for collective Jacobi relaxation and collective Gauss-Seidel iteration.

4.1 Smoothing Property: Collective Jacobi Relaxation

In the smoothing step we are concerned with the high-frequency parts of the error. Consequently, if we replace $\cos\theta$ by a real variable c, then the condition $\theta \in \Theta^{(high)}$ translates to $-1 \le c \le 0$.

With this substitution in the case of a collective Jacobi relaxation σ is given by $\sigma(\theta, h_k, \alpha, \tau) = \widetilde{\sigma}(\cos\theta, h_k, \alpha, \tau)$, where

$$\widetilde{\sigma}^2(c, h_k, \alpha, \tau) := \frac{h_k^4((c+2)\tau - 2)^2 + 36\alpha((c-1)\tau + 1)^2}{4\left(h_k^4 + 9\alpha\right)}.$$

With this rewriting the condition in formula (10) has become a purely polynomial inequality and can invoke CAD to determine $q(\tau)$. For this purpose we used the CAD-implementation in *Mathematica*. The subscript 2 for σ below indicates that we are dealing with the square of the actual expression.

In[8]:= $\boldsymbol{\sigma_2 = \dfrac{h^4((c+2)\tau - 2)^2 + 36\alpha((c-1)\tau + 1)^2}{4\,(h^4 + 9\alpha)};}$

In[9]:= **Resolve[ForAll[**$c, -1 \leq c \leq 0$**, ForAll[**$h, h > 0$**, ForAll[**$\alpha, \alpha > 0, \sigma_2 \leq \lambda$**]]], {**$\tau, \lambda$**}, Reals]**

Out[9]= $\left(\tau \leq 0 \&\& \lambda \geq 4\tau^2 - 4\tau + 1\right) \| \left(0 < \tau \leq \frac{4}{5} \&\& \lambda \geq \frac{1}{4}\left(\tau^2 - 4\tau + 4\right)\right) \|$

$\left(\tau > \frac{4}{5} \&\& \lambda \geq 4\tau^2 - 4\tau + 1\right)$

The computation takes about one second and the result is a quantifier-free formula equivalent to the quantified formula given in (10). Note also that this is again a statement formulated in terms of polynomial inequalities. It is normalized in the sense that the parameter τ is assumed to be on the bottom level, which is indicated by the order of variables within the "Resolve"-command. Thus the output is sorted in a way that the conditions on τ are inequalities comparing to (algebraic) numbers, whereas the conditions on λ on the next higher level are formulated in terms of τ.

So for every τ the function value $q^2(\tau)$ is the smallest λ fulfilling Out[2]. For instance, if we plug $\tau = \frac{1}{2}$ into Out[2], the formula reduces to:

$$\text{False} \vee \left(\text{True} \wedge \lambda \geq \frac{3}{4}\right) \vee \text{False}.$$

As $q^2\left(\frac{1}{2}\right)$ is the smallest λ fulfilling the inequality, we have $q^2\left(\frac{1}{2}\right) = \frac{3}{4}$. Guided by this example we read off the general form for $q^2(\tau)$ which is a piecewise quadratic function given by

$$q^2(\tau) = \begin{cases} 4\tau^2 - 4\tau + 1 & \text{for} \quad \tau \leq 0 \\ \frac{1}{4}\left(\tau^2 - 4\tau + 4\right) & \text{for} \ 0 < \tau \leq \frac{4}{5} \\ 4\tau^2 - 4\tau + 1 & \text{for} \ \frac{4}{5} < \tau \end{cases}. \tag{11}$$

Summarizing we have determined the supremum in (7) and therefore the smoothing rate $q(\tau)$. If we take the square root of (11) and restrict ourselves to the relevant range $\tau \in [0, 1]$, we obtain the smoothing rates for the collective Jacobi relaxation:

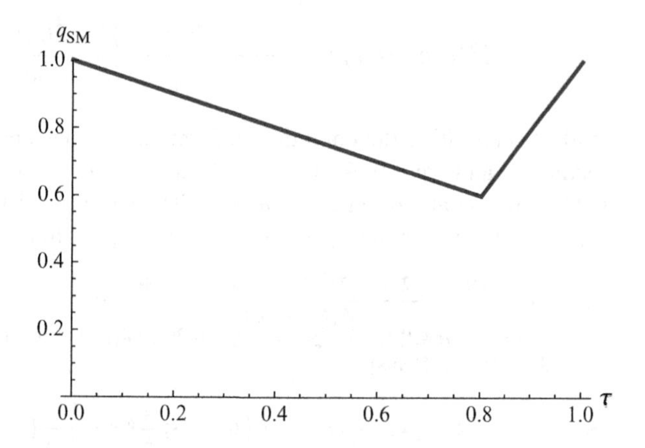

Fig. 1 Smoothing factor depending on damping parameter τ

$$q(\tau) = \begin{cases} \frac{1}{2}(2-\tau) & \text{for } 0 \le \tau \le \frac{4}{5} \\ 2\tau - 1 & \text{for } \frac{4}{5} < \tau \le 1 \end{cases}. \tag{12}$$

Since our method gives an equivalent reformulation, we know that these bounds on the smoothing rate are sharp. The graph of the function q can be seen in Fig. 1. From this we see that $q(\tau)$ takes its minimum for $\tau = \frac{4}{5}$ with value $q\left(\frac{4}{5}\right) = \frac{3}{5}$. For the canonical choice $\tau = \frac{1}{2}$, we obtain $q\left(\frac{1}{2}\right) = \frac{3}{4}$.

A smoothing analysis in a similar setting has been carried out in [1], where the authors obtain estimates for smoothing and convergence rates using numerical interpolation. To the knowledge of the authors (12) and (13) below provide the first rigorously proven sharp bounds for the smoothing rate.

4.2 *Smoothing Property: Collective Gauss-Seidel Iteration*

In this subsection we carry over the smoothing analysis which we have done for the collective Jacobi relaxation to the collective Gauss-Seidel iteration. Again we can determine the symbol of the involved preconditioning matrices:

$$\hat{M}_k^{(gs)} \underline{\varphi_k(\theta)} = \underbrace{\frac{(4+e^{-i\theta})h_k}{6}}_{\hat{M}_k^{(gs)}(\theta):=} \underline{\varphi_k(\theta)} \quad \text{and} \quad \hat{K}_k^{(gs)} \underline{\varphi_k(\theta)} = \underbrace{\frac{2-e^{-i\theta}}{h_k}}_{\hat{K}_k^{(gs)}(\theta):=} \underline{\varphi_k(\theta)}.$$

The procedure for determining $\sigma(\theta, h, \alpha)$ (which now is independent of τ) is completely analogous to the previous case. Again, by our choice of matrix norm $\|\cdot\|_{\hat{X}}$, the same value for σ is obtained no matter whether we consider the spectral radius or the norm.

In order to have a purely polynomial input for the CAD computations, we simplify the formula for σ: the occurrences of $\cos\theta$ and $\sin\theta$ are replaced by c and s, respectively. Moreover we expand numerator and denominator and replace all occurrences of s^2 by $1 - c^2$ thus arriving at numerator and denominator being linear in s. After these simple rewriting steps we obtain $\sigma(\theta, h_k, \alpha) = \widetilde{\sigma}(\sin\theta, \cos\theta, h_k, \alpha)$, where

$$\widetilde{\sigma}^2(s, c, h_k, \alpha) := \frac{\left(h_k^4 + 36\alpha\right)\left((17 + 8c)h_k^4 + 72h_k^2\alpha^{1/2}|s| + 36(5 - 4c)\alpha\right)}{(17 + 8c)^2 h_k^8 + 72(40c^2 - 28c + 13)h_k^4\alpha + 1296(5 - 4c)^2\alpha^2}.$$

The smoothing rate q is again the supremum over all high frequencies, grid sizes and choices of the parameter α and is given by

$$q^2 = \sup_{(s,c)\in D} \sup_{h_k>0} \sup_{\alpha>0} \widetilde{\sigma}^2(s, c, h_k, \alpha),$$

where $D := \{(s, c) \in \mathbb{R}^2 \ : \ s^2 + c^2 = 1, \ c \le 0\}$.

Note that in this definition above still the absolute value of s and a term $\alpha^{1/2}$ occurs. Before we can invoke CAD-computations, we have to rewrite $\widetilde{\sigma}$ as rational function. A first simplification is that as $\widetilde{\sigma}$ does not depend on the sign of s, we can restrict ourselves to assuming only non-negative values and thus replace $|s|$ by $s \ge 0$. To eliminate also $\alpha^{1/2}$ in the numerator, we replace α by $\widetilde{\alpha}^2$, where $\widetilde{\alpha} > 0$. Having completed these rewritings the final formula for q reads as

$$q^2 = \sup_{(s,c)\in\widetilde{D}} \sup_{h_k>0} \sup_{\widetilde{\alpha}>0} \frac{\left(h_k^4 + 36\widetilde{\alpha}^2\right)\left((17 + 8c)h_k^4 + 72h_k^2\widetilde{\alpha}s + 36(5 - 4c)\widetilde{\alpha}^2\right)}{(17 + 8c)^2 h_k^8 + 72(40c^2 - 28c + 13)h_k^4\widetilde{\alpha}^2 + 1296(5 - 4c)^2\widetilde{\alpha}^4},$$

where $\widetilde{D} := \{(s, c) \in \mathbb{R}^2 \ : \ s^2 + c^2 = 1, \ c \le 0, \ s \ge 0\}$. We can again rewrite the supremum as quantified expression where the quantifiers can be eliminated with the help of a CAD computation. With *Mathematica's* quantifier elimination algorithm, we obtain the smoothing rate for the collective Gauss-Seidel iteration after about twenty minutes:

$$q = \tfrac{1}{7}(3 + \sqrt{2}) \approx 0.63. \tag{13}$$

Even though twenty minutes are not a very long time to wait for a result that needs to be obtained only once, it still seems too long for such a simple formula. We can speed up the calculation significantly by reducing both the number of variables and the degrees of the polynomials by substitution using the variable $\eta := h_k^2/\widetilde{\alpha} > 0$. This substitution reduces the formula for q to

$$q^2 = \sup_{(s,c)\in\widetilde{D}} \sup_{\eta>0} \frac{\left(\eta^2 + 36\right)\left((17 + 8c)\eta^2 + 72\eta s + 36(5 - 4c)\right)}{(17 + 8c)^2\eta^4 + 72(40c^2 - 28c + 13)\eta^2 + 1296(5 - 4c)^2}.$$

Based on this representation *Mathematica's* quantifier elimination algorithm is able to derive q within about twenty seconds.

5 Concluding Remarks

In this paper we have shown a strategy to compute the smoothing rate for a multigrid method using collective Jacobi relaxation or Gauss-Seidel iteration by means of symbolic computation in an entirely automatic manner. The proposed strategy strongly relies on the fact that local Fourier analysis is a systematic machinery which is applied to the problem and the given numerical method. Typically this approach leads to determining the supremum of an explicitly given term.

On the one hand, the smoothing rates we obtained this way may be viewed as an interesting result on their own. On the other hand, these rates will also enter a full two- or multigrid analysis which again can be done using local Fourier analysis.

Also for the full analysis, or the extension to higher dimensional cases, Fourier analysis leads to an expression that in the first step is a rational function in the mesh size h_k, the regularization parameter α, the damping parameter τ, and trigonometric expressions of the frequencies θ. This is in particular the case for the model problem described in this paper for the above mentioned generalizations.

Theoretical results guarantee that also these problems can be solved with the methods applied in this work. To obtain the full results in reasonable time, it is necessary to apply proper strategies to reduce the complexity of the problems in the formulation of the input which is ongoing work.

Acknowledgements We thank the anonymous referees for their remarks that have significantly improved the quality of this paper. This work is supported by the Austrian Science Fund (FWF) under grants W1214/DK12 and DK6.

References

1. Borzi, A., Kunisch, K., Kwak, D.Y.: Accuracy and convergence properties of the finite difference multigrid solution of an optimal control optimality system. SIAM J. Control Optim **41**(5), 1477–1497 (2003)
2. Brandt, A.: Multi-level adaptive solutions to boundary-value problems. Math. Comp. **31**, 333–390 (1977)
3. Brandt, A.: Rigorous quantitative analysis of multigrid, I: Constant coefficients two-level cycle with L_2-norm. SIAM J. Numer. Anal. **31**(6), 1695–1730 (1994)
4. Brown, C.W.: QEPCAD B – a program for computing with semi-algebraic sets. Sigsam Bulletin **37**(4), 97–108 (2003)
5. Collins, G.E.: Quantifier elimination for real closed fields by cylindrical algebraic decomposition. In Automata theory and formal languages (Second GI Conference, Kaiserslautern, 1975), Lecture Notes in Computer Science, Vol. 33, pp. 134–183. Springer, Berlin (1975)
6. Hackbusch, W.: Multi-Grid Methods and Applications. Springer, Berlin (1985)
7. Holmes, R.B.: A formula for the spectral radius of an operator. Am. Math. Mon. **75**(2), 163–166 (1968)

8. Hong, H., Liska, R., Steinberg, S.: Testing stability by quantifier elimination. J. Symbolic Comput. **24**(2), 161–187 (1997); Applications of quantifier elimination (Albuquerque, NM, 1995).
9. Lass, O., Vallejos, M., Borzi, A., Douglas, C.C.: Implementation and analysis of multigrid schemes with finite elements for elliptic optimal control problems. Computing **84**(1–2), 27–48 (2009)
10. Schöberl, J., Simon, R., Zulehner, W.: A Robust Multigrid Method for Elliptic Optimal Control Problems. SIAM J. Numer. Anal. **49**, 1482 (2011)
11. Seidl, A., Sturm, T.: A generic projection operator for partial cylindrical algebraic decomposition. In Proceedings of the 2003 International Symposium on Symbolic and Algebraic Computation, pp 240–247 (electronic), ACM Press, New York (2003)
12. Strzeboński, A.: Solving systems of strict polynomial inequalities. J. Symbolic Comput. **29**(3), 471–480 (2000)
13. Tarski, A.: A decision method for elementary algebra and geometry. 2nd ed. University of California Press, Berkeley and Los Angeles, California (1951)
14. Trottenberg, U., Oosterlee, C., Schüller, A.: Multigrid. Academic Press, London (2001)
15. Vanka, S.P.: Block-implicit multigrid solution of Navier-Stokes equations in primitive variables. J. Comput. Phys. **65**, 138–158 (1986)
16. Wienands, R., and Joppich, W. Practical Fourier analysis for multigrid methods. Chapman & Hall, CRC (2005)

Analytical Evaluations of Double Integral Expressions Related to Total Variation

Carsten Pontow and Otmar Scherzer

Abstract In this paper, for certain classes of functions, we present analytical evaluations of double integral expressions representing approximations of the total variation seminorm. These calculations are carried out by using the Maple computer algebra software package in a sophisticated way.

The derived expressions can be used for approximations of the total variation seminorm, and, in particular, can be used for efficient numerical total variation regularization energy minimization.

1 Introduction

Let $N \in \{1, 2\}$ and let Ω denote a bounded connected subset of $\mathbb{R}^N$ with Lipschitz-continuous boundary. In this paper we address the problem of evaluating double integrals of the kind

$$\mathcal{R}_n(f) := \int_\Omega \int_\Omega \frac{|f(x) - f(y)|}{|x - y|} \varphi_n(x - y)\, dx\, dy.$$

where f is a measurable real function defined on Ω and (φ_n) denotes a sequence of non-negative, radially symmetric, and radially decreasing functions from $L^1(\mathbb{R}^N)$ satisfying for every $\delta > 0$

O. Scherzer (✉)
RICAM, Austrian Academy of Sciences & Computational Science Center, University of Vienna, Altenbergerstraße 69, 4040 Linz, Austria
e-mail: otmar.scherzer@univie.ac.at

C. Pontow
Computational Science Center, University of Vienna, Nordbergstraße 15, 1090 Wien, Austria
e-mail: carsten.pontow@univie.ac.at

U. Langer and P. Paule (eds.), *Numerical and Symbolic Scientific Computing*, Texts and Monographs in Symbolic Computation, DOI 10.1007/978-3-7091-0794-2_10,

$$\lim_{n\to\infty} \int_{\{x:|x|>\delta\}} \varphi_n(x)\, dx = 0 \tag{1}$$

and for all $n \in \mathbb{N}$

$$\int_\Omega \varphi_n(x)\, dx = 1. \tag{2}$$

Conditions (1) and (2) imply that the unit mass of the functions φ_n concentrates around the origin as n strives to infinity. The properties of (φ_n) imply further that this sequence represents an approximation of the identity.

The integral expression above occurred in papers by Bourgain et al. [3] and Dávila [6], providing new characterizations of the total variation seminorm. In fact, they proved that for every function f of bounded variation

$$\frac{1}{K_{1,N}} \lim_{n\to\infty} \mathcal{R}_n(f) = |Df| \tag{3}$$

and that the above limit diverges to infinity if f is not a function of bounded variation. In the above equation the constant $K_{1,N}$ is a real constant depending only on dimension N and $|Df|$ denotes the total variation seminorm of f.

Prominent applications of the total variation seminorm lie in the field of image processing, where the seminorm represents a usefool tool for measuring the amount of variation within an image. For example, total variation regularization, i.e., the search for a minimum of functionals of the type (Rudion, Osher, Fatemi functional [12])

$$\mathcal{F}(f) := \frac{1}{2} \int_\Omega (f - f^\delta)^2(x)\, dx + \alpha |Df|$$

is a prominent method for the denoising of images. Here f^δ is an image corrupted by noise and α represents a positive real constant. The first summand of the functional above is called the *fidelity* term and penalizes the deviation of an image f from the data f^δ. The second summand is named *regularization term* and penalizes the rate of variation within f. While the first summand provides that the outcome of the minimization process (the denoised image) preserves similarity to f^δ, the second term is intended to reduce the oscillations within the argument f, in order to generate an approximation to f^δ that contains a small amount of inherent noise. In fact, this denoising strategy has proven to be successful and even more, the total variation seminorm has proven to be superior to other regularization terms in the sense that edges within the image are preserved.

Since in many cases the total variation seminorm of a function cannot be determined analytically, appropriate simplifications are introduced when such an evaluation has to be done. A first step lies in the discretization of the function. Typically, a mesh is introduced on Ω and the functions f is approximated by linear combinations of piecewise constant or piecewise linear ansatz functions defined with respect to that mesh. For example, a one-dimensional function $f:(0,1) \to \mathbb{R}$ can be approximated by a linear combination of indicator functions

$\sum_{k=1}^{n} a_k \chi_{(\frac{k-1}{n},\frac{k}{n})}$ where the coefficients a_k represent mean values of f over the intervals $(\frac{k-1}{n},\frac{k}{n})$. The total variation seminorm of this discrete approximation of f equals $\sum_{k=1}^{n} |a_k - a_{k-1}|$. This approximation in turn can be used in standard numerical procedures for problems involving total variation. In the case of total variation regularization such procedures include the method of steepest descent [7, 8] and other algorithms like [4, 11]. In this paper, we discuss whether other discrete schemes can also be justified as approximations of the total variation seminorm: We base this discussion on the limit relation (3), which opens up new possibilities for computing approximations of the total variation seminorm for special functions. Instead of computing the total variation seminorm of a finite dimensional approximation f_d of f, the integral expressions $\mathcal{R}_n$ may be considered for the evaluation of f_d and in fact the main subject of this work is an investigation of such analytical evaluations.

In a standard numerical setting one would try to approximate the integral expressions within $\mathcal{R}_n(f_d)$ by some quadrature rules associated with the mesh underlying f_d. However, the singularity within the integrand in $\mathcal{R}_n(f)$ causes difficulties when treated with standard numerical techniques and thus we decided to follow a different approach, which to our knowledge has not been applied before in this setting, namely to evaluate $\mathcal{R}_n(f_d)$ analytically for special functions. It became soon obvious that without the power of computer algebra systems in many cases such evaluations are close to being impossible because of the size of the algebraic terms that occur in the computations of $\mathcal{R}_n(f_d)$. However, with the help of the Maple software program we could actually do some evaluations for some special, practically important, cases of f_d. We present the respective results here together with some descriptions of how we proceeded.

In terms of total variation regularization our results may be particularly useful because it can be shown that indeed minimizers of functionals

$$\mathcal{F}_n(f) := \frac{1}{2}\int_\Omega (f - f^\delta)^2(x)\,dx + \frac{\alpha}{K_{1,N}}\mathcal{R}_n(f). \tag{4}$$

are converging to minimizers of the total variation regularization functional

$$\mathcal{F}(f) := \frac{1}{2}\int_\Omega (f - f^\delta)^2(x)\,dx + \alpha|Df|. \tag{5}$$

We give a short review about this result in Sect. 3.

However,we want to state as well that we will not give a detailed convergence analysis of the resulting discrete functionals, and we will not explore all degrees of freedom associated with our approach of the analytical evaluation of functionals $\mathcal{R}_n(f_d)$. We just want to demonstrate here that this approach actually can be carried out successfully and give some hints that these evaluations can provide some new tools for the numerical treatment of functionals involving the total variation seminorm, in particular total variation regularization.

Notations. We summarize some further general assumptions and notations that will be used throughout the rest of this paper.

For $p \geq 1$ the space of L^p-functions on Ω with mean value zero is symbolized by

$$L^p_\diamond(\Omega) := \left\{ f \in L^p(\Omega) \ : \ \int_\Omega f(x)\, dx = 0 \right\}.$$

All occurring functions in this paper defined on Ω are supposed to be real-valued.

Let $C_c^\infty(\Omega)$ be the space of infinitely differentiable functions from Ω to $\mathbb{R}$ with compact support. For the total variation seminorm of a locally integrable function f we use the symbol

$$|Df| = \sup \left\{ \int_\Omega f(x) \nabla \cdot \psi(x) dx : \psi \in C_c^\infty(\Omega; \mathbb{R}^N), |\psi(x)| \leq 1 \text{for all } x \in \Omega \right\};$$

Here $|\psi(x)|$ denotes the Euclidean distance of the vector $\psi(x)$. The space of functions of bounded variation on Ω is the set

$$BV(\Omega) := \{ f \in L^1(\Omega) \ : \ |Df| < \infty \}.$$

The constants $K_{1,N}$ are defined by

$$K_{1,N} = \begin{cases} \frac{2}{\pi} \text{ if } N = 2, \\ 1 \text{ if } N = 1. \end{cases}$$

2 Some Examples of Evaluations of $\mathcal{R}_n(f_d)$

In this section we present some analytical evaluations of integral expressions $\mathcal{R}_n(f_d)$ where f_d is a linear combination of piecewise constant or piecewise linear functions representing a discrete approximation to some continuous or integrable function f as sketched in the introduction. For these evaluations we had to choose some domain Ω, some approximation of identity (φ_n) and some linear combinations f_d of the kind stated above. Since we could not find any examples to follow in literature we decided to start with choices as simple and intuitive as possible and then gradually to increase the complexity of the setting.

For the sake of simplicity, we chose generally for Ω in the one-dimensional case the open interval $(0, 1)$ and in the two-dimensional case the open square $(0, 1) \times (0, 1)$. For the sake of intuiveness, we chose for (φ_n) approximations of the identity that comply with the discretization f_d in the sense that f_d depends on the index n, too, i.e., we treat evaluations of type $\mathcal{R}_n(f_n)$ where f_n is of the form

$$f_n = \sum_{k=1}^{n} a_k g_k$$

where for $k \in \{1, \ldots, n\}$ the symbols a_k denote real constants and the symbols g_k stand for elements of a familiy of piecewise constant or piecewise linear functions associated to a regular mesh on Ω that itself depends on n. As we will see below in this way the standard discretization of the total variation seminorm in dimension one is recoverable.

2.1 The One-Dimensional Case

We work on the domain

$$\Omega := (0, 1).$$

and consider evaluation of $\mathcal{R}_n$ with piecewise constant or piecewise linear functions defined on a regular mesh.

1. The first two schemes are for piecewise constant functions. They can be computed manually rather easily:

   a. We use the sequence of kernel functions (φ_n) defined by

   $$\varphi_n := \frac{n}{2} \chi_{(-\frac{1}{n}, \frac{1}{n})}.$$

   Let $a_1, \ldots, a_n \in \mathbb{R}$. Evaluating the one-dimensional piecewise constant function

   $$f_n := \sum_{i=1}^{n} a_i \chi_{(-\frac{1}{n}, \frac{1}{n})}$$

   with $\mathcal{R}_n$ yields the standard total variation seminorm of f_n:

   $$\mathcal{R}_n(f_n) = \sum_{i=2}^{n} |a_i - a_{i-1}| = |Df_n|.$$

   b. Using instead of (φ_n) the family of kernels $(\varphi_n^{(2)})$ defined by

   $$\varphi_n^{(2)} := \frac{n}{4} \chi_{(-\frac{2}{n}, \frac{2}{n})}$$

   yields

   $$\mathcal{R}_n(f_n) = \sum_{i=2}^{n-1} \frac{1 - \ln(2)}{2} |a_{i+1} - a_{i-1}| + \sum_{i=2}^{n} \ln(2) |a_i - a_{i-1}|. \tag{6}$$

   We recall that $\ln(2) \approx 0.7$.

2. Now we consider piecewise linear functions for evaluation. Let $a_0, \ldots, a_n \in \mathbb{R}$ and f_n be the piecewise linear spline interpolating the nodes $(\frac{k}{n}, a_k), k = 0 \ldots n$, i.e.,

$$f_n := \sum_{i=0}^{n} a_i g_i$$

where

$$g_i(x) := \max\left(1 - n\left|x - \frac{i}{n}\right|, 0\right).$$

Inserting f_n in $\mathcal{R}_n$ and using the kernel functions φ_n yields

$$\mathcal{R}_n(f_n) = \sum_{i=1}^{n} \frac{|a_i - a_{i-1}|}{2} + \sum_{i=1}^{n-1} t(a_{i-1}, a_i, a_{i+1})$$

where

$$t(a_{i-1}, a_i, a_{i+1}) = \begin{cases} \frac{|a_{i+1}-a_{i-1}|}{4} & \text{if } \operatorname{sgn}(a_{i-1} - a_i) = \operatorname{sgn}(a_i - a_{i+1}), \\ \frac{(a_i-a_{i-1})^2+(a_i-a_{i+1})^2}{4(|a_i-a_{i-1}|+|a_i-a_{i+1}|)} & \text{if } \operatorname{sgn}(a_{i-1} - a_i) \neq \operatorname{sgn}(a_i - a_{i+1}). \end{cases}$$

The change from piecewise constant to piecewise linear ansatz functions results in a huge increase of the complexity of the evaluation. Thus, we applied the computer algebra program Maple for this evaluation. We provide a sketch of some parts of the computation for the case $n \geq 2$.

Before we start we would like to note that Maple is far from being capable of evaluating an expression $\mathcal{R}_n(f_n)$ on its own; the user has to design a strategy for the evaluation, has to divide the problem in appropriate substeps that Maple can treat, has to keep track of the many occurring case distinctions and in particular has to check whether the program misses some simplifications or even delivers wrong results in some cases.

We want to compute

$$\int_0^1 \int_0^1 \frac{\left|\sum_{i=0}^n a_i g_i(x) - \sum_{i=0}^n a_i g_i(y)\right|}{|x-y|} \varphi_n(x-y)\, dxdy.$$

which is equal to

$$\frac{n}{2} \sum_{k=1}^{n} \int_{\frac{k-1}{n}}^{\frac{k}{n}} \int_{\max(0,y-\frac{1}{n})}^{\min(1,y+\frac{1}{n})} \frac{\left|\sum_{i=0}^n a_i g_i(x) - \sum_{i=0}^n a_i g_i(y)\right|}{|x-y|}\, dxdy.$$

Looking at the supports of the functions g_i we realize that the latter double integral equals

$$\frac{n}{2}\left(\int_0^{\frac{1}{n}}\int_0^{y+\frac{1}{n}}\frac{\left|\sum_{i=0}^{2}a_i g_i(x)-\sum_{j=0}^{1}a_j g_j(y)\right|}{|x-y|}\,dxdy\right.$$

$$+\sum_{k=2}^{n-1}\int_{\frac{k-1}{n}}^{\frac{k}{n}}\int_{y-\frac{1}{n}}^{y+\frac{1}{n}}\frac{\left|\sum_{i=k-2}^{k+1}a_i g_i(x)-\sum_{j=k-1}^{k}a_j g_j(y)\right|}{|x-y|}\,dxdy$$

$$\left.+\int_{\frac{n-1}{n}}^{1}\int_{y-\frac{1}{n}}^{1}\frac{\left|\sum_{i=n-2}^{n}a_i g_i(x)-\sum_{j=n-1}^{n}a_j g_j(y)\right|}{|x-y|}\,dxdy\right).$$

We only treat the second double integral, the other two are evaluated analogously. Let $2 \leq k \leq n-1$.

Then

$$\int_{\frac{k-1}{n}}^{\frac{k}{n}}\int_{y-\frac{1}{n}}^{y+\frac{1}{n}}\frac{\left|\sum_{i=k-2}^{k+1}a_i g_i(x)-\sum_{j=k-1}^{k}a_j g_j(y)\right|}{|x-y|}\,dxdy$$

can be decomposed into the sum

$$\int_{\frac{k-1}{n}}^{\frac{k}{n}}\int_{y-\frac{1}{n}}^{\frac{k-1}{n}}\frac{\left|\sum_{i=k-2}^{k-1}a_i g_i(x)-\sum_{j=k-1}^{k}a_j g_j(y)\right|}{y-x}\,dxdy$$

$$+\int_{\frac{k-1}{n}}^{\frac{k}{n}}\int_{\frac{k-1}{n}}^{\frac{k}{n}}\frac{\left|\sum_{i=k-1}^{k}a_i g_i(x)-\sum_{j=k-1}^{k}a_j g_j(y)\right|}{|x-y|}\,dxdy$$

$$+\int_{\frac{k-1}{n}}^{\frac{k}{n}}\int_{\frac{k}{n}}^{y+\frac{1}{n}}\frac{\left|\sum_{i=k}^{k+1}a_i g_i(x)-\sum_{j=k-1}^{k}a_j g_j(y)\right|}{x-y}\,dxdy.$$

whose summands we denote by $I_{k,-}$, I_k and $I_{k,+}$, respectively. The integrand of I_k is of the simple form

$$I_k = \frac{|a_k - a_{k-1}|}{n}.$$

By an application of Fubini's theorem

$$I_{k,-} = I_{k-1,+}.$$

Thus, it suffices to treat the evaluation of $I_{k,-}$ whose integrand $J(x,y)$ is reshaped as follows:

$$\frac{1}{y-x}\Big|(a_{k-1}-a_k)\,ny + (a_{k-1}-a_{k-2})\,nx$$

$$+ (a_{k-2}+a_k-2\,a_{k-1})\,k - a_{k-2} + 2\,a_{k-1} - a_k.\Big|$$

From this representation it is already visible that the evaluation will be dependent from the sign of the differences

$$\Delta_k := a_k - a_{k-1} \quad \text{and} \quad \Delta_{k-1} = a_{k-1} - a_{k-2}.$$

We treat here one instance of the more complex case when

$$\operatorname{sgn}(\Delta_k) \neq \operatorname{sgn}(\Delta_{k-1}),$$

namely the subcase where the middle coefficient a_{k-1} is the maximum of the three coefficients; the other subcase where a_{k-1} is the minimum can be treated just the same. The less complex case where a_{k-1} lies between a_{k-2} and a_k needs fewer case distinctions but apart from that can be treated analogously. Note that in the chosen subcase the second difference $\Delta_k^2 = a_{k-2} - 2\,a_{k-1} + a_k$ is negative.

The numerator $J_N(x,y)$ of $J(x,y)$ now reads

$$J_N(x,y) = \left|-\Delta_k ny + \Delta_{k-1} nx + \Delta_k^2 (k-1)\right|$$

and is positive if and only if

$$x > f_N(y) := \frac{\Delta_k ny - \Delta_k^2 (k-1)}{\Delta_{k-1} n}.$$

Thus, to evaluate the inner integral of $I_{k,-}$ we have to determine the intersections of its integration domain $(y - \frac{1}{n}, \frac{k-1}{n})$ with the intervals $(f_N(y), +\infty)$ and $(-\infty, f_N(y))$, respectively, for all $y \in (\frac{k-1}{n}, \frac{k}{n})$.

We get that

$$f_N(y) < \frac{k-1}{n} \iff \frac{k-1}{n} < y$$

and

$$y - \frac{1}{n} < f_N(y) \iff y < \frac{1}{n}\left(k - \frac{\Delta_k}{\Delta_k^2}\right).$$

Let $C := \frac{1}{n}\left(k - \frac{\Delta_k}{\Delta_k^2}\right)$. Note that $C < \frac{k}{n}$. Then

$$I_{k,-} = -\int_{\frac{k-1}{n}}^{C} \int_{y-\frac{1}{n}}^{f_N(y)} J(x,y)\,dxdy + \int_{\frac{k-1}{n}}^{C} \int_{f_N(y)}^{\frac{k-1}{n}} J(x,y)\,dxdy$$

$$+\int_{C}^{\frac{k}{n}}\int_{y-\frac{1}{n}}^{\frac{k-1}{n}} J(x,y)\,dxdy.$$

Integrating $J(x,y)$ with respect to x yields the primitive function

$$K(x,y) = \left((ny-k+1)\Delta_k^2\right)\ln(y-x) - xn\Delta_{k-1}.$$

Inserting the limits of the inner integral of the third summand yields

$$K\left(y-\frac{1}{n},y\right) = \left((k-1-ny)\Delta_k^2\right)\ln(n) - (ny-1)\Delta_{k-1},$$

$$K\left(\frac{k-1}{n},y\right) := \left((ny-k+1)\Delta_k^2\right)\ln\left(y-\frac{k-1}{n}\right) - (k-1)\Delta_{k-1}$$

and

$$\begin{aligned}L_3(y) &:= \int_{y-\frac{1}{n}}^{\frac{k-1}{n}} J(x,y)\,dx\\ &= \left((ny-k+1)\Delta_k^2\right)\ln(ny-k+1) + (ny-k)\Delta_{k-1}.\end{aligned}$$

A primitive function for L_3 is

$$\begin{aligned}M_3(y) := &\frac{(ny-k+1)^2\,\Delta_k^2}{2n}\ln(ny-k+1)\\ &-\frac{1}{4n}(ny-k+1)^2\,\Delta_k^2 + \frac{1}{2}y(nk-2y)\Delta_{k-1}.\end{aligned}$$

Analogous computations for the first and second summand yield the functions

$$L_2(y) = (ny+1-k)\left(\Delta_k^2\ln\left(-\frac{\Delta_{k-1}}{\Delta_k^2}\right) + \Delta_k\right)$$

and

$$M_2(y) = \frac{1}{2}\,y(ny-2k+2)\left(\Delta_k^2\ln\left(-\frac{\Delta_{k-1}}{\Delta_k^2}\right) + \Delta_k\right)$$

and

$$L_1(y) = \left(-\Delta_k^2\right)(ny-k+1)\left(\ln\left(-\frac{\Delta_k^2(ny-k+1)}{\Delta_{k-1}}\right) - 1\right) + \Delta_{k-1}$$

and

$$
\begin{aligned}
M_1(y) = - \frac{\Delta_k^2}{2}\Bigg(& y\,(-ny - 2 + 2k) \ln\left(-\frac{\Delta_{k-1}}{\Delta_k^2}\right) \\
& + \frac{(ny - k + 1)^2 \ln(ny - k + 1)}{n} \\
& - \frac{(ny - k + 1)^2 + 2ny\,(ny - 2k)}{2n}\Bigg) \\
& + (a_k - a_{k-1})\, y,
\end{aligned}
$$

respectively.

Now

$$
\begin{aligned}
I_{k,-} = & M_1\left(\frac{k-1}{n}\right) - M_1(C) + M_2(C) - M_2\left(\frac{k-1}{n}\right) \\
& + M_3\left(\frac{k}{n}\right) - M_3(C) \\
= & \frac{(a_{k-1} - a_k)^2 + (a_{k-1} - a_{k-2})^2}{4n\Delta_k{}^2}.
\end{aligned}
$$

From this result and the results from above the final result follows easily.

3. The results of the following example gives some hints that even the evaluation of very simple piecewise constant functions with $\mathcal{R}_n$ can get quite complicated if the size of the supports of the piecewise constant functions on the one hand and the size of the supports of the functions (φ_n) on the other hand are not proportional to each other.

   Here we fix the ansatz functions and φ_n for a fixed n. In this example we evaluate $\mathcal{R}_n$ for the Haar-functions $h_j^{(k)}$ using the kernel functions (φ_n). For $j = k = 0$ the function $h_j^{(k)}$ is defined by

$$
h_0^{(0)}(x) := 1
$$

for all $x \in \Omega$. For $k \in \mathbb{N}_0, 1 \leq j \leq 2^k$ we have

$$
h_j^{(k)}(x) := \begin{cases} \sqrt{2^k} & \text{if } x \in \left(\frac{2j-2}{2^{k+1}}, \frac{2j-1}{2^{k+1}}\right), \\ -\sqrt{2^k} & \text{if } x \in \left(\frac{2j-1}{2^{k+1}}, \frac{2j}{2^{k+1}}\right), \\ 0 & \text{otherwise.} \end{cases}
$$

Since $h_0^{(0)}$ is constant

$$
\mathcal{R}_n\left(h_0^{(0)}\right) = 0.
$$

For $h_1^{(0)}$ we get

$$\mathcal{R}_n\left(h_0^{(1)}\right) = \begin{cases} 2\ln(2) \text{ if } n = 1, \\ 2 \quad \text{if } n > 1. \end{cases}$$

Note that by symmetry for $k \geq 1$ and $1 \leq j \leq 2^{k-1}$

$$\mathcal{R}_n\left(h_j^{(k)}\right) = \mathcal{R}_n\left(h_{2^k-j+1}^{(k)}\right). \tag{7}$$

For $k \geq 1$, $\mathcal{R}_n\left(h_1^{(k)}\right)$ equals

$$\begin{cases} \sqrt{2^k}\left(\left(k + \frac{1}{2^{k-1}}\right)\ln(2) - \left(1 - \frac{1}{2^k}\right)\ln\left(2^k - 1\right)\right) & \text{if } n = 1, \\ \frac{n}{\sqrt{2^k}}\left((k+2)\ln(2) - \ln(n) + 1\right) & \text{if } 2 \leq n \leq 2^k, \\ \sqrt{2^k}n\left(\frac{k+1}{2^{k-1}}\ln(2) - \frac{1}{2^{k-1}}\ln(n) + \frac{1}{2^{k-1}} - \frac{1}{n}\right) & \text{if } 2^k \leq n \leq 2^{k+1}, \\ 3\sqrt{2^k} & \text{if } n \geq 2^{k+1}. \end{cases}$$

For $k \geq 2$ and $j = 2, \ldots, 2^{k-1}$ the inner functions $h_j^{(k)}$, evaluate to

$$\mathcal{R}_n\left(h_j^{(k)}\right) = \begin{cases} \sqrt{2^k}\left(\frac{j\ln(j)}{2^k} - \frac{(j-1)\ln(j-1)}{2^k}\right. & \\ -\left(1 - \frac{j}{2^k}\right)\ln\left(2^k - j\right) & \\ \left.+\left(1 - \frac{j-1}{2^k}\right)\ln\left(2^k - j + 1\right) + \frac{\ln(2)}{2^{k-1}}\right) & \text{if } n = 1, \\ \frac{n}{\sqrt{2^k}}\left(j\ln(j) + (j-1)\ln(j-1)\right. & \\ \left.+(k+2)\ln(2) - \ln(n) + 1\right) & \text{if } \frac{2^k}{2^k - j} \leq n \leq \frac{2^k}{j}, \\ n\sqrt{2^k}\left(\frac{(j-1)\ln(j-1)}{2^k} - \frac{(j+1)\ln(n)}{2^k} -\right. & \\ \left.\frac{(kj+k+2)\ln(2)}{2^k} + \frac{j+1}{2^k} - \frac{1}{n}\right) & \text{if } \frac{2^k}{j} \leq n \leq \frac{2^k}{j-1}, \\ \frac{2n}{\sqrt{2^k}}\left((k+1)\ln(2) - \ln(n) + 1\right) & \text{if } \frac{2^k}{j-1} \leq n \leq 2^{k+1}, \\ 4\sqrt{2^k} & \text{if } n \geq 2^{k+1}. \end{cases}$$

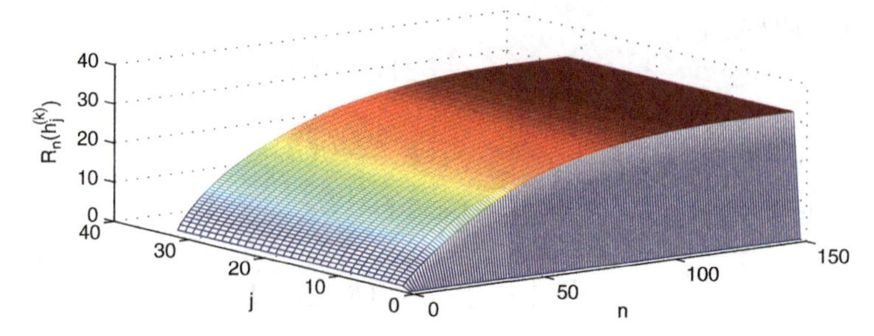

Fig. 1 Visualization of the size of $\mathcal{R}_n\left(h_j^{(k)}\right)$ for $k = 6, n = 1, \dots, 150, k = 2, \dots, 32$

The evaluation of $h_j^{(k)}$ for $j = 2^{k-1} + 1, \dots, 2^k$ is reduced to the evaluations directly above via (7).

Some of the results for $k = 6$ are visualized by Fig. 1. Note that despite the many case distinctions the size of the results $\mathcal{R}_n\left(h_j^{(k)}\right)$ does not vary very much over j. For fixed j the size of the results increases concavely for growing n until the total variation of $h_j^{(k)}$ is obtained for $n \geq 128$.

2.2 *The Two-Dimensional Case*

We now switch to the two-dimensional case and evaluate $\mathcal{R}_n$ for a piecewise constant function defined on a subset of $\mathbb{R}^2$. In detail, let Ω be chosen as the open square $(0, 1) \times (0, 1)$ and f be defined on Ω via

$$f(\vec{v}) := \sum_{i,j=1}^{n} a_{i,j}\, \chi_{I_{i,j}}(\vec{v})$$

for all $\vec{v} \in \Omega$ where $I_{k,l} := I_k \times I_l$ with $I_k := (\frac{k-1}{n}, \frac{k}{n})$ for all $k, l \in \{1, \dots, n\}$.

The following computations are much more complicated than those of the one-dimensional case. The sheer size alone of the occurring expressions makes them extremely difficult to handle, and a manual evaluation seems close to being impossible. Certainly, the same considerations concerning the use of the Maple software as already stated in point two of the previous subsection apply to the cases treated here, too. A priori it was everything but clear whether those integrals below could be evaluated at all using Maple, and we ran more than one time into dead-ends during the computations while trying to divide the problem in subproblems that we supposed to be treatable by Maple. Sometimes transformations between different representations of the occurring terms were needed in order to allow Maple to do some crucial simplification or integration steps, or the order of computation steps did matter. Thus, the successful approaches presented below may not be the most elegant ones from an analytical point of view. Our excuse for their choice is simply that they work at all.

In the following we investigate two cases of piecewise constant kernel functions, which differ by the ratio between the size of the support and the grid size. At the present stage of research it is not clear what the optimal ratio actually is.

2.2.1 Piecewise Constant Ansatz I

We choose the kernel functions

$$\varphi_n(\vec{v}) := \frac{n^2}{\pi}\chi_{B(0,\frac{1}{n})}(\vec{v})$$

for all $\vec{v}$ in $\mathbb{R}^2$ where $B\left(0,\frac{1}{n}\right)$ denotes the ball around the origin with radius $\frac{1}{n}$. The sequence (φ_n) satisfies all conditions stated in the introduction. We note that for two points (x,y) and (w,z) from Ω we have $(x,y)-(w,z)\in B\left(0,\frac{1}{n}\right)$ if and only if

$$(w,z)\in S_\Omega\left(x,y,\frac{1}{n}\right) := ((0,1)\times(0,1))\cap B\left((x,y),\frac{1}{n}\right).$$

We further define the intersection of the circle $B((x,y),\frac{1}{n})$ with the square $I_{k,l}$ by

$$S_{k,l}\left(x,y,\frac{1}{n}\right) := I_{k,l}\cap B\left((x,y),\frac{1}{n}\right).$$

We have to evaluate

$$\mathcal{R}_n^1(f) = \int_\Omega\int_\Omega \frac{|f(x,y)-f(w,z)|}{|(x,y)-(w,z)|}\varphi((x,y)-(w,z))\,d(w,z)\,d(x,y)$$

$$=\frac{n^2}{\pi}\int_0^1\int_0^1\int\int_{S_\Omega(x,y,\frac{1}{n})}\frac{\left|\sum_{i,j=1}^n a_{i,j}\chi_{I_{i,j}}(x,y)-\sum_{k,l=1}^n a_{k,l}\chi_{I_{k,l}}(w,z)\right|}{|(x-w,y-z)|}$$

$$d(w,z)\,dydx.$$

The occurring quadruple integral can be rewritten as follows:

$$\sum_{i,j=1}^n\sum_{k,l=1}^n\int_{\frac{i-1}{n}}^{\frac{i}{n}}\int_{\frac{j-1}{n}}^{\frac{j}{n}}\int\int_{S_{k,l}(x,y,\frac{1}{n})}\frac{|a_{i,j}-a_{k,l}|}{|(x-w,y-z)|}\,d(w,z)\,dydx.$$

For fixed $1\le i,j\le n$ and fixed $x,y\in I_{i,j}$ the domain $S_{k,l}\left(x,y,\frac{1}{n}\right)$ of the inner double integral is empty if $|i-k|\ge 2$ or $|j-l|\ge 2$. Thus, it suffices to evaluate those summands of the inner sum above that fulfill $|i - k| \leq 1$ and $|j - l| \leq 1$. However, in the case where $i = k$ and $j = l$ the integrand of the corresponding summand vanishes such that this case may be left out, too.

Given a pair of indices (i, j) let

$$\mathcal{I}_{i,j} := \{(k,l) \in \{1, \ldots, n\}^2 \; : \; |i - k| \leq 1, |j - l| \leq 1, (k,l) \neq (i,j)\}.$$

denote the set of pairs of indices for which the corresponding summands of the inner sum in the integral expression above do not vanish generally.

Then the above quadruple integral equals

$$\sum_{i,j=1}^{n} \sum_{(k,l) \in \mathcal{I}_{i,j}} |a_{i,j} - a_{k,l}| \int_{\frac{i-1}{n}}^{\frac{i}{n}} \int_{\frac{j-1}{n}}^{\frac{j}{n}} \int\int_{S_{k,l}(x,y,\frac{1}{n})} \frac{1}{|(x - w, y - z)|} \, d(w,z) \, dy dx.$$

We denote for all $1 \leq i, j \leq n$ and all $(k,l) \in \mathcal{I}_{i,j}$ the quadruple integral expression on the right hand side above with $J_{i,j}^{k,l}$, i.e.

$$J_{i,j}^{k,l} := \int_{\frac{i-1}{n}}^{\frac{i}{n}} \int_{\frac{j-1}{n}}^{\frac{j}{n}} \int\int_{S_{k,l}(x,y,\frac{1}{n})} \frac{1}{|(x - w, y - z)|} \, d(w,z) \, dy dx.$$

Again, let $1 \leq i, j \leq n$ be fixed. The set of pairs of indices with non-vanishing summands $\mathcal{I}_{i,j}$ may be partitioned into the sets

$$\mathcal{I}_{i,j}^{d} := \{(k,l) \in \mathcal{I}_{i,j} \; : \; k \neq i \text{ and } l \neq j\}$$

of pairs of indices marking squares diagonally adjacent to the square $I_{i,j}$ and

$$\mathcal{I}_{i,j}^{l} := \{(k,l) \in \mathcal{I}_{i,j} \; : \; k = i \text{ or } l = j\}$$

collecting pairs of indices that denote squares laterally adjacent to $I_{i,j}$. By simple transformations of the kind $(x, w) \mapsto (x \pm \frac{1}{n}, w \pm \frac{1}{n})$ etc. and applications of Fubini's theorem (as in the one-dimensional case) (or by geometric insight) we realize that the integrals $J_{i,j}^{k,l}$ are equal for all $(k,l) \in \mathcal{I}_{i,j}^{d}$ and the same holds true for all $(k,l) \in \mathcal{I}_{i,j}^{l}$. Further, the respective values of the two evaluations are independent of i and j.

Thus, it suffices to compute the values of $J_{i_*,j_*}^{i_*-1,j_*-1}$ and $J_{i_*,j_*}^{i_*,j_*-1}$ for some fixed $2 \leq i_*, j_* \leq n$, and the final result will be

$$\begin{aligned}
&\frac{\pi}{2n^2} \mathcal{R}_n^1(f) \\
&= \left(\sum_{i=2}^{n} \sum_{j=2}^{n} |a_{i,j} - a_{i-1,j-1}| + \sum_{i=1}^{n-1} \sum_{j=2}^{n} |a_{i,j} - a_{i+1,j-1}| \right) J_{i_*,j_*}^{i_*-1,j_*-1}
\end{aligned} \tag{8}$$

$$+\left(\sum_{i=1}^{n}\sum_{j=2}^{n}\left|a_{i,j}-a_{i,j-1}\right|+\sum_{i=2}^{n}\sum_{j=1}^{n}\left|a_{i,j}-a_{i-1,j}\right|\right)J_{i*,j*}^{i*,j*-1}.$$

We begin with the more complex case of laterally adjacent squares and evaluate

$$J_{i,j}^{i,j-1}=\int_{\frac{i-1}{n}}^{\frac{i}{n}}\int_{\frac{j-1}{n}}^{\frac{j}{n}}\int\int_{S_{i,j-1}(x,y,\frac{1}{n})}\frac{1}{|(x-w,y-z)|}\,d(w,z)\,dy dx$$

for some fixed $2\leq i,j\leq n$.

We first point out that

$$J_{\frac{1}{2}}:=\int_{\frac{i-1}{n}+\frac{1}{2n}}^{\frac{i}{n}}\int_{\frac{j-1}{n}}^{\frac{j}{n}}\int\int_{S_{i,j-1}(x,y,\frac{1}{n})}\frac{1}{|(x-w,y-z)|}\,d(w,z)\,dy dx=$$

$$\int_{\frac{i-1}{n}}^{\frac{i-1}{n}+\frac{1}{2n}}\int_{\frac{j-1}{n}}^{\frac{j}{n}}\int\int_{S_{i,j-1}(x,y,\frac{1}{n})}\frac{1}{|(x-w,y-z)|}\,d(w,z)\,dy dx$$

such that

$$J_{i,j}^{i,j-1}=2J_{\frac{1}{2}}.$$

The above can be established by application of the transformations $(x,w)\mapsto(\frac{2i-1}{n}-x,\frac{2i-1}{n}-w)$.

Let (x,y) be chosen from $I_{i,j}$ with $x\geq\frac{i-1}{n}+\frac{1}{2n}$. We analyze the inner double integral

$$\int\int_{S_{i,j-1}(x,y,\frac{1}{n})}\frac{1}{|(x-w,y-z)|}\,d(w,z)$$

of $J_{\frac{1}{2}}$. We use the abbreviations

$$a:=x-\frac{i-1}{n},\qquad b:=y-\frac{j-1}{n}\qquad\text{and}\qquad d:=x-\frac{i}{n}.$$

for the distances of x and y to some of the nodes. Note that by our choice of x and y we have the inequalities $a,b>0$, $d<0$ and, in particular, $a>|d|$.

Let $(w,z)\in S_{i,j-1}\left(x,y,\frac{1}{n}\right)$. It follows that

$$(x-w)^2<\frac{1}{n^2}-(y-z)^2.$$

and therefore,

$$z>y-\frac{1}{n}\qquad\text{and}\qquad x-\sqrt{\frac{1}{n^2}-(y-z)^2}<w<x+\sqrt{\frac{1}{n^2}-(y-z)^2}.$$

Thus, $z \in \left(y - \frac{1}{n}, \frac{j-1}{n}\right)$ and we have to analyze the intersection of intervals

$$I_w := \left(\frac{i-1}{n}, \frac{i}{n}\right) \cap \left(x - \sqrt{\frac{1}{n^2} - (y-z)^2}, x + \sqrt{\frac{1}{n^2} - (y-z)^2}\right). \quad (9)$$

(The index w in the symbol I_w is just used as a symbol to indicate that we are dealing with the integration domain of the variable w but does not stand for the values of w. The same applies to I_z etc. below.) We first point out that

$$\frac{i-1}{n} < x - \sqrt{\frac{1}{n^2} - (y-z)^2} \Longleftrightarrow z < y - \sqrt{\frac{1}{n^2} - a^2}. \quad (10)$$

Thus the result of (9) is dependent from the intersection

$$I_z := \left(y - \frac{1}{n}, \frac{j-1}{n}\right) \cap \left(-\infty, y - \sqrt{\frac{1}{n^2} - a^2}\right). \quad (11)$$

While it is clear that $y - \sqrt{\frac{1}{n^2} - a^2} > y - \frac{1}{n}$ we have

$$y - \sqrt{\frac{1}{n^2} - a^2} < \frac{j-1}{n} \Longleftrightarrow y < \frac{j-1}{n} + \sqrt{\frac{1}{n^2} - a^2} \quad (12)$$

with $\frac{j-1}{n} < \frac{j-1}{n} + \sqrt{\frac{1}{n^2} - a^2} < \frac{j}{n}$.

We first consider the case that $y \in I_y := \left(\frac{j-1}{n}, \frac{j-1}{n} + \sqrt{\frac{1}{n^2} - a^2}\right)$. Then $I_z = \left(y - \frac{1}{n}, y - \sqrt{\frac{1}{n^2} - a^2}\right)$.

We look at the subcase $z \in I_z$. It follows by (10) that now the lower bound of I_w is $x - \sqrt{\frac{1}{n^2} - (y-z)^2}$.

Considering its upper bound we have to find the minimum of $x + \sqrt{\frac{1}{n^2} - (y-z)^2}$ and $\frac{i}{n}$. Similarly to (10) we get that

$$\frac{i}{n} > x + \sqrt{\frac{1}{n^2} - (y-z)^2} \Longleftrightarrow z < y - \sqrt{\frac{1}{n^2} - d^2}. \quad (13)$$

Since $a^2 > d^2$ it is true that $y - \sqrt{\frac{1}{n^2} - d^2} \in I_z$ implying that the treated subcase has two more subsubcases: $z \in I_{z,1} := \left(y - \frac{1}{n}, y - \sqrt{\frac{1}{n^2} - d^2}\right)$ and $z \in I_{z,2} := \left(y - \sqrt{\frac{1}{n^2} - d^2}, y - \sqrt{\frac{1}{n^2} - a^2}\right)$.

By (13) and (9) the upper bound of I_w is $x+\sqrt{\frac{1}{n^2}-(y-z)^2}$ if $z \in I_{z,1}$ and equals $\frac{i}{n}$ if $z \in I_{z,2}$. Thus, the treated subcase gives rise to the following two quadruple integrals

$$\int_{\frac{i-1}{n}+\frac{1}{2n}}^{\frac{i}{n}} \int_{\frac{j-1}{n}}^{\frac{j-1}{n}+\sqrt{\frac{1}{n^2}-a^2}} \int_{y-\frac{1}{n}}^{y-\sqrt{\frac{1}{n^2}-d^2}} \int_{x-\sqrt{\frac{1}{n^2}-(y-z)^2}}^{x+\sqrt{\frac{1}{n^2}-(y-z)^2}} \frac{1}{|(x-w,y-z)|} dwdzdydx$$

$$\int_{\frac{i-1}{n}+\frac{1}{2n}}^{\frac{i}{n}} \int_{\frac{j-1}{n}}^{\frac{j-1}{n}+\sqrt{\frac{1}{n^2}-a^2}} \int_{y-\sqrt{\frac{1}{n^2}-d^2}}^{y-\sqrt{\frac{1}{n^2}-a^2}} \int_{x-\sqrt{\frac{1}{n^2}-(y-z)^2}}^{\frac{i}{n}} \frac{1}{|(x-w,y-z)|} dwdzdydx$$

which we denote by K_1 and K_2, respectively.

We turn to the subcase that $z \notin I_z$, i.e. $z \in \left(y-\sqrt{\frac{1}{n^2}-a^2}, \frac{j-1}{n}\right)$. By (10) the lower bound of I_w is in this subcase $\frac{i-1}{n}$. Since $y-\sqrt{\frac{1}{n^2}-a^2} > y-\sqrt{\frac{1}{n^2}-d^2}$ by (13) in this subcase the upper bound of I_w is $\frac{i}{n}$. This subcase yields the quadruple integral

$$K_3 := \int_{\frac{i-1}{n}+\frac{1}{2n}}^{\frac{i}{n}} \int_{\frac{j-1}{n}}^{\frac{j-1}{n}+\sqrt{\frac{1}{n^2}-a^2}} \int_{y-\sqrt{\frac{1}{n^2}-a^2}}^{\frac{j-1}{n}} \int_{\frac{i-1}{n}}^{\frac{i}{n}} \frac{1}{|(x-w,y-z)|} dwdzdydx.$$

We still need to analyze the case $y \notin I_y$, that is, $y \in \left(\frac{j-1}{n}+\sqrt{\frac{1}{n^2}-a^2}, \frac{j}{n}\right)$. In this case by (12) and (11) the interval I_z equals $\left(y-\frac{j-1}{n}, \frac{j-1}{n}\right)$. By (12) and (11) it is clear that the lower bound of I_w is $x-\sqrt{\frac{1}{n^2}-(y-z)^2}$. The determination of the upper bound is a little more intricate including two subcases concerning the choice of the domain of y, one of which generating two subsubcases concerning the domain of z. However, its computation processes similarly enough to the computations in the first case that we skip it here and just state the resulting quadruple integrals which we name K_4, K_5 and K_6, respectively.

$$\int_{\frac{i-1}{n}+\frac{1}{2n}}^{\frac{i}{n}} \int_{\frac{j-1}{n}+\sqrt{\frac{1}{n^2}-a^2}}^{\frac{j-1}{n}+\sqrt{\frac{1}{n^2}-d^2}} \int_{y-\frac{1}{n}}^{y-\sqrt{\frac{1}{n^2}-d^2}} \int_{x-\sqrt{\frac{1}{n^2}-(y-z)^2}}^{x+\sqrt{\frac{1}{n^2}-(y-z)^2}} \frac{1}{|(x-w,y-z)|} dwdzdydx,$$

$$\int_{\frac{i-1}{n}+\frac{1}{2n}}^{\frac{i}{n}} \int_{\frac{j-1}{n}+\sqrt{\frac{1}{n^2}-a^2}}^{\frac{j-1}{n}+\sqrt{\frac{1}{n^2}-d^2}} \int_{y-\sqrt{\frac{1}{n^2}-d^2}}^{\frac{j-1}{n}} \int_{x-\sqrt{\frac{1}{n^2}-(y-z)^2}}^{\frac{i}{n}} \frac{1}{|(x-w,y-z)|} dwdzdydx,$$

$$\int_{\frac{i-1}{n}+\frac{1}{2n}}^{\frac{i}{n}} \int_{\frac{j-1}{n}+\sqrt{\frac{1}{n^2}-d^2}}^{\frac{j}{n}} \int_{y-\frac{1}{n}}^{\frac{j-1}{n}} \int_{x-\sqrt{\frac{1}{n^2}-(y-z)^2}}^{x+\sqrt{\frac{1}{n^2}-(y-z)^2}} \frac{1}{|(x-w,y-z)|}\, dwdzdydx.$$

Altogether,

$$J_{\frac{1}{2}} = \sum_{i=1}^{6} K_i.$$

The evaluation of the six quadruple integrals K_i involves the transformation of the respective inner double integrals to polar coordinates. In order to simplify this procedure we first translate the integration domain of the respective inner double integral to the rectangle $(-\frac{1}{n}, \frac{1}{n}) \times (0, \frac{1}{n})$. In all six cases given a point (x, y) from the domain of the respective outer double integral this is done by application of the transformation $(w, z) \mapsto (x - w, y - z)$. Let L_i be the result of this application to K_i. Then

$$L_1 = \int_{\frac{i-1}{n}+\frac{1}{2n}}^{\frac{i}{n}} \int_{\frac{j-1}{n}}^{\frac{j-1}{n}+\sqrt{\frac{1}{n^2}-a^2}} \int_{\sqrt{\frac{1}{n^2}-d^2}}^{\frac{1}{n}} \int_{-\sqrt{\frac{1}{n^2}-z^2}}^{\sqrt{\frac{1}{n^2}-z^2}} \frac{1}{|(w,z)|}\, dwdzdydx,$$

$$L_2 = \int_{\frac{i-1}{n}+\frac{1}{2n}}^{\frac{i}{n}} \int_{\frac{j-1}{n}}^{\frac{j-1}{n}+\sqrt{\frac{1}{n^2}-a^2}} \int_{\sqrt{\frac{1}{n^2}-a^2}}^{\sqrt{\frac{1}{n^2}-d^2}} \int_{d}^{\sqrt{\frac{1}{n^2}-z^2}} \frac{1}{|(w,z)|}\, dwdzdydx,$$

$$L_3 = \int_{\frac{i-1}{n}+\frac{1}{2n}}^{\frac{i}{n}} \int_{\frac{j-1}{n}}^{\frac{j-1}{n}+\sqrt{\frac{1}{n^2}-a^2}} \int_{b}^{\sqrt{\frac{1}{n^2}-a^2}} \int_{d}^{a} \frac{1}{|(w,z)|}\, dwdzdydx,$$

$$L_4 = \int_{\frac{i-1}{n}+\frac{1}{2n}}^{\frac{i}{n}} \int_{\frac{j-1}{n}+\sqrt{\frac{1}{n^2}-a^2}}^{\frac{j-1}{n}+\sqrt{\frac{1}{n^2}-d^2}} \int_{\sqrt{\frac{1}{n^2}-d^2}}^{\frac{1}{n}} \int_{-\sqrt{\frac{1}{n^2}-z^2}}^{\sqrt{\frac{1}{n^2}-z^2}} \frac{1}{|(w,z)|}\, dwdzdydx,$$

$$L_5 = \int_{\frac{i-1}{n}+\frac{1}{2n}}^{\frac{i}{n}} \int_{\frac{j-1}{n}+\sqrt{\frac{1}{n^2}-a^2}}^{\frac{j-1}{n}+\sqrt{\frac{1}{n^2}-d^2}} \int_{b}^{\sqrt{\frac{1}{n^2}-d^2}} \int_{d}^{\sqrt{\frac{1}{n^2}-z^2}} \frac{1}{|(w,z)|}\, dwdzdydx,$$

$$L_6 = \int_{\frac{i-1}{n}+\frac{1}{2n}}^{\frac{i}{n}} \int_{\frac{j-1}{n}+\sqrt{\frac{1}{n^2}-d^2}}^{\frac{j}{n}} \int_{b}^{\frac{1}{n}} \int_{-\sqrt{\frac{1}{n^2}-z^2}}^{\sqrt{\frac{1}{n^2}-z^2}} \frac{1}{|(w,z)|}\, dwdzdydx.$$

Let for $1 \le i \le 6$ the function F_i be the evaluation function of the inner double integral of L_i defined on the domain of the outer double integral of L_i.

In L_1, L_4 and L_6 the integration domain of the inner double integral is a segment of the circle $B(0, \frac{1}{n})$ that results from the intersection of that circle with a parallel to the x-axis. A straight-forward transformation to polar coordinates (r, ϕ) yields for example for F_1:

$$F_1(x,y) = \int_{\sqrt{\frac{1}{n^2}-d^2}}^{\frac{1}{n}} \int_{\arcsin\left(\frac{\sqrt{\frac{1}{n^2}-d^2}}{r}\right)}^{\arcsin\left(-\frac{\sqrt{\frac{1}{n^2}-d^2}}{r}\right)+\pi} d\phi dr$$

$$= \int_{\sqrt{\frac{1}{n^2}-d^2}}^{\frac{1}{n}} \pi - 2\arcsin\left(-\frac{\sqrt{\frac{1}{n^2}-d^2}}{r}\right) dr$$

$$= \pi r - 2r\arcsin\left(\frac{\sqrt{\frac{1}{n^2}-d^2}}{r}\right)$$

$$-2\sqrt{\frac{1}{n^2}-d^2}\,\operatorname{arcoth}\left(\frac{r}{\sqrt{r^2-\frac{1}{n^2}+d^2}}\right)\Bigg|_{\sqrt{\frac{1}{n^2}-d^2}}^{\frac{1}{n}}$$

$$= \frac{\pi}{n} - 2\frac{\arcsin\left(\sqrt{\frac{1}{n^2}-d^2}\,n\right)}{n} +$$

$$\sqrt{\frac{1}{n^2}-d^2}\left(\ln\left(\frac{1}{n}+d\right) - \ln\left(\frac{1}{n}-d\right)\right),$$

and F_4 and F_6 are treated analogously.

The integration domain of F_3 is a rectangle with edges parallel to the axes stretching across both quadrants of the upper half plane. In order to transform this domain to polar coordinates we split it along the y-axis in two axis-parallel rectangles that reside in the second and first quadrant, respectively,

$$F_3(x,y) = \int_b^{\sqrt{\frac{1}{n^2}-a^2}} \int_d^0 \frac{1}{|(w,z)|}\, dw dz + \int_b^{\sqrt{\frac{1}{n^2}-a^2}} \int_0^a \frac{1}{|(w,z)|}\, dw dz,$$

and call the resulting double integrals $A(x,y)$ and $B(x,y)$.

We turn to the computation of $A(x,y)$. The transformation of a rectangle domain located in the second quadrant to polar coordinates depends on whether its bottom left vertex or its top right vertex is more distant from the origin. In the case of $A(x,y)$ this conditions reads

$$|(d,b)| < \left|\left(0, \frac{1}{n^2} - a^2\right)\right|. \tag{14}$$

In the case where (14) holds true the transformation to polar coordinates yields

$$A(x,y) = \int_b^{|(d,b)|} \int_{\frac{\pi}{2}}^{-\arcsin\left(\frac{b}{r}\right)+\pi} d\phi dr + \int_{|(d,b)|}^{\left|\left(0,\frac{1}{n^2}-a^2\right)\right|} \int_{\frac{\pi}{2}}^{\arccos\left(\frac{d}{r}\right)} d\phi dr$$

$$+ \int_{\left|\left(0,\frac{1}{n^2}-a^2\right)\right|}^{\left|\left(d,\sqrt{\frac{1}{n^2}-a^2}\right)\right|} \int_{-\arcsin\left(\frac{\sqrt{\frac{1}{n^2}-a^2}}{r}\right)+\pi}^{\arccos\left(\frac{d}{r}\right)} d\phi dr,$$

in the opposite case $A(x, y)$ equals

$$\int_b^{\left|\left(0,\frac{1}{n^2}-a^2\right)\right|} \int_{\frac{\pi}{2}}^{-\arcsin\left(\frac{b}{r}\right)+\pi} d\phi dr + \int_{\left|\left(0,\frac{1}{n^2}-a^2\right)\right|}^{|(d,b)|} \int_{-\arcsin\left(\frac{\sqrt{\frac{1}{n^2}-a^2}}{r}\right)+\pi}^{-\arcsin\left(\frac{b}{r}\right)+\pi} d\phi dr$$

$$+ \int_{|(d,b)|}^{\left|\left(d,\sqrt{\frac{1}{n^2}-a^2}\right)\right|} \int_{-\arcsin\left(\frac{\sqrt{\frac{1}{n^2}-a^2}}{r}\right)+\pi}^{\arccos\left(\frac{d}{r}\right)} d\phi dr.$$

In both cases the occurring three double integrals can be evaluated similarly like $F_1(x, y)$ above. Summing together the respective three results yields in both cases the same result: a sum consisting of summands that are of one of the following three types: binary products where one factor is a logarithmic expression, binary products where one factor is an arcsin- or arccos-expression or binary products of a square root and π. By use of the appropriate transformation rules for arcus-expressions the two latter groups of binary products cancel each other out. Therefore, in both cases $A(x, y)$ equals

$$\frac{1}{2}\Bigg(b\left(\ln\left(\sqrt{d^2+b^2}+d\right)-\ln\left(\sqrt{d^2+b^2}-d\right)\right)$$

$$+ d\left(\ln\left(\sqrt{d^2+b^2}+b\right)-\ln\left(\sqrt{d^2+b^2}-b\right)\right.$$

$$\left.+\ln\left(\sqrt{\frac{1}{n^2}-a^2+d^2}-\sqrt{\frac{1}{n^2}-a^2}\right)-\ln\left(\sqrt{\frac{1}{n^2}-a^2+d^2}+\sqrt{\frac{1}{n^2}-a^2}\right)\right)$$

$$+\sqrt{\frac{1}{n^2}-a^2}\left(\ln\left(\sqrt{\frac{1}{n^2}-a^2+d^2}-d\right)-\ln\left(\sqrt{\frac{1}{n^2}-a^2+d^2}+d\right)\right)\Bigg).$$

The double integral $B(x, y)$ is evaluated in a completely analogous fashion.

The integration domains of F_2 and F_5 have a similar geometric structure. We give a short overview of the evaluation of F_2. As in the case of F_3 we split the integration domain along the y-axis in order to have less case distinction when transforming to polar coordinates:

$$F_2(x,y) := \int_{\sqrt{\frac{1}{n^2}-a^2}}^{\sqrt{\frac{1}{n^2}-d^2}} \int_d^0 \frac{1}{|(w,z)|}\, dwdz + \int_{\sqrt{\frac{1}{n^2}-a^2}}^{\sqrt{\frac{1}{n^2}-d^2}} \int_0^{\sqrt{\frac{1}{n^2}-z^2}} \frac{1}{|(w,z)|}\, dydx.$$

The first double integral has an axis-parallel rectangle domain located in the second quadrant and is treated like $A(x,y)$. The domain of the second double integral is the intersection of two shapes: an axis-parallel rectangle domain located in the first quadrant whose lower right vertex B lies on the circle $B(0,\frac{1}{n})$, and the circle $B(0,\frac{1}{n})$ itself. This means that by the intersection the right edge of the rectangle and parts of its top edge are exchanged with a circular arc around zero with radius $\frac{1}{n}$. The double integral is translated to polar coordinates as follows:

$$\int_{\sqrt{\frac{1}{n^2}-a^2}}^{\sqrt{\frac{1}{n^2}-d^2}} \int_{\arcsin\left(\frac{\sqrt{\frac{1}{n^2}-a^2}}{r}\right)}^{\frac{\pi}{2}} d\phi dr + \int_{\sqrt{\frac{1}{n^2}-d^2}}^{\frac{1}{n}} \int_{\arcsin\left(\frac{\sqrt{\frac{1}{n^2}-a^2}}{r}\right)}^{\arcsin\left(\frac{\sqrt{\frac{1}{n^2}-d^2}}{r}\right)} d\phi dr.$$

For the evaluation of F_5 proceed as for F_2. The only difference to F_2 lies in the fact that by the intersection with the circle $B(0,\frac{1}{n})$ also parts of the bottom line of the corresponding underlying rectangle are removed. In symbols this is reflected by exchanging every occurrence of the term $\sqrt{\frac{1}{n^2}-a^2}$ with b in the double integral directly above.

The integration of the functions F_i with respect to y can be executed in all cases by standard means. Note that the double integrals related to F_1, F_2 and F_4 do not depend on y such that those integrations are mere multiplications of the respective functions with the difference between the limits of the respective integrals. After all the resulting functions of the variable x have been summed up the fourth integration can be carried out yielding

$$J_{\frac{1}{2}} = \sum_{i=1}^{6} K_i = \sum_{i=1}^{6} L_i = \frac{1}{3n^3} \quad \text{and} \quad J_{i,j}^{i,j-1} = 2J_{\frac{1}{2}} = \frac{2}{3n^3}.$$

This solves the case of laterally adjacent squares

The case of diagonally adjacent squares is much simpler. For $2 \leq i,j \leq n$ we have to compute

$$J_{i,j}^{i-1,j-1} = \int_{\frac{i-1}{n}}^{\frac{i}{n}} \int_{\frac{j-1}{n}}^{\frac{j}{n}} \int\int_{S_{i-1,j-1}(x,y,\frac{1}{n})} \frac{1}{|(x-w,y-z)|}\, d(w,z)\, dydx.$$

where

$$S_{i-1,j-1}\left(x,y,\frac{1}{n}\right) = \left(\left(\frac{i-2}{n},\frac{i-1}{n}\right) \times \left(\frac{j-2}{n},\frac{j-1}{n}\right)\right) \cap B\left((x,y),\frac{1}{n}\right).$$

An easy computation shows that the latter set is empty if and only if $|(a,b)| \geq \frac{1}{n}$. Therefore, for the computation of the inner double integral of $J_{i,j}^{i-1,j-1}$ we may restrict ourselves to points $(x,y) \in I_{i,j}$ that satisfy $|(a,b)| < \frac{1}{n}$. By similar reasoning as in the lateral case we infer that $J_{i,j}^{i-1,j-1}$ equals

$$\int_{\frac{i-1}{n}}^{\frac{i}{n}} \int_{\frac{j-1}{n}}^{\frac{j-1}{n}+\sqrt{\frac{1}{n^2}-a^2}} \int_{y-\sqrt{\frac{1}{n^2}-a^2}}^{\frac{j-1}{n}} \int_{x-\sqrt{\frac{1}{n^2}-(z-y)^2}}^{\frac{i-1}{n}} \frac{1}{|(x-w,y-z)|}\, dw dz dy dx.$$

By applying the transformation $(w,z) \mapsto (x-w, y-z)$ this quadruple integral transforms to

$$\int_{\frac{i-1}{n}}^{\frac{i}{n}} \int_{\frac{j-1}{n}}^{\frac{j-1}{n}+\sqrt{\frac{1}{n^2}-a^2}} \int_{b}^{\sqrt{\frac{1}{n^2}-a^2}} \int_{a}^{\sqrt{\frac{1}{n^2}-z^2}} \frac{1}{|(w,z)|}\, dw dz dy dx,$$

and by changing the inner double integral to polar coordinates we get

$$\begin{aligned} J_{i,j}^{i-1,j-1} &= \int_{\frac{i-1}{n}}^{\frac{i}{n}} \int_{\frac{j-1}{n}}^{\frac{j-1}{n}+\sqrt{\frac{1}{n^2}-a^2}} \int_{|(a,b)|}^{\frac{1}{n}} \int_{\arcsin(\frac{b}{r})}^{\arccos(\frac{a}{r})} \frac{1}{|(w,z)|}\, dw dz dy dx \\ &= \frac{1}{6n^3}. \end{aligned}$$

Altogether, by (8) the final result is

$$\begin{aligned} \mathcal{R}_n^1(f) = {} & \frac{1}{3\pi n}\left(\sum_{i=2}^{n}\sum_{j=2}^{n}\left|a_{i,j}-a_{i-1,j-1}\right| + \sum_{i=1}^{n-1}\sum_{j=2}^{n}\left|a_{i,j}-a_{i+1,j-1}\right|\right) \\ & + \frac{4}{3\pi n}\left(\sum_{i=1}^{n}\sum_{j=2}^{n}\left|a_{i,j}-a_{i,j-1}\right| + \sum_{i=2}^{n}\sum_{j=1}^{n}\left|a_{i,j}-a_{i-1,j}\right|\right). \end{aligned}$$

2.2.2 Piecewise Constant Ansatz II

With f as above we also evaluated $\mathcal{R}_n^1(f)$ with the kernel functions

$$\varphi_n(\vec{v}) := \frac{n^2}{4}\chi_{(-\frac{1}{n},\frac{1}{n})\times(-\frac{1}{n},\frac{1}{n})}(\vec{v})$$

for all $\vec{v}$ in $\mathbb{R}^2$. Note that these kernel functions are not radial such that in this case the assumptions on (φ_n) from the introduction are not fully satisfied.

The evaluation proceeds similar to the above one. As a result we get

$$\mathcal{R}_n(f) = \frac{1}{3}\frac{\sqrt{2}-1}{n}\left(\sum_{i=2}^{n}\sum_{j=2}^{n}\left|a_{i,j}-a_{i-1,j-1}\right| + \sum_{i=1}^{n-1}\sum_{j=2}^{n}\left|a_{i,j}-a_{i+1,j-1}\right|\right)$$

$$+\frac{1}{12}\frac{3\ln\left(\sqrt{2}+1\right)-3\ln\left(\sqrt{2}-1\right)-2(\sqrt{2}-1)}{n}$$

$$\times\left(\sum_{i=1}^{n}\sum_{j=2}^{n}\left|a_{i,j}-a_{i,j-1}\right| + \sum_{i=2}^{n}\sum_{j=1}^{n}\left|a_{i,j}-a_{i-1,j}\right|\right).$$

3 Connection to Total Variation Regularization

In this section we give a short summary of the theoretical background on approximation properties of the minimizers of (4) to the minimizer of (5). This justifies the minimizers of the Galerkin approximations, where the calculations of the previous section have to be used, as approximations of the minimizer of the total variation functional. In the following we review existence and uniqueness of a minimizer of the functional $\mathcal{F}_n$ defined in (4) as in [1].

The first stated result is on weak lower semicontinuity of the functional $\mathcal{R}_n$. The proof uses standard techniques from convex analysis and functional analysis and is therefore omitted.

Lemma 1. *Let $1 \leq q < \infty$. For all $n \in \mathbb{N}$, $\mathcal{R}_n$ is weakly lower semicontinuous on $L^q(\Omega)$, that is,*

$$\mathcal{R}_n(g) \leq \liminf_{k\to\infty} \mathcal{R}_n(g_k)$$

for every sequence $(g_k) \in L^q(\Omega)$ that converges weakly with respect to the L^q-topology to a function $g \in L^q(\Omega)$.

Moreover, by using the following variant of compactness results established by Bourgain et al. [3] and Ponce [9]), we can prove existence and uniqueness of minimizers of $\mathcal{F}$ in a very similar manner as in [1].

Theorem 2. *Assume that (g_n) is a sequence of functions in $L^1_\diamond(\Omega)$ such that $\mathcal{R}_n(g_n)$ is uniformly bounded. Then the sequence (g_n) is relatively compact in $L^1_\diamond(\Omega)$ and has a subsequence (g_{n_k}) converging (in the L^1-norm) to a limit function g that lies in $BV(\Omega)$.*

We note that the regularization term of the functional $\mathcal{F}$ depends exclusively on a derivative. Therefore, for data f^δ and $f^\delta + C$, where C is a constant, the according minimizers of the functional $\mathcal{F}$ also differs by C. The same holds true for the functionals $\mathcal{F}_n$. Thus, we may reduce our investigations to the case where we assume that the mean of f^δ is zero, and consequently, also the mean of the minimizers of $\mathcal{F}_n$ and $\mathcal{F}$ is zero.

Proposition 3. *Let $\alpha > 0$ and assume that $f^\delta \in L^2_\diamond(\Omega)$.*

1. *Then the functional $\mathcal{F}_n$ attains a unique minimizer f_n over $L^2_\diamond(\Omega)$ that also belongs to $L^1_\diamond(\Omega)$.*
2. *The function f_n is also a minimizer of $\mathcal{F}_n$ over $L^1(\Omega)$.*
3. *The sequence of numbers $(\mathcal{R}_n(f_n))$ is uniformly bounded over $n \in \mathbb{N}$, and the sequence f_n is relatively compact in $L^1_\diamond(\Omega)$ and has a convergent subsequence whose limit f is an element of $BV(\Omega)$. With the help of Γ-convergence (s. below) it can be established that in fact, the whole sequence (f_n) is converging to f.*

It remains to clarify whether the limit function f is a minimum of the respective limit functional $\mathcal{F}$. The concerning questions are answered to a large extent by a result of Ponce [10] in terms of Γ-convergence. This technique has also been used in [1]

We recall the notion of Γ-convergence in $L^1(\Omega)$ [5]. Let (F_n) denote a sequence of lower semicontinuous functionals mapping functions from $L^1(\Omega)$ to the set of extended real numbers $\bar{\mathbb{R}}$, and let F be a functional of this kind, too. Then the sequence (F_n) Γ-converges to F with respect to the $L^1(\Omega)$-topology if and only if the following two conditions are satisfied:

- For every $g \in L^1(\Omega)$ and for every sequence (g_n) in $L^1(\Omega)$ converging to g in the L^1-norm we have
$$F(g) \le \liminf_{n\to\infty} F_n(g_n);$$
- For every $g \in L^1(\Omega)$ there exists a sequence (g_n) in $L^1(\Omega)$ converging to g in the L^1-norm with
$$F(g) = \lim_{n\to\infty} F_n(g_n).$$

In this case we write
$$\Gamma_{L^1(\Omega)}\text{-}\lim_{n\to\infty} F_n = F.$$

We note that all of the functionals $\mathcal{F}_n$ and $\mathcal{F}$ are lower semicontinuous with respect to the topology of $L^1(\Omega)$.

Theorem 4. *The sequence of functionals $(\mathcal{F}_n)$ converges in the $\Gamma_{L^1(\Omega)}$-sense to the limit functional $\mathcal{F}$.*

Corollary 5. *The limit function f of the sequence of minimizers (f_n) of $\mathcal{F}_n$ is the unique minimum of the limit functional $\mathcal{F}$ over $L^1(\Omega)$. The minimum f also belongs to the space $L^2_\diamond(\Omega) \cap BV(\Omega)$.*

4 Discussion and Outlook

In Sect. 2 of this paper we have shown that analytical evaluations of double integrals $\mathcal{R}_n(f_d)$ are possible for some types of discrete approximations f_d to a continuous or integrable function f by using the functionalities of computer algebra software systems. By such evaluations, we have produced new candidates for approximation schemes of the total variation seminorm of a function, which can be implemented in standard numerical algorithms for solving problems like total variation regularization.

In Sect. 3 we reviewed that in fact solutions of total variation regularization can be approximated by minimizers of the functional that results from the total variation regularization functional by exchanging the total variation seminorm with a double integral functional $\mathcal{R}_n$.

We have already started to investigate the usefulness of the new approximation schemes in numerical applications. For example, approximation scheme (6) has been applied to total variation regularization within the framework of a gradient descent algorithm, and indeed, promising results have been obtained this way [2].

However, from the theoretical point of view there are many important points open, for example a systematic analysis of the integral expression $\mathcal{R}_n(f_d)$ in terms of analytical evaluation exploring all its degrees of freedom, a detailed convergence analysis for the resulting discrete functionals and further numerical tests for appropriate applications involving the evaluation of the total variation seminorm.

Altogether, we have presented a new approach to the treatment of the total variation seminorm in numerical applications. Our contribution might pave the way for the development of more sophisticated tools for the treatment of double integrals of the kind $\mathcal{R}_n(f_d)$ as well as the development of new approximation schemes to the total variation seminorm.

Acknowledgements The authors would like to express their gratitude to Paul F. X. Müller for introducing us to the recent work on the new characterizations of Sobolev spaces and BV and some stimulating discussions. This work has been supported by the Austrian Science Fund (FWF) within the national research networks Industrial Geometry, project 9203-N12, and Photoacoustic Imaging in Biology and Medicine, project S10505-N20. Moreover, the authors thank a referee for the careful reading of the manuscript and many detailed comments.

References

1. Aubert, G., Kornprobst, P.: Can the nonlocal characterization of Sobolev spaces by Bourgain et al. be useful for solving variational problems? SIAM J. Numer. Anal. **47**(2), 844–860 (2009)
2. Boulanger, J., Elbau, P, Pontow, C, Scherzer, O.: Non-local functionals for imaging. In: Bauschke, H.H., Burachik, R.S., Combettes, P.L., Elser, V., Luke, D.R., Wolkowicz, H. (eds.) Springer Optimization and Its Applications. Fixed-Point Algorithms for Inverse Problems in Science and Engineering, vol. 49, 1st edn., pp. 131–154. Springer, New York (2011). ISBN 978-1-4419-9568-1
3. Bourgain, J., Brézis, H., Mironescu, P.: Another look at Sobolev spaces. In: Menaldi, J.L., Rofman, E., Sulem, A. (eds.) Optimal Control and Partial Differential Equations-Innovations & Applications: In honor of Professor Alain Bensoussan's 60th anniversary, pp. 439–455. IOS press, Amsterdam (2000)
4. Chambolle, A.: An algorithm for total variation minimization and applications. J. Math. Imaging Vis. **20**(1–2), 89–97 (2004)

5. Dal Maso, G.: An Introduction to Γ-Convergence, volume 8 of Progress in Nonlinear Differential Equations and their Applications. Birkhäuser (1993)
6. Dávila, J.: On an open question about functions of bounded variation. Calc. Var. Partial Differ. Equ. **15**(4), 519–527 (2002)
7. Dobson, D., Scherzer, O.: Analysis of regularized total variation penalty methods for denoising. Inverse Probl. **12**(5), 601–617 (1996)
8. Dobson, D.C., Vogel, C.R.: Convergence of an iterative method for total variation denoising. SIAM J. Numer. Anal. **34**(5), 1779–1791 (1997)
9. Ponce, A.: An estimate in the spirit of Poincaré's inequality. J. Eur. Math. Soc. (JEMS) **6**(1), 1–15 (2004)
10. Ponce, A.: A new approach to Sobolev spaces and connections to Γ-convergence. Calc. Var. Partial Differ. Equ. **19**, 229–255 (2004)
11. Vogel, C.R.: Computational Methods for Inverse Problems, volume 23 of Frontiers in Applied Mathematics. SIAM, Philadelphia (2002)
12. Rudin, L.I., Osher, S., Fatemi, E.: Nonlinear total variation based noise removal algorithms. Phys. D. Nonlinear Phenom. **60**(1–4), 259–268 (1992)

Sound and Complete Verification Condition Generator for Functional Recursive Programs

Nikolaj Popov and Tudor Jebelean

Abstract We present a method for verifying recursive functional programs by defining a verification condition generator (VCG) which covers the most frequent type of recursive programs. These programs may operate on arbitrary domains. We prove *soundness* and *completeness* of the VCG and this provides a warranty that any system based on our results will be sound.

We introduce here the notion of *completeness* of a VCG as a duality of *soundness*. It is important for the following two reasons: theoretically, it is the dual of *soundness* and practically, it helps debugging. Any counterexample for the failing verification condition will carry over to a counterexample for the given program and specification. Moreover, the failing proof gives information about the place of the bug.

Furthermore, we introduce a specialized strategy for termination. The termination problem is reduced to the termination of a simplified version of the program. The conditions for the simplified versions are identical for special classes of functional programs, thus they are highly reusable.

1 Introduction

Since the beginning of program verification back in the 1950-s, a good deal of theoretical and practical results have been achieved in research, and the concrete application of these in industrial software development is slowly but steadily progressing. We are convinced that in order to increase the quality of the software production, program verification and formal methods should play a bigger role during the process of software design and composition.

N. Popov (✉) · T. Jebelean
Research Institute for Symbolic Computation, Johannes Kepler University, Linz, Austria
e-mail: popov@risc.uni-linz.ac.at; jebelean@risc.uni-linz.ac.at

U. Langer and P. Paule (eds.), *Numerical and Symbolic Scientific Computing*, Texts and Monographs in Symbolic Computation, DOI 10.1007/978-3-7091-0794-2_11,

The research on program verification presented here is dedicated to the study and development of a relevant theory, which may serve as a basis for the practical need of proving program correctness in an automatic manner. Additionally this basis can be used in a tutorial way for the introduction of formal verification techniques to the students in computer science.

We are primarily concerned with the generation of verification conditions, while the actual proving of these verification conditions is subject to further research, in particular in the frame of the *Theorema* project (`www.theorema.org`). One important purpose of this research is to create a mechanism for the generation of the verification conditions which is *very simple*, but still provably correct and complete in the context of predicate logic.

Program specification (or formal specification of a program) is the definition of what a program is expected to do. Normally, it does not describe, and it should not, how the program is implemented. The specification is usually provided by logical formulas describing a relationship between input and output parameters. We consider specifications which are pairs, containing a precondition (input condition) and a postcondition (output condition).

Given such a specification, it is possible to use formal verification techniques to demonstrate that a program is correct with respect to the specification.

A precondition (or input predicate) of a program is a condition that must always be true just prior to the execution of that program. It is expressed by a predicate on the input of the program. If a precondition is violated, the effect of the program becomes undefined and thus may or may not carry out its intended work. For example: the factorial is only defined for integers greater than or equal to zero. So a program that calculates the factorial of an input number would have preconditions that the number be an integer and that it be greater than or equal to zero.

A postcondition (or output predicate) of a program is a condition that must always be true just after the execution of that program. It is expressed by a predicate on the input and the output of the program.

We do not consider informal specifications, which are normally written as comments between the lines of code.

Formal verification is, in general, the act of proving mathematically the correctness of a program with respect to a certain formal specification. Software testing, in contrast to verification, cannot prove that a system does not contain any defects, neither that it has a certain property, e.g., correctness with respect to a specification. Only the process of formal verification can prove that a system does not have a certain defect or does have a certain property.

The problem of verifying programs is usually split into two subproblems: generate verification conditions which are sufficient for the program to be correct and prove the verification conditions, within the theory of the domain for which the program is defined. In this work we will concentrate on the generation of verification conditions.

A verification condition generator (VCG) is a device – normally implemented by a program – which takes a program, actually its source code, and the specification, and produces verification conditions. These verification conditions do not contain any part of the program text, and are expressed in some logical formalism.

Let us say, the program is F and the specification I_F (input predicate), and O_F (output predicate) is provided. The verification conditions generated by VCG are: $VC_1, VC_2, \ldots, VC_n$. After having the verification conditions at hand, one has to prove them as logical formulas in the theory of the domain $\mathbb{D}_F$ on which the program F is defined, e.g., integers, reals, etc.

Normally, these conditions are given to an automatic or semi-automatic theorem prover. If all of them hold, then the program is correct with respect to its specification. The latter statement we call *soundness* of the VCG.

Formally, soundness is expressed as follows: *Given a program F and a specification $\langle I_F, O_F \rangle$, if the verification conditions generated by the VCG hold as logical formulas, then the program F is correct with respect to the specification $\langle I_F, O_F \rangle$.*

It is clear that whenever one defines a VCG, the first task to be done is proving its soundness statement.

Completing the notion of *soundness* of a VCG, we introduce its dual notion – *completeness*. The respective *completeness* statement of the VCG is :

Given a program F and a specification $\langle I_F, O_F \rangle$, if the program F is correct with respect to the specification $\langle I_F, O_F \rangle$, then the verification conditions generated by the VCG hold as logical formulas.

The notion of *completeness* of a VCG is important for the following two reasons: theoretically, it is the dual of *soundness* and practically, it helps debugging. Any counterexample for the failing verification condition would carry over to a counterexample for the program and the specification, and thus give a hint on "what is wrong".

Indeed, most books about program verification present methods for verifying correct programs. However, in practical situations, it is the failure which occurs more often until the program and the specification are completely debugged.

A distinction is made between total correctness, which additionally requires that the program terminates, and partial correctness, which simply requires that if an answer is returned (that is, the program terminates) it will be correct.

For example, if we are successively searching through integers $1, 2, 3, \ldots$ to see if we can find an example of some phenomenon – say an odd perfect number – it is quite easy to write a partially correct program (use integer factorization to check n as perfect or not). But to say this program is totally correct would be to assert something currently not known in number theory.

The relation between partial and total correctness is informally given by:

$$\text{Total Correctness} = \text{Partial Correctness} + \text{Termination}.$$

The precise definition is as follows: Let us consider the program F (which takes an input and should produce an output) and the specification $\langle I_F, O_F \rangle$. The restriction to *one* input and *one* output is not important, because these could also be vectors of values. We denote by $F[x]$ the output of the program. Since when the program does not terminate this output does not exists, one should use $F[x]$ with a certain care. The input condition I_F is a unary predicate expressing the restrictions on the possible inputs; the output condition O_F is a binary predicate expressing the relationship between the input and the output of F.

Moreover, by $\downarrow$ we denote the predicate expressing termination. We will write $F[x] \downarrow$ and say "F terminates on x". The notation $F[x] \downarrow$ is used in most papers and textbooks, so we also use it here in order to avoid confusion. However, from the logical point of view, the correct notation should be $F \downarrow x$. Partial correctness of F is expressed by the formula:

$$(\forall x : I_F[x])(F[x] \downarrow \Longrightarrow O_F[x, F[x]]). \tag{1}$$

Termination of F is expressed by:

$$(\forall x : I_F[x]) F[x] \downarrow, \tag{2}$$

and total correctness of F is respectively:

$$(\forall x : I_F[x])(F[x] \downarrow \wedge O_F[x, F[x]]). \tag{3}$$

Logically, it is clear that partial correctness (1) and termination (2) imply total correctness (3).

The above considerations apply to all type of programs which have input and output, however in the current work we concentrate on *functional programs*. From our point of view, a logical program for a function F is collection of logical formulas of the type:

$$\forall x \varphi \Longrightarrow F[x] = \tau,$$

where φ is a formula and τ is a term using the predicate and the function symbols from the underlying theory of the domain[s] of the objects manipulated by the program. It is easy to see that the usual programming construct **If-then-else** can be used to abbreviate such a collection of formulas into an expression like e.g., (17). Thus, an important feature of our approach is that *the programs are logical formulas in the signature of the underlying theory of the objects manipulated by the program.*[1] This simplifies the process of reasoning about programs, because there is no need of translation from the programming language into the language of logic, and also no need to define the semantics of the programs, besides the semantics of the logic language itself.

Note that F is used both for denoting the program itself as well as the function implemented by the program, but the different meanings are easy to differentiate by the context.

[1]This approach is in fact borrowed from the *Theorema* system.

Automatic theorem proving, and more generally, automated reasoning is a border area of computer science and mathematics dedicated to understanding different aspects of reasoning in a way that allows the creation of software which makes computers to reason completely or nearly completely automatically. Automatic theorem proving is, in particular, the proving of mathematical theorems by an algorithm. In contrast to proof checking, where an existing proof for a theorem is certified valid, automatic theorem provers generate the proofs themselves. A recent and relatively comprehensive overview on that area may be found in [40].

The research presented in this paper is performed in the frame of the *Theorema* system [12], a mathematical computer assistant which aims at supporting the entire process of mathematical theory exploration: invention of mathematical concepts, invention and verification (proof) of propositions about concepts, invention of problems formulated in terms of concepts, invention and verification (proof of correctness) of algorithms, and storage and retrieval of the formulas invented and verified during this process. The system includes a collection of general as well as specific provers for various interesting domains (e.g., integers, sets, reals, tuples, etc.). More details about *Theorema* are available at `www.theorema.org`. The papers [11–13] are surveys, and point to earlier relevant papers.

1.1 Related Research

There is a huge amount of literature on program verification and a comprehensive overview on the topic may evolve into PhD thesis itself. However, summarizing, there are two main types of presentations:

- Classical approaches which are concerned more with the theory of computation and less with possible implementation of verification systems.
- Practical approaches to proving program correctness automatically or semi-automatically, which are less concerned with the theoretical foundations.

1.1.1 Theoretical Approaches

One of the first approaches to program verification was the axiomatic reasoning, which was initially developed by Floyd [16] for the verifications of flowcharts. This method was then further developed by Hoare [20] for dealing with *while*-programs and became known as Hoare Logic. The method provides a set of logical rules allowing to reason about the correctness of (imperative) computer programs with the methods of mathematical logic.

The central feature of Hoare logic is the Hoare triple. It describes how the execution of a piece of code changes the state of the computation. A Hoare triple is of the form:

$$\{P\}C\{Q\}, \qquad (4)$$

where P and Q are assertions (normally given by logical formulas) and C is a command, or a program. In the literature, P is called the precondition and Q the postcondition. Hoare logic has axioms and inference rules for all the constructs of simple imperative programming languages.

Although P and Q are very similar to the pre– and post–condition used in our approach, one should note that they are specifically designed for imperative constructs, since C is a *piece of program* (succession of imperative statements). In our approach the pre– and post–condition refer to the *whole* program, more specifically to the function implemented by the program.

Fixpoint induction or Scott induction is a proof principle due to Dana Scott [2]. It is useful for proving continuous properties of least fixed points of continuous operators. Scott semantics treats programs as continuous operators, and the computable functions they define are the least fixpoints of the operators [29]. Using Scott induction one may prove continuous properties (e.g., partial correctness) of programs, and therefore applying Scott induction for proving properties of programs is a very powerful technique. However, in automated reasoning proving by induction can get computationally very expensive, much more than proving predicate logic formulas.

Other important techniques exposed in classical books (e.g., [27, 29]) are very comprehensive, however, their orientation is theoretical rather than practical and mechanized. Verification in that context is normally a process in which the reader is required to understand the concept and perform creatively.

Furthermore, in order to perform verification, one uses a certain model of computation, which significantly increases the proving effort.

1.1.2 Practical Approaches

In contrast to classical books, practical computer aided verification is oriented towards verification of practical and popular types of programs (like in Java, C, Lisp) implementing primitive recursive functions, mutual recursive functions, etc. Performing creatively there is normally not required and the aim is to speed up the verification of relatively large programs.

In the PVS system [35] the approach is type theoretical and relies on exploration of certain subtyping properties. The realization is based on Church's higher-order logic.

The HOL system [21], originally constructed by Gordon, is also based on generalization of Church's higher-order logic. It mainly deals with primitive recursive functions, however, there is a very interesting work dedicated to transforming non-primitive recursive to primitive recursive functions [38]. There are various versions of HOL – in [17] one may find how it evolved over the years.

The Coq system [4] is based on a framework called "Calculus of Inductive Constructions" that is both a logic and a functional programming language. Coq has relatively big library with theories (e.g., $\mathbb{N}$, $\mathbb{Z}$, $\mathbb{Q}$, lists, etc.) where the individual proofs of the verification conditions may be carried over [5].

The KeY system [3] is not a classical verification condition generator, but a theorem prover for program logic. It is distinguished from most other deductive verification systems in that symbolic execution of programs, first-order reasoning, arithmetic simplification, external decision procedures, and symbolic state simplification are interleaved. KeY is mainly dedicated to support object-oriented models, where for loop- and recursion-free programs, symbolic execution is performed in an automated manner.

The Sunrise system [37] contains embedding of an imperative language within the HOL theorem prover. A very specific feature of the system is that its VCG is verified as sound, and that soundness proof is checked by the HOL system. The programming language containing assignments, conditionals, and *while* commands, and also mutually recursive procedures, however, all variables have the type $\mathbb{N}$.

The ACL2 system [1] is, in our opinion, one of the most comprehensive systems for program verification. It contains a programming language, an extensible theory in a first-order logic, and a theorem prover. The language and implementation of ACL2 are built on Common Lisp. ACL2 is intended to be an industrial strength version of the Boyer-Moore theorem prover NQTHM [6], however its logical basis remains the same.

Furthermore, in [30] it is shown how a theorem prover may be used directly on the operational semantics to generate verification conditions. Thus no separate VCG is necessary, and the theorem prover can be employed both to generate and to prove the verification conditions.

All these practical systems are very important and interesting, however they do not investigate the theoretical basis of verification, thus are not so relevant from the point of view of the research presented here. Moreover most of these practical approaches address programs written in imperative languages (see also some recent results in [15, 26, 41]), where the verification process has a quite different flavor.

1.2 Summary of Main Results

- We define a verification condition generator (VCG) which covers the most frequent type of recursive programs operating on arbitrary domains.
- As a distinctive feature of our method, the verification conditions do not refer to a theoretical model for program semantics or program execution, but only to the theory of the domain used in the program. This is very important for the automatic verification, because any additional theory present in the system would significantly increase the proving effort.
- We introduce here the notion of *completeness* of a VCG as a duality of *soundness*. It is important for the following two reasons: theoretically, it is the dual of *soundness* and practically, it helps debugging. Any counterexample for the failing verification condition will carry over to a counterexample for the given program and specification. Moreover, the failing proof gives information about the place of the bug.

- We prove *soundness* and *completeness* of the VCG and this provides a warranty that any system based on our results will be sound and will detect all bugs.
- We introduce a specialized strategy for termination. The termination problem is reduced to the termination of a simplified version of the program. The conditions for the simplified versions are identical for special classes of functional programs, thus they are highly reusable.

2 Automation of the Verification: VCG

In this section we develop a theoretical framework whose results are then used for automatic verification.

In the literature, there is a variety of strategies for obtaining proof rules. However, some of them have been discovered to be unsound [22]. Soundness of verification condition generators is automatically assumed, however many of them have not been proven sound. This implies that any of the programs which were verified by the help of an unsound VCG may, in fact, be incorrect.

In this paper we define necessary and also sufficient conditions for a program (of certain kind) to be totaly correct. We then construct a VCG which generates these conditions.

We prove soundness and completeness of the respective verification conditions. This implies that the validity of the verification conditions is necessary and sufficient to verify the total correctness of the program under consideration.

These proofs of soundness and completeness form the basis of an implementation of the VCG that ensures the verification of concrete programs.

2.1 Program Schemata

Considering program schemata [28] instead of concrete programs has a relatively long tradition. Early surveys on the theory of program schemata can be found in [18], and in more general splitting programs into types of programs is well studied in [33].

More generally, the use of schemata (axiom schemata, proposition schemata, problem schemata, and algorithm schemata) plays a very important role for algorithm-supported mathematical theory exploration [7, 10].

Program schemata are (almost) programs where the concrete constants, functions and predicates are replaced by symbolic expressions.

When investigating program schemata instead of concrete programs, one may derive properties which concern not just one concrete program, but many similar programs, more generally – a whole class of programs – those which fit into the schema.

Moreover, for a given schema, each concrete program can be obtained from it by an instantiation which gives concrete meanings to the constant, function and predicate symbols in the schema.

Smith proposed the use of schemata for synthesis of functional programs [36]. In fact, his work spans over more than two decades, and has produced some of the more important results in practical program synthesis.

A recent result on the application of program schemata to program synthesis is available at [7, 10]. There one may find how even non-trivial algorithms, e.g., Buchberger's algorithm for Gröbner bases [8, 9] may be synthesized fully automatically starting from the specification and the schema.

We approach the problem of program verification by studying one concrete program schema. When deriving necessary (and also sufficient) conditions for program correctness, we actually prove at the meta-level that for any program of that class (defined by the schema) it suffices to check only the respective verification conditions. This is very important for the automation of the whole process, because the production of the verification conditions is not expensive from the computational point of view.

The following example will give more intuition on the notions of program schemata and concrete programs. Let us consider the schema defining *simple recursive programs* :

$$F[x] = \mathbf{If}\ Q[x]\ \mathbf{then}\ S[x]\ \mathbf{else}\ C[x, F[R[x]]], \tag{5}$$

where Q is a predicate and S, C, R are auxiliary functions.

Consider also, the program $Fact$ for computing the *factorial* function:

$$Fact[n] = \mathbf{If}\ n = 0\ \mathbf{then}\ 1\ \mathbf{else}\ Fact[n-1]. \tag{6}$$

It is now obvious, that the program $Fact$ fits to the simple recursive program schema. In order to automate the process of reasoning about programs like $Fact$ we reason at the meta-level about their schemata.

2.2 Coherent Programs

Here we state the general principles we use for writing coherent programs with the aim of building up a non-contradictory system of verified programs. Although, these principles are not our invention (similar ideas appear in [24]), we state them here because we want to emphasize on and later formalize them. Similar to these ideas appear also in software engineering – they are called there *design by contract* or *programming by contract* [31].

We build our system such that it preserves the modularity principle, that is, a subprogram may be replaced by any other program that satisfies the same specification.

Building up correct programs: Firstly, we want to ensure that our system of coherent programs would contain only correct (verified) programs. This we achieve, by:

- Start from basic (trustful) functions e.g., addition, multiplication, etc.;
- Define each new function in terms of already known (defined previously) functions by giving its source text, the specification (input and output predicates) and prove their total correctness with respect to the specification.

This simple inductively defined principle would guarantee that no wrong program may enter our system. Next we want to ensure is the easy exchange (mobility) of our program implementations. This principle is usually referred as:

Modularity: Once we define the new function and prove its correctness, we "forbid" using any knowledge concerning the concrete function definition. The only knowledge we may use is the specification. This gives the possibility of easy replacement of existing functions. For example we have a powering function P, with the following program definition (implementation):

$$P[x,n] = \textbf{If } n = 0 \textbf{ then } 1 \textbf{ else } P[x,n-1] * x. \tag{7}$$

The specification of P is: the domain $\mathbb{D} = \mathbb{R}^2$, the precondition $I_P[x,n] \iff n \in \mathbb{N}$ and the postcondition $O_P[x,n,y] \iff y = x^n$.

Additionally, we have proven the correctness of P. Later, after using the powering function P for defining other functions, we decide to replace its definition (implementation) by another one, however, keeping the same specification. In this situation, the only thing we should do (besides preserving the name) is to prove that the new definition (implementation) of P meets the specification as defined initially.

In order to achieve the modularity, we need to ensure that when defining a new program, all the calls made to the existing (already defined) programs obey the input restrictions of that programs – we call this principle *the appropriate values for the function calls principle.*

We now define naturally the class of coherent programs as those which obey the *appropriate values to the function calls* principle. The general definition comes in two parts: for functions defined by superposition and for functions defined by the **If-then-else** construct.

Definition 1. Let F be obtained from $H, G_1, \ldots, G_n$ by superposition:

$$F[x] = H[G_1[x], \ldots, G_n[x]]. \tag{8}$$

The program F with the specification $\langle I_F, O_F \rangle$ is *coherent* with respect to its auxiliary functions H, G_i and their specifications $\langle I_H, O_H \rangle$, $\langle I_{G_i}, O_{G_i} \rangle$ *if and only if*

$$(\forall x : I_F[x])(I_{G_1}[x] \wedge \ldots \wedge I_{G_n}[x]) \tag{9}$$

and

$$(\forall x : I_F[x])(\forall y_1 \dots y_n)(O_{G_1}[x, y_1] \wedge \dots \wedge O_{G_n}[x, y_n] \Longrightarrow I_H[y_1, \dots y_n]). \quad (10)$$

Definition 2. Let F be obtained from H and G by the **If-then-else** construct:

$$F[x] = \textbf{If } Q[x] \textbf{ then } H[x] \textbf{ else } G[x]. \quad (11)$$

The program F with the specification $\langle I_F, O_F \rangle$ is *coherent* with respect to its auxiliary functions H, G and their specifications $\langle I_H, O_H \rangle$, $\langle I_G, O_G \rangle$ *if and only if*

$$(\forall x : I_F[x])(Q[x] \Longrightarrow I_H[x]) \quad (12)$$

$$\wedge$$

$$(\forall x : I_F[x])(\neg Q[x] \Longrightarrow I_G[x]).$$

Throughout this paper we deal mainly with coherent functions. As a first step of the verification process we check if the program is coherent. Incoherent programs need not be incorrect. However, if we want to achieve the modularity of our system, we need to restrict ourselves to deal with coherent programs only.

In order to demonstrate the importance of the coherence we give an example of a function GM, defined in terms of an auxiliary function M. Initially GM is correct with respect to its specification but it is not coherent. However, after a slight modification of the implementation of M, GM will not be correct any more and therefore the modularity principle is not met.

Let us have GM and M with the following program definitions (implementations):

$$GM[x, y] = Sqrt[M[x, y]] \quad (13)$$

$$M[x, y] = |x * y|.$$

The specifications of GM, M and $Sqrt$ are: the domain for the first two is $\mathbb{R}^2$, and the domain for the $Sqrt$ is $\mathbb{R}$. The precondition of GM is $I_{GM}[x, y] \iff \mathbb{T}$, the postcondition of GM is $O_{GM}[x, y, z] \iff z^4 = (x.y)^2$, the precondition of M is $I_M[x, y] \iff x \geq 0 \wedge y \geq 0$, the postcondition of M is $O_M[x, y, z] \iff z = x.y$, the precondition of $Sqrt$ is $I_{Sqrt}[x] \iff x \geq 0$ and the postcondition of $Sqrt$ is $O_{Sqrt}[x, y] \iff y^2 = x \wedge y \geq 0$.

The function GM is expected for positive numbers to compute their geometric means, but it is also defined for any combination of positive and negative numbers. The function M is computing the absolute value of the multiplication of two numbers.

It is easy to see that GM and M are correct with respect to their specifications. Moreover, we do not concentrate here on the implementation of $Sqrt$ – we just assume it is correct with respect to the given specification. We will however show that GM is not coherent. The coherence conditions are:

$$(\forall x, y : \mathbb{T})\ (x \geq 0 \wedge y \geq 0) \tag{14}$$

and

$$(\forall x, y : \mathbb{T})\ (\forall z)\ (z = x.y \Longrightarrow z > 0). \tag{15}$$

As we can see both conditions for GM to be coherent are violated, but the program GM is correct. Now comes the question: "Why do we need to require our functions to be coherent if non-coherent ones may still be correct?". In fact, the function GM is correct due to the special behavior of the function M – it makes sure that its output is nonnegative. Let us now change the implementation of M and remove this special feature (which is not needed, because it is not required by its specification). The new definition is:

$$M[x, y] = x * y. \tag{16}$$

We preserve however its specification unchanged, i.e., the precondition of M ($I_M[x, y] \iff x \geq 0 \wedge y \geq 0$), and the postcondition of M ($O_M[x, y, z] \iff z = x.y$).

It is easy to check that the new implementation of M meets its specification. If we now execute the function GM on one positive and one negative numbers, say 2 and -2, M will return a negative output ($M[2, -2] = -4$) and therefore we may not expect a correct output from the function *Sqrt* executed on a negative input.

3 General Recursive Programs

In this section we study the class of general recursive programs (more precisely: recursive programs with multiple recursive calls and multiple conditional branches) and we extract the purely logical conditions which are sufficient for the program correctness. These are inferred using Scott induction [27, 29] and induction on natural numbers in the fixpoint theory of functions and constitute a meta-theorem which is proven once for the whole class. The concrete verification conditions for each program are then provable without having to use the fixpoint theory.

We approach the correctness problem by splitting it into two parts: *partial correctness* (prove that the program satisfies the specification provided the program terminates), and *termination* (prove that the program always terminates).

General recursive programs are programs of the form:

$$\begin{aligned} F[x] = &\ \textbf{If } Q_0[x] \textbf{ then } S[x] \\ &\ \textbf{elseif } Q_1[x] \textbf{ then } C_1[x, F[R_{1,1}[x]], \ldots, F[R_{1,k_1}[x]]] \\ &\ \ldots \\ &\ \textbf{elseif } Q_n[x] \textbf{ then } C_n[x, F[R_{n,1}[x]], \ldots, F[R_{n,k_n}[x]]], \end{aligned} \tag{17}$$

where Q_i are predicates and $S, C_i, R_{i,j}$ are auxiliary functions ($S[x]$ is a "simple" function (the bottom of the recursion), $C_i[x, y_1, \ldots, y_{k_i}]$ are "combinator" functions, and $R_{i,j}[x]$ are "reduction" functions).

We assume that the functions S, C_i, and $R_{i,j}$ satisfy their specifications given by $\langle I_S[x], O_S[x, y]\rangle$, $\langle I_{C_i}[x, y_1, \ldots, y_{k_i}], O_{C_i}[x, y_1, \ldots, y_{k_i}, z]\rangle$, $\langle I_{R_{i,j}}[x], O_{R_{i,j}}[x, y]\rangle$.

Without loss of generality, we may assume that the Q_i predicates are exhaustive and mutually disjoint, that is:

$$Q_i \Rightarrow \neg Q_j, \quad \text{for each } i \neq j,$$

and:

$$Q_0 \vee \cdots \vee Q_n.$$

As an important note, we point out that functions with multiple arguments also fall into this scheme, because the arguments x, y, z could be vectors (tuples).

In practice Q_i may also be implemented by programs, and they may also have input conditions, but we do not want to complicate the present discussion by including this aspect, which has a special flavor.

Type (or domain) information does not appear explicitly in this formulation, however it may be included in the input conditions.

Note that the "programming language" used here contains only the construct **If-then-else** in addition to the language of first order predicate logic.

One may also use some additional restrictions on the shape of the definitions of Q_i, S, C_i, and $R_{i,j}$ (e.g., that they do not contain quantifiers) in order to make the program "easy" to execute. However, this depends on the complexity of the "interpreter" ("compiler") and does not influence the actual generation of the verification conditions. In general, the auxiliary functions may already be defined in the underlying theory, or by other programs (that includes logical terms).

3.1 Coherent General Recursive Multiple Conditional Programs

As already discussed, we first check if the program is coherent, that is, all function calls are applied to arguments obeying the respective input specifications.

The corresponding conditions for this class of programs, which are derived from the definition of coherent programs (1) and (2), are:

Definition 3. Let for all i, j, the functions S, C_i, and $R_{i,j}$ be such that they satisfy their specifications $\langle I_S, O_S\rangle$, $\langle I_{C_i}, O_{C_i}\rangle$, and $\langle I_{R_{i,j}}, O_{R_{i,j}}\rangle$. Then the program F as defined in (17) with its specification $\langle I_F, O_F\rangle$ is coherent with respect to S, C_i, $R_{i,j}$, and their specifications, if and only if the following conditions hold:

$$(\forall x : I_F[x])(Q_0[x] \Longrightarrow I_S[x]) \tag{18}$$

$$(\forall x : I_F[x])(Q_1[x] \Longrightarrow I_F[R_{1,1}[x]] \wedge \cdots \wedge I_F[R_{1,k_1}[x]]) \quad (19)$$

$$\cdots$$

$$(\forall x : I_F[x])(Q_n[x] \Longrightarrow I_F[R_{n,1}[x]] \wedge \cdots \wedge I_F[R_{n,k_n}[x]]) \quad (20)$$

$$(\forall x : I_F[x])(Q_1[x] \Longrightarrow I_{R_{1,1}}[x] \wedge \cdots \wedge I_{R_{1,k_1}}[x]) \quad (21)$$

$$\cdots$$

$$(\forall x : I_F[x])(Q_n[x] \Longrightarrow I_{R_{n,1}}[x] \wedge \cdots \wedge I_{R_{n,k_n}}[x]) \quad (22)$$

$$(\forall x, y_1, \ldots, y_{k_1} : I_F[x])(Q_1[x] \wedge O_F[R_{1,1}[x], y_1] \wedge \cdots \wedge O_F[R_{1,k_1}[x], y_{k_1}] \quad (23)$$

$$\Longrightarrow$$

$$I_{C_1}[x, y_1, \ldots, y_{k_1}])$$

$$\cdots$$

$$(\forall x, y_1, \ldots, y_{k_n} : I_F[x])(Q_n[x] \wedge O_F[R_{n,1}[x], y_1] \wedge \cdots \wedge O_F[R_{n,k_n}[x], y_{k_n}] \quad (24)$$

$$\Longrightarrow$$

$$I_{C_1}[x, y_1, \ldots, y_{k_n}])$$

Again we see that the respective conditions for coherence correspond very much to our intuition about coherent programs, namely:

- (18) treats the base case, that is, $Q_0[x]$ holds and no recursion is applied, thus the input x must fulfill the precondition of S.
- (19) – (20) treat the general case, that is, $\neg Q_0[x]$, and say $Q_i[x]$ holds and recursion is applied, thus all the new inputs $R_{i,1}[x]$, …, $R_{i,k_i}[x]$ must fulfill the precondition of the main function F.
- (21) – (22) treat the general case, that is, $\neg Q_0[x]$, and say $Q_i[x]$ holds and recursion is applied, thus the input x must fulfill the preconditions of the reduction functions $R_{i,1}, \ldots, R_{i,k_i}$.
- (23) – (24) treat the general case, that is, $\neg Q_0[x]$, and say $Q_i[x]$ holds and recursion is applied, thus the input x, together with any y_1, …, y_{k_i} (where for each j, y_j is a possible output $F[R_{i,j}[x]]$) must fulfill the precondition of the function C_i.

3.2 Verification Conditions and Their Soundness

As we already discussed, in order to be sure that a program is correctly proven to be correct, one has to formally rely on the technique used for verification. Thus we formulate here a *soundness* theorem, for the class of coherent general recursive programs.

Theorem 1. *Let for each i, j: S, C_i, and $R_{i,j}$ be functions which satisfy their specifications $\langle I_S, O_S\rangle$, $\langle I_{C_i}, O'_{C_i} rangle$, and $\langle I_{R_{i,j}}, O_{R_{i,j}}\rangle$. Let also the general recursive program F as defined in* (17) *with its specification $\langle I_F, O_F\rangle$ be coherent with respect to S, C_i, $R_{i,j}$, and their specifications. Then F is totally correct with respect to $\langle I_F, O_F\rangle$ if the following verification conditions hold:*

$$(\forall x : I_F[x])(Q_0[x] \Longrightarrow O_F[x, S[x]]) \tag{25}$$

$$(\forall x, y_1, \ldots, y_{k_1} : I_F[x])(Q_1[x] \wedge O_F[R_{1,1}[x], y_1] \wedge \cdots \wedge O_F[R_{1,k_1}[x], y_{k_1}] \tag{26}$$

$$\Longrightarrow$$

$$O_F[x, C_1[x, y_1, \ldots, y_{k_1}]])$$

$$\ldots$$

$$(\forall x, y_1, \ldots, y_{k_n} : I_F[x])(Q_n[x] \wedge \; O_F[R_{n,1}[x], y_1] \wedge \cdots \wedge O_F[R_{n,k_n}[x], y_{k_n}] \tag{27}$$

$$\Longrightarrow$$

$$O_F[x, C_n[x, y_1, \ldots, y_{k_n}]])$$

$$(\forall x : I_F[x])(F'[x] = \mathbb{T}) \tag{28}$$

where:

$$\begin{aligned} F'[x] = \; & \textbf{If } Q_0[x] \textbf{ then } \mathbb{T} \\ & \textbf{elseif } Q_1[x] \textbf{ then } F'[R_{1,1}[x]] \wedge \cdots \wedge F'[R_{1,k_1}[x]] \\ & \ldots \\ & \textbf{elseif } Q_n[x] \textbf{ then } F'[R_{n,1}[x]] \wedge \cdots \wedge F'[R_{n,k_n}[x]]. \end{aligned} \tag{29}$$

The above conditions constitute the following principle:

- (25) prove that the base case is correct.
- (26) – (27) for any *else* branch, prove that the recursive expression is correct under the assumption that the reduced calls are correct.
- (28) prove that a simplified version F' of the initial program F terminates.

Proof. The proof of the *soundness* statement is split into two major parts:

- **(A)** – prove termination.
- **(B)** prove partial correctness using Scott induction.

(A): From the assumption that for all i: S, C_i, and $R_{i,j}$ are totally correct (with respect to I_S, I_{C_i}, and $I_{R_{i,j}}$) by the coherence of F, we obtain that all the calls to the auxiliary functions S, C_i, and $R_{i,j}$ will terminate.

Take arbitrarily x and assume $I_F[x]$. From (28), we obtain that $F'[x] = \mathbb{T}$.

Now we construct the recursive tree $RT_{F'}[x]$ of F', starting form x in the following way:

- x is the root of the tree, that is, the uppermost node.
- For any node u, if $Q_0[u]$ holds, then stop further construction on that branch, and put the symbol $\overline{\top}$.
- For any node u, if $Q_i[u]$ holds, for some $i \neq 0$, then construct all the k_i descendent nodes $R_{i,1}[u], \ldots, R_{i,k_i}[u]$.

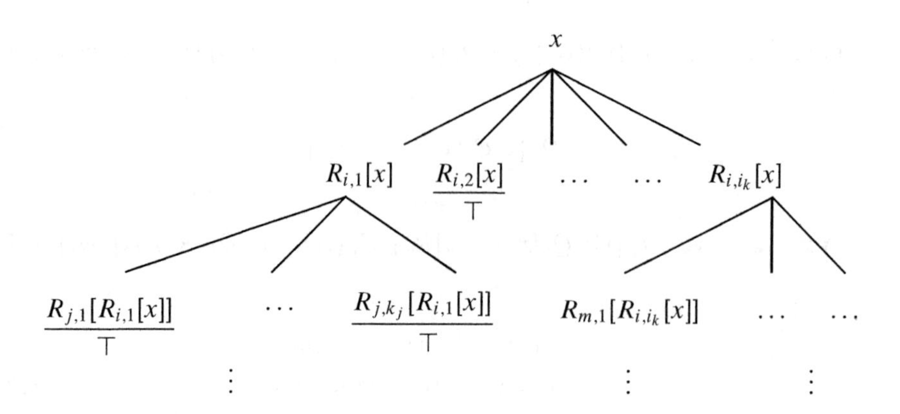

$Q_i[x]$

$Q_j[R_{i,1}[x]]$ $\quad$ $Q_0[R_{i,2}[x]] \rightsquigarrow \overline{\top}$ $\quad$ $Q_m[R_{i,i_k}[x]]$

$Q_0[R_{j,1}[R_{i,1}[x]]] \rightsquigarrow \overline{\top}$ $\quad$ $Q_0[R_{j,k_j}[R_{i,1}[x]]] \rightsquigarrow \overline{\top}$

We first show that $RT_{F'}[x]$ is finite.

We prove this statement by contradiction, i.e. assume $RT_{F'}[x]$ is infinite. Hence, there exists an infinite path $(\langle i_1, j_1\rangle, \langle i_2, j_2\rangle, \ldots, \langle i_l, j_l\rangle \ldots)$, such that:

$$\begin{aligned}
&\neg Q_0[x] \text{ but:} Q_{i_1}[x] \\
&\neg Q_0[R_{i_1,j_1}[x]] \text{ but: } Q_{i_2}[R_{i_1,j_1}[x]] \\
&\ldots \\
&\neg Q_0[R_{i_l,j_l}[\ldots[R_{i_1,j_1}[x]]]] \text{ but: } Q_{i_{l+1}}[R_{i_l,j_l}[\ldots[R_{i_1,j_1}[x]]]] \\
&\ldots.
\end{aligned} \tag{30}$$

Now, we look at the construction of F' as being the least fixpoint of the operator F' as defined in (29).

Let us denote the nowhere defined function by Ω ($\Omega = \lambda x.\bot$). Let $f_0, f_1, \ldots$ $f_m, \ldots$ be the finite approximations of F' obtained in the following way:

$$f_0[x] = \Omega[x]$$

$$\begin{aligned} f_{m+1}[x] = &\ \textbf{If } Q_0[x] \textbf{ then } \mathbb{T} \\ &\ \textbf{elseif } Q_1[x] \textbf{ then } f_m[R_{1,1}[x]] \wedge \cdots \wedge f_m[R_{1,k_1}[x]] \\ &\ \ldots \\ &\ \textbf{elseif } Q_n[x] \textbf{ then } f_m[R_{n,1}[x]] \wedge \cdots \wedge f_m[R_{n,k_n}[x]]. \end{aligned}$$

The computable function F', corresponding to (29) is defined as

$$F' = \bigcup_m f_m,$$

that is, the least fixpoint of (29).

Since for our particular x (it was taken arbitrarily) we have $F'(x) = \mathbb{T}$, there must exist a finite approximation f_m, such that:

$$f_m[x] = \mathbb{T}.$$

If $m = 0$, then $f_0[x] = \mathbb{T}$, but on the other hand, by its definition $f_0 = \Omega$, thus this is not a case. Hence, we conclude that $m > 0$.

From the assumption (30), and in particular $\neg Q_0[x]$ and $Q_{i_1}[x]$, by the definition of f_m we obtain:

$$f_m[x] = f_{m-1}[R_{i_1,1}[x]] \wedge \cdots \wedge f_{m-1}[R_{i_1,k_{i_1}}[x]].$$

From here, and $f_m[x] = \mathbb{T}$ we obtain that:

$$f_{m-1}[R_{i_1,1}[x]] = \mathbb{T} \wedge \cdots \wedge f_{m-1}[R_{i_1,k_{i_1}}[x]] = \mathbb{T},$$

and hence $f_{m-1}[R_{i_1,j_1}[x]] = \mathbb{T}$.

By repeating the same kind of reasoning m times (in fact, formally it is done by induction), we obtain that:

$$f_0[R_{i_m,j_m}[\ldots[R_{i_1,j_m}[x]]]] = \mathbb{T}.$$

On the other hand $f_0 = \Omega$ and hence:

$$f_0[R_{i_m,j_m}[\ldots[R_{i_1,j_1}[x]]]] = \bot.$$

This is the desired contradiction, and hence, we have proven that the recursive tree $RT_{F'}[x]$ is finite.

Now we continue the proof of the termination of F. We prove this statement by contradiction, i.e. assume $RT_{F'}[x]$ is finite and $F[x] = \bot$.

For our particular x (it was taken arbitrary but fixed, $I_F[x]$), we consider the following two cases:

- Case 1: $Q_0[x]$.
  Now by the definition of F, we have $F[x] = S[x]$. We chose x such that $I_F[x]$, and by (18) we obtain that $S[x] \downarrow$ and hence $F[x] \downarrow$ and thus we obtain a contradiction.
- Case 2: $\neg Q_0[x]$, and assume $Q_{i_1}[x]$. Now, by following the definition of F, we have,

$$F[x] = C_{i_1}[x, F[R_{i_1,1}[x], \ldots, R_{i_1,k_{i_1}}[x]]],$$

  and since F is coherent, we have $I_{R_{i_1,1}}[x]$, $I_{R_{i_1,2}}[x]$, and $I_{R_{i_1,k_{i_1}}}[x]$, and $I_C[x, y_1, \ldots, y_{k_{i_1}}]$. Thus there exist j_1, such that $R_{i_1,j_1}[x] = \bot$.
  Applying the same kind of reasoning we obtain the infinite path
  $(\langle i_1, j_1\rangle, \langle i_2, j_2\rangle, \ldots, \langle i_l, j_l\rangle \ldots)$, that is:
  $\neg Q_0[x]$ but: $Q_{i_1}[x]$
  $\neg Q_0[R_{i_1,j_1}[x]]$ but: $Q_{i_2}[R_{i_1,j_1}[x]]$
  ...
  $\neg Q_0[R_{i_l,j_l}[\ldots[R_{i_1,j_1}[x]]]]$ but: $Q_{i_{l+1}}[R_{i_l,j_l}[\ldots[R_{i_1,j_1}[x]]]]$
  ....

This implies that the three $RT_{F'}[x]$ is infinite, which is the desired contradiction.

(B): Using Scott induction, we will show that F is partially correct with respect to its specification, namely:

$$(\forall x : I_F[x])(F[x] \downarrow \Longrightarrow O_F[x, F[x]]). \tag{31}$$

As it is well known (e.g., [27, 29]), not every property is admissible and may be proven by Scott induction. However, properties which express partial correctness are known to be admissible.

Let us remind the definition of these properties: A property ϕ is said to be a partial correctness property if and only if there are predicates I and O, such that:

$$(\forall f)(\phi[f] \Longleftrightarrow (\forall a)\ (f[a] \downarrow \wedge I[a] \Longrightarrow O[a, f[a]])). \tag{32}$$

We now consider the following partial correctness property ϕ:

$$(\forall f)(\phi[f] \Longleftrightarrow (\forall a)\ (f[a] \downarrow \wedge I_F[a] \Longrightarrow O_F[a, f[a]])).$$

The first step in Scott induction is to show that ϕ holds for Ω. By the definition of ϕ we obtain:

$$\phi[\Omega] \Longleftrightarrow (\forall a)\ (\Omega[a] \downarrow \wedge I_F[a] \Longrightarrow\ O_F[a, \Omega[a]]),$$

and so, $\phi[\Omega]$ holds, since $\Omega[a] \downarrow$ never holds.

In the second step of Scott induction, we assume $\phi[f]$ holds for some f:

$$(\forall a)(f[a] \downarrow \wedge I_F[a] \Longrightarrow\ O_F[a, f[a]]), \tag{33}$$

and show $\phi[f_{new}]$, where f_{new} is obtained from f by the main program (17) as follows:

$$\begin{aligned} f_{new}[x] = &\ \textbf{If}\ Q_0[x]\ \ \textbf{then}\ S[x] \\ &\ \textbf{elseif}\ Q_1[x]\ \textbf{then}\ C_1[x, f[R_{1,1}[x]], \ldots, f[R_{1,k_1}[x]]] \\ &\ \ldots \\ &\ \textbf{elseif}\ Q_n[x]\ \textbf{then}\ C_n[x, f[R_{n,1}[x]], \ldots, f[R_{n,k_n}[x]]], \end{aligned}$$

Now, we need to show now that for an arbitrary a,

$$f_{new}[a] \downarrow \wedge I_F[a] \Longrightarrow\ O_F[a, f_{new}[a]].$$

Assume $f_{new}[a] \downarrow$ and $I_F[a]$. We have now the following two cases:

- Case 1: $Q_0[a]$.

  By the definition of f_{new} we obtain $f_{new}[a] = S[a]$ and since $f_{new}[a] \downarrow$, we obtain that $S[a]$ must terminate as well, that is $S[a] \downarrow$. Now using verification condition (25) we may conclude $O_F[a, S[a]]$ and hence $O_F[a, f_{new}[a]]$.
- Case 2: $Q_i[a]$ for some i, $1 \leq i \leq n$.

  By the definition of f_{new} we obtain:

  $$f_{new}[a] = C_i[a, f[R_{i,1}[a]], \ldots, f[R_{i,k_i}[a]]]$$

  and since $f_{new}[a] \downarrow$, we conclude that all the others involved in this computation must also terminate, that is:

  $$C_i[a, f[R_{i,1}[a]], \ldots, f[R_{i,k_i}[a]]] \downarrow,$$

  $$f[R_{i,1}[a]] \downarrow, \ldots, f[R_{i,k_i}[a]] \downarrow$$

  and

  $$R_{i,1}[a] \downarrow, \ldots, R_{i,k_i}[a] \downarrow .$$

  Since F is coherent, namely from $I_F[a]$, by (19)–(20), we obtain:

  $$I_F[R_{i,1}[x]] \wedge \cdots \wedge I_F[R_{i,k_1}[x]].$$

Knowing that for each j: $f[R_{i,j}[a]] \downarrow$, by the induction hypothesis (33) we obtain $O_F[R_{i,j}[a], f[R_{i,j}[a]]]$.

Considering the appropriate i^{th} verification condition (26)–(27), note that all the assumptions from the left part of the implication are at hand and thus we can conclude:

$$O_F[a, C_i[a, f[R_{i,1}[a]], \ldots, f[R_{i,k_i}[a]]]],$$

which is

$$O_F[a, f_{new}[a]].$$

Now we conclude that the property ϕ holds for the least fixpoint of (17) and hence, ϕ holds for the function computed by (17), which completes the proof of Theorem 1.

3.3 Completeness of the Verification Conditions

Completing the notion of *soundness*, we introduce its dual notion – *completeness*.

As we already mentioned, after generating the verification conditions, one has to prove them as logical formulas. If all of them hold, then the program is correct with respect to its specification – Theorem 1.

Now, we formulate the *completeness* theorem for the class of coherent general recursive programs.

Theorem 2. *Let for any i, j the functions S, C_i, and $R_{i,j}$ satisfy their specifications $\langle I_S, O_S \rangle$, $\langle I_C, O_C \rangle$, and $\langle I_{R_{i,j}}, O_{R_{i,j}} \rangle$. Let also the general recursive program F (17) with its specification $\langle I_F, O_F \rangle$ be coherent with respect to S, C_i, $R_{i,j}$, and their specifications, and the output specification of F, (O_F) is functional one.*

Then if F is totally correct with respect to $\langle I_F, O_F \rangle$ then the following verification conditions hold:

$$(\forall x : I_F[x])(Q_0[x] \Longrightarrow O_F[x, S[x]]) \tag{34}$$

$$(\forall x, y_1, \ldots, y_{k_1} : I_F[x])(Q_1[x] \wedge O_F[R_{1,1}[x], y_1] \wedge \cdots \wedge O_F[R_{1,k_1}[x], y_{k_1}] \tag{35}$$

$$\Longrightarrow$$

$$O_F[x, C_1[x, y_1, \ldots, y_{k_1}]])$$

$$\ldots$$

$$(\forall x, y_1, \ldots, y_{k_n} : I_F[x])(Q_n[x] \wedge O_F[R_{n,1}[x], y_1] \wedge \cdots \wedge O_F[R_{n,k_n}[x], y_{k_n}] \tag{36}$$

$$\Longrightarrow$$

$$O_F[x, C_n[x, y_1, \ldots, y_{k_n}]])$$

$$(\forall x : I_F[x])(F'[x] = \mathbb{T}) \tag{37}$$

where:

$$F'[x] = \textbf{If } Q_0[x] \textbf{ then } \mathbb{T} \tag{38}$$
$$\textbf{elseif} Q_1[x] \textbf{ then } F'[R_{1,1}[x]] \wedge \cdots \wedge F'[R_{1,k_1}[x]]$$
$$\ldots$$
$$\textbf{elseif } Q_n[x] \textbf{ then } F'[R_{n,1}[x]] \wedge \cdots \wedge F'[R_{n,k_n}[x]],$$

which are the same as (25), (26)–(27), (28), *and* (29) *from Theorem 1.*

Proof. We assume now that:

- For all i, j the functions S, C_i, and $R_{i,j}$ are totally correct with respect to their specifications $\langle I_S, O_S \rangle$, $\langle I_{C_i}, O_{C_i} \rangle$, and $\langle I_{R_{i,j}}, O_{R_{i,j}} \rangle$.
- The program F (17) with its specification $\langle I_F, O_F \rangle$ is coherent.
- The output specification of F, O_F is functional one, that is:

$$(\forall x : I_F[x])(\exists! y)(O_F[x, y]).$$

- The program F (17) is correct with respect to its specification, that is, the total correctness formula holds:

$$(\forall x : I_F[x])(F[x] \downarrow \wedge O_F[x, F[x]]). \tag{39}$$

We show that (34), (35) – (36), and (37) hold as logical formulas.
We start now with proving (34) and (35) – (36) simultaneously.
Take arbitrarily x and assume $I_F[x]$. We consider the following two cases:

- Case 1: $Q_0[x]$
  By the definition of F, we have $F[x] = S[x]$, and by using the correctness formula (39) of F, we conclude (34) holds. The formulas (35) – (36) hold, because the predicates Q are consistent and noncontradictory, and hence $\neg Q_i[x]$ for all i, $1 \le i \le n$.
- Case 2: $Q_i[x]$ for some i, $1 \le i \le n$.
  Now, the formulas (34) and all except one of (35) – (36) hold trivially, because at the left hand side of the implication we have $\neg Q_i[x]$.
  Assume $y_1, \ldots, y_{k_i}$ are such that:

$$O_F[R_{i,1}[x], y_1], \ldots, O_F[R_{i,k_i}[x], y_{k_i}].$$

Since F is correct, we obtain that:

$$y_1 = F[R_{i,1}[x]], \ldots, y_{k_i} = F[R_{i,k_i}[x]]$$

because O_F is a functional predicate.

On the other hand, by the definition of F, we have:
$F[x] = C_i[a, F[R_{i,1}[a]], \ldots, F[R_{i,k_i}[a]]]$ and hence $F[x] = C_i[x, y_1, \ldots, y_{k_i}]$. Again, from the correctness of F, we obtain $O_F[x, C_i[x, y_1, \ \ldots, y_{k_i}]]$, which had to be proven.

Now, we show that the simplified version $F'[x] = \mathbb{T}$. Moreover, F' terminates if F terminates, which is equivalent to $F'[x] = \mathbb{T}$.

Take arbitrarily x and assume $I_F[x]$.

Now we construct the recursive tree $RT_F[x]$ of F, starting form x in the following way:

- x is the root of the tree, that is, the uppermost node.
- For any node u, if $Q_0[u]$ holds, then stop further construction on that branch, and put the symbol $\top$.
- For any node u, if $Q_i[u]$ holds, for some $i \neq 0$, then construct all the k_i descendent nodes $R_{i,1}[u], \ldots, R_{i,k_i}[u]$.

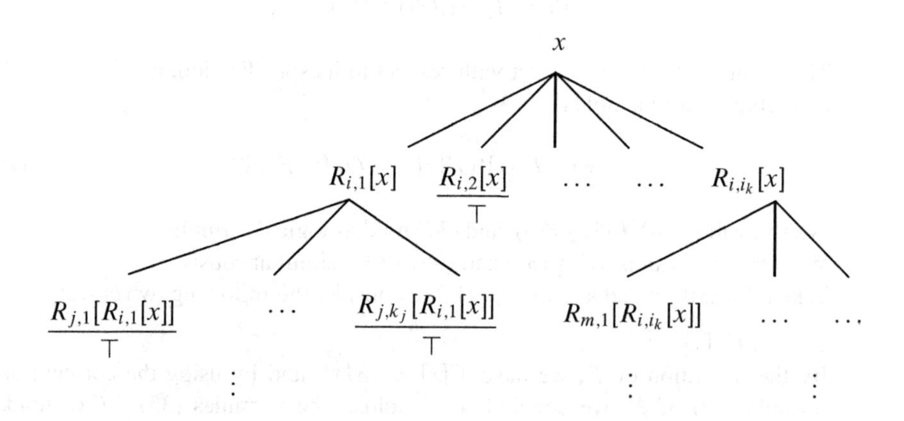

$Q_i[x]$

$Q_j[R_{i,1}[x]]$ $\qquad$ $Q_0[R_{i,2}[x]] \rightsquigarrow \top$ $\qquad$ $Q_m[R_{i,i_k}[x]]$

$Q_0[R_{j,1}[R_{i,1}[x]]] \rightsquigarrow \top$ $\qquad$ $Q_0[R_{j,k_j}[R_{i,1}[x]]] \rightsquigarrow \top$

Note that the recursive tree of F, $RT_F[x]$ is the same as the recursive tree of F', $RT_{F'}[x]$. Thus $RT_F[x]$ is finite.

Now we need to show termination of F'. For our particular x (it was taken arbitrary but fixed, $I_F[x]$), we consider the following two cases:

- Case 1: $Q_0[x]$.
  Now by the definition of F', we have $F[x] = \mathbb{T}$ and hence $F'[x] \downarrow$.

- Case 2: $\neg Q_0[x]$, and say $Q_i[x]$. Now, by following the definition of F', we have,

$$F'[x] = F'[R_{i,1}[x]] \wedge \cdots \wedge F'[R_{i,k_i}[x]].$$

We need to apply the same kind of reasoning to all the nodes of the recursive tree $RT_F[x]$. Since the tree is finite, after unfolding finitely many times we reach the leaves. Moreover, for each leaf we arrive at the Case 1 and thus $F'[x] \downarrow$.

By this we completed our proof of the *completeness* theorem.

4 Two Relevant Examples

4.1 Binary Powering

In order to make clear our contributions, we first consider a relatively simple example, namely an implementation of the *binary powering* algorithm:

$$\begin{aligned} P[x,n] = \; & \textbf{If } n = 0 \textbf{ then } 1 \\ & \textbf{elseif } \text{Even}[n] \textbf{ then } P[x * x, n/2] \\ & \textbf{else } x * P[x * x, (n-1)/2]. \end{aligned}$$

This program is in the context of the theory of real numbers, and in the following formulas, all variables are implicitly assumed to be real. Additional type information (e.g., $n \in \mathbb{N}$) may be explicitly included in some formulas.

The specification is:

$$\langle I_P[x,n] \iff n \in \mathbb{N}, O_P[x,n,y] \iff y = x^n \rangle. \tag{40}$$

The conditions for **coherence** are:

$$(\forall x, n : n \in \mathbb{N})\ (n = 0 \Rightarrow \mathbb{T}) \tag{41}$$

$$(\forall x, n : n \in \mathbb{N})(n \neq 0 \wedge \text{Even}[n] \Rightarrow \text{Even}[n]) \tag{42}$$

$$(\forall x, n : n \in \mathbb{N})(n \neq 0 \wedge \neg\text{Even}[n] \Rightarrow \text{Odd}[n]) \tag{43}$$

$$(\forall x, n, m : n \in \mathbb{N})(n \neq 0 \wedge \text{Even}[n] \wedge m = (x * x)^{n/2} \Rightarrow \mathbb{T}) \tag{44}$$

$$(\forall x, n, m : n \in \mathbb{N})(n \neq 0 \wedge \neg\text{Even}[n] \wedge m = (x * x)^{(n-1)/2} \Rightarrow \mathbb{T}) \tag{45}$$

$$(\forall x, n : n \in \mathbb{N})(n \neq 0 \wedge \text{Even}[n] \Rightarrow n/2 \in \mathbb{N}) \tag{46}$$

$$(\forall x, n : n \in \mathbb{N})(n \neq 0 \wedge \neg\text{Even}[n] \Rightarrow (n-1)/2 \in \mathbb{N}) \tag{47}$$

One sees that the formulas (41), (44) and (45) are trivially valid, because we have the logical constant $\mathbb{T}$ at the right side of an implication. The origin of these $\mathbb{T}$ come from the preconditions of the 1 *constant-function-one* and the $*$ *multiplication*.

The formulas (42), (43), (46) and (47) are easy consequences of the elementary theory of reals and naturals. For the further check of **correctness** the generated conditions are:

$$(\forall x, n : n \in \mathbb{N})(n = 0 \Rightarrow 1 = x^n) \tag{48}$$

$$(\forall x, n, m : n \in \mathbb{N})(n \neq 0 \wedge \text{Even}[n] \wedge m = (x * x)^{n/2} \Rightarrow m = x^n) \tag{49}$$

$$(\forall x, n, m : n \in \mathbb{N})(n \neq 0 \wedge \neg\text{Even}[n] \wedge m = (x * x)^{(n-1)/2} \Rightarrow x * m = x^n) \tag{50}$$

$$(\forall x, n : n \in \mathbb{N}) P'[x, n] = \mathbb{T}, \tag{51}$$

where

$$\begin{aligned} P'[x, n] = \; & \textbf{If } n = 0 \textbf{ then } \mathbb{T} \\ & \textbf{elseif } \text{Even}[n] \textbf{ then } P'[x * x, n/2] \\ & \textbf{else } P'[x * x, (n-1)/2]. \end{aligned}$$

The proofs of these verification conditions are straightforward.

Now comes the question: What if the program is not correctly written? Thus, we introduce now a bug. The program P is now almost the same as the previous one, but in the base case (when $n = 0$) the return value is 0.

$$\begin{aligned} P[x, n] = \; & \textbf{If } n = 0 \textbf{ then } 0 \\ & \textbf{elseif } \text{Even}[n] \textbf{ then } P[x * x, n/2] \\ & \textbf{else } x * P[x * x, (n-1)/2]. \end{aligned}$$

Now, for this buggy version of P we may see that all the respective verification conditions remain the same except one, namely, (48) is now:

$$(\forall x, n : n \in \mathbb{N})(n = 0 \Rightarrow 0 = x^n) \tag{52}$$

which itself reduces to:

$$0 = 1$$

(because we consider a theory where $0^0 = 1$).

Therefore, according to the *completeness* of the method, we conclude that the program P does not satisfy its specification. Moreover, the failed proof gives a hint for "debugging": we need to change the return value in the case $n = 0$ to 1.

Furthermore, in order to demonstrate how a bug might be located, we construct one more "buggy" example where in the "Even" branch of the program we have $P[x, n/2]$ instead of $P[x * x, n/2]$:

$$
\begin{aligned}
P[x,n] = \textbf{If } n = 0 \textbf{ then } 1 \\
&\textbf{elseif } \text{Even}[n] \textbf{ then } P[x,n/2] \\
&\textbf{else } x * P[x * x, (n-1)/2].
\end{aligned}
$$

Now, we may see again that all the respective verification conditions remain the same as in the original one, except one, namely, (49) is now:

$$(\forall x, n, m : n \in \mathbb{N})(n \neq 0 \wedge \text{Even}[n] \wedge m = (x)^{n/2} \Rightarrow m = x^n) \tag{53}$$

which itself reduces to:

$$m = x^{n/2} \Rightarrow m = x^n$$

From here, we see that the "Even" branch of the program is problematic and one should satisfy the implication. The most natural candidate would be:

$$m = (x^2)^{n/2} \Rightarrow m = x^n$$

which finally leads to the correct version of P.

4.2 Neville's Algorithm

Neville's algorithm [19], [34] constructs the polynomial of degree $n-1$ which passes through given n different points.

The original problem is as follows:

- Given a field K and two non-empty tuples $\overline{x}$ and $\overline{a}$ over K of same length n, such that
$$(\forall i, j : i, j = 1, \ldots, n)(i \neq j \Rightarrow x_i \neq x_j),$$
that is, no two x_i from $\overline{x}$ are the same.
- Find a polynomial p over the field K, such that
  - $Deg[p] \leq n - 1$ and
  - $(\forall i : i = 1, \ldots, n)(Eval[p, x_i] = a_i)$,

  where the $Eval$ function evaluates a polynomial p at value x_i.

This original problem, as stated here, was solved by E. H. Neville [25] by inventing an algorithm for the construction of such a polynomial [32]. The algorithm itself may be formulated as follows:

$$
\begin{aligned}
P[\overline{x}, \overline{a}] = &\textbf{ If } \|\overline{a}\| \leq 1 \textbf{ then } First[\overline{a}] \\
\textbf{else} &\frac{(\mathcal{X} - First[\overline{x}])(P[Tail[\overline{x}], Tail[\overline{a}]]) - (\mathcal{X} - Last[\overline{x}])(P[Bgn[\overline{x}], Bgn[\overline{a}]])}{Last[\overline{x}] - First[\overline{x}]},
\end{aligned}
\tag{54}
$$

where we use the following notation:

- $\|\overline{a}\|$ gives the number n of elements of $\overline{a}$,
- *First* $[\overline{a}]$ gives the first element of $\overline{a}$,
- *Last* $[\overline{a}]$ gives the last element of $\overline{a}$,
- *Tail* $[\overline{a}]$ gives $\overline{a}$ without its first element,
- *Bgn* $[\overline{a}]$ gives $\overline{a}$ without its last element, and
- $\mathcal{X}$ is a constant expressing the single polynomial of degree 1, leading coefficient 1 and free coefficient 0.

In fact, in abstract algebra $\mathcal{X}$ may also be interpreted as an indeterminate of the polynomials, that is, the variable in polynomial functions. This is a discussion which is not relevant for this presentation, however, it is very important when constructing the theory of polynomials, on which the verification conditions would have to be proven.

In order to illustrate how Neville's algorithm works, we consider the following example: $\overline{x} = \langle -1, 0, 1\rangle$ and $\overline{a} = \langle 3, 4, 7\rangle$. After executing (54), we obtain:

$$P[\langle -1,0,1\rangle, \langle 3,4,7\rangle] = \cdots = \mathcal{X}^2 + 2\mathcal{X} + 4.$$

This polynomial has a degree 2, as expected, and if we now evaluate it at the values -1, 0, and 1, we obtain:

$$Eval[\mathcal{X}^2 + 2\mathcal{X} + 4, -1] = 3,$$

$$Eval[\mathcal{X}^2 + 2\mathcal{X} + 4, 0] = 4,$$

and

$$Eval[\mathcal{X}^2 + 2\mathcal{X} + 4, 1] = 7,$$

which corresponds to the initial $\overline{a}$.

In order to verify (54), we first formalize the specification, and then produce the respective verification conditions.

We give here some notations which we use for the formalization of the specification:

- $\|\overline{a}\|_i$ gives the ith element a_i of a tuple $\overline{a}$. Sometimes, a_i is used as an abbreviation for $\|\overline{a}\|_i$. In addition to it, we have the restriction $1 \leq i \leq \|\overline{a}\|$.
- *IsPoly*$[poly]$ is a predicate standing that the expression $poly$ is a polynomial. For example $IsPoly[\mathcal{X}^2 + 2\mathcal{X} + 4]$.
- *IsTuple* $[\overline{a}]$ is a predicate standing that the expression $\overline{a}$ is a tuple. For example *IsTuple*$[\langle 3, 4, 7\rangle]$.
- *Deg[poly]* gives the degree of the polynomial $poly$. For example $Deg[\mathcal{X}^2 + 2\mathcal{X} + 4] = 2$.
- *Eval[poly, x]* evaluates a polynomial $poly$ at value x.

The preconditions of the functions used for the definition of (54) are as follows:

- *First*: $I_{First}\,[\overline{a}] \iff IsTuple\,[\overline{a}] \wedge \|\overline{a}\| \geq 1$
- *Last*: $I_{Last}\,[\overline{a}] \iff IsTuple\,[\overline{a}] \wedge \|\overline{a}\| \geq 1$
- *Tail*: $I_{Tail}\,[\overline{a}] \iff IsTuple\,[\overline{a}] \wedge \|\overline{a}\| \geq 1$
- *Bgn*: $I_{Bgn}\,[\overline{a}] \iff IsTuple\,[\overline{a}] \wedge \|\overline{a}\| \geq 1$
- $\|\overline{a}\|_i$: $I_{Projection}[\overline{a}, i] \iff IsTuple\,[\overline{a}] \wedge 1 \leq i \leq \|\overline{a}\|$
- $\frac{u}{v}$: $I_{Div}[u, v] \iff v \neq 0$
- uv: $I_{Mult}[u, v] \iff \mathbb{T}$
- $u + v$: $I_{Add}[u, v] \iff \mathbb{T}$
- $u - v$: $I_{Sub}[u, v] \iff \mathbb{T}$

We are now ready to give the formal specification of (54). The precondition is:

$$(\forall \overline{x}, \overline{a})(I_p[\overline{x}, \overline{a}] \iff \tag{55}$$

$$\iff IsTuple\,[\overline{x}] \wedge\ IsTuple\,[\overline{a}] \wedge \|\overline{x}\| = \|\overline{a}\| \wedge \|\overline{a}\| \geq 1 \wedge$$
$$\wedge((\forall i, j : i, j \in \mathrm{N})\ (1 \leq i, j \leq \|\overline{a}\| \wedge i \neq j \implies \|\overline{x}\|_i \neq \|\overline{x}\|_j))),$$

and the postcondition is:

$$(\forall \overline{x}, \overline{a})\ (O_p[\overline{x}, \overline{a}, p] \iff \tag{56}$$

$$\iff IsPoly\,[p] \wedge Deg[p] \leq \|\overline{a}\| - 1 \wedge$$
$$\wedge((\forall i : i \in \mathrm{N})(1 \leq i \leq \|\overline{a}\| \wedge i \neq j \implies Eval[p, x_i] = a_i))).$$

The conditions for coherence are:

$$(\forall \overline{x}, \overline{a})(IsTuple\,[\overline{x}] \wedge IsTuple\,[\overline{a}] \wedge \|\overline{x}\| = \|\overline{a}\| \wedge \|\overline{a}\| \geq 1 \wedge \tag{57}$$

$$\wedge((\forall i, j : i, j \in \mathrm{N})\ (1 \leq i, j \leq \|\overline{a}\| \wedge i \neq j \implies \|\overline{x}\|_i \neq \|\overline{x}\|_j)) \wedge \|\overline{a}\| \leq 1$$

$$\implies$$

$$IsTuple\,[\overline{a}] \wedge \|\overline{a}\| \geq 1)$$

$$(\forall \overline{x}, \overline{a})(IsTuple\,[\overline{x}] \wedge IsTuple\,[\overline{a}] \wedge \|\overline{x}\| = \|\overline{a}\| \wedge \|\overline{a}\| \geq 1 \wedge \tag{58}$$
$$\wedge((\forall i, j : i, j \in \mathrm{N})\ (1 \leq i, j \leq \|\overline{a}\| \wedge i \neq j \implies \|\overline{x}\|_i \neq \|\overline{x}\|_j)) \wedge \neg(\|\overline{a}\| \leq 1)$$

$$\implies$$

$$(IsTuple\,[Tail[\overline{x}]] \wedge\ IsTuple\,[Tail\,[\overline{a}]] \wedge \|Tail\,[\overline{x}]\| = \|Tail\,[\overline{a}]\| \wedge \|Tail\,[\overline{a}]\| \geq 1 \wedge$$
$$\wedge((\forall i, j : i, j \in \mathrm{N})\ (1 \leq i, j \leq \|Tail\,[\overline{a}]\| \wedge i \neq j \implies \|Tail\,[\overline{x}]\|_i \neq \|Tail\,[\overline{x}]\|_j))$$

$$(\forall \overline{x}, \overline{a})(IsTuple\,[\overline{x}] \wedge IsTuple\,[\overline{a}] \wedge \|\overline{x}\| = \|\overline{a}\| \wedge \|\overline{a}\| \geq 1 \wedge \tag{59}$$

$$\wedge((\forall i, j : i, j \in \mathrm{N})(1 \leq i, j \leq \|\overline{a}\| \wedge i \neq j \implies \|\overline{x}\|_i \neq \|\overline{x}\|_j)) \wedge \neg(\|\overline{a}\| \leq 1)$$

$$\implies$$

$$(IsTuple\,[Bgn\,[\overline{x}]] \wedge IsTuple\,[Bgn\,[\overline{a}]] \wedge \|Bgn[\overline{x}]\| = \|Bgn\,[\overline{a}]\| \wedge \|Bgn\,[\overline{a}]\| \geq 1 \wedge$$

$$\wedge((\forall i,j: i,j \in \mathbb{N})\ (1 \leq i,j \leq \|Bgn[\overline{a}]\| \wedge i \neq j \Longrightarrow \|Bgn[\overline{x}]\|_i \neq \|Bgn[\overline{x}]\|_j))))$$

$$(\forall \overline{x}, \overline{a})(IsTuple[\overline{x}] \wedge IsTuple[\overline{a}] \wedge \|\overline{x}\| = \|\overline{a}\| \wedge \|\overline{a}\| \geq 1 \wedge \tag{60}$$
$$\wedge((\forall i,j: i,j \in \mathbb{N})\ (1 \leq i,j \leq \|\overline{a}\| \wedge i \neq j \Longrightarrow \|\overline{x}\|_i \neq \|\overline{x}\|_j)) \wedge \neg(\|\overline{a}\| \leq 1)$$
$$\Longrightarrow$$
$$(IsTuple[\overline{x}] \wedge \|\overline{x}\| \geq 1 \wedge IsTuple[\overline{a}] \wedge \|\overline{a}\| \geq 1))$$

$$(\forall \overline{x}, \overline{a})(IsTuple[\overline{x}] \wedge IsTuple[\overline{a}] \wedge \|\overline{x}\| = \|\overline{a}\| \wedge \|\overline{a}\| \geq 1 \wedge \tag{61}$$
$$\wedge((\forall i,j: i,j \in \mathbb{N})\ (1 \leq i,j \leq \|\overline{a}\| \wedge i \neq j \Longrightarrow \|\overline{x}\|_i \neq \|\overline{x}\|_j)) \wedge \neg(\|\overline{a}\| \leq 1)$$
$$\Longrightarrow$$
$$(IsTuple[\overline{x}] \wedge \|\overline{x}\| \geq 1 \wedge IsTuple[\overline{a}] \wedge \|\overline{a}\| \geq 1))$$

$$(\forall \overline{x}, \overline{a}, p_1, p_2)(IsTuple[\overline{x}] \wedge IsTuple[\overline{a}] \wedge \|\overline{x}\| = \|\overline{a}\| \wedge \|\overline{a}\| \geq 1 \wedge$$
$$\wedge((\forall i,j: i,j \in \mathbb{N})\ (1 \leq i,j \leq \|\overline{a}\| \wedge i \neq j \Longrightarrow \|\overline{x}\|_i \neq \|\overline{x}\|_j)) \wedge \neg(\|\overline{a}\| \leq 1) \wedge$$
$$\wedge IsPoly[p_1] \wedge ((\forall i: i \in \mathbb{N})(1 \leq i \leq \|Tail[\overline{x}]\| \Longrightarrow Eval[p_1, \|Tail[\overline{x}]\|_i] = \|Tail[\overline{a}]\|_i] \wedge$$
$$\wedge Deg[p_1] \leq \|Tail[\overline{a}]\| - 1 \wedge$$
$$\wedge IsPoly[p_2] \wedge ((\forall i: i \in \mathbb{N})\ (1 \leq i \leq \|Bgn[\overline{x}]\| \Longrightarrow Eval[p_2, \|Bgn[\overline{x}]\|_i] = \|Bgn[\overline{a}]\|_i] \wedge$$
$$\wedge Deg[p_2] \leq \|Bgn[\overline{a}]\| - 1$$
$$\Longrightarrow$$
$$(Last[\overline{x}] - First[\overline{x}] \neq 0) \wedge IsTuple[\overline{x}] \wedge \|\overline{x}\| \geq 1))). \tag{62}$$

At the first side, the formulas look very complicated, however, they are almost trivial to prove.

For example in (57) the outermost symbol is "$\Longrightarrow$" and, at the right-hand-side, we have to prove:

$$IsTuple[\overline{a}] \wedge \|\overline{a}\| \geq 1,$$

which is assumed at the left-hand-side. Thus, the formula holds.

After proving that the algorithm is coherent, we generate the verification conditions which would ensure the total correctness of the algorithm.

The condition treating the base case, that is, the bottom of the recursion is:

$$(\forall \overline{x}, \overline{a})(IsTuple[\overline{x}] \wedge IsTuple[\overline{a}] \wedge \|\overline{x}\| = \|\overline{a}\| \wedge \|\overline{a}\| \geq 1 \wedge$$
$$\wedge((\forall i,j: i,j \in \mathbb{N})\ (1 \leq i,j \leq \|\overline{a}\| \wedge i \neq j \Longrightarrow \|\overline{x}\|_i \neq \|\overline{x}\|_j)) \wedge \|\overline{a}\| \leq 1$$

$$\Longrightarrow$$

$$IsPoly[First[\overline{a}]]\wedge$$
$$\wedge((\forall i : i \in \mathbb{N})(1 \leq i \leq \|\overline{a}\| \Longrightarrow Eval[First[\overline{a}], \|\overline{x}\|_i] = \|\overline{a}\|_i) \wedge$$
$$\wedge Deg[First[\overline{a}]] \leq \|\overline{a}\| - 1)). \tag{63}$$

The condition treating the general case, that is, the recursive calls is:

$$(\forall \overline{x}, \overline{a}, p_1, p_2)(IsTuple\ [\overline{x}] \wedge IsTuple\ [\overline{a}] \wedge\ \|\overline{x}\| = \|\overline{a}\| \wedge \|\overline{a}\| \geq 1\wedge$$
$$\wedge((\forall i, j : i, j \in \mathbb{N})\ (1 \leq i, j \leq \|\overline{a}\| \wedge i \neq j \Longrightarrow\ \|\overline{x}\|_i \neq \|\overline{x}\|_j)) \wedge \neg(\|\overline{a}\| \leq 1)\wedge$$
$$\wedge IsPoly[p_1] \wedge ((\forall i : i \in \mathbb{N})\ (1 \leq i \leq \|Tail[\overline{x}]\| \Longrightarrow Eval[p_1, \|Tail[\overline{x}]\|_i] = \|Tail[\overline{a}]\|_i])\wedge$$
$$\wedge Deg[p_1] \leq \|Tail[\overline{a}]\| - 1\wedge$$
$$\wedge IsPoly[p_2] \wedge\ ((\forall i : i \in \mathbb{N})\ (1 \leq i \leq \|Bgn[\overline{x}]\| \Longrightarrow Eval[p_2, \|Bgn[\overline{x}]\|_i] = \|Bgn[\overline{a}]\|_i]\wedge$$
$$\wedge Deg[p_2] \leq \|Bgn[\overline{a}]\| - 1$$

$$\Longrightarrow$$

$$IsPoly[\frac{(\mathcal{X} - First[\overline{x}])p_1 - (\mathcal{X} - Last[\overline{x}])p_2}{Last[\overline{x}] - First[\overline{x}]}]\wedge$$
$$\wedge(\forall i : i \in \mathbb{N})(1 \leq i \leq \|\overline{x}\| \Longrightarrow Eval[\frac{(\mathcal{X} - First[\overline{x}])p_1 - (\mathcal{X} - Last[\overline{x}])p_2}{Last[\overline{x}] - First[\overline{x}]}, \|x\|_i] = \|a\|_i)\wedge$$
$$\wedge Deg[\frac{(\mathcal{X} - First[\overline{x}])p_1 - (\mathcal{X} - Last[\overline{x}])p_2}{Last[\overline{x}] - First[\overline{x}]}] \leq \|\overline{a}\| - 1)). \tag{64}$$

5 Termination

In this section we present a specialized strategy for proving termination of recursive functional programs. The detailed termination proofs may in many cases be skipped, because the termination conditions are reusable and thus collected in specialized libraries. Enlargement of the libraries is possible by proving termination of each candidate, but also by taking new elements directly from existing libraries.

Termination proofs of individual programs are, in general, expensive from the automatic theorem proving point of view – they normally involve induction and thus an induction prover must be applied. In some cases, program termination, however, may be ensured – and this is the main contribution of this section – by matching against *simplified versions* (of programs) collected in specialized libraries.

As we already saw, proving total correctness of a program is split into three distinct steps: first – proving coherence, second – proving partial correctness, and third – proving termination.

Furthermore, partial correctness and termination, expressed as verification conditions which themselves may be proven without taking into account their order. Moreover, as we have shown in the previous sections, a coherent program (of a certain recursive type) is totaly correct if and only if its verification conditions hold as logical formulas.

Proving any of the three kinds of verification conditions has its own difficulty, however, our experience shows that proving coherence is relatively easy, proving partial correctness is more difficult and proving the termination verification condition (it is only one condition) is in general the most difficult one.

The proof typically needs an induction prover and the induction step may sometimes be difficult to find. Fortunately, due to the specific structure, the proof is not always necessary, and this is what we discuss here.

5.1 Libraries of Terminating Programs

In this subsection we describe the idea of proving termination of recursive programs by creating and exploring libraries of terminating programs, and thus avoiding redundancy of induction proofs. The core idea is that different recursive programs may have the same *simplified version.*

Let us reconsider the following very simple recursive program for computing the factorial function:

$$Fact[n] = \textbf{If } n = 0 \textbf{ then } 1 \textbf{ else } n * Fact[n-1], \tag{65}$$

with the specification of $Fact$, input:

$$\forall n (I_{Fact}[n] \Longleftrightarrow n \in \mathrm{N}) \tag{66}$$

and Output:

$$\forall n, m\, (O_{Fact}[n, m] \Longleftrightarrow n! = m)\,. \tag{67}$$

The verification condition for the termination of $Fact$ is expressed using a *simplified version* of the initial function:

$$Fact'[n] = \textbf{If } n = 0 \textbf{ then } \mathbb{T} \textbf{ else } Fact'[n-1], \tag{68}$$

namely, the verification condition is

$$(\forall n : n \in \mathrm{N})(Fact'[n] = \mathbb{T}), \tag{69}$$

where $\mathbb{T}$ expresses the logical constant *true*.

Note, that different recursive programs may have the same *simplified version.* Let us now consider another very simple recursive program for computing the sum function:

$$Sum[n] = \textbf{If } n = 0 \textbf{ then } 0 \textbf{ else } n + Sum[n-1], \tag{70}$$

with the specification of Sum, input:

$$\forall n (I_{Sum}[n] \Longleftrightarrow n \in \mathrm{N}) \tag{71}$$

and Output:

$$\forall n, m \left(O_{Sum}[n, m] \Longleftrightarrow \frac{n * (n+1)}{2} = m \right). \tag{72}$$

The verification condition for the termination of Sum is expressed using a *simplified version* of the initial function:

$$Sum'[n] = \textbf{If } n = 0 \textbf{ then } \mathbb{T} \textbf{ else } Sum'[n-1], \tag{73}$$

namely, the verification condition is

$$(\forall n : n \in \mathrm{N})(Sum'[n] = \mathbb{T}). \tag{74}$$

Notably, the termination verification conditions (69) and (74) of the programs (68) and (73) are the same.

Primitive recursive functions (75) comprise a very large and powerful class of functions [23]. It is well known that they always terminate [39]. The schemata of the recursive part of a primitive recursive function is:

$$Prim[n] = \textbf{If } n = 0 \textbf{ then } S[n] \textbf{ else } C[n, Prim[n-1]], \tag{75}$$

Now, the simplified version of (75) is:

$$Prim'[n] = \textbf{If } n = 0 \textbf{ then } \mathbb{T} \textbf{ else } Prim'[n-1], \tag{76}$$

namely, the verification condition is

$$(\forall n : n \in \mathrm{N})(Prim'[n] = \mathbb{T}). \tag{77}$$

which is the same as (69).

For serving the termination proofs, we are now creating libraries containing *simplified versions* together with their input conditions, whose termination is proven. The proof of the termination may now be skipped if the *simplified version* is already in the library and this membership check is much easier than an induction proof – it only involves matching against simplified versions.

Starting from a small library – actually it is not only one, but more, because each recursive schema has several domain based libraries – we intend to enlarge it. One way of doing so is by carrying over the whole proof of any new candidate, appearing during a verification process.

5.2 Enlargement Within libraries

Enlargement within a library is also possible by applying special knowledge retrieval. As we have seen, termination depends on the *simplified version* F' and on the input condition I_F. Considering again the factorial example (65), in order to prove its termination we need to prove (69). Assume, now the pair (68), (69) is in our library. We may now strengthen the input condition I_{Fact} and actually produce a new one:

$$I_{F-new}[n] \Longleftrightarrow (n \in \mathrm{N} \wedge n \geq 100).$$

The *simplified version* $Fact'$ remains the same (68) – we did not change the initial program (65), however, the termination condition becomes:

$$(\forall n : n \in \mathrm{N} \wedge n \geq 100)(Fact'[n] = \mathbb{T}), \tag{78}$$

and (after proving them) we add it to the library. It is easy to see that any new version of a simplified program which is obtained by strengthening the input condition can also be included in the library without further proof. Assume

$$(\forall x : I_F[x])(F'[x] = \mathbb{T})$$

is a member of a library. Then for any "stronger" input condition $I_{F-strng}$, we have:

$$I_{F-strng}[x] \Longrightarrow I_F[x],$$

and thus

$$(\forall x : I_{F-strng}[x])(F'[x] = \mathbb{T}).$$

This is of course not the case for weakening the input condition. Consider the following weakening of I_{Fact}:

$$I_{F-real}[n] \Longleftrightarrow (n \in \mathrm{R}),$$

which leads to nontermination of our $Fact'$ (69), that is:

$$(\forall n : n \in \mathrm{R})(Fact'[n] = \mathbb{T}),$$

which does not hold.

Strengthening of input conditions leads to preserving the termination properties and thus enlarging a library without additional proof is possible. However, for a fixed *simplified version*, keeping (and collecting in some cases) the weakest input condition is the most efficient strategy, because then proving the implication from stronger to weaker condition is relatively easier.

5.3 A Note on the Termination of Fibonacci-like Programs

In this subsection we share some thoughts about proving termination of Fibonacci-like programs. In fact, we show that certain simplification is not possible by constructing a counterexample.

Consider the following program (already a simplified version) for computing F:

$$F[x] = \textbf{If } Q[x] \textbf{ then } \mathbb{T} \textbf{ else } F[R_1] \wedge F[R_2]. \tag{79}$$

The question we want to ask is the following: In order to prove termination of F, would it be sufficient to prove termination of a split of F, namely to prove termination of F_1 and F_2, where:

$$F_1[x] = \textbf{If} Q[x] \textbf{ then } \mathbb{T}\textbf{else} F[R_1], \tag{80}$$

$$F_2[x] = \textbf{If } Q[x] \textbf{ then } \mathbb{T} \textbf{ else} F[R_2]. \tag{81}$$

We do not want to go into discussions on *if this were so*. We give an example in order to show that this is not a case. Let us have:

$$Q[x] \Longleftrightarrow x = 0$$

$$R_1[0] = 1, R_1[1] = 2, R_1[2] = 0$$

$$R_2[0] = 2, R_2[1] = 0, R_2[2] = 1$$

$$I_F[x] \Longleftrightarrow I_{R_1}[x] \Longleftrightarrow I_{R_2}[x] \Longleftrightarrow x = 0 \vee x = 1 \vee x = 2.$$

First check if the program F is coherent. In order to perform the coherence check, we instantiate the relevant conditions:

$$(\forall x : x = 0 \vee x = 1 \vee x = 2)(x = 0 \Longrightarrow x = 0 \vee x = 1 \vee x = 2)$$

$$(\forall x : x = 0 \vee x = 1 \vee x = 2)(x \neq 0 \Longrightarrow R_1[x] = 0 \vee R_1[x] = 1 \vee R_1[x] = 2)$$

$$(\forall x : x = 0 \vee x = 1 \vee x = 2)(x \neq 0 \Longrightarrow R_2[x] = 0 \vee R_2[x] = 1 \vee R_2[x] = 2)$$

$$(\forall x : x = 0 \vee x = 1 \vee x = 2)(x \neq 0 \Longrightarrow x = 0 \vee x = 1 \vee x = 2)$$

$$(\forall x : x = 0 \vee x = 1 \vee x = 2)(x \neq 0 \Longrightarrow x = 0 \vee x = 1 \vee x = 2)$$

$$(\forall x : x = 0 \vee x = 1 \vee x = 2)(x \neq 0 \wedge \ldots \Longrightarrow \mathbb{T}).$$

After we are convinced that F is coherent, we first observe that its split F_1 and F_2 both terminate.

For F_1, all the possibilities are:

$$F_1[0] = \mathbb{T}$$

$$F_1[1] = F_1[2] = F_1[0] = \mathbb{T}$$

$$F_1[2] = F_1[0] = \mathbb{T},$$

and for F_2:

$$F_2[0] = \mathbb{T}$$

$$F_2[1] = F_2[0] = \mathbb{T}$$

$$F_2[2] = F_2[1] = F_2[0] = \mathbb{T}.$$

Now we will show that $F[1]$ does not terminate. Indeed:

$$F[1] = F[R_1[1]] \wedge F[R_2[1]] = F[2] \wedge F[0] = F[R_1[2]] \wedge F[R_2[2]] =$$

$$= F[0] \wedge F[1] = \mathbb{T} \wedge F[1].$$

Thus, proving termination of Fibonacci-like programs requires proving termination of the whole simplified version, and, in general, no split into parts is possible.

5.4 Further Examples

Let us recall the definition of the binary powering algorithm analysed in subsection 4.1:

$$\begin{aligned} P[x, n] = {} & \textbf{If } n = 0 \textbf{ then } 1 \\ & \textbf{elseif } \text{Even}[n] \textbf{ then } P[x * x, n/2] \\ & \textbf{else } x * P[x * x, (n-1)/2]. \end{aligned}$$

In this case the simplified function is:

$$\begin{aligned} P'[n] = {} & \textbf{If } n = 0 \textbf{ then } \mathbb{T} \\ & \textbf{elseif } \text{Even}[n] \textbf{ then } P'[n/2] \\ & \textbf{else } P'[(n-1)/2]. \end{aligned}$$

Note that the first argument x of P is removed, because it does not occur in the conditions. For termination one needs to show that $P'[n] = \mathbb{T}$ for all naturals n, which can be easily done by induction on n.

Likewise, the definition of the Neville's algorithm analtyzed in subsection 4.2 was:

$$P[\overline{x},\overline{a}] = \textbf{If } \|\overline{a}\| \leq 1 \textbf{ then } \mathit{First}[\overline{a}] \\ \textbf{else} \frac{(\mathcal{X} - \mathit{First}[\overline{x}])(P[\mathit{Tail}[\overline{x}],\mathit{Tail}[\overline{a}]]) - (\mathcal{X} - \mathit{Last}[\overline{x}])(P[\mathit{Bgn}[\overline{x}],\mathit{Bgn}[\overline{a}]])}{\mathit{Last}[\overline{x}] - \mathit{First}[\overline{x}]}, \tag{82}$$

Correspondingly, the simplified function is:

$$P'[\overline{a}] = \textbf{If } \|\overline{a}\| \leq 1 \textbf{ then } \mathbb{T} \textbf{ else } P'[\mathit{Tail}[\overline{a}]] \wedge P'[\mathit{Bgn}[\overline{a}]]. \tag{83}$$

The termination condition is: $(\forall \overline{a})(P'[\overline{a}] = \mathbb{T})$, which is provable by a simple induction on the lenght of $\overline{a}$.

6 Conclusions and Further Work

Our theoretical framework for the verification of functional programs is relatively simple, although sound and complete.

In contrast to most approaches which expose methods for verifying correct programs, we put a special emphasis on falsifying incorrect programs.

We first perform a check whether the program under consideration is *coherent* with respect to its specification, that is, each function call is applied to arguments obeying the respective input specification. The completeness of the method relies on the coherence of the program.

The program correctness is then transformed into a set of first-order predicate logic formulas by a verification condition generator (VCG) – a device, which takes the program (its source code) and the specification (precondition and postcondition) and produces several verification conditions, which themselves, do not refer to any theoretical model for program semantics or program execution, but only to the theory of the domain used in the program.

For coherent programs we are able to define a necessary and sufficient set of verification conditions, thus our condition generator is not only *sound*, but also *complete*. This distinctive feature of our method is very useful in practice for program debugging.

We would like to address not only logicians (interested on program verification and automatic theorem proving), but also mathematicians, physicists and engineers who are inventing algorithms for solving concrete problems. On one hand, the help comes with the automatically obtained correctness proof. On the other hand, the inventor may try to prove the correctness of any conjecture, and in case of a failure obtain a counterexample, which may eventually help making a new conjecture.

The approach to program verification presented here is a result of an theoretical work with the long term aim of practical verification of recursive programs. Although the examples presented here appear to be relatively simple, they already demonstrate the usefulness of our approach in the general case. We aim at extending these experiments to more practical examples, because these, usually, are not more complex from the mathematical point of view. Furthermore we aim at improving the education of future software engineers by exposing them to successful examples of using formal methods (and in particular automated reasoning) for the verification and the debugging of concrete programs.

Another possible direction of our further work is the development of methods for proving total correctness of tail recursive programs. More precisely, methods for programs having a specific structure in which an auxiliary tail recursive function is driven by a main nonrecursive function, and only the specification of the main function is provided.

The difficulty there is that it is impossible to find automatically, in general, verification conditions for an arbitrary tail recursive function without knowing its specification. However, in many particular cases this is, nevertheless, possible. The specification of the auxiliary function could be obtained automatically, for example by solving coupled linear recursive sequences with constant coefficients.

References

1. ACL2. http://www.cs.utexas.edu/users/moore/acl2/
2. de Bakker, J.W., Scott, D.: A Theory of Programs. In IBM Seminar, Vienna, Austria (1969)
3. Bernhard, B., Reiner, H., Peter, H.S. (ed.): Verification of Object-Oriented Software: The KeY Approach. vol. 4334, LNCS, Springer (2007)
4. Bertot, Y., Casteran, P.: Interactive Theorem Proving and Program Development. Coq'Art: The Calculus of Inductive Constructions. Springer, Berlin (2004)
5. Blanqui, F., Hinderer, S., Coupet-Grimal, S., Delobel, W., Kroprowski, A.: CoLoR, a Coq Library on Rewriting and Termination. In: Geser, A., Søndergaard, H. (eds.) Proceedings of 8th International Workshop on Termination, Seattle, WA, USA (2006)
6. Boyer, R.S., Moore, J.S.: A Computational Logic Handbook. Academic Press Professional, Incorporation, San Diego, CA, USA (1988)
7. Buchberger, B.: Algorithm Supported Mathematical Theory Exploration: A Personal View and Stragegy. In: Buchberger, B. John Campbell, (eds.) Proceedings of AISC 2004 (7th International Conference on Artificial Intelligence and Symbolic Computation), Springer Lecture Notes in Artificial Intelligence, vol. 3249, pp. 236–250. Springer, Berlin 22–24 (2004)
8. Buchberger, B.: Ein Algorithmus zum Auffinden der Basiselemente des Restklassenringes nach einem nulldimensionalen Polynomideal (An Algorithm for Finding the Basis Elements in the Residue Class Ring Modulo a Zero Dimensional Polynomial Ideal). PhD thesis, Mathematical Institute, University of Innsbruck, Austria (1965). (English translation Journal of Symbolic Computation 41 (2006) 475–511).
9. Buchberger, B.: Gröbner-Bases: An Algorithmic Method in Polynomial Ideal Theory. In: Bose, N.K. (eds.) Multidimensional Systems Theory – Progress, Directions and Open Problems in Multidimensional Systems, pp. 184–232. Reidel Publishing Company, Dodrecht, Boston, Lancaster (1985)

10. Buchberger, B.: Towards the Automated Synthesis of a Groebner Bases Algorithm. RACSAM – Revista de la Real Academia de Ciencias (Review of the Spanish Royal Academy of Science), Serie A: Mathematicas, **98**(1), 65–75 (2004)
11. Buchberger, B., Affenzeller, M., Ferscha, A., Haller, M., Jebelean, T., Klement, E.P., Paule, P., Pomberger, G., Schreiner, W., Stubenrauch, R., Wagner, R., Weiß, G., Windsteiger, W.: Hagenberg Research. Springer, 1st edition (2009)
12. Buchberger, B., Craciun, A., Jebelean, T., Kovacs, L., Kutsia, T., Nakagawa, K., Piroi, F., Popov, N., Robu, J., Rosenkranz, M., Windsteiger, W.: Theorema: Towards Computer-Aided Mathematical Theory Exploration. J. Appl. Logic pp. 470–504 (2006)
13. Buchberger, B., Dupre, C., Jebelean, T., Kriftner, F., Nakagawa, K., Vasaru, D., Windsteiger, W.: The Theorema Project: A Progress Report. In Calculemus 2000: Integration of Symbolic Computation and Mechanized Reasoning, Calculemus (2000)
14. Buchberger, B., Lichtenberger, F.: Mathematics for Computer Science I – The Method of Mathematics (in German). Springer, 2nd edition (1981)
15. Chlipala, A.: A Verified Compiler for an Impure Functional Language. In Proceedings of the 37th ACM SIGPLAN-SIGACT Symposium on Principles of Programming Languages (POPL'10), pp. 237–248. New York, NY, USA (2010)
16. Floyd, R.W.: Assigning Meanings to Programs. In Proceedings of Symphosia in Applied Mathematics 19, pp. 19–37 (1967)
17. Gordon, M.: From LCF to HOL: A Short History. pp. 169–185. MIT Press, MA, USA (2000)
18. Greibach, S.A.: Theory of Program Structures: Schemes, Semantics, Verification. Springer, New York, Secaucus, NJ, USA (1985)
19. Hildebrand, B.F.: Introduction to Numerical Analysis: 2nd Edition. Dover Publications, New York, NY, USA (1987)
20. Hoare, C.A.R.: An axiomatic basis for computer programming. Comm. ACM **12**(10), 576–580 (1969)
21. HOL. http://hol.sourceforge.net/
22. Homeier, P.V., Martin, D.F.: Secure Mechanical Verification of Mutually Recursive Procedures. Inf. Comput. **187**(1), 1–19 (2003)
23. Neil, I.: Computability and Complexity. In: Edward, N. Zalta, (eds) The Stanford Encyclopedia of Philosophy (Fall 2008 Edition). http://plato.stanford.edu/archives/fall2008/entries/computability
24. Kaufmann, M., Moore, J.S.: An Industrial Strength Theorem Prover for a Logic Based on Common Lisp. Software Eng. **23**(4), 203–213 (1997)
25. Langford, W.J., Broadbent, T.A.A., Goodstein, R.L.: Obituary: Professor Eric Harold Neville. Math. Gaz. **48**(364), 131–145 (1964)
26. Rustan, K., Leino, M., Peter, M., Jan, S.: Verification of Concurrent Programs with Chalice. In: Aldini, A., Barthe, G., Gorrieri, R. (eds.) Foundations of Security Analysis and Design V, pp. 195–222. Springer, New York (2009)
27. Loeckx, J., Sieber, K.: The Foundations of Program Verification. Teubner, 2nd edition (1987)
28. Luckham, D.C., Park, D.M.R., Paterson, M.: On Formalised Computer Programs. J. Comput. Syst. Sci. **4**(3), 220–249 (1970)
29. Manna, Z.: Mathematical Theory of Computation. McGraw-Hill, New York (1974)
30. Matthews, J., Moore, J.S., Ray, S., Vroon, D.: Verification Condition Generation Via Theorem Proving. In: Hermann, M., Voronkov, A. (eds.) In Proceedings of the 13th International Conference on Logic for Programming, Artificial Intelligence, and Reasoning (LPAR 2006), LNCS, vol. 4246, pp. 362–376. Phnom Penh, Cambodia (2006)
31. Meyer, B.: Applying design by contract. Computer **25**(10), 40–51 (1992)
32. Neville, E.H.: Iterative Interpolation. J. Indian Math. Soc. **20**, 87–120 (1934)
33. Peter, R.: Rekursive Funktionen in der Komputer-Theorie. Verlag d. ungarisch. Akademie d. Wiss., Budapest (1976)
34. Press, W.H., Teukolsky, S.A., Vetterling, W.T., Flannery, B.P.: Numerical Recipes in C: The Art of Scientific Computing. Cambridge University Press, 2 edition (1992)
35. PVS group: PVS Specification and Verification System. http://pvs.csl.sri.com (2004)

36. Smith, D.R.: Top-Down Synthesis of Divide-and-Conquer algorithms. Artif. Intell. **27**(1), 43–96 (1985)
37. Sunrise: http://www.cis.upenn.edu/~hol/sunrise/
38. van der Voort, M.: Introducing Well-founded Function Definitions in HOL. In Higher Order Logic Theorem Proving and its Applications: Proceedings of the IFIP TC10/WG10.2 Workshop, pp. 117–132 (1992)
39. Walter, S.B., Lawrence, H.L.: Theory of Computation. Wiley, New York, NY, USA (1974)
40. Wiedijk, F. (ed.): The Seventeen Provers of the World, Foreword by Dana, S. Scott, Lecture Notes in Computer Science. vol. 3600, Springer (2006)
41. Zee, K., Kuncak, V., Rinard, M.C.: Full Functional Verification of Linked Data Structures. In Proceedings of the 30th ACM Conference on Programming Language Design and Implementation, pp. 349–361 (2008)

An Introduction to Automated Discovery in Geometry through Symbolic Computation

Tomas Recio and María P. Vélez

Abstract In this chapter we will present, for the novice, an introduction to the automated discovery of theorems in elementary geometry. Here the emphasis is on the rationale behind different possible formulations of goals for discovery.

1 Introduction

This paper rises from the survey lecture given by the first author at the SNSC-project final Conference, that took place at RISC-Linz, July 2008. It aims to introduce the novice (and curious) reader to automatic *discovery* of elementary geometry theorems, by means of the algebraic geometry approach that has already shown its success for automatic theorem *proving* (see Sect. 2 below).

Different approaches to discovery have already been presented by the authors ([10, 13, 15]), all of them illustrated with many examples. The emphasis here is on building up and discussing the rationale behind the potential alternative formulations of goals and methods for discovery. We refer the reader interested in details about each one of the protocols, to the above articles. Parts of of them have been summarily sketched in this chapter.

T. Recio (✉)
Universidad de Cantabria, Avda. Los Castros, Santander, Spain
e-mail: tomas.recio@unican.es

M.P. Vélez
Universidad Antonio de Nebrija, Pirineos, 55, Madrid, Spain

U. Langer and P. Paule (eds.), *Numerical and Symbolic Scientific Computing*,
Texts and Monographs in Symbolic Computation, DOI 10.1007/978-3-7091-0794-2_12,

2 Automatic Proving

Automatic proving of elementary geometry theorems through symbolic computation has reached a certain mature status. More than thirty years after the foundational paper by Wu "On the decision problem and the mechanization of theorem-proving in elementary geometry" [20], and over twenty years after the popular book of Chou "Mechanical geometry theorem proving" [7], the topic seems active enough to deserve the publication of new books, such as [14] or [21], and continues gathering the international community of researchers at the series of biennial Automated Deduction in Geometry (ADG) conferences[1]. We refer the reader to the impressive bibliography on the subject kept by Prof. D. Wang in [19].

The goal of this particular approach to automatic proving through symbolic computation is to provide algorithms, using computer algebra methods, for confirming (or refuting) the truth of some given geometric statement. More precisely, it aims to decide if a given statement is true (except for some degenerate cases, to be described by the algorithm). Hundreds of non-elementary theorems in elementary geometry have been successfully –and almost instantaneously– verified by a variety of symbolic computation methods. Different collections of examples are presented on the books we referred to in the precedent paragraph.

Briefly, this particular automatic proving approach proceeds by translating a geometric statement $\mathcal{T}\{H \Rightarrow T\}$ into algebraic terms, after adopting a coordinate system. More precisely, geometric instances verifying the hypotheses H (respectively, the theses T) can be expressed as the set of solutions[2] of a system of polynomial equations $H = \{h_1 = 0, \ldots, h_r = 0\}$ ($T = \{t_1 = 0, \ldots, t_s = 0\}$, respectively). As it is usual in algebraic geometry we will denote by $V(H)$ or $V(T)$ the solution set of the corresponding system of equations. At this point it seems reasonable to say that our statement $\mathcal{T}$ is true if and only if $V(H) \subseteq V(T)$.

But this interpretation is, in some sense, too strong, because it requires that all instances verifying the polynomial system H should satisfy, as well, the polynomial system T. In fact, it often happens that the algebraic formulation of the given hypotheses includes, for instance, degenerate cases of the proposed geometric configuration (triangles that collapse to a line, parallel lines that coincide, etc.), that should not be considered for the validity of our statement. Therefore, it seems more convenient to rephrase our formulation of the truth of a statement, by just requiring that a (Zariski) open subset of $V(H)$ is included in $V(T)$. Thus, we are thinking of an algebraic procedure that automatically generates a set $\{f_1 \neq 0 \vee \cdots \vee f_s \neq 0\}$ of inequations, the complement of the closed set defined by $R = \{f_1 = 0, \ldots, f_s = 0\}$, such that, by avoiding *degeneracy conditions*,

[1] Visit https://lsiit-cnrs.unistra.fr/adg2010/index.php/Main_Page for information about the ADG-2010 conference, with links to the URLs of previous meetings.

[2] To be considered over a suitable field. There will be different interpretations for different choices of this field. Here we will assume to be in the algebraically closed case, so we will miss, for instance, oriented geometry.

$V(H)\backslash V(R) \subseteq V(T)$. That is, our original statement now merely claims that $H \wedge \neg R \Rightarrow T$, for some suitable collection of degeneracy conditions R that should be obtained by algebraic manipulation from the given theses and hypotheses.

But, even in this summary and rough description, the reader should be warned that it does not reflects the existence of some subtle, but serious, difficulties. Diverse foundational problems may arise in the geometric/algebraic translation process (see the references to this issue that appear, for instance, at the book [7], or at the papers [1, 8, 15], or in the introduction to [2]). In particular, we should point out that different ideas of truth, diverse protocols to grasp it and several methods to perform them, have been considered in this algebraic geometry approach to automated theorem proving (in particular, see the general discussion on the relative concept of truth in this context, included in the papers [5, 9]).

3 Automatic Discovery

A closely related, yet different, issue is that of the automatic *discovery* of theorems. While automatic *proving* deals with establishing that some statement holds in most instances, automatic *discovery* addresses the case of statements that are false in most relevant cases. In fact, it aims to produce, automatically, additional hypotheses for the statement to be correct. In other words, when proving fails, we might try discovering why...

Let us consider the following example from [13]: we draw a triangle and, then, the feet of the corresponding altitudes. These feet are the vertices of a new triangle, the so called *orthic* triangle for the given triangle. We state that this orthic triangle is isosceles, but it is not so, in general. That is, we fail proving that the orthic triangle of any given triangle is isosceles. It seems quite obvious that our statement holds if the original triangle is itself isosceles, but, only in this case? Searching for other possibilities is the task of the automatic discovery of theorems protocols (see Fig. 1 below).

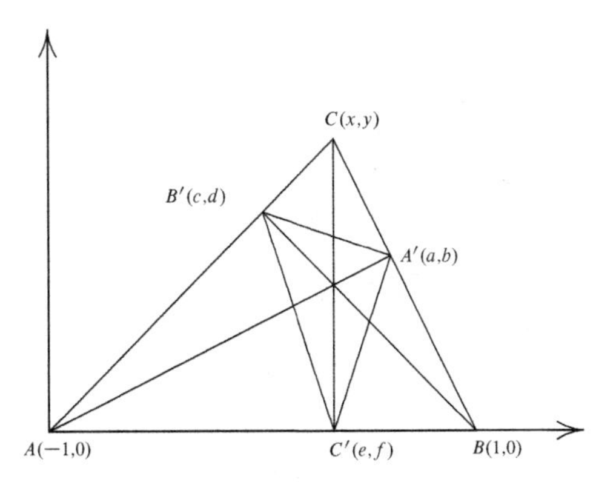

Fig. 1 The triangle ABC and its orthic triangle $A'B'C'$

The interest of developing such automatized discovery procedure is quite obvious. First, it could help finding out missing cases in intriguing classical theorems. For instance, searching for the conditions that a triangle has to verify in order to have two or three equal length internal or external bisectors, corresponding to different vertices (extending the Theorem of Steiner-Lehmus, see [4, 17, 18] or [12] and the references thereof). Second, in the context of CAD, automatic discovery could be used as an auxiliary tool for determining further constraints verified by the elements of a given sketch if the user imposes among them some geometric restrictions (an observation already remarked in [11]). For instance, we draw a triangle, then its orthic triangle, but we want to make our sketch so that the later triangle is isosceles. Then we should be warned that this obliges the first triangle to be drawn with some special features (and the CAD program should provide information about all of them).

Finally, automated discovery could be also useful in the educational context, since it would allow a dynamic geometry program (provided with a link to a computer algebra program, as shown in [3] or [16]) to act as an intelligent agent, being able to *know* in advance the response for most (right or wrong) conjectures made by a user attempting to construct a certain figure on the screen; in this way, the dynamic geometry program could act as a tutor, guiding in the right direction the efforts of the user towards the assigned task. Suppose the student is given a triangle and one arbitrary point P and is asked to determine the locus of P so that the three symmetrical points P_1, P_2, P_3 of P, with respect to the three sides of the triangle, are aligned. After dragging P around for a while, the student finds some positions where P_1, P_2, P_3 are close to be on a line, but it could happen that he/she does not have any further insight on this problem (Fig. 2).

The student could ask the computer for a (partial or total) answer to the query. The machine, provided with an automatic discovery tool, will "know" the locus of P is, precisely, the circle through the three vertices of the given triangle [15], and –if adequately programmed– could present to the student different hints towards finding the solution. A project regarding the implementation of proving and discovering features on GeoGebra, a popular dynamic geometry program for mathematics education, is being currently considered (see http://www.ciem.unican.es/proving2010)

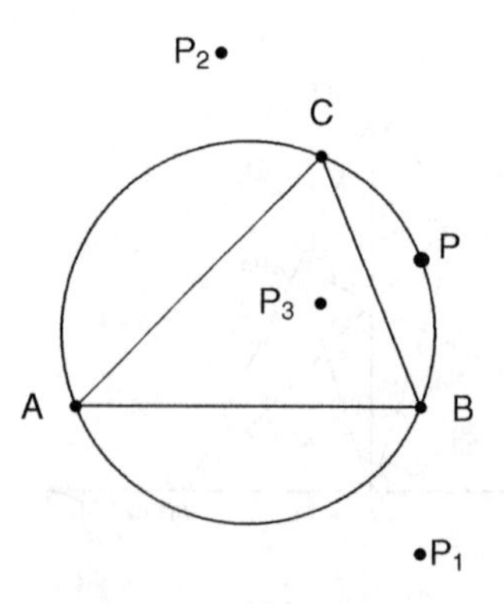

Fig. 2 Locus of P for the alignment of its reflections on the three sides of a triangle

4 Stating Our Goal

As in the automatic proving context, we start considering an algebraically translated statement of the kind $\{H \Rightarrow T\}$ (such as: *for any given triangle, if we construct its orthic triangle then... it is isosceles*, or: *given any triangle and a point P and its reflections with respect to the triangle's sides, then... they are aligned*), where H stands for the equations describing the construction (orthic triangle, symmetries, etc.) and T describes the desired property (isosceles, aligned, etc.). By abuse of notation, we will denote also by H and T the ideals generated by the polynomials involved in the equations describing the statement. Suppose that $H = (h_1, \ldots, h_r)$ and $T = (t_1, \ldots, t_s))$ are these ideals of polynomials in a ring $K[X]$, $X = \{x_1, \ldots, x_n\}$, over a field K, with algebraic closure $\overline{K}$. Then, the geometric instances verifying the hypotheses (respectively, the theses) of the statement are the algebraic sets $V(H)$ (respectively, $V(T)$) over the affine space $\overline{K}^n$.

Following the tradition of automatic theorem proving it is also natural here to search for complementary hypotheses of inequality type, taking account of degeneracy conditions. But, since it is quite reasonable to assume that the given discovery statement is generally false, we should also search for complementary hypotheses of equality type (such as: *P is on the circle described by the three vertices*), so that, adding them to the given hypotheses and avoiding degeneracies, the given statement becomes true. Let us denote by R' the collection of polynomials describing these equality type conditions and by R'' the system of polynomials representing degeneracy conditions. Thus, a natural goal for discovering theorems could be finding sets R', R'', such that $\{(H \wedge R' \wedge \neg R'') \Rightarrow T\}$, assuming $\{H \wedge R' \wedge \neg R''\}$ is not empty. But this formulation is, in fact, too slack to be of any use. In fact, by taking $R' = T$ and a trivial R'', we will obtain, in general, a (useless) solution to our goal (such as: *for any given triangle, if we construct its orthic triangle and it is isosceles then... it is isosceles*).

Therefore we should reconsider the formulation of the goal, taking into account that we actually need to find the complementary hypotheses in terms of some specific set of variables ruling our statement, as the following example shows.

Example 1. Suppose we are searching conditions for the orthic triangle of a given one to be isosceles. That is, we want to find new hypotheses to achieve this property, and these hypotheses should be expressed in terms of the variables assigned to the vertices of the given triangle, and not, for example, in terms of the variables naming the vertices of the orthic triangle.

Without loss of generality let us assume that, up to a change of coordinates, the vertices of the triangle are $A = (-1, 0)$, $B = (1, 0)$ and $C = (x, y)$. Denote by $A' = (a, b)$, $B' = (c, d)$ and $C' = (e, f)$ the corresponding vertices of the orthic triangle. Then, the given construction is described by the following six equations in eight variables (H):

$$(a-1)y - b(x-1) = 0, \quad (c+1)y - d(x+1) = 0, \quad f = 0,$$
$$(a+1)(x-1) + by = 0, \quad (c-1)(x+1) + dy = 0, \quad e - x = 0,$$

where first line states that vertices A', B' and C' belong to the corresponding side of the given triangle and the second line expresses these vertices are also points lying in the corresponding altitude.

Now we would like to know when this orthic triangle is isosceles. Remark that there are three possibilities for a triangle to be isosceles depending on which couple of sides is to be considered having equal length. Again, for simplicity, let us analyze here only one of these cases, for example, when the two sides meeting at vertex C' are equal, yielding to the following thesis

$$T: \ (e-c)^2 + (d-f)^2 - (e-a)^2 - (b-f)^2 = 0.$$

In order to search for the extra constraints yielding to an isosceles orthic triangle it seems obvious we should proceed finding R', R'' as polynomials in variables x, y. In fact, the system H has two degrees of freedom, as expected, because variables x, y have been freely chosen and variables a, b, c, d, e, f depend on them. Now, let us remark that the mere consideration of the algebraic system does not allow to highlight a meaningful set of variables –such as x, y– carrying the relevant geometric information (to build the orthic triangle). In this system H there are, as well, other sets of two free variables. For instance, we could express all involved variables in terms of a, c.

Different and more complicated examples show that it is impossible to determine a set of meaningful variables in an automatic way, even relying on heuristics, such as considering those variables that are not involved in the thesis, etc. It should be human intuition (i.e. the user) who has to point out the concrete collection of variables that will turn meaningful the discovery process. Thus, our goal, as stated above, should be modified by referring to some specific set of variables for the complementary hypotheses.

But this requirement is not enough. In fact, we should notice that, once the equality type extra hypotheses R' are found, the degenerate conditions R'' should be expressed in terms of some subset of the selected variables, since the whole construction, after adding R', could possess, then, less degrees of freedom (for instance, in the example above, if R' is found and it states –say– that the given triangle must be isosceles and, therefore, that $x = 0$, then the degree of freedom, of the new system of hypotheses, will be reduced from two to one). Bearing this in mind, our goal should be reformulated to look for the existence of two subsets of variables $U' \subseteq U \subseteq X$, and two ideals (R', R''), in $K[U]$ and $K[U']$, respectively, such that $(H \wedge R' \wedge \neg R'') \Rightarrow T$ and $\{H \wedge R' \wedge \neg R''\}$ is not contradictory.

Now, a more subtle consideration must be taken into account. It is true that, if we could find a couple (R', R'') verifying the above conditions, we would have a true statement, keeping the given theses T and adding some extra hypotheses $\{H \wedge R' \wedge \neg R''\}$. But nothing guarantees that such statement really covers *all* possibilities related to the given statement $H \Rightarrow T$.

In the example above, imagine that some R' is found expressing that the given triangle should be isosceles; then it will yield to a true statement (in fact, the orthic triangle of an isosceles triangle is also isosceles), but there are other restrictions on

ABC (far less evident) for the given statement to hold. That the given triangle is isosceles is, indeed, a sufficient condition for the orthic triangle to be isosceles, but it is not a necessary condition. So, if we want to avoid discovering just some trivial statements, what we really need to find out is a collection of non contradictory (i.e. such that there is at least one instance of the given hypotheses were they actually hold) extra hypotheses R', R'',

(a) Expressed in the right variables, $(R' \subset K[U], R'' \subset K[U']$, with $U' \subseteq U \subseteq X)$.
(b) Which are, when added to H, sufficient for T, so that $\{(H \wedge R' \wedge \neg R'') \Rightarrow T\}$.
(c) Which are as well necessary for the thesis T to hold under the given hypotheses H, so that $\{(T \wedge H) \Rightarrow (H \wedge R' \wedge \neg R'')\}$.

Coming back to the above example, what we wish to obtain is that the orthic triangle of the given one is isosceles if and only if one of the following conditions hold for the vertex (x, y):

- $x = 0$, that is, the triangle is isosceles, as expected, or
- $x^2 + y^2 - 1 = 0$ and $(x, y) \neq (\pm 1, 0)$, (a degenerate case), or
- $x^2 - y^2 - 1 = 0$ and $(x, y) \neq (\pm 1, 0)$.

5 Refining Our Goal

It seems we have achieved a sound description of our automatic discovery goal. Yet, some more difficulties arise. In fact, assume we have found R', R'' verifying the above conditions. This equivalent to $\{H \wedge T\} \equiv \{H \wedge R' \wedge \neg R''\}$. Now consider the projection $\pi(V(H) \cap V(T))$ of $V(H) \cap V(T)$ over the affine space described by the variables U. It is easy to show that $\{H \wedge T\} \equiv \{H \wedge R' \wedge \neg R''\}$ implies

$$\pi(V(H) \cap V(T)) = \pi(V(H) \cap (V(R') \backslash V(R''))$$

Moreover, since we would have chosen the variables U as those freely ruling our construction $V(H)$, in many statements every assignment of the U variables should yield to at least one instance on $V(H)$. That is, in the standard case when $V(H)$ projects onto the whole U-variables affine space, the projection of $V(H) \cap V(T)$ will be equal to $V(R') \backslash V(R'')$, i.e. to the difference of two algebraic sets.

But the projection of an algebraic variety is a general constructible set, that is, a finite union of sets, each one being the intersection of an algebraic variety and the complement of another one, such as $V(R') \backslash V(R'')$. It is a finite union, and not, in general, just one of the terms of such union[3], as we have concluded from our

[3]For instance, the constructible set $(V(R_1') \setminus V(R_1'')) \cup V(R_2')$, where $V(R_1') =$ a plane, $V(R_1'') =$ a line on the plane, $V(R_2') =$ a point on this line, can not be expressed as $V(R_3') \backslash V(R_3'')$, for whatever sets of polynomials R_3', R_3''.

assumption about the existence of R', R'' verifying the three conditions $a), b), c)$ above. This means that declaring these three conditions as the goal for the discovery of theorems would yield to failure (no couple R', R'' would exist) in most instances, due to the lack of an appropiate language in our setting to express all necessary and sufficient conditions.

At this point two possibilities arise. One, that of reformulating the whole approach to discovery, allowing, from the beginning, the introduction of a finite union of collections of equations R'_i in the U-variables, and inequations R''_i (some of them in the variables U, to take care of the possible degenerate cases of the free variables for H, and some in the U'-variables, to consider the possible degenerate cases after including the new hypotheses R'_i), which would provide:

- When added to H, sufficient conditions for T, so that $\{(H \wedge (\vee_i(R'_i \wedge \neg R''_i))) \Rightarrow T\}$.
- Which are as well necessary, so that $\{(T \wedge H) \Rightarrow (H \wedge (\vee_i(R'_i \wedge \neg R''_i)))\}$.

It is, obviously, a more complex (albeit more complete) approach. We will deal with it in the next Sect. 6.

A second possibility is that of weakening condition $c)$ in such a way that the discovery goal is redirected to finding out a collection of non contradictory extra hypotheses R', R'',

(a) Expressed in the right variables, $(R' \subset K[U], R'' \subset K[U']$, with $U' \subseteq U \subseteq X)$.
(b) Which are, when added to H, sufficient for T, so that $\{(H \wedge R' \wedge \neg R'') \Rightarrow T\}$.
(c) And verifying that R' is necessary for T to hold on H, i.e. $\{(T \wedge H) \Rightarrow R'\}$.

Notice that we have deleted the reference to R'' in the last item, at the risk of losing some necessary inequality-type conditions. Since these conditions, in general, only describe the degeneracy cases that should be avoided for the statement to become true, we think it is quite safe to keep the new formulation of condition $c)$ in our approach, as we will not miss any interesting results just because of not paying attention to some degenerate cases. We will study this approach in Sect. 7.

6 Comprehensive Bases

In this Section let us assume that we had settled our discovery goal to finding a finite union of collections of equations R'_i in U, and inequations R''_i, some in U and some in U', so that $\{(H \wedge (\vee_i(R'_i \wedge \neg R''_i)))$ is not empty, $\{(H \wedge (\vee_i(R'_i \wedge \neg R''_i))) \Rightarrow T\}$ and $\{(T \wedge H) \Rightarrow (H \wedge (\vee_i(R'_i \wedge \neg R''_i)))\}$. That is, $V(H) \cap V(T) = V(H) \cap \bigcup_i (V(R'_i) \setminus V(R''_i))$. Then, as argued above, the projection over the U variables of $V(H) \cap V(T)$ will be equal to the projection of $V(H)$ intersected with $\bigcup_i (V(R'_i) \setminus V(R''_i))$.

Let the projection of $V(H)$ over the U-affine space be described by another finite union of collections of equations P'_j and inequations P''_j, in the U variables, as $\bigcup_j (V(P'_j)\backslash V(P''_j))$.

Then we remark that if we have succeeded finding R'_i, R''_i verifying the above conditions, it will also hold that

$$\{(H \wedge (\vee_i (R'_i \wedge \neg R''_i)) \wedge (\vee_j (P'_i \wedge \neg P''_i))) \Rightarrow T\}$$

and

$$\{(T \wedge H) \Rightarrow (H \wedge (\vee_i (R'_i \wedge \neg R''_i)) \wedge (\vee_j (P'_i \wedge \neg P''_i)))\}$$

But $(\vee_i (R'_i \wedge \neg R''_i)) \wedge (\vee_j (P'_i \wedge \neg P''_i))$ can be as well explicitly expressed as $(\vee_k (M'_k \wedge \neg M''_k)))$ for some new set of equations and inequations M'_k, M''_k. Thus, for this particular set we would have $\{(H \wedge (\vee_k (M'_k \wedge \neg M''_k))) \Rightarrow T\}$, $\{(T \wedge H) \Rightarrow (H \wedge (\vee_k (M'_k \wedge \neg M''_k)))\}$ and the projection over the U variables of $V(H) \cap V(T)$ will be exactly equal to $\bigcup_k (V(M'_k)\backslash V(M''_k))$.

Therefore, setting our discovery goal in such broad way (i.e. allowing unions of basic constructible sets in its formulation), we are driven to describing the projection of $V(H) \cap V(T)$. More precisely, it is easy to prove that if there is a finite union of collections of equations and inequations R'_i, R''_i fulfilling our discovery goal, then there will be another finite union of sets, described by analogous equations and inequations, M'_k, M''_k, accomplishing the goal and yielding directly the projection of $V(H) \cap V(T)$. Conversely, if there is a finite union of collections of equations and inequations R'_i, R''_i fulfilling our discovery goal, any description $\bigcup_k (V(M'_k)\backslash V(M''_k))$ of the projection of $V(H) \cap V(T)$ will satisfy as well the discovery goal.

Thus, in this setting, we face two main issues:

- Deciding if there is a collection of couples R'_i, R''_i, verifying the conditions (as stated at the beginning of this Section) for discovery.
- And, in the affirmative case, computing a description of the $V(H) \cap V(T)$ projection.

Let C be the cylinder over the projection $V(H) \cap V(T)$ onto the U-variables. That is, the set of points in the X- affine space that project over $V(H) \cap V(T)$.

Proposition 1. *There is a collection of couples R'_i, R''_i, verifying the conditions for discovery, if and only if $V(H) \cap C = V(H) \cap V(T)$.*

Proof. First remark that always $V(H) \cap C \supseteq V(H) \cap V(T)$. Now, as we have already shown, there is a collection R'_i, R''_i of couples holding the discovery conditions, if and only if those M'_k, M''_k describing the projection of $V(H) \cap V(T)$ verify as well these conditions. But, then, the equality $V(H) \cap C = V(H) \cap V(T)$ is just a reformulation of $V(H) \cap V(T) = V(H) \cap \bigcup_k (V(M'_k)\backslash V(M''_k))$, i.e. accomplishing the discovery goal. □

Let $H' = (H+T) \cap K[U]$, i.e. H' is the elimination ideal of $H+T$ for the U-variables. Then consider $H + H'^e$, where H'^e provides the extension of H' to $K[X]$. Its variety $V(H + H'^e)$ is the intersection of $V(H)$ with the cylinder $V(H'^e)$ over the Zariski closure of the projection of $V(H) \cap V(T)$. It might be strictly larger than C (think of $H = \{(xu-1)(u) = 0\}, T = \{(xu-1) = 0\}$, then $U = \{u\}$ is a free variable for H, $H' = (0)$, $H + H'^e = H$, but the projection of $V(H) \cap V(T)$ over the u-line is $u \neq 0$ and $V(H) \cap C = V(T)$ which is strictly smaller than $V(H)$).

Proposition 2. *If T is contained in all the minimal primes of $H + H'^e$, there is a collection of couples R'_i, R''_i, verifying the conditions for discovery for all $U' \subseteq U$.*

Proof. It always holds that $H \subseteq H + H'^e \subseteq H + T$. Thus

$$V(H) \cap V(T) \subseteq V(H) \cap V(H'^e) \subseteq V(H)$$

But if T is contained in all the minimal primes of $H + H'^e$, we have also $V(H) \cap V(H'^e) \subseteq V(T)$, so $V(H) \cap V(H'^e) = V(H) \cap V(T)$. As a side remark, conversely, if this equality holds, then, $\sqrt{H + H'^e} = \sqrt{(H+T)} \supseteq T$, and T is contained in all the minimal primes of $H + H'^e$.

Now, since it always hods that $V(H) \cap V(T) \subseteq V(H) \cap C \subseteq V(H) \cap V(H'^e)$, the equality $V(H) \cap V(H'^e) = V(H) \cap V(T)$ implies $V(H) \cap C = V(H) \cap V(T)$ and thus, by the above proposition, there is a collection of couples of ideals verifying the discovery conditions. □

The converse does not hold, and the previous example $H = \{(xu-1)(u) = 0\}$, $T = \{(xu-1) = 0\}$, with $U = \{u\}$ shows that there are cases with suitable discovery conditions but where T does not vanish over all the minimal primes of $H + H'^e$.

Finally, remark that, even if there is not a finite union of collections of equations and inequations R'_i, R''_i fulfilling our discovery goal, any set $(V(M'_k) \backslash V(M''_k))$ being part of the projection of $V(H) \cap V(T)$, will provide a necessary condition for $H \Rightarrow T$.

Therefore, a reasonable way to proceed in order to find R'_i, R''_i verifying the above conditions consists in computing the projection of $V(H) \cap V(T)$ and express it as $\bigcup_i (V(R'_i) \backslash V(R''_i))$. Then, we should check if this set of equations and inequations are sufficient for T.

In some sense, this is what has been achieved in [13] or [6] and can be seen as quite close to performing a certain kind of quantifier elimination procedure. But let us remark that, in the theorem *proving* context, this formulation (i.e. requiring that $\{H \wedge T\} \Rightarrow \{\neg R''\}$ for non-degeneracy conditions) has not been followed in most works, perhaps due to its complexity. But we refer to [13] for a large collection of discovery results using this protocol.

7 FSDIC

As mentioned above, a second procedure for automatic discovery is that of [10], related to finding a *Full Set of (Discovering) Interesting Conditions (FSDIC)*. Assume we are setting our goal to find out a collection of non contradictory extra hypotheses R', R'':

(a) Expressed in the right variables, ($R' \subset K[U], R'' \subset K[U']$, with $U' \subseteq U \subseteq X$).
(b) Which are, when added to H, sufficient for T, so that $\{(H \wedge R' \wedge \neg R'') \Rightarrow T\}$.
(c) And verifying that R' is necessary for T to hold on H, i.e. $\{(T \wedge H) \Rightarrow R'\}$.

This can be formally translated into the following definition. First, some notation. Let us consider some subsets of a main set of variables $X = \{x_1, \ldots, x_n\}$, namely $U' \subseteq U \subseteq X$. Then, we will often deal with the extension $K[U'] \hookrightarrow K[U] \hookrightarrow K[X]$ of polynomial rings on the corresponding variables, with coefficients in a fixed field K. Let A be an ideal in $K[U']$, B an ideal in $K[U]$, and C an ideal in $K[X]$. We will denote –as it is standard in Commutative Algebra– by $A^{e'} = AK[U]$, the extended ideal; by $A^e = AK[X]$, and by $B^e = BK[X]$. Clearly $(A^{e'})^e = A^e$. Moreover we will denote by $C^{c'} = C \cap K[U']$, its contraction ideal; by $C^c = C \cap K[U]$, and by $B^{c'} = B \cap K[U']$. Again, it is clear that $(C^c)^{c'} = C^{c'}$. Finally, if I is an ideal in $K[X]$, we will denote by $V(I) = \{(x_1, \ldots, x_n) \in \overline{K}^n \mid f(x_1, \ldots, x_n) = 0, \ \forall f \in I\}$ the algebraic set defined by I in $\overline{K}^n$, where $\overline{K}$ is the algebraic closure of K.

Definition 1. Let $\mathcal{T}$ be a statement, of the kind $H \Rightarrow T$, where the ideals $H, T \subseteq K[x_1, \ldots, x_n]$ will be the corresponding hypothesis ideal and thesis ideal. Let $U' \subseteq U \subseteq \{x_1, \ldots, x_n\} = X$.
Then a couple (R', R'') of ideals, respectively in $K[U]$ and $K[U']$, will be called a **Full Set of (Discovering) Interesting Conditions (FSDIC)** for $\mathcal{T}$ with respect to U and U' if the following conditions hold:

(a) $R' \subseteq K[U]$ and $R'' \subseteq K[U']$.
(b) $V(H + R'^e) \backslash V(R''^e) \subseteq V(T)$.
(c) $V(H + T) \subseteq V(R'^e)$.
(d) If $f \in K[U']$ is such that $V(H + R'^e) \backslash V((f)^e) \subseteq V(T)$, then $f \in \sqrt{R''}$.
(e) $V(H + R'^e) \backslash V(R''^e) \neq \emptyset$.

Remark 1. Condition $d)$ is equivalent to the following:

$d')$ if $R''' \subseteq K[U']$ is an ideal such that $V(H + (R')^e) \backslash V((R''')^e) \subseteq V(T)$, then $\overline{K}^n \backslash V((R''')^e) \subseteq \overline{K}^n \backslash V((R'')^e)$. See [10] for details.

Following [10], we address now two issues: when such pair of ideals R', R'' exist and how can we compute them. The following propositions give us a complete answer to these questions (we refer the reader to [10] for proofs):

Theorem 1. *Let $H' = (H + T) \cap K[U]$ and $H'' = ((H + H'^e) : (T)^\infty) \cap K[U']$ (the saturation by T). Then there exist two ideals R', R'' such that (R', R'') is an*

FSDIC *for $\mathcal{T}$ with respect to U and U' if and only if (H', H'') is* **FSDIC** *for $\mathcal{T}$ with respect to U and U'.*

Remark 2. If (R', R'') is an **FSDIC**, we have that

$$V(H + R'^{e})\backslash V(R''^{e}) = V(H + H'^{e})\backslash V(R''^{e})$$

Remark 3. Suppose, as motivated at the end of the previous Sect. 5, that an alternate definition of **FSDIC** is given, in which we drop property d). Then the statement of the theorem above will still hold. Thus the existence of an **FSDIC**, with or without condition d), is always equivalent to (H', H'') being an **FSDIC** in the stronger sense we have formally introduced in Definition 1.

The above theorem tells us that, if an **FSDIC** exists, then the couple (H', H'') is indeed one such full set of conditions, providing an extra algebraic set of equality-type constraints that is the smallest one in terms of the variety given by the first ideal of the couple (since $V(H'^{e}) \subseteq V(R'^{e})$, see Remark 2) and also providing the largest set of non degeneracy conditions in terms of the complement of the variety given by the second ideal of the couple (as we have shown in the proof that always $V(R''^{e}) \supseteq V(H''^{e})$).

Moreover, the above Remark 2 shows that the hypotheses of equality type $H + R'^{e}$ arising from whatever **FSDIC** will be always geometrically equivalent to $H + H'^{e}$ (after adding the non-degeneracy hypotheses), and in this sense we can conclude that our protocol yields, essentially, to a unique solution (when it exists one) on the additional hypotheses of equality-type for the statement to become true.

Now let us see describe here some further results of [10] for the existence of an an **FSDIC**, determining some necessary and sufficient algorithmic conditions for (H', H'') to be an **FSDIC**.

Theorem 2. *(H', H'') is* **FSDIC** *for $\mathcal{T}$ with respect to U and U' if and only if $1 \notin (H')^{c'} : H''^{\infty}$ (equivalently, iff $H'' \nsubseteq \sqrt{(H')^{c'}}$).*

Corollary 1. *Moreover, if U' is a set of algebraically independent variables for H', then (H', H'') is an* **FSDIC** *for $\mathcal{T}$ with respect to U and U' if and only if $H'' \neq (0)$.*

Proposition 3. *Notation as in the previous section. Suppose that $U' \subset U$ is a set of algebraically independent variables for $H + H'^{e}$. Then T is contained in all the minimal primes of $H + H'^{e}$ where U' are independent if and only if $1 \notin (H')^{c'} : H''^{\infty}$ (and this is equivalent to the couple (H', H'') being an* **FSDIC***).*

Remark 4. Compare to Proposition 2, for the Comprehensive bases approach. It shows, in some sense, that the existence of couples verifying the *FSDIC* protocol implies the existence of such couples for the Comprehensive bases setting, but not conversely, according to the counterexample after Proposition 2.

The above proposition can be refined in a quite useful sense (see example 2)

Proposition 4. *Suppose that $U' \subset U \subset X$ is a set of algebraically independent variables for $H + H'^{e}$ and, moreover, suppose it is maximal among the subsets of X with this property (ie. $K[\tilde{U}] \cap (H + H^{e}) \neq (0)$ for any $U' \subset \tilde{U} \subset X$).*

Then, the couple (H', H'') is not an **FSDIC** *is equivalent to the fact that T is not contained in all the minimal primes of $H + H'^{e}$ where U' are independent, but also that it is contained in at least one of them.*

Finally, let us introduce an example from [10] showing that, even when there is no **FSDIC**, we know –sometimes– it is due to the fact that the thesis holds over some relevant component (but not over all) and this information can be the clue to discover a statement. Further examples can be consulted at [10].

Example 2. The example deals with a generalization of the Steiner-Lehmus Theorem on the equality of lengths of the angle bisectors on a given triangle, an issue which has attracted along the years a considerable interest (see references at Sect. 3).

Without loss of generality we will consider a triangle of vertices $A(0,0)$, $B(1,0)$, $C(x, y)$. Then at each vertex we can determine two bisectors (one internal, another one external) for the angles described by the lines supporting the sides of the triangle meeting at that vertex. We want to discover what kind of triangle has, say, one bisector at vertex A and one bisector at vertex B, of equal length. Recall that the Steiner-Lehmus Theorem states that this is the case, for internal bisectors, if and only if the triangle is isosceles. So the question here is about the equality of lengths when we consider external bisectors, too.

Algebraically we translate the construction of a bisector, say, at vertex A, as follows. We take a point (p,q) at the same distance as $C = (x, y)$ from A, so it verifies $p^2 + q^2 - (x^2 + y^2) = 0$. Then, we place this point at the line AB, by adding the equation $q = 0$. Then the midpoint from (p,q) and C will be $((x + p)/2, (y + q)/2)$ and the line defined by A and by this midpoint intersects the opposite side BC (or its prolongation) at point (a, b), verifying $\{p^2 + q^2 - (x^2 + y^2) = 0, q = 0, -a(y + q)/2 + b(x + p)/2 = 0, -ay + b(x - 1) + y = 0\}$. Finally, distance from (a, b) to A is given as $a^2 + b^2$, and this quantity provides the length of the bisector(s) associated to A. Notice that by placing (p, q) at different positions in the line AB, the previous construction provides both the internal and the external bisector through A. There is no way of distinguishing both bisectors, without introducing inequalities, something alien to our setting (since we work on algebraically closed fields).

Likewise, we associate a set of equations to determine the length of the bisector(s) at B, introducing a point (r, s) in the line AB, so that its distance to B is equal to that of vertex C. Then we consider the midpoint of (r, s) and C and place a line through it and B. This line intersects side AC at a point (m, n), which is defined by the following set of equations: $\{(r - 1)^2 + s^2 - ((x - 1)^2 + y^2) = 0$, $s = 0, -m((y+s)/2) + n((x+r)/2 - 1) + (y+s)/2 = 0, -my + nx = 0\}$. The length of this bisector will be $(m - 1)^2 + n^2$.

Finally, we apply our discovery protocol to the hypotheses H given by the two sets of equations and having as thesis T the equality $(a^2+b^2)-((m-1)^2+n^2) = 0$.

It is clear the that the only two (geometrically meaningful for the construction) independent variables are $\{x, y\}$, so we eliminate in $H + T$ all variables except these two, getting in this way the ideal H'. The result is a polynomial that factors as the product of y^3 (a degenerate case), $2x-1$ (triangle is isosceles) and the degree 10 polynomial $14x^2y^4+y^2+246y^2x^6+76x^8-y^6+8x^{10}+9y^{10}-164y^2x^5+12y^4x-10x^2y^2-4x^4-44y^8x-136y^4x^3+278y^4x^4-64x^7-164x^7y^2+122y^6x^2-6y^4+8x^5-36y^6x+20y^2x^3+84y^4x^6+86x^4y^6+44x^2y^8+16x^6+41y^2x^8+31y^2x^4-40x^9-252y^4x^5-172y^6x^3+14y^8$ (cf.[18], page 150, also [4] for a picture of the curve given by this polynomial).

Next, in order to compute H'' we must choose one of the variable x, y, say, variable x, and eliminate y in the saturation of $H + H'$ by T. The result is (0), so there is no **FSDIC**, according to Corollary 1. In fact it is hard to expect that for almost all triangles with vertex C placed at the locus of H' and for any interpretation of the bisectors at A and B, they will all have simultaneously an equal length. But it also means (by Proposition 4) that adding H' to the set of hypotheses, for instance, placing vertex C at any point on the degree 10 curve, there will be an interpretation for the bisectors such that the equality of lengths follow. It is easy to deduce that this is so (except for some degenerate cases) considering internal/external, external/internal and external/external bisectors (since the internal/internal case holds only for isosceles triangles). Moreover, intersecting this curve with the line $2x - 1 = 0$ we can find out two points $x = 1/2, y = (1/2)RootOf(-1 + 3Z^2)$ (aprox. $x = 0.5000000000, y = +-0.2886751346$) where all four bisectors (the internal and external ones of A and B) have equal length. The other two points of intersection correspond to the case of equilateral triangles, where the two internal bisectors and the two infinite external bisectors of A, B have pairwise equal length, but the length is not equal for the internal and external bisectors.

Acknowledgements First author supported by grant "Algoritmos en Geometría Algebraica de Curvas y Superficies" (MTM2008-04699-C03-03) from the Spanish MICINN. Second author supported by grant "Geometría Algebraica y Analítica Real" (UCM-910444). Thanks also to the Austrian FWF Special Research Program SFB F013 "Numerical and Symbolic Scientific Computing" for the invitation to present this talk and paper.

References

1. Bazzotti, L., Dalzotto, G., Robbiano, L.: Remarks on geometric theorem proving", in automated deduction in geometry (Zurich, 2000), Lecture Notes in Computer Science 2061, 104–128, Springer, Berlin (2001)
2. Beltrán, C., Dalzotto, G., Recio, T.: The moment of truth in automatic theorem proving in elementary geometry, in Proceedings ADG 2006 (extended abstracts). Botana, F., Roanes-Lozano, E. (eds.) Universidad de Vigo (2006)
3. Botana, F., Recio, T.: Towards solving the dynamic geometry bottleneck via a symbolic approach in Proceedings ADG (Automatic Deduction in Geometry) 2004. Springer. Lec. Not. Artificial Intelligence LNAI 3763, pp. 761–771 (2005)

4. Botana, F.: Bringing more intelligence to dynamic geometry by using symbolic computation, in Symbolic Computation and Education. Edited by Shangzhi. L., Dongming, W., Jing-Zhong, Z. World Scientific, pp. 136–150 (2007)
5. Bulmer, M., Fearnley-Sander, D., Stokes, T.: The Kinds of Truth of Geometric Theorems, in Automated Deduction in Geometry (Zurich, 2000), Lecture Notes in Computer Science 2061, 129–142, Springer, Berlin (2001)
6. Chen, X.F., Li, P., Lin, L., Wang, D.K.: Proving Geometric Theorems by Partitioned - Parametric Groebner Bases, Automated Deduction in Geometry, LNAI 3763, 34–43, Springer, Berlin (2006)
7. Chou, S.-C.: Mechanical Geometry Theorem Proving, in Mathematics and its Applications, D. Reidel Publ. Comp. (1987)
8. Conti, P., Traverso, C.: A case of automatic theorem proving in Euclidean geometry: the Maclane 8_3 theorem, in Applied Algebra, Algebraic Algorithms and Error-Correcting Codes (Paris, 1995), Lecture Notes in Computer Science, 948, 183–193, Springer, Berlin (1995)
9. Conti, P., Traverso, C.: Algebraic and Semialgebraic Proofs: Methods and Paradoxes, in Automated deduction in geometry (Zurich, 2000), Lecture Notes in Comput. Sci. 2061, 83–103, Springer, Berlin (2001)
10. Dalzotto, G., Recio, T.: On Protocols for the Automated Discovery of Theorems in Elementary Geometry. J. Autom. Reasoning **43**, 203–236 (2009)
11. Kapur, D.: Wu's method and its application to perspective viewing. In: Geometric Reasoning. Kapur, D. Mundy, J.L. (eds.) The MIT press, Cambridge, MA (1989)
12. Losada, R., Recio, T., Valcarce, J.L.: Sobre el descubrimiento automático de diversas generalizaciones del Teorema de Steiner-Lehmus, Boletín de la Sociedad Puig Adam, **82**, 53–76 (2009) (in Spanish)
13. Montes, A., Recio, T.: Automatic discovery of geometry theorems using minimal canonical comprehensive Groebner systems, In Botana, F., Recio, T. (eds.) Automated Deduction in Geometry, LNAI (Lect. Notes Artificial Intelligence) 4869, pp. 113–139, Springer, Berlin (2007)
14. Pech, P.: Selected topics in geometry with classical vs. computer proving. World Scientific Publishing Company (2007)
15. Recio, T., Pilar Vélez, M.: Automatic Discovery of Theorems in Elementary Geometry, J. Autom. Reasoning **23**, 63–82 (1999)
16. Recio, T., Botana, F.: Where the truth lies (in automatic theorem proving in elementary geometry), in Proceedings ICCSA (International Conference on Computational Science and its Applications) 2004. Springer. Lec. Not. Com. Sci. 3044, pp. 761–771 (2004)
17. http://www.mathematik.uni-bielefeld.de/~sillke/PUZZLES/steiner-lehmus
18. Wang, D.: Elimination practice: software tools and applications, Imperial College Press, London (2004)
19. http://www-calfor.lip6.fr/~wang/
20. Wen-Tsün, W.: On the decision problem and the mechanization of theorem-proving in elementary geometry. Sci. Sinica **21**, 159–172 (1978); Also in: Automated theorem proving: After 25 years (Bledsoe, W. W., Loveland, D. W., eds.), AMS, Providence, pp. 213–234 (1984)
21. Zeliberger, D.: Plane geometry: an elementary textbook by Shalosh B. Ekhad, XIV, (Circa 2050), http://www.math.rutgers.edu/~zeilberg/PG/gt.html

Symbolic Analysis for Boundary Problems: From Rewriting to Parametrized Gröbner Bases

Markus Rosenkranz, Georg Regensburger, Loredana Tec, and Bruno Buchberger

Abstract We review our algebraic framework for linear boundary problems (concentrating on ordinary differential equations). Its starting point is an appropriate algebraization of the domain of functions, which we have named integro-differential algebras. The algebraic treatment of boundary problems brings up two new algebraic structures whose symbolic representation and computational realization is based on canonical forms in certain commutative and noncommutative polynomial domains. The first of these, the ring of integro-differential operators, is used for both stating and solving linear boundary problems. The other structure, called integro-differential polynomials, is the key tool for describing extensions of integro-differential algebras. We use the canonical simplifier for integro-differential polynomials for generating an automated proof establishing a canonical simplifier for integro-differential operators. Our approach is fully implemented in the Theorema system; some code fragments and sample computations are included.

M. Rosenkranz (✉)
School of Mathematics, Statistics and Actuarial Science (SMSAS), University of Kent, Canterbury CT2 7NF, UK
e-mail: m.rosenkranz@kent.ac.uk

G. Regensburger
Johann Radon Institute for Computational and Applied Mathematics (RICAM), Austrian Academy of Sciences, 4040 Linz, Austria
INRIA Saclay – Île de France, Project DISCO, L2S, Supélec, 91192 Gif-sur-Yvette Cedex, France
e-mail: georg.regensburger@ricam.oeaw.ac.at

L. Tec · B. Buchberger
Research Institute for Symbolic Computation (RISC), Johannes Kepler University, 4032 Hagenberg, Austria
e-mail: ltec@risc.uni-linz.ac.at; bruno.buchberger@risc.uni-linz.ac.at

U. Langer and P. Paule (eds.), *Numerical and Symbolic Scientific Computing*, Texts and Monographs in Symbolic Computation, DOI 10.1007/978-3-7091-0794-2_13,

1 Introduction

1.1 Overall View

When problems from Analysis – notably differential equations – are treated by methods from Symbolic Computation, one speaks of *Symbolic Analysis*, as in the eponymous workshops of the FoCM conference series [34]. Symbolic Analysis is based on algebraic structures, as all other symbolic branches, but its special flavor comes from its connection with analytic and numeric techniques. As most differential equations arising in the applications can only be solved numerically, this connection is absolutely vital.

If symbolic techniques cannot solve "most" differential equations, *what else* can they do? The answers are very diverse (reductions, normal forms, symmetry groups, singularity analysis, triangularization etc), and in the frame of this paper we can only point to surveys like [79] and [36, §2.11]. In fact, even the notion of "solving" is quite subtle and can be made precise in various ways. Often a symbolic method will not provide the "solution" in itself but valuable information about it to be exploited for subsequent numerical simulation.

Our own approach takes a somewhat intermediate position while diverging radically in another respect: Unlike most other symbolic methods known to us, we consider differential equations along with their *boundary conditions*. This is not only crucial for many applications, it is also advantageous from an algebraic point of view: It allows to define a linear operator, called the Green's operator, that maps the so-called forcing function on the right-hand side of an equation to the unique solution determined by the boundary conditions. This gives rise to an interesting structure on Green's operators and on boundary problems (Sect. 5). Algebraically, the consequence is that we have to generalize the common structure of differential algebras to what we have called integro-differential algebras (Sect. 3).

Regarding the *solvability issues*, the advantage of this approach is that it uncouples the task of finding an algebraic representation of the Green's operator from that of carrying out the quadratures involved in applying the Green's operator to a forcing function. While the latter may be infeasible in a symbolic manner, the former can be done by our approach (with numerical quadratures for integrating forcing functions).

The *research program* just outlined has been pursued in the course of the SFB project F013 (see below for a brief chronology), and the results have been reported elsewhere [66, 70, 72]. For the time being, we have restricted ourselves to linear boundary problems, but the structure of integro-differential polynomials [73] may be a first stepping stone towards nonlinear Green's operators. Since the algebraic machinery for Green's operators is very young, our strategy was to concentrate first on boundary problems for ordinary differential equations (ODEs), with some first steps towards partial differential equations (PDEs) undertaken more recently [74]. For an application of our methods in the context of actuarial mathematics, we refer to [2], for a more algebraic picture from the skew-polynomial perspective see [67].

1.2 New Results

In the present paper, we will present a *new confluence proof* for the central data structure used in our approach: As the algebraic language for Green's operators, the integro-differential operators (Sect. 4) are defined as a ring of noncommutative polynomials in infinitely many variables, modulo an infinitely generated ideal. While the indeterminates represent the basic operations of analysis (differentiation, integration, extraction of boundary values and multiplication by one of infinitely many coefficient functions), this ideal specifies their interaction (e.g. the fundamental theorem of calculus describing how differentiation and integration interact). Our new proof is fully automated within the TH∃OREM∀ system (Sect. 2), using a generic noncommutative polynomial reduction based on a noncommutative adaption of reduction rings [22]; see also [83] for a short outline of the proof.

In a way, the new proof completes the circle started with the ad-hoc confluence proof in [69]. For the latter, no algebraic structure was available for coping with certain expressions that arise in the proof because they involved generic coefficient functions along with their integrals and derivatives (rather than the operator indeterminates modeling integration and differentiation!), while this structure is now provided by the afore-mentioned *integro-differential polynomials* (Sect. 6). Roughly speaking, this means within the spectrum between rewrite systems (completion by the Knuth-Bendix procedure) and Gröbner bases (completion by Buchberger's algorithm), we have moved away from the former towards the latter [18]. We will come back to this point later (Sect. 7).

Moreover, the paper includes the following *improvements and innovations*: The setting for Gröbner bases and the Buchberger algorithm are introduced generically for commutative and noncommutative rings (allowing infinitely many variables and generators), based on reduction rings and implemented in the TH∃OREM∀ system (Sect. 2). The presentation of integro-differential algebras is streamlined and generalized (Sect. 3). For both of the main computational domains – integro-differential operators and integro-differential polynomials – we have a basis free description while a choice of basis is only need for deciding equality (Sects. 4, 6). The construction of integro-differential polynomials, which was sketched in [73], is carried out in detail (Sect. 6). In particular, a complete proof of the crucial result on canonical forms (Theorem 42) is now given.

1.3 Chronological Outline

As indicated above, this paper may be seen as a kind of target line for the research that we have carried out within Project F1322 of the SFB F013 supported by the Austrian Science Fund (FWF). We have already pointed out the crucial role of analysis/numerics in providing the right inspirations for the workings of Symbolic

Analysis. The development of this project is an illuminating and pleasant case in point. It was initiated by the stimulating series of *Hilbert Seminars* conducted jointly by Bruno Buchberger and Heinz W. Engl from October 2001 to July 2002, leading to the genesis of Project F1322 as a spin-off from Projects F1302 (Buchberger) and F1308 (Engl). Triggered by the paper [42], the idea of symbolic operator algebras emerged as a common leading theme. It engendered a vision of transplanting certain ideas like the Moore-Penrose inverse on Hilbert spaces from their homeground in functional analysis into a new domain within Symbolic Analysis, where powerful algebraic tools like Gröbner bases are available [9, 19, 20, 24]. This vision eventually crystallized in the algebraic machinery for computing Green's operators as described before.

In the early stage of the project, those two main tools from analysis (Moore-Penrose inverse) and algebra (Gröbner bases) were welded together in a rather ad-hoc manner, but it did provide a new tool for solving boundary problems [71]. In the course of the *dissertation* [69], a finer analysis led to a substantial simplification where the Moore-Penrose inverse was superseded by a purely algebraic formulation in terms of one-sided inverses and the expensive computation of a new noncommutative Gröbner basis for each boundary problem was replaced by plain reduction modulo a fixed Gröbner basis for modeling the essential operator relations. The resulting quotient algebra (called "Green's polynomials" at that time) is the precursor of the integro-differential operators described below (Sect. 4). The final step towards the current setup was the reformulation and generalization in a differential algebra setting [72] and in an abstract linear algebra setting [66].

The advances on the theoretical side were paralleled by an early *implementation* of the algorithm for computing Green's operators. While the ad-hoc approach with computing Gröbner bases per-problem was carried out by the help of NCAlgebra, a dedicated Mathematica package for noncommutative algebra [42], the fixed Gröbner basis for simplifying Green's operator was implemented in the THƎOREM∀ system [26]; see Sect. 2 for a general outline of this system. As the new differential algebra setting emerged, however, it became necessary to supplant this implementation by a new one. It was again integrated in the THƎOREM∀ system, but now in a much more intimate sense: Instead of using a custom-tailored interface as in [69], the new package was coded directly in the THƎOREM∀ language using the elegant structuring constructs of functors [25]. Since this language is also the object language of the provers, this accomplishes the old ideal of integrating computation and deduction.

The presentation of several parts of this paper – notably Sects. 3–5 – benefited greatly from a *lecture* given in the academic year 2009/10 on Symbolic Integral Operators and Boundary Problems by the first two authors. The lecture was associated with the Doctoral Program "Computational Mathematics: Numerical Analysis and Symbolic Computation" (W1214), which is a follow-up program to the SFB F013. We would like to thank our students for the lively discussions and valuable comments.

1.4 Overview of the Paper

We commence by having a closer look at the TH∃OREM∀ system (Sect. 2), which will also be used in all sample computations presented in subsequent sections; both the sample computations and the TH∃OREM∀ program code is available in an executable Mathematica notebook from www.theorema.org. We discuss canonical simplifiers for quotient structures and Gröbner bases in reduction rings, and we give a short overview of the functors used in building up the hierarchy of the algebraic structures used in the computations. The main structure among these is that of an integro-differential algebra (Sect. 3), which is the starting point for the integro-differential operators as well as the integro-differential polynomials. Since the former are, in turn, the foundation for computing Green's operators for boundary problems, we will next summarize the construction of integro-differential operators and their basic properties (Sect. 4), while the algorithms for solving and factoring boundary problems are explained and exemplified thereafter (Sect. 5). Driving towards the focus point of this paper, we describe then the algebra of integro-differential polynomials (Sect. 6), which will be the key tool to be employed for the confluence proof. Since this proof is reduced to a computation in TH∃OREM∀, we will only explain the main philosophy and show some representative fragments (Sect. 7). We wind up with some thoughts about open problems and future work (Sect. 8).

2 Data Structures for Polynomials in Theorema

2.1 The Theorema Functor Language

The TH∃OREM∀ system [26] was designed by B. Buchberger as an integrated environment for proving, solving and computing in various domains of mathematics. Implemented on top of Mathematica, its core language is a version of higher-order predicate logic that contains a natural programming language such that algorithms can be coded and verified in a unified formal frame. In this logic-internal programming language, functors are a powerful tool for building up *hierarchical domains* in a modular and generic way. They were introduced and first implemented in TH∃OREM∀ by B. Buchberger. The general idea – and its use for structuring those domains in which Gröbner bases can be computed – is described in [22, 25], where one can also find references to pertinent early papers by B. Buchberger. See also [87] for some implementation aspects of functor programming.

The notion of functor in TH∃OREM∀ is akin to functors in ML, not to be confused with the functors of category theory. From a computational point of view, a TH∃OREM∀ functor is a higher-order function that produces a *new domain from given domains*, where each domain is considered as a bundle of operations (including relations qua boolean-valued operations – in particular also carrier predicates). Operations in the new domain are defined in terms of operations in the underlying domains.

Apart from this computational aspect, functors also have an important reasoning aspect – a functor transports properties of the input domains to properties of the output domain, typical examples being the various "preservation theorems" in mathematics: "If R is a ring, then $R[x]$ is also a ring". This means the functor $R \mapsto R[x]$ preserves the property of being a ring, in other words: it goes from the "category of rings" to itself. In this context, a *category* is simply a collection of domains characterized by a common property (a higher-order predicate on domains).

See below for an example of a functor named `LexWords`. It takes a linearly ordered alphabet `L` as input domain and builds the *word monoid* over this alphabet:

$$\textbf{Definition}\Big[\text{"Word Monoid"},\ \textbf{any}[L],$$

$$\textbf{LexWords}[L] = \textbf{Functor}\Big[W,\ \textbf{any}\big[v,\ w,\ \xi,\ \eta,\ \bar{\xi},\ \bar{\eta}\big],$$

$$\frac{s = \langle\rangle}{\underset{W}{\in}[w] \Leftrightarrow \left(\bigwedge \begin{cases} \textbf{is-tuple}[w] \\ \underset{i=1,\ldots,|w|}{\forall} \underset{L}{\in}[w_i] \end{cases}\right)}$$

$$\underset{W}{\square} = \langle\rangle$$

$$v \underset{W}{*} w = v \asymp w$$

$$\left(\langle \eta, \bar{\eta}\rangle \underset{W}{>} \langle\rangle\right) \Leftrightarrow \textbf{True}$$

$$\left(\langle\rangle \underset{W}{>} \langle \bar{\eta}\rangle\right) \Leftrightarrow \textbf{False}$$

$$\left(\langle \eta, \bar{\eta}\rangle \underset{W}{>} \langle \xi, \bar{\xi}\rangle\right) \Leftrightarrow \left(\bigvee \begin{cases} \eta \underset{L}{>} \xi \\ (\eta = \xi) \bigwedge \langle \bar{\eta}\rangle \underset{W}{>} \langle \bar{\xi}\rangle \end{cases}\right) \Big]\Big]$$

Here $\bar{\xi}$, $\bar{\eta}$ are sequence variables, i.e. they can be instantiated with finite sequences of terms. The new domain `W` has the following operations: `W`[$\in$] denotes the carrier predicate, the neutral element is given by `W`[$\square$], the multiplication `W`[$*$] is defined as concatenation, and `W`[$>$] defines the lexicographic ordering on `W`.

In the following code fragments, we illustrate one way of *building up polynomials* in THEOREM∀ starting from the base categories of fields with ordering and ordered monoids. Via the functor `FreeModule`, we construct first the free vector space `V` over a field `K` generated by the set of words in an ordered monoid `W`. The elements of `V` are described by `V`[$\in$] as lists of pairs, each pair containing one (non-zero) coefficient from `K` and one basis vector from `W`, where the basis vectors are ordered according to the ordering on `W`. The operations of addition, subtraction and scalar multiplication are defined recursively, using the operations on `K` and `W`:

$$\mathbf{Definition}\Big[\text{"Free Module"},\ \mathrm{any}[K, W],$$

$$\mathrm{FreeModule}[K, W] = \mathrm{Functor}\Big[V,\ \mathrm{any}[c, d, x, y, \xi, \eta, A, \bar{x}, \bar{y}],$$

$$s = \langle\rangle$$

$$\underset{V}{\in}[x] \Leftrightarrow \left(\mathrm{where}\Big[z = |x|,\ \bigwedge\begin{cases}\text{is-tuple}[x] \\ \underset{i=1,\dots,z}{\forall}\left(\bigwedge\begin{cases}\text{is-tuple}[x_i] \\ |x_i| = 2 \\ \underset{V}{\text{is-coeff}}[(x_i)_1] \\ \underset{V}{\text{is-bvec}}[(x_i)_2]\end{cases}\right) \\ \underset{i=1,\dots,z-1}{\forall}\ (x_i)_2 \underset{W}{>} (x_{i+1})_2\end{cases}\Big]\right)$$

$$\underset{V}{\text{is-bvec}}[\xi] \Leftrightarrow \underset{W}{\in}[\xi]$$

$$\underset{V}{\text{is-coeff}}[c] \Leftrightarrow \left(\underset{K}{\in}[c] \bigwedge c \neq \underset{K}{0}\right)$$

$$\underset{V}{0} = \langle\rangle$$

...

$$\langle\langle c, \xi\rangle, \bar{x}\rangle \underset{V}{+} \langle\langle d, \eta\rangle, \bar{y}\rangle = \begin{cases}\langle c, \xi\rangle \smile \left(\langle\bar{x}\rangle \underset{V}{+} \langle\langle d, \eta\rangle, \bar{y}\rangle\right) & \Leftarrow \xi \underset{W}{>} \eta \\ \langle d, \eta\rangle \smile \left(\langle\langle c, \xi\rangle, \bar{x}\rangle \underset{V}{+} \langle\bar{y}\rangle\right) & \Leftarrow \eta \underset{W}{>} \xi \\ \left\langle c \underset{K}{+} d, \xi\right\rangle \smile \left(\langle\bar{x}\rangle \underset{V}{+} \langle\bar{y}\rangle\right) & \Leftarrow (\xi = \eta) \bigwedge c \underset{K}{+} d \neq \underset{K}{0} \\ \langle\bar{x}\rangle \underset{V}{+} \langle\bar{y}\rangle & \Leftarrow \text{otherwise}\end{cases}$$

...

$$x \underset{V}{\cdot} 0 = \langle\rangle$$

$$0 \underset{V}{\cdot} y = \langle\rangle$$

$$\dots \qquad \Big]\Big]$$

$$c \underset{V}{\cdot} \langle\langle d, \eta\rangle, \bar{y}\rangle = \left\langle c \underset{K}{*} d, \eta\right\rangle \smile c \underset{V}{\cdot} \langle\bar{y}\rangle$$

$$\langle\langle c, \xi\rangle, \bar{x}\rangle \underset{V}{\cdot} d = \left\langle c \underset{K}{*} d, \xi\right\rangle \smile \langle\bar{x}\rangle \underset{V}{\cdot} d$$

By the `MonoidAlgebra` functor we extend this domain, introducing a multiplication using the corresponding operations in K and W:

$$\mathrm{MonoidAlgebra}[K, W] = \mathrm{where}\Big[V = \mathrm{FreeModule}[K, W],$$

$$\mathrm{Functor}\Big[P,\ \mathrm{any}[c, d, f, g, \xi, \eta, \bar{m}, \bar{n}],$$

$$s = \langle\rangle$$

...(* *linear operations from V* *)

(* **multiplication** *)

$$\langle\rangle \underset{P}{*} g = \langle\rangle \qquad \Big]\Big]$$

$$f \underset{P}{*} \langle\rangle = \langle\rangle$$

$$\langle\langle c, \xi\rangle, \bar{m}\rangle \underset{P}{*} \langle\langle d, \eta\rangle, \bar{n}\rangle = \left\langle\left\langle c \underset{K}{*} d, \xi \underset{W}{*} \eta\right\rangle\right\rangle \underset{P}{+} \langle\langle c, \xi\rangle\rangle \underset{P}{*} \langle\bar{n}\rangle \underset{P}{+} \langle\bar{m}\rangle \underset{P}{*} \langle\langle d, \eta\rangle, \bar{n}\rangle$$

The new domain inherits the structure on the elements of V.

The main advantage of the above construction is that it is fully *generic*: Not only can it be instantiated for different coefficient rings (or fields) and different sets of indeterminates, it comprises also the commutative and noncommutative case (where `W` is instantiated respectively by a commutative and noncommutative monoid).

2.2 Quotient Structures and Canonical Simplifiers

In algebra (and also in the rest of mathematics), one encounters quotient structures on many occasions. The general setting is a set A with various operations (an algebra in the general sense used in Sect. 6) and a congruence relation $\equiv$ on A, meaning an equivalence relation that is compatible with all the operations on A. Then one may form the quotient $A/{\equiv}$, which will typically inherit some properties of A. For example, $A/{\equiv}$ belongs to the category of rings if A does, so we can view the *quotient construction* $A \mapsto A/{\equiv}$ as a functor on the category of rings.

But for computational purposes, the usual set-theoretic description of $A/{\equiv}$ as a set of equivalence classes is not suitable (since each such class is typically uncountably infinite). We will therefore use an alternative approach that was introduced in [27] as a general framework for symbolic representations. The starting point is a *canonical simplifier* for $A/{\equiv}$, meaning a map $\sigma\colon A \to A$ such that

$$\sigma(a) \equiv a \quad \text{and} \quad \sigma(a) = \sigma(a') \text{ whenever } a \equiv a'. \tag{1}$$

The set $\tilde{A} = \sigma(A)$ is called the associated *system of canonical forms* for $A/{\equiv}$.

Clearly canonical simplifiers exist for every quotient $A/{\equiv}$, but for computational purposes the crucial question is whether σ is *algorithmic*. Depending on $A/{\equiv}$, it may be easy or difficult or even impossible to construct a computable $\sigma\colon A \to A$. In the examples that we will treat, canonical simplifiers are indeed available.

Canonical simplifiers are also important because they allow us to *compute in the quotient structure*. More precisely, one can transplant the operations on A to $\tilde{A}$ by defining $\omega(a_1, \ldots, a_n) = \sigma(\omega(a_1, \ldots, a_n))$ for every operation ω on A. With these new operations, one may easily see that $\tilde{A}$ is isomorphic to the quotient $A/{\equiv}$; see the Theorem "Canonical simplification and computation" in [27, p. 13].

There is an intimate relation between canonical forms and *normal forms* for rewrite systems (Sect. 4 contains some basic terminology and references). In fact, every rewrite system $\to$ on an algebraic structure A creates an equivalence relation $\equiv$, the symmetric closure of $\overset{*}{\to}$. Thus $a \equiv a'$ if and only if a and a' can be connected by an equational chain (using the rewrite rules in either direction). Typically, the relation $\equiv$ will actually be a congruence on A, so that the quotient $A/{\equiv}$ has a well-defined algebraic structure. Provided the rewrite system is noetherian, the normal forms of $\to$ are then also canonical forms for $A/{\equiv}$. Hence we will often identify these terms in a rewriting context.

For our *implementation*, we use canonical simplifiers extensively. In fact, the observation made above about computing in the quotient structure is realized by a

THEOREM∀ functor, which is applied at various different places. Here A is typically a K-algebra, with the ground field K being $\mathbb{Q}$ or computable subfields of $\mathbb{R}$ and $\mathbb{C}$.

2.3 Reduction Rings and Gröbner Bases

For defining reduction on polynomials, we use the *reduction ring* approach in the sense of [17, 22]. For commutative reduction rings, see also [81, 82]; for another noncommutative approach we refer to [57–59].

To put it simply, a reduction ring is a ring in which Gröbner bases can be done. A full *axiomatization* for the commutative case is given in [17]. If such rings satisfy certain additional axioms (defining the category of so-called "Gröbner rings"), then Gröbner bases can be computed by iterated S-polynomial reduction in the given ring – this is the Gröbner Ring Extension Theorem, stated and proved in [17].

A detailed presentation of their *construction* in the THEOREM∀ setting was given in [21, 23]; it is the starting point for our current work. At this point we do not give an axiomatic characterization for noncommutative reduction rings, but we do use a construction that is similar to the commutative setting. Thus we endow a polynomial domain `P`, built via the `MonoidAlgebra` functor with word monoid `W` and field `K`, with the following three operations: a noetherian (partial) ordering, a binary operation *least common reducible*, and a binary operation *reduction multiplier*. The noetherian ordering is defined in the usual way in terms of the given orderings on `K` and `W`.

The basic idea of *reduction multipliers* is to answer the question: "With which monomial do I have to multiply a given polynomial so that it cancels the leading term of another given polynomial?" In the noncommutative case, the corresponding operation `rdm` splits into left reduction multiplier `lrdm` and its right counterpart `rrdm` defined as follows:

$$\underset{P}{\mathtt{lrdm}}[\langle\langle c,\xi\rangle,\bar{m}\rangle,\langle\langle d,\eta\rangle,\bar{n}\rangle]=\begin{cases}\left\langle\left\langle \underset{K}{1},\ \underset{W}{\mathtt{lquot}}[\xi,\eta]\right\rangle\right\rangle & \Leftarrow\ \underset{K}{\mathtt{rdm}}[c,d]\neq\underset{K}{0}\bigwedge \eta\underset{W}{\mid}\xi\\ \underset{P}{0} & \Leftarrow\ \mathtt{otherwise}\end{cases}$$

$$\underset{P}{\mathtt{rrdm}}[\langle\langle c,\xi\rangle,\bar{m}\rangle,\langle\langle d,\eta\rangle,\bar{n}\rangle]=\begin{cases}\left\langle\left\langle \underset{K}{\mathtt{rdm}}[c,d],\ \underset{W}{\mathtt{rquot}}[\xi,\eta]\right\rangle\right\rangle & \Leftarrow\ \underset{K}{\mathtt{rdm}}[c,d]\neq\underset{K}{0}\bigwedge \eta\underset{W}{\mid}\xi\\ \underset{P}{0} & \Leftarrow\ \mathtt{otherwise}\end{cases}$$

Here the divisibility relation | on `W` checks whether a given word occurs within another word, and the corresponding quotients `lquot` and `rquot` yield the word segments respectively to the left and to the right of this occurrence. Since the scalars from `K` commute with the words, it is an arbitrary decision whether one includes it in the right (as here) or left reduction multiplier. In typical cases, this scalar factor is just `rdm[c,d]` $=$ `c/d`.

The operations relating Gröbner bases are introduced via a functor which is called `GroebnerExtension`. It defines *polynomial reduction* using reduction multipliers (note that this includes also the commutative case, where one actually needs only one reduction multiplier, the other one being unity):

$$\mathtt{hred}_{\mathtt{G}}[\mathtt{f},\ \mathtt{g}] = \mathtt{f} -_{\mathtt{P}} \mathtt{lrdm}_{\mathtt{P}}[\mathtt{f},\ \mathtt{g}] *_{\mathtt{P}} \mathtt{g} *_{\mathtt{P}} \mathtt{rrdm}_{\mathtt{P}}[\mathtt{f},\ \mathtt{g}]$$

The next step is to introduce reduction modulo a system of polynomials. For some applications (like the integro-differential operators described in Sect. 4), it is necessary to deal with *infinite reduction systems*: the polynomial ring contains infinitely many indeterminates, and reduction is applied modulo an infinite set of polynomials. In other words, we want to deal with an infinitely generated ideal in an infinitely generated algebra.

This is a broad topic, and we cannot hope to cover it in the present scope. In general one must distinguish situations where both the generators of the ideal and the algebra are parametrized by finitely many families involving finitely many parameters and more general algebras/ideals where this is not so. In the latter case, one works with finite subsets, and all computations are approximate: one never catches the whole algebraic picture. Fortunately, the applications we have in mind – in particular the integro-differential operators – are of the first type where *full algorithmic control* can be achieved. However, most of the common packages implementing noncommutative Gröbner bases do not support such cases [55, 56]. For some recent advances, we refer the reader to [3, 14, 43, 51] as well as Ufnarovski's extensive survey chapter [86].

Let us point out just one important class of decidable reductions in infinitely generated algebras – if an infinite set of (positively weighted) *homogeneous polynomials* is given, which is known to be complete for each given degree (see [51] for the proof) since one can compute a truncated Gröbner basis of such a graded ideal, which is finite up to a given degree. But if the given set is not homogeneous or cannot be clearly presented degree by degree, basically nothing can be claimed in general. Unfortunately, the applications we have in mind seem to be of this type.

In our setting, infinitely generated ideals are handled by an algorithmic operation for instantiating reduction rules. The *reduction* of polynomial `f` modulo a system `S` is realized thus:

$$\mathtt{hredp}_{\mathtt{G}}[\mathtt{f},\ \mathtt{l},\ \mathtt{g},\ \mathtt{r}] = \mathtt{f} -_{\mathtt{P}} \mathtt{l} *_{\mathtt{P}} \mathtt{g} *_{\mathtt{P}} \mathtt{r}$$

$$\mathtt{hred}_{\mathtt{G}}[\mathtt{f},\ \mathtt{S}] = \mathtt{where}\Big[\mathtt{q} = \mathtt{S}[\mathtt{f}],\ \mathtt{hredp}_{\mathtt{G}}[\mathtt{f},\ \mathtt{q}_1,\ \mathtt{q}_2,\ \mathtt{q}_3]\Big]$$

where `S[f]` is the operation that decides if there exists `g` modulo which `f` can be reduced, and it returns a triple containing the `g` and the left/right reduction multipliers needed for performing the reduction.

The main tool for the Gröbner bases construction, namely the notion of S-polynomial, can now be defined in terms of the *least common reducible*:

$$\mathtt{spol}_{\mathtt{G}}[\mathtt{f},\ \mathtt{g}] = \mathtt{where}\Big[\mathtt{L} = \mathtt{lcrd}_{\mathtt{P}}[\mathtt{f},\ \mathtt{g}],\ \mathtt{hredp}_{\mathtt{G}}[\mathtt{L},\ \mathtt{f}] -_{\mathtt{G}} \mathtt{hredp}_{\mathtt{G}}[\mathtt{L},\ \mathtt{g}]\Big]$$

Here `lcrd[f,g]` represents the smallest monomial that can be reduced both modulo `f` and modulo `g`, built from the least common reducible of the corresponding coefficients in `K` and the least common multiple of the words in `W`:

$$\underset{P}{\texttt{lcrd}}[\langle\langle c,\, \xi\rangle,\, \bar{m}\rangle,\, \langle\langle d,\, \eta\rangle,\, \bar{n}\rangle] = \left\langle\left\langle \underset{K}{\texttt{lcrd}}[c,\, d],\, \underset{W}{\texttt{lcm}}[\xi,\, \eta]\right\rangle\right\rangle$$

In our setting, the `lcrd[c,d]` can of course be chosen as unity since we work over a field `K`, but in rings like $\mathbb{Z}$ one would have to use the least common multiple.

Finally, *Gröbner bases* are computed by the usual accumulation of S-polynomials reduction, via the following version of Buchberger algorithm [24]:

$$\underset{G}{\texttt{Gb}}[R,\, S] = \texttt{where}\left[\texttt{pairs} = \left\langle \langle R_i,\, R_j\rangle \underset{\substack{i=1,\dots,|R| \\ j=1,\dots,|R|}}{|} R_i \neq R_j\right\rangle,\, \underset{G}{\texttt{Gb}}[R,\, \texttt{pairs},\, S]\right]$$

$$\underset{G}{\texttt{Gb}}[R,\, \langle\rangle,\, S] = R$$

$$\underset{G}{\texttt{Gb}}[R,\, \langle\langle f,\, g\rangle,\, \bar{m}\rangle,\, S] = \texttt{where}\Big[h = \underset{G}{\texttt{tred}}\Big[\underset{G}{\texttt{spol}}[f,\, g],\, S\Big],$$

$$\left\{\begin{array}{lcl} \underset{G}{\texttt{Gb}}[R,\, \langle\bar{m}\rangle,\, S] & \Leftarrow & h \underset{P}{=} 0 \\ \underset{G}{\texttt{Gb}}[R \frown h,\, ((h \twoheadrightarrow R) \asymp \langle\bar{m}\rangle) \asymp (R \twoheadleftarrow h),\, S] & \Leftarrow & \texttt{otherwise} \end{array}\right.$$

$$\Big]$$

Total reduction modulo a system, denoted here by `tred`, is computed by iteratively performing reductions, until no more reduction is possible. The above implementation of Buchberger's algorithm is again *generic* since it can be used in both commutative and noncommutative settings. For finitely many indeterminates, the algorithm always terminates in the commutative case (by Dickson's Lemma); in the noncommutative setting, this cannot be guaranteed in general. For our applications we also have to be careful to ensure that the reduction systems we use are indeed noetherian (Sect. 4).

3 Integro-Differential Algebras

For working with boundary problems in a symbolic way, we first need an algebraic structure having *differentiation along with integration*. In the following definitions, one may think of our standard example $\mathcal{F} = C^\infty(\mathbb{R})$, where $\partial = {}'$ is the usual derivation and $\int$ the integral operator

$$f \mapsto \int_a^x f(\xi)\, d\xi$$

for a fixed $a \in \mathbb{R}$.

3.1 Axioms and Basic Properties

Let K be a commutative ring. We first recall that $(\mathcal{F}, \partial)$ is a *differential K-algebra* if $\partial\colon \mathcal{F} \to \mathcal{F}$ is a K-linear map satisfying the *Leibniz rule*

$$\partial(fg) = \partial(f)\, g + f\, \partial(g). \tag{2}$$

For convenience, we may assume $K \leq \mathcal{F}$, and we write f' as a shorthand for $\partial(f)$. The following definition [72] captures the algebraic properties of the Fundamental Theorem of Calculus and Integration by Parts.

Definition 1. We call $(\mathcal{F}, \partial, \textstyle\int)$ an *integro-differential algebra* if $(\mathcal{F}, \partial)$ is a commutative differential K-algebra and $\textstyle\int$ is a K-linear section (right inverse) of ∂, i.e.

$$(\textstyle\int f)' = f, \tag{3}$$

such that the *differential Baxter axiom*

$$(\textstyle\int f')(\textstyle\int g') + \textstyle\int (fg)' = (\textstyle\int f')g + f(\textstyle\int g') \tag{4}$$

holds.

We refer to ∂ and $\textstyle\int$ respectively as the *derivation* and *integral* of $\mathcal{F}$ and to (3) as *section axiom*. Moreover, we call a section $\textstyle\int$ of ∂ an integral for ∂ if it satisfies (4). For the similar notion of differential Rota-Baxter algebras, we refer to [39] but see also below.

Note that we have applied *operator notation* for the integral; otherwise, for example, the section axiom (3) would read $(\textstyle\int(f))' = f$, which is quite unusual at least for an analyst. We will likewise often use operator notation for the derivation, so the Leibniz rule (2) can also be written as $\partial fg = (\partial f)g + f(\partial g)$. For the future we also introduce the following convention for saving parentheses: Multiplication has precedence over integration, so $\textstyle\int f \int g$ is to be parsed as $\textstyle\int (f \int g)$.

Let us also remark that Definition 1 can be generalized: First, no changes are needed for the *noncommutative case* (meaning $\mathcal{F}$ is noncommutative). This would for example be an appropriate setting for matrices with entries in $\mathcal{F} = C^{\infty}[a,b]$, providing an algebraic framework for the results on linear systems of ODEs. Second, one may add a *nonzero weight* in the Leibniz axiom, thus incorporating also discrete models where ∂ is the difference operator defined by $(\partial f)_k = f_{k+1} - f_k$. The nice thing is that all other axioms remain unchanged. For both generalizations confer also to [39].

We study first some direct consequences of the section axiom (3). For further details on linear left and right inverses, we refer for example to [13, p. 211] or to [63] in the context of generalized inverses. We also introduce the following names for the *projectors and modules* associated with a section of a derivation.

Definition 2. Let $(\mathcal{F}, \partial)$ be a differential K-algebra and $\int$ a K-linear section of ∂. Then we call the projectors

$$\mathrm{J} = \textstyle\int \circ \partial \qquad \text{and} \qquad \mathrm{E} = 1 - \textstyle\int \circ \partial$$

respectively the *initialization* and the *evaluation* of $\mathcal{F}$. Moreover, we refer to

$$\mathcal{C} = Ker(\partial) = Ker(\mathrm{J}) = \mathrm{Im}(\mathrm{E}) \quad \text{and} \quad \mathcal{I} = \mathrm{Im}(\textstyle\int) = \mathrm{Im}(\mathrm{J}) = Ker(\mathrm{E})$$

as the submodules of respectively *constant* and *initialized functions*.

Note that they are indeed projectors since $\mathrm{J} \circ \mathrm{J} = \int \circ (\partial \circ \int) \circ \partial = \mathrm{J}$ by (3), which implies $\mathrm{E} \circ \mathrm{E} = 1 - \mathrm{J} - \mathrm{J} + \mathrm{J} \circ \mathrm{J} = \mathrm{E}$. As is well known [13, p. 209], every projector is characterized by its kernel and image – they form a direct decomposition of the module into two submodules, and every such decomposition corresponds to a unique projector. We have therefore a *canonical decomposition*

$$\mathcal{F} = \mathcal{C} \dotplus \mathcal{I},$$

which allows to split off the "constant part" of every "function" in $\mathcal{F}$.

Before turning to the other axioms, let us check what all this means in the *standard example* $\mathcal{F} = C^\infty(\mathbb{R})$ with $\partial = \frac{d}{dx}$ and $\int = \int_a^x$. Obviously, the elements of $\mathcal{C}$ are then indeed the constant functions $f(x) = c$, while $\mathcal{I}$ consists of those functions that satisfy the homogeneous initial condition $f(a) = 0$. This also explains the terminology for the projectors: Here $\mathrm{E} f = f(a)$ evaluates f at the initialization point a, and $\mathrm{J} f = f - f(a)$ enforces the initial condition. Note that in this example the evaluation E is multiplicative; we will show below that this holds in any integro-differential algebra.

The Leibniz rule (2) and the differential Baxter axiom (4) entail interesting properties of the two submodules $\mathcal{C}$ and $\mathcal{I}$. For understanding these, it is more economic to forget for a moment about integro-differential algebras and turn to the following general observation about *projectors on an algebra*. We use again operator notation, giving precedence to multiplication over the linear operators.

Lemma 3. *Let E and J be projectors on a K-algebra with $E + J = 1$, set*

$$C = \mathrm{Im}(E) = Ker(J) \quad \textit{and} \quad I = Ker(E) = \mathrm{Im}(J).$$

Then the following statements are equivalent:

1. *The projector E is multiplicative, meaning $Efg = (Ef)(Eg)$.*
2. *The projector J satisfies the identity $(Jf)(Jg) + Jfg = (Jf)g + f(Jg)$.*
3. *The submodule C is a subalgebra and the submodule I an ideal.*

Proof. 1. ⇔ 2. Multiplicativity of $E = 1 - J$ just means

$$fg - Jfg = fg - (Jf)g - f(Jg) + (Jf)(Jg).$$

<u>1. ⇒ 3.</u> This follows immediately because C is the image and I the kernel of the algebra endomorphism E.
<u>3. ⇒ 1.</u> Let f, g be arbitrary. Since the given K-algebra is a direct sum of C and I, we have $f = f_C + f_I$ and $g = g_C + g_I$ for $f_C = Ef, g_C = Eg \in C$ and $f_I = Jf, g_I = Jg \in I$. Then

$$Efg = Ef_C g_C + Ef_C g_I + Ef_I g_C + Ef_I g_I$$

Since I is an ideal, the last three summands vanish. Furthermore, C is a subalgebra, so $f_C g_C \in C$. This implies $Ef_C g_C = f_C \, g_C$ because E is a projector onto C. □

This lemma is obviously applicable to integro-differential algebras $\mathcal{F}$ with the projectors $E = \mathtt{E}$ and $J = \mathtt{J}$ and with the submodules $C = \mathcal{C}$ and $I = \mathcal{I}$ because the differential Baxter axiom (4) is exactly condition 2. From now on, we will therefore refer to $\mathcal{C}$ as the *algebra of constant functions* and to $\mathcal{I}$ as the *ideal of initialized functions*. Moreover, we note that in any integro-differential algebra the evaluation $\mathtt{E} = 1 - \int \circ \partial$ is multiplicative, meaning

$$\mathtt{E} fg = (\mathtt{E} f)(\mathtt{E} g). \tag{5}$$

Altogether we obtain now the following characterization of integrals (note that the requirement that $\mathcal{C}$ be a subalgebra already follows from the Leibniz axiom).

Corollary 4. *Let $(\mathcal{F}, \partial)$ be a differential algebra. Then a section $\int$ of ∂ is an integral if and only if its evaluation $\mathtt{E} = 1 - \int \circ \partial$ is multiplicative, and if and only if $\mathcal{I} = \operatorname{Im}(\int)$ is an ideal.*

Note that the ideal $\mathcal{I}$ corresponding to an integral is in general *not a differential ideal* of $\mathcal{F}$. We can see this already in the standard example $C^\infty[0,1]$, where $\mathcal{I}$ consists of all $f \in C^\infty[0,1]$ with $f(0) = 0$. Obviously $\mathcal{I}$ is not differentially closed since $x \in \mathcal{I}$ but $x' = 1 \notin \mathcal{I}$.

The above corollary implies immediately that an integro-differential algebra $\mathcal{F}$ can *never be a field* since then the only possibilities for $\mathcal{I}$ would be 0 and $\mathcal{F}$. The former case is excluded since it means that $Ker(\partial) = \mathcal{F}$, contradicting the surjectivity of ∂. The latter case corresponds to $Ker(\partial) = 0$, which is not possible because $\partial 1 = 0$.

Corollary 5. *An integro-differential algebra is never a field.*

In some sense, this observation ensures that all integro-differential algebras are *fairly complicated.* The next result points in the same direction, excluding finite-dimensional algebras.

Proposition 6. *The iterated integrals $1, \int 1, \int\int 1, \ldots$ are all linearly independent over K. In particular, every integro-differential algebra is infinite-dimensional.*

Proof. Let (u_n) be the sequence of iterated integrals of 1. We prove by induction on n that $u_0, u_1, \ldots, u_n$ are linearly independent. The base case $n = 0$ is trivial. For the induction step from n to $n+1$, assume $c_0 u_0 + \cdots + c_{n+1} u_{n+1} = 0$. Applying ∂^{n+1}

yields $c_{n+1} = 0$. But by the induction hypothesis, we have then also $c_0 = \cdots = c_n = 0$. Hence $u_0, \ldots, u_{n+1}$ are linearly independent. □

Let us now return to our discussion of the *differential Baxter axiom* (4). We will offer an equivalent description that is closer to analysis. It is more compact but less symmetric. (In the noncommutative case one has to add the opposite version – reversing all products – for obtaining equivalence.)

Proposition 7. *The differential Baxter axiom* (4) *is equivalent to*

$$f \textstyle\int g = \int fg + \int f' \int g, \tag{6}$$

in the presence of the Leibniz axiom (2) *and the section axiom* (3).

Proof. For proving (6) note that since $\mathscr{I}$ is an ideal, $f \int g$ is invariant under the projector J and thus equal to $\int (f \int g)' = \int f' \int g + \int fg$ by the Leibniz axiom (2) and the section axiom (3). Alternatively, one can also obtain (6) from (4) if one replaces g by $\int g$ in (4). Conversely, assuming (6) we see that $\mathscr{I}$ is an ideal of $\mathscr{F}$, so Corollary 4 implies that $\int$ satisfies the differential Baxter axiom (4). □

For obvious reasons, we refer to (6) as *integration by parts*. The usual formulation $\int fG' = fG - \int f'G$ is only satisfied "up to a constant", or if one restricts G to $\mathrm{Im}(\int)$. Substituting $G = \int g$ then leads to (6). But note that we have now a more algebraic perspective on this well-known identity of Calculus: It tells us how $\mathscr{I}$ is realized as an ideal of $\mathscr{F}$.

Sometimes a variation of (6) is useful. Applying $\int$ to the Leibniz axiom (2) and using the fact that $\mathrm{E} = 1 - \mathrm{J}$ is multiplicative (5), we obtain

$$\textstyle\int fg' = fg - \int f'g - (\mathrm{E} f)(\mathrm{E} g), \tag{7}$$

which we call the *evaluation variant* of integration by parts (a form that is also used in Calculus). Observe that, we regain integration by parts (6) upon replacing g by $\int g$ in (7) since $\mathrm{E} \int g = 0$.

Note that in general one cannot extend a given differential algebra to an integro-differential algebra since the latter requires a *surjective derivation*. For example, in $(K[x^2], x\partial)$ the image of ∂ does not contain 1. As another example (cf. Sect. 6), the algebra of differential polynomials $\mathscr{F} = K\{u\}$ does not admit an integral in the sense of Definition 1 since the image of ∂ does not contain u.

How can we isolate the *integro part* of an integro-differential algebra? The disadvantage (and also advantage!) of the differential Baxter axiom (4) is that it entangles derivation and integral. So how can one express "integration by parts" without referring to the derivation?

Definition 8. Let $\mathscr{F}$ be a K-algebra and $\int$ a K-linear operation satisfying

$$(\textstyle\int f)(\int g) = \int f \int g + \int g \int f. \tag{8}$$

Then $(\mathscr{F}, \int)$ is called a *Rota-Baxter algebra* (of weight zero).

Rota-Baxter algebras are named after Glen Baxter [7] and Gian-Carlo Rota [75]; see also [37,38] for further details. In the following, we refer to (8) as *Baxter axiom*; in contrast to the differential Baxter axiom (4), we will sometimes also call it the *pure Baxter axiom*.

One might now think that an integro-differential algebra $(\mathcal{F}, \partial, \smallint)$ is a differential algebra $(\mathcal{F}, \partial)$ combined with a Rota-Baxter algebra $(\mathcal{F}, \smallint)$ such that the section axiom (3) is satisfied. In fact, such a structure was introduced independently by Guo and Keigher [39] under the name *differential Rota-Baxter algebras*. But we will see that an integro-differential algebra is a little bit more – this is why we also refer to (8) as "weak Baxter axiom" and to (4) and (6) as "strong Baxter axioms".

Proposition 9. *Let $(\mathcal{F}, \partial)$ be a differential algebra and $\smallint$ a section for ∂. Then $\smallint$ satisfies the pure Baxter axiom* (8) *if and only if $\mathcal{I} = \operatorname{Im}(\smallint)$ is a subalgebra of $\mathcal{F}$. In particular, $(\mathcal{F}, \smallint)$ is a Rota-Baxter algebra for any integro-differential algebra $(\mathcal{F}, \partial, \smallint)$.*

Proof. Clearly (8) implies that $\mathcal{I}$ is a subalgebra of $\mathcal{F}$. Conversely, if $(\smallint f)(\smallint g)$ is contained in $\mathcal{I}$, it is invariant under the projector J and must therefore be equal to $\smallint \partial\,(\smallint f)(\smallint g) = \smallint f \smallint g + \smallint g \smallint f$ by the Leibniz axiom (2). □

So the strong Baxter axiom (4) requires that $\mathcal{I}$ be an ideal, the weak Baxter axiom (8) only that it be a subalgebra. We will soon give a counterexample for making sure that (4) is indeed asking for more than (8), see Example 14. But before this we want to express the difference between the two axioms in terms of a *linearity property*. Recall that both ∂ and $\smallint$ were introduced as K-linear operations on $\mathcal{F}$. Using the Leibniz axiom (2), one sees immediately that ∂ is even $\mathcal{C}$-linear. It is natural to expect the same from $\smallint$, but this is exactly the difference between (4) and (8).

Proposition 10. *Let $(\mathcal{F}, \partial)$ be a differential algebra and $\smallint$ a section for ∂. Then $\smallint$ satisfies the differential Baxter axiom* (4) *if and only if it satisfies the pure Baxter axiom* (8) *and is $\mathcal{C}$-linear.*

Proof. Assume first that $\smallint$ satisfies the differential Baxter axiom (4). Then the pure Baxter axiom (8) holds by Proposition 9. For proving $\smallint cg = c \smallint g$ for all $c \in \mathcal{C}$ and $g \in \mathcal{F}$, we use the integration-by-parts formula (6) and $c' = 0$.

Conversely, assume the pure Baxter axiom (8) is satisfied and $\smallint$ is $\mathcal{C}$-linear. By Proposition 7 it suffices to prove the integration-by-parts formula (6) for $f, g \in \mathcal{F}$. Since $\mathcal{F} = \mathcal{C} \dotplus \mathcal{I}$, we may first consider the case $f \in \mathcal{C}$ and then the case $f \in \mathcal{I}$. But the first case follows from $\mathcal{C}$-linearity; the second case means $f = \smallint \tilde{f}$ for $\tilde{f} \in \mathcal{F}$, and (6) becomes the pure Baxter axiom (8) for $\tilde{f}$ and g. □

Let us now look at some natural *examples of integro-differential algebras*, in addition to our standard examples $C^\infty(\mathbb{R})$ and $C^\infty[a, b]$.

Example 11. The *analytic functions* on the real interval $[a, b]$ form an integro-differential subalgebra $C^\omega[a, b]$ of $C^\infty[a, b]$ over $K = \mathbb{R}$ or $K = \mathbb{C}$. It contains in turn the integro-differential algebra $K[x, e^{Kx}]$ of *exponential polynomials*, defined as the space of all K-linear combinations of $x^n e^{\lambda x}$, with $n \in \mathbb{N}$ and $\lambda \in K$. Finally, the algebra of *ordinary polynomials* $K[x]$ is an integro-differential subalgebra in all cases.

All the three examples above have *algebraic analogs*, with integro-differential structures defined in the expected way.

Example 12. For a field K of characteristic zero, the *formal power series* $K[[x]]$ are an integro-differential algebra. One sets $\partial x^k = kx^{k-1}$ and $\int x^k = x^{k+1}/(k+1)$; note that the latter needs characteristic zero. The formal power series contain a highly interesting and important integro-differential subalgebra: the *holonomic power series*, defined as those whose derivatives span a finite-dimensional K-vector space [29, 77].

Of course $K[[x]]$ also contains (an isomorphic copy of) the integro-differential algebra of *exponential polynomials*. In fact, one can define $K[x, e^{Kx}]$ algebraically as a quotient of the free algebra generated by the symbols x^k and $e^{\lambda x}$, with λ ranging over K. Derivation and integration are then defined in the obvious way. The exponential polynomials contain the *polynomial ring* $K[x]$ as an integro-differential subalgebra. When $K = \mathbb{R}$ or $K = \mathbb{C}$, we use the notation $K[x]$ and $K[x, e^{Kx}]$ both for the analytic and the algebraic object since they are isomorphic.

The following example is a clever way of transferring the previous example to coefficient fields of *positive characteristic*.

Example 13. Let K be an arbitrary field (having zero or positive characteristic). Then the algebra $H(K)$ of *Hurwitz series* [46] over K is defined as the K-vector space of infinite K-sequences with the multiplication defined as

$$(a_n) \cdot (b_n) = \left(\sum_{i=0}^{n} \binom{n}{i} a_i b_{n-i} \right)_n$$

for all $(a_n), (b_n) \in H(K)$. If one introduces derivation and integration by

$$\partial\,(a_0, a_1, a_2, \ldots) = (a_1, a_2, \ldots),$$
$$\textstyle\int (a_0, a_1, \ldots) = (0, a_0, a_1, \ldots),$$

the Hurwitz series form an integro-differential algebra $(H(K), \partial, \int)$, as explained by [47] and [37]. Note that as an additive group, $H(K)$ coincides with the formal power series $K[[z]]$, but its multiplicative structure differs: We have an isomorphism

$$\sum_{n=0}^{\infty} a_n z^n \mapsto (n!\, a_n)$$

from $K[[z]]$ to $H(K)$ if and only if K has characteristic zero. The point is that one can integrate every element of $H(K)$, whereas the formal power series z^{p-1} does not have an antiderivative in $K[[z]]$ if K has characteristic $p > 0$. □

Now for the promised *counterexample* to the claim that the section axiom would suffice for merging a differential algebra $(\mathcal{F}, \partial)$ and a Rota-Baxter algebra $(\mathcal{F}, {\textstyle\int})$ into an integro-differential algebra $(\mathcal{F}, \partial, {\textstyle\int})$.

Example 14. Set $R = K[y]/(y^4)$ for K a field of characteristic zero and define ∂ on $\mathcal{F} = R[x]$ as usual. Then $(\mathcal{F}, \partial)$ is a differential algebra. Let us define a K-linear map ${\textstyle\int}$ on $\mathcal{F}$ by

$$\textstyle\int f = \int^* f + f(0,0)\, y^2, \tag{9}$$

where ${\textstyle\int}^*$ is the usual integral on $R[x]$ with $x^k \mapsto x^{k+1}/(k+1)$. Since the second term vanishes under ∂, we see immediately that ${\textstyle\int}$ is a section of ∂. For verifying the pure Baxter axiom (8), we compute

$$(\textstyle\int f)(\int g) = (\int^* f)(\int^* g) + y^2 \int^* \big(g(0,0)\, f + f(0,0)\, g\big) + f(0,0)\, g(0,0)\, y^4,$$
$$\textstyle\int f \int g = \int f \big(\int^* g + g(0,0)\, y^2\big) = \int^* f \int^* g + g(0,0)\, y^2 \int^* f.$$

Since $y^4 \equiv 0$ and the ordinary integral ${\textstyle\int}^*$ fulfills the pure Baxter axiom (8), this implies immediately that ${\textstyle\int}$ does also. However, it does not fulfill the differential Baxter axiom (4) because it is not $\mathcal{C}$-linear: Observe that $\mathcal{C}$ is here $Ker(\partial) = R$, so in particular we should have ${\textstyle\int}(y \cdot 1) = y \cdot {\textstyle\int} 1$. But one checks immediately that the left-hand side yields xy, while the right-hand side yields $xy + y^3$. □

3.2 Ordinary Integro-Differential Algebras

The following example shows that our current notion of integro-differential algebra includes also algebras of "multivariate functions".

Example 15. Consider $\mathcal{F} = C^\infty(\mathbb{R}^2)$ with the derivation $\partial u = u_x + u_y$. Finding sections for ∂ means solving the *partial differential equation* $u_x + u_y = f$. Its general solution is given by

$$u(x, y) = \int_a^x f(t, y - x + t)\, dt + g(y - x),$$

where $g \in C^\infty(\mathbb{R})$ and $a \in \mathbb{R}$ are arbitrary. Let us choose $a = 0$ for simplicity. In order to ensure a linear section, one has to choose $g = 0$, arriving at

$$\textstyle\int f = \displaystyle\int_0^x f(t, y - x + t)\, dt,$$

Using a change of variables, one may verify that ${\textstyle\int}$ satisfies the pure Baxter axiom (8), so $(\mathcal{F}, {\textstyle\int})$ is a Rota-Baxter algebra.

We see that the *constant functions* $\mathcal{C} = Ker(\partial)$ are given by $(x, y) \mapsto c(x - y)$ with arbitrary $c \in C^\infty(\mathbb{R})$, while the *initialized functions* $\mathcal{I} = \mathrm{Im}(\smallint)$ are those $F \in \mathcal{F}$ that satisfy $F(0, y) = 0$ for all $y \in \mathbb{R}$. In other words, $\mathcal{C}$ consists of all functions constant on the characteristic lines $x - y = \text{const}$, and $\mathcal{I}$ of those satisfying the homogeneous initial condition on the vertical axis (which plays the role of a "noncharacteristic initial manifold"). This is to be expected since $\smallint$ integrates along the characteristic lines starting from the initial manifold. The *evaluation* $\mathrm{E}: \mathcal{F} \to \mathcal{F}$ maps a function f to the function $(x, y) \mapsto f(0, y - x)$. This means that f is "sampled" only on the initial manifold, effectively becoming a univariate function: the general point (x, y) is projected along the characteristics to the initial point $(0, y - x)$.

Since E is multiplicative on $\mathcal{F}$, Lemma 3 tells us that $(\mathcal{F}, \partial, \smallint)$ is in fact an *integro-differential algebra*. Alternatively, note that $\mathcal{I}$ is an ideal and that $\smallint$ is $\mathcal{C}$-linear. Furthermore, we observe that here the polynomials are given by $K[x]$. □

In the following, we want to restrict ourselves to boundary problems for *ordinary differential equations*. Hence we want to rule out cases like Example 15. The most natural way for distinguishing ordinary from partial differential operators is to look at their kernels since only the former have finite-dimensional ones. Note that in the following definition we deviate from the standard terminology in differential algebra [48, p. 58], where ordinary only refers to having a single derivation.

From now on, we restrict the ground ring K to a *field*. We can now characterize when a differential algebra is ordinary by requiring that $\mathcal{C}$ be one-dimensional over K, meaning $\mathcal{C} = K$.

Definition 16. A differential algebra $(\mathcal{F}, \partial)$ is called *ordinary* if $\dim_K Ker(\partial) = 1$.

Note that except for Example 15 all our examples have been ordinary integro-differential algebras. The requirement of ordinariness has a number of pleasant consequences. First of all, the somewhat tedious distinction between the weak and strong *Baxter axioms* disappears since now $\mathcal{F}$ is an algebra over its own field of constants $K = \mathcal{C}$. Hence $\smallint$ is by definition $\mathcal{C}$-linear, and Lemma 10 ensures that the pure Baxer axiom (8) is equivalent to the differential Baxter axiom (4). Let us summarize this.

Corollary 17. *In an ordinary integro-differential algebra, the constant functions coincide with the ground field, and the strong and weak Baxter axioms are equivalent.*

Recall that a *character* on an algebra (or group) is a multiplicative linear functional; this may be seen as a special case of the notion of character in representation theory, namely the case when the representation is one-dimensional. In our context, a character on an integro-differential algebra $\mathcal{F}$, is a K-linear map $\varphi: \mathcal{F} \to K$ satisfying $\varphi(fg) = \varphi(f)\,\varphi(g)$ and a fortiori also $\varphi(1) = 1$. So we just require φ to be a K-algebra homomorphism, as for example in [52, p. 407].

Ordinary integro-differential algebras will always have at least one character, namely the *evaluation*: One knows from Linear Algebra that a projector P onto a one-dimensional subspace $[w]$ of a K-vector space V can be written as $P(v) = \varphi(v)\, w$, where $\varphi\colon V \to K$ is the unique functional with $\varphi(w) = 1$. If V is moreover a K-algebra, a projector onto $K = [1]$ is canonically described by the functional φ with normalization $\varphi(1) = 1$. Hence multiplicative projectors like $\mathtt{E}$ can be viewed as characters. In the next section, we consider other characters on $\mathcal{F}$; for the moment let us note $\mathtt{E}$ as a distinguished character. We write $\mathcal{F}^\bullet$ for the set of all nonzero characters on a K-algebra $\mathcal{F}$, in other words all algebra homomorphisms $\mathcal{F} \to K$.

One calls a K-algebra *augmented* if there exists a character on it. Its kernel $\mathcal{I}$ is then known as an *augmentation ideal* and forms a direct summand of K; see for example [33, p. 132]. Augmentation ideals are always maximal ideals (generalizing the $C^\infty[a,b]$ case) since the direct sum $\mathcal{F} = K \dotplus \mathcal{I}$ induces a ring isomorphism $\mathcal{F}/\mathcal{I} \cong K$. Corollary 4 immediately translates to the following characterization of integrals in ordinary differential algebras.

Corollary 18. *In an ordinary differential algebra $(\mathcal{F}, \partial)$, a section $\textstyle\int$ of ∂ is an integral if and only if its evaluation is a character if and only if $\mathcal{I} = \mathrm{Im}(\textstyle\int)$ is an augmentation ideal.*

3.3 Initial Value Problems

It is clear that in general we cannot assume that the solutions of a differential equation with coefficients in $\mathcal{F}$ are again in $\mathcal{F}$. For example, in $\mathcal{F} = K[x]$, the differential equation $u' - u = 0$ has no solution. In fact, its "actual" solution space is spanned by $u(x) = e^x$ if $K = \mathbb{R}$ or $K = \mathbb{C}$. So in this case we should have taken the exponential polynomials $\mathcal{F} = K[x, e^{Kx}]$ for ensuring that $u \in \mathcal{F}$. But if this is the case, we can also solve the *inhomogeneous differential equation* $u' - u = f$ whose general solution is $Ke^x + e^x \textstyle\int e^{-x} f$, with $\textstyle\int = \int_0^x$ as usual. Of course we can also incorporate the initial condition $u(0) = 0$, which leads to $u = e^x \textstyle\int e^{-x} f$.

This observation generalizes: Whenever we can solve the homogeneous differential equation within $\mathcal{F}$, we can also solve the initial value problem for the corresponding inhomogeneous problem. The classical tool for achieving this explicitly is the *variation-of-constants formula* [30, p. 74], whose abstract formulation is given in Theorem 20 below.

As usual [64], we will write $\mathcal{F}[\partial]$ for the ring of differential operators with coefficients in $\mathcal{F}$, see also Sect. 4. Let

$$T = \partial^n + c_{n-1}\partial^{n-1} + \cdots + c_0$$

be a monic (i.e. having leading coefficient 1) differential operator in $\mathcal{F}[\partial]$ of degree n. Then we call $u_1, \ldots, u_n \in \mathcal{F}$ a *fundamental system* for T if it is a K-basis for *Ker*(T), so it yields the right number of solutions for the homogeneous differential equation $Tu = 0$. A fundamental system will be called *regular* if its associated Wronskian matrix

$$W(u_1, \ldots, u_n) = \begin{pmatrix} u_1 & \cdots & u_n \\ u_1' & \cdots & u_n' \\ \vdots & \ddots & \vdots \\ u_1^{(n-1)} & \cdots & u_n^{(n-1)} \end{pmatrix}$$

is invertible in $\mathcal{F}^{n \times n}$ or equivalently if its Wronskian $\det W(u_1, \ldots, u_n)$ is invertible in $\mathcal{F}$. Of course this alone implies already that $u_1, \ldots, u_n$ are linearly independent.

Definition 19. A monic differential operator $T \in \mathcal{F}[\partial]$ is called *regular* if it has a regular fundamental system.

For such differential operators, variation of constants goes through – the canonical initial value problem can be solved uniquely. This means in particular that regular differential operators are always *surjective*.

Theorem 20. *Let $(\mathcal{F}, \partial, \int)$ be an ordinary integro-differential algebra. Given a regular differential operator $T \in \mathcal{F}[\partial]$ with $\deg T = n$ and a regular fundamental system $u_1, \ldots, u_n \in \mathcal{F}$, the canonical initial value problem*

$$\boxed{\begin{array}{l} Tu = f \\ \mathtt{E}u = \mathtt{E}u' = \cdots = \mathtt{E}u^{(n-1)} = 0 \end{array}} \tag{10}$$

has the unique solution

$$u = \sum_{i=1}^{n} u_i \textstyle\int d^{-1} d_i f \tag{11}$$

for every $f \in \mathcal{F}$, where $d = \det W(u_1, \ldots, u_n)$, and d_i is the determinant of the matrix W_i obtained from W by replacing the i-th column by the n-th unit vector.

Proof. We can use the usual technique of reformulating $Tu = f$ as a system of linear first-order differential equations with companion matrix $A \in \mathcal{F}^{n \times n}$. We extend the action of the operators $\int, \partial, \mathtt{E}$ componentwise to $\mathcal{F}^n$. Setting now

$$\hat{u} = W \textstyle\int W^{-1} \hat{f}$$

with $\hat{f} = (0, \ldots, 0, f)^\top \in \mathcal{F}^n$, we check that $\hat{u} \in \mathcal{F}^n$ is a solution of the first-order system $\hat{u}' = A\hat{u} + \hat{f}$ with initial condition $\mathtt{E}(\hat{u}) = 0$. Indeed we have $\hat{u}' = W' \int W^{-1} \hat{f} + W W^{-1} \hat{f}$ by the Leibniz rule and $AW = W'$ since $u_1, \ldots, u_n$ are solutions of $Tu = 0$; so the differential system is verified. For checking the initial condition, note that $\mathtt{E} \int W^{-1} \hat{f}$ is already the zero vector, so we have also $\mathtt{E}(\hat{u}) = 0$ since $\mathtt{E}$ is multiplicative.

Writing u for the first component of $\hat{u}$, we obtain a solution of the initial value problem (10), due to the construction of the companion matrix. Let us now compute $\hat{g} = W^{-1} \hat{f}$. Obviously $\hat{g}$ is the solution of the linear equation system $W\hat{g} = \hat{f}$.

Hence Cramer's rule, which is also applicable for matrices over rings [53, p. 513], yields $\hat{g}_i$ as $d^{-1} d_i f$ and hence

$$u = (W \textstyle\int \hat{g})_1 = \displaystyle\sum_{i=1}^{n} u_i \textstyle\int d^{-1} \, d_i \, f$$

since the first row of W is $(u_1, \ldots, u_n)$.

For proving uniqueness, it suffices to show that the homogeneous initial value problem only has the trivial solution. So assume u solves (10) with $f = 0$ and choose coefficients $c_1, \ldots, c_n \in K$ such that

$$u = c_1 u_1 + \cdots + c_n u_n.$$

Then the initial conditions yield $\mathtt{E}(Wc) = 0$ with $c = (c_1, \ldots, c_n)^\top \in K^n$. But we have also $\mathtt{E}(Wc) = (\mathtt{E}W)c$ because $\mathtt{E}$ is linear, and $\det \mathtt{E}W = \mathtt{E}(\det W)$ because it is moreover multiplicative. Since $\det W \in \mathcal{F}$ is invertible, $\mathtt{E}W \in K^{n \times n}$ is regular, so $c = (\mathtt{E}W)^{-1} 0 = 0$ and $u = 0$. □

4 Integro-Differential Operators

With integro-differential algebras, we have algebraized the *functions* to be used in differential equations and boundary problems, but we must also algebraize the *operators* inherent in both – the differential operators on the left-hand side of the former, and the integral operators constituting the solution of the latter. As the name suggests, the integro-differential operators provide a data structure that contains both of these operator species. In addition, it has as a third species the boundary operators needed for describing (global as well as local) boundary conditions of any given boundary problem for a LODE.

4.1 Definition

The basic idea is similar to the construction of the algebra of *differential operators* $\mathcal{F}[\partial]$ for a given differential algebra $(\mathcal{F}, \partial)$. But we are now starting from an ordinary integro-differential algebra $(\mathcal{F}, \partial, \textstyle\int)$, and the resulting algebra of integro-differential operators will accordingly be denoted by $\mathcal{F}[\partial, \textstyle\int]$. Recall that $\mathcal{F}[\partial]$ can be seen as the quotient of the free algebra generated by ∂ and $f \in \mathcal{F}$, modulo the ideal generated by the Leibniz rule $\partial f = f\partial + f'$. For $\mathcal{F}[\partial, \textstyle\int]$, we do the same but with more generators and more relations. In the following, all integro-differential algebras are assumed to be ordinary.

Apart from $\int$, we will also allow a collection of "point evaluations" as new generators since they are needed for the specification of boundary problems. For example, the local boundary condition $u(1) = 0$ on a function $u \in \mathcal{F} = C^\infty[0,1]$ gives rise to the functional $\mathrm{E}_1 \in \mathcal{F}^*$ defined by $u \mapsto u(1)$. As one sees immediately, E_1 is a *character* on $\mathcal{F}$, meaning $\mathrm{E}_1(uv) = \mathrm{E}_1(u)\,\mathrm{E}_1(v)$ for all $u, v \in \mathcal{F}$. This observation is the key for algebraizing "point evaluations" to an arbitrary integro-differential algebra where one cannot evaluate elements as in $C^\infty[0,1]$. We will see later how the characters serve as the basic building blocks for general local conditions like $3u(\pi) - 2u(0)$ or global ones like $\int_0^1 \xi u(\xi)\,d\xi$. Recall that we write $\mathcal{F}^\bullet$ for the set of all characters on integro-differential algebra $\mathcal{F}$. In Sect. 3 we have seen that every integro-differential algebra $(\mathcal{F}, \partial, \int)$ contains at least one character, namely the evaluation $\mathrm{E} = 1 - \int\partial$ associated with the integral. Depending on the application, one may add other characters.

Definition 21. Let $(\mathcal{F}, \partial, \int)$ be an ordinary integro-differential algebra over a field K and $\Phi \subseteq \mathcal{F}^\bullet$. The *integro-differential operators* $\mathcal{F}_\Phi[\partial, \int]$ are defined as the free K-algebra generated by ∂, and $\int$, the "functions" $f \in \mathcal{F}$, and the characters $\varphi \in \Phi \cup \{\mathrm{E}\}$, modulo the rewrite rules in Table 1. If Φ is understood, we write $\mathcal{F}[\partial, \int]$.

The notation $U \cdot f$, used in the right-hand side of some of the rules above, refers to the *action* of $U \in \mathcal{F}\langle\partial, \int\rangle$ on a function $f \in \mathcal{F}$; in particular, $f \cdot g$ denotes the product of two functions $f, g \in \mathcal{F}$. It is an easy matter to check that the rewrite rules of Table 1 are fulfilled in $(\mathcal{F}, \partial, \int)$, so we may regard $\cdot$ as an action of $\mathcal{F}[\partial, \int]$ on $\mathcal{F}$. Thus every element $T \in \mathcal{F}[\partial, \int]$ acts as a map $T\colon \mathcal{F} \to \mathcal{F}$.

We have given the relations as a *rewrite system*, but their algebraic meaning is also clear: If in the free algebra $\mathcal{F}\langle\partial, \int\rangle$ of Definition 21 we form the two-sided ideal $\mathfrak{g}$ generated by the left-hand side minus right-hand side for each rule, then $\mathcal{F}_\Phi[\partial, \int] = \mathcal{F}\langle\partial, \int\rangle/\mathfrak{g}$. Note that there are infinitely many such rules since each choice of $f, g \in \mathcal{F}$ and $\varphi, \psi \in \Phi$ yields a different instance (there may be just finitely many characters in Φ but the coefficient algebra $\mathcal{F}$ is always infinite), so $\mathfrak{g}$ is an infinitely generated ideal (it was called the "Green's ideal" in [70] in a slightly more special setting). Note that one gets back the rewrite system of Table 1 if one uses the implied set of generators and a suitable ordering (see Sect. 7).

The reason for specifying $\mathfrak{g}$ via a rewrite system is of course that we may use it for generating a canonical simplifier for $\mathcal{F}[\partial, \int]$. This can be seen either from the term rewriting or from the Gröbner basis perspective: In the former case, we see Table 1 as a confluent and terminating rewrite system (modulo the ring axioms); in the latter case, as a *noncommutative Gröbner basis with noetherian reduction* (its elements are of course the left-hand side minus right-hand side for each rule). While

Table 1 Rewrite rules for integro-differential operators

$fg \to f \cdot g$	$\partial f \to f\partial + \partial \cdot f$	$\int f \int \to (\int \cdot f)\int - \int(\int \cdot f)$
$\varphi\psi \to \psi$	$\partial\varphi \to 0$	$\int f \partial \to f - \int(\partial \cdot f) - (\mathrm{E} \cdot f)\,\mathrm{E}$
$\varphi f \to (\varphi \cdot f)\varphi$	$\partial\int \to 1$	$\int f \varphi \to (\int \cdot f)\varphi$

we cannot give a detailed account of these issues here, we will briefly outline the Gröbner basis setting since our new proof in Sect. 7 will rely on it.

4.2 Noncommutative Gröbner Bases

As detailed in Sect. 2, it is necessary for our application to deal with *infinitely generated ideals* and an *arbitrary set of indeterminates*. The following description of such a noncommutative Gröbner basis setting is based on the somewhat dated but still highly readable Bergman paper [9]; for a summary see [28, §3.3]. For other approaches we refer the reader to [61, 62, 85, 86].

Let us first recall some notions for abstract reduction relations [4]. We consider a relation $\rightarrow \subseteq A \times A$ for a set A; typically $\rightarrow$ realizes a single step in a simplification process like the transformation of integro-differential operators according to Table 1. The transitive closure of $\rightarrow$ is denoted by $\overset{+}{\rightarrow}$, its reflexive-transitive closure by $\overset{*}{\rightarrow}$. We call $a \in A$ *irreducible* if there is no $a_0 \in A$ with $a \rightarrow a_0$; we write $A_\downarrow$ for the set of all irreducible elements. If $a \overset{*}{\rightarrow} a_0$ with $a_0 \in A_\downarrow$, we call a_0 a *normal form* of a, denoted by $\downarrow a = a_0$ in case it is unique.

If all elements are to have a unique normal form, we have to impose two conditions: termination for banning infinite reductions and confluence reuniting forks. More precisely, $\rightarrow$ is called *terminating* if there are no infinite chains $a_1 \rightarrow a_2 \rightarrow \ldots$ and *confluent* if for all $a, a_1, a_2 \in A$ the fork $a_1 \overset{*}{\leftarrow} a \overset{*}{\rightarrow} a_2$ finds a reunion $a_1 \overset{*}{\rightarrow} a_0 \overset{*}{\leftarrow} a_2$ for some $a_0 \in A$. If $\rightarrow$ is both terminating and confluent, it is called *convergent*.

Turning to noncommutative Gröbner bases theory, we focus on reduction relations on the free algebra $K\langle X\rangle$ over a commutative ring K in an arbitrary set of indeterminates X; the corresponding monomials form the free monoid $\langle X\rangle$. Then a *reduction system* for $K\langle X\rangle$ is a set $\Sigma \subseteq \langle X\rangle \times K\langle X\rangle$ whose elements are called rules. For a rule $\sigma = (W, f)$ and monomials $A, B \in \langle X\rangle$, the K-module endomorphism of $K\langle X\rangle$ that fixes all elements of $\langle X\rangle$ except sending AWB to AfB is denoted by ${}_A\sigma_B$ and called a reduction. It is said to act trivially on $a \in K\langle X\rangle$ if the coefficient of AWB in a is zero.

Every reduction system Σ induces the relation $\rightarrow_\Sigma \subseteq K\langle X\rangle \times K\langle X\rangle$ defined by setting $a \rightarrow_\Sigma b$ if and only if $r(a) = b$ for some reduction acting nontrivially on a. We call its reflexive-transitive closure $\overset{*}{\rightarrow}_\Sigma$ the *reduction relation* induced by Σ, and we say that a reduces to b when $a \overset{*}{\rightarrow}_\Sigma b$. Accordingly we can speak of irreducible elements, normal forms, termination and confluence of Σ.

For ensuring termination, one can impose a noetherian *monoid ordering* on $\langle X\rangle$, meaning a partial ordering such that $1 < A$ for all $A \in \langle X\rangle$ and such that $B < B'$ implies $ABC < AB'C$ for $A, B, B', C \in \langle X\rangle$. Recall that for partial (i.e. not necessarily total) orderings, noetherianity means that there are no infinite descending chains or equivalently that every nonempty set has a minimal element [8, p. 156]. Note that in a noetherian monoid ordering (like the divisibility relation on natural numbers), elements are not always comparable.

Now if one has a noetherian monoid ordering on $\langle X \rangle$, then Σ will be *terminating* provided it respects $<$ in the sense that $W' < W$ for every rule $(W, f) \in \Sigma$ and every nonzero monomial W' of f. (Let us also remark that the condition $1 < A$ from above might as well be dropped, as in [9]: The given rewrite system cannot contain a rule $1 \to f$ since then $W < 1$ for at least one nonzero monomial W of f, so $1 > W > WW > \cdots$ would yield an infinite descending chain. Such rules precluded, it is not stringent that constants in K be comparable with the elements in X. But since it is nevertheless very natural and not at all restrictive, we stick to the monoid orderings as given above.)

It is typically more difficult to ensure confluence of a reduction system Σ. According to the definition, we would have to investigate all forks $a_1 \overset{*}{\leftarrow} a \overset{*}{\rightarrow} a_2$, which are usually infinite in number. The key idea for a practically useful criterion is to consider only certain *minimal forks* (called ambiguities below, following Bergman's terminology) and see whether their difference eventually becomes zero. This was first described by Buchberger in [24] for the commutative case; see also [19, 20]. The general intuition behind minimal forks is analyzed in [18], where Gröbner bases are compared with Knuth-Bendix completion and Robinson's resolution principle.

An *overlap ambiguity* of Σ is given by a quintuple (σ, τ, A, B, C) with Σ-rules $\sigma = (W, f)$, $\tau = (V, g)$ and monomials $A, B, C \in \langle X \rangle \setminus \{1\}$ such that $W = AB$ and $V = BC$. Its associated S-polynomial is defined as $fC - Ag$, and the ambiguity is called resolvable if the S-polynomial reduces to zero. (In general one may also have so-called inclusion ambiguities, but it turns out that one can always remove them without changing the resulting normal forms [9, §5.1]. Since the reduction system of Table 1 does not have inclusion ambiguities, we will not discuss them here.)

For making the connection to ideal theory, we observe that every reduction system Σ gives rise to a two-sided *ideal* I_Σ generated by all elements $W - f$ for $(W, f) \in \Sigma$; we have already seen this connection for the special case of the integro-differential operators. Note that $a \overset{*}{\rightarrow}_\Sigma 0$ is equivalent to $a \in I_\Sigma$.

In the given setting, the task of proving convergence can then be attacked by the so-called *Diamond Lemma for Ring Theory*, presented as Theorem 1.2 in Bergman's homonymous paper [9]; see also Theorem 3.21 in [28]. It is the noncommutative analog of Buchberger's criterion [19] for infinitely generated ideals. (In the version given below, we have omitted a fourth equivalent condition that is irrelevant for our present purposes.)

Theorem 22. *Let Σ be a reduction system for $K\langle X \rangle$ and $\leq$ a noetherian monoid ordering that respects Σ. Then the following conditions are equivalent:*

- *All ambiguities of Σ are resolvable.*
- *The reduction relation $\overset{*}{\rightarrow}_\Sigma$ is convergent.*
- *We have the direct decomposition $K\langle X \rangle = K\langle X \rangle_\downarrow \dotplus I_\Sigma$ as K-modules.*

When these conditions hold, the quotient algebra $K\langle X \rangle / I_\Sigma$ may be identified with the K-module $K\langle X \rangle_\downarrow$, having the multiplication $a \cdot b = {\downarrow} ab$.

We will apply Theorem 22 in Sect. 7 for proving that Table 1 constitutes a Gröbner basis for the ideal $\mathfrak{g}$. Hence we may conclude that $\mathcal{F}[\partial, \int]$ can be identified with the algebra $\mathcal{F}\langle\partial, \int\rangle_{\downarrow}$ of *normal forms*, and this is what gives us an algorithmic handle on the integro-differential operators. It is thus worth investigating these normal forms in some more detail.

4.3 Normal Forms

We start by describing a set of *generators*, which will subsequently be narrowed to the normal forms of $\mathcal{F}_\Phi[\partial, \int]$. The key observation is that in any monomial we never need more than one integration while all the derivatives can be collected at its end.

Lemma 23. *Every integro-differential operator in $\mathcal{F}_\Phi[\partial, \int]$ can be reduced to a linear combination of monomials $f \varphi \int g \psi \partial^i$, where $i \geq 0$ and each of $f, \varphi, \int, g, \psi$ may also be absent.*

Proof. Call a monomial consisting only of functions and functionals "algebraic". Using the left column of Table 1, it is immediately clear that all such monomials can be reduced to f or φ or $f\varphi$. Now let w be an arbitrary monomial in the generators of $\mathcal{F}_\Phi[\partial, \int]$. By using the middle column of Table 1, we may assume that all occurrences of ∂ are moved to the right, so that all monomials have the form $w = w_1 \cdots w_n \partial^i$ with $i \geq 0$ and each of $w_1, \ldots, w_n$ either a function, a functional or $\int$. We may further assume that there is at most one occurrence of $\int$ among the $w_1, \ldots, w_n$. Otherwise the monomials $w_1 \cdots w_n$ contain $\int \tilde{w} \int$, where each $\tilde{w} = f\varphi$ is an algebraic monomial. But then we can reduce

$$\int \tilde{w} \int = (\int f\varphi) \int = (\int \cdot f)\varphi \int$$

by using the corresponding rule of Table 1. Applying these rules repeatedly, we arrive at algebraic monomials left and right of $\int$ (or just a single algebraic monomial if $\int$ is absent). □

In TH∃OREM∀, the *integro-differential operators* over an integro-differential algebra $\mathcal{F}$ of coefficient functions are built up by FreeIntDiffOp[$\mathcal{F}$,K]. This functor constructs an instance of the monoid algebra with the word monoid over the infinite alphabet consisting of the letters ∂ and $\int$ along with a basis of $\mathcal{F}$ and with all multiplicative characters induced by evaluations at points in K:

```
Definition["IntDiffOp", any[ℱ, K],
 IntDiffOp[ℱ, K] = where[𝒜 = FreeIntDiffOp[ℱ, K], 𝒢 = GreenSystem[ℱ, K]
    QuotAlg[GBNF[𝒜, 𝒢]]]
]
```

In this code fragment, the `GreenSystem` functor contains the encoding of the aforementioned rewrite system (Table 1), here understood as a noncommutative Gröbner basis. Normal forms for total reduction modulo infinite Gröbner bases are created by the `GBNF` functor, while the `QuotAlg` functor constructs the quotient algebra from the corresponding canonical simplifier (see Sect. 2 for details). For instance, multiplying the integral operator $\int$ by itself triggers an application of the Baxter rule:

```
Compute[⟨⟨1, ⟨"∫"⟩⟩⟩ ⋆_I ⟨⟨1, ⟨"∫"⟩⟩⟩] // FormatP

-A x + x A
```

Here integral operators are denoted by `A`, following the notation in the older implementation [70].

We turn now to the normal forms of *boundary conditions*. Since they are intended to induce mappings $\mathcal{F} \to K$, it is natural to define them as those integro-differential operators that "end" in a character $\varphi \in \Phi$. For example, if φ is the point evaluation $\mathtt{E}_1$ considered before, the composition $\mathtt{E}_1\partial$ describes the local condition $u'(1) = 0$, the composition $\mathtt{E}_1\int$ the global condition $\int_0^1 u(\xi)\, d\xi = 0$. In general, boundary conditions may be arbitrary linear combinations of such composites; they are known as "Stieltjes conditions" in the literature [15, 16].

Definition 24. The elements of the right ideal

$$|\Phi) = \Phi \cdot \mathcal{F}_\Phi[\partial, \textstyle\int]$$

are called *boundary conditions* over $\mathcal{F}$.

It turns out that their *normal forms* are exactly the linear combinations of local and global conditions as in the example mentioned above. As a typical representative over $\mathcal{F} = C^\infty(\mathbb{R})$, one may think of an element like

$$\mathtt{E}_0\partial^2 + 3\,\mathtt{E}_\pi - 2\,\mathtt{E}_{2\pi}\textstyle\int \sin x,$$

written as $u''(0) + 3\,u(\pi) - 2\int_0^{2\pi} \sin\xi\, u(\xi)\, d\xi$ in traditional notation.

Proposition 25. *Every boundary condition of* $|\Phi)$ *has the normal form*

$$\sum_{\varphi\in\Phi} \left(\sum_{i\in\mathbb{N}} a_{\varphi,i}\, \varphi\partial^i + \varphi \textstyle\int f_\varphi \right), \tag{12}$$

with only finitely many $a_{\varphi,i} \in K$ *and* $f_\varphi \in \mathcal{F}$ *nonzero.*

Proof. By Lemma 23, every boundary condition of $|\Phi)$ is a linear combination of monomials having the form

$$w = \chi f \varphi {\textstyle\int} g \psi \partial^i \quad \text{or} \quad w = \chi f \varphi \partial^i \tag{13}$$

where each of f, g, φ, ψ may also be missing. Using the left column of Table 1, the prefix $\chi f \varphi$ can be reduced to a scalar multiple of a functional, so we may as well assume that f and φ are not present; this finishes the right-hand case of (13). For the remaining case $w = \chi \int g \psi \partial^i$, assume first that ψ is present. Then we have

$$\chi\,({\textstyle\int} g\psi) = \chi\,({\textstyle\int}\cdot g)\,\psi = (\chi{\textstyle\int}\cdot g)\,\chi\psi = (\chi{\textstyle\int}\cdot g)\,\psi,$$

so w is again a scalar multiple of $\psi\partial^i$, and we are done. Finally, assume we have $w = \chi \int g \partial^i$. If $i = 0$, this is already a normal form. Otherwise we obtain

$$w = \chi\,({\textstyle\int} g\partial)\,\partial^{i-1} = (\chi \cdot g)\,\chi\partial^{i-1} - \chi{\textstyle\int} g'\partial^{i-1} - (\mathtt{E}\cdot g)\,\mathtt{E}\partial^{i-1},$$

where the first and the last summand are in the required normal form, while the middle summand is to be reduced recursively, eventually leading to a middle term in normal form $\pm\chi\int g'\partial^0 = \pm\chi\int g'$. □

Most expositions of boundary problems – both analytic and numeric ones – restrict their attention to local conditions, even more specifically to those with just two point evaluation (so-called two-point boundary problems). While this is doubtless the most important case, there are at least three reasons for considering *Stieltjes boundary conditions* of the form (12).

- They arise in certain applications (e.g. heat radiated through a boundary) and in treating ill-posed problems by generalized Green's functions [70, p. 191].
- As we shall see (Sect. 5), they are needed for factoring boundary problems.
- Their algebraic description as a right ideal is very natural.

Hence we shall always mean all of $|\Phi)$ when we speak of boundary conditions.

Let us now turn to the other two ingredients of integro-differential operators: We have already mentioned the *differential operators* $\mathcal{F}[\partial]$, but we can now see them as a subalgebra of $\mathcal{F}_\Phi[\partial, \int]$. They have the usual normal forms since the Leibniz rule is part of the rewrite system. Analogously, we introduce the subalgebra of *integral operators* generated by the functions and $\int$. Using Lemma 23, it is clear that the normal forms of integral operators are $\mathcal{F}$ itself and linear combinations of $f \int g$, and the only rule applicable to them is $\int f \int \to \cdots$ in Table 1. Since we have already included $\mathcal{F}$ in $\mathcal{F}[\partial]$, we introduce $\mathcal{F}[\int]$ as the $\mathcal{F}$-bimodule generated by $\int$ so that it contains only the monomials of the form $f \int g$.

Finally, we must consider the two-sided ideal (Φ) of $\mathcal{F}_\Phi[\partial, \int]$ generated by Φ whose elements are called *boundary operators*. A more economical description of (Φ) is as the left $\mathcal{F}$-submodule generated by $|\Phi)$ because by Lemma 23 any $w\chi\tilde{w}$ with $w, \tilde{w} \in \mathcal{F}[\partial, \int]$ can be reduced to $f\varphi\int g\psi\partial^i\chi\tilde{w}$. Note that (Φ) includes all finite dimensional projectors P along Stieltjes boundary conditions. Any such projector can be described in the following way: If $u_1, \ldots, u_n \in \mathcal{F}$ and $\beta_1, \ldots, \beta_n \in |\Phi)$ are biorthogonal (meaning $\beta_i(u_j) = \delta_{ij}$), then

$$P = \sum_{i=1}^{n} u_i \, \beta_i \, : \, \mathscr{F} \to \mathscr{F} \tag{14}$$

is the projector onto $[u_1, \ldots, u_n]$ whose kernel is the subspace of all $u \in \mathscr{F}$ such that $\beta(u) = 0$ for all $\beta \in [\beta_1, \ldots, \beta_n]$. See for example [50, p. 71] or [66] for details. Note that all elements of (Φ) have the normal form (14), except that the (u_j) need not be biorthogonal to the (β_i).

We can now characterize the normal forms of $\mathscr{F}_\Phi[\partial, \int]$ in a very straightforward and intuitive manner: Every monomial is either a *differential operator* or an *integral operator* or a *boundary operator*. Hence every element of $\mathscr{F}_\Phi[\partial, \int]$ can be written uniquely as a sum $T + G + B$, with a differential operator $T \in \mathscr{F}[\partial]$, an integral operator $G \in \mathscr{F}[\int]$, and a boundary operator $B \in (\Phi)$.

Proposition 26. *For an integro-differential algebra $\mathscr{F}$ and characters $\Phi \subseteq \mathscr{F}^\bullet$, we have the direct decomposition $\mathscr{F}_\Phi[\partial, \int] = \mathscr{F}[\partial] \dotplus \mathscr{F}[\int] \dotplus (\Phi)$.*

Proof. Inspection of Table 1 confirms that all integro-differential operators having the described sum representation $T + G + P$ are indeed in normal form. Let us now prove that every integro-differential operator of $\mathscr{F}_\Phi[\partial, \int]$ has such a representation. It is sufficient to consider its monomials w. If w starts with a functional, we obtain a boundary condition by Proposition 25; so assume this is not the case. From Lemma 23 we know that

$$w = f \varphi \textstyle\int g \psi \partial^i \quad \text{or} \quad w = f \varphi \partial^i,$$

where each of φ, g, ψ may be absent. But $w \in (\Phi)$ unless φ is absent, so we may actually assume

$$w = f \textstyle\int g \psi \partial^i \quad \text{or} \quad w = f \partial^i.$$

The right-hand case yields $w \in \mathscr{F}[\partial]$. If ψ is present in the other case, we may reduce $\int g \psi$ to $(\int \cdot g)\, \psi$, and we obtain again $w \in (\Phi)$. Hence we are left with $w = f \int g \partial^i$, and we may assume $i > 0$ since otherwise we have $w \in \mathscr{F}[\int]$ immediately. But then we can reduce

$$\begin{aligned} w &= f\, (\textstyle\int g \partial)\, \partial^{i-1} = f \Big(g - \textstyle\int (\partial \cdot g) - (\mathrm{E} \cdot g)\, \mathrm{E} \Big) \partial^{i-1} \\ &= (fg)\, \partial^{i-1} - f \textstyle\int (\partial \cdot g)\, \partial^{i-1} - (\mathrm{E} \cdot g)\, f\, \mathrm{E}\, \partial^{i-1}, \end{aligned}$$

where the first term is obviously in $\mathscr{F}[\partial]$ and the last one in (Φ). The middle term may be reduced recursively until the exponent of ∂ has dropped to zero, leading to a term in $\mathscr{F}[\int]$. □

We can observe the direct decomposition $\mathscr{F}_\Phi[\partial, \int] = \mathscr{F}[\partial] \dotplus \mathscr{F}[\int] \dotplus (\Phi)$ in the following *sample multiplication* of $\int \partial$ and $\partial \partial x e^x \int$:

```
Compute[⟨⟨1, ⟨"∫", "∂"⟩⟩⟩ ⋆ ⟨⟨1, ⟨"∂", "∂", ⟨"[]", ⟨1, 1⟩⟩, "∫"⟩⟩⟩] // FormatP
```

```
-2 E + 2 eˣ + 2 eˣ x + 2 eˣ A + eˣ x A + eˣ x D
```

As in the previous computation, `A` stands for the integral $\int$, moreover `D` for the derivation ∂, and `E` for the evaluation. As we can see, the sum is composed of one differential operator (the last summand), two integral operators (in the middle), and three boundary operators (the first summands). Observe also that the input operators are not in normal form but the output operator is.

4.4 Basis Expansion

Regarding the canonical forms for $\mathcal{F}[\partial, \int]$, there is one more issue that we have so far swept under the rug. The problem is that in the current setup elements like $x\partial + 3x^2\partial$ and $(x + 3x^2)\partial$ are considered distinct normal forms. More generally, if $f + g = h$ holds in $\mathcal{F}$, there is no rule that allows us to recognize that $f + g \in \mathcal{F}[\partial, \int]$ and $h \in \mathcal{F}[\partial, \int]$ are the same. Analogously, if $\lambda \tilde{f} = \tilde{g}$ holds in $\mathcal{F}$ with $\lambda \in K$, then $\lambda \tilde{f}$ and $\tilde{g}$ are still considered to be different in $\mathcal{F}[\partial, \int]$. A slightly less trivial example is when $f = (\cos x)(\cos^2 x^2)$ and $g = -(\sin x)(\sin x^2)$ so that $h = \cos(x + x^2)$. What is needed in general is obviously a *choice of basis* for $\mathcal{F}$. But since such a choice is always to some degree arbitrary, we would like to postpone it as much as possible.

An unbiased way of introducing all K-linear relations in $\mathcal{F}$ is simply to collect them in all in the two-sided ideal

$$\mathfrak{l} = (f + g - h,\, \lambda \tilde{f} - \tilde{g} \mid f + g = h \text{ and } \lambda \tilde{f} = \tilde{g} \text{ in } \mathcal{F}) \subseteq \mathcal{F}\langle \partial, \textstyle\int \rangle,$$

which we shall call the *linear ideal*. Since $\mathfrak{l} + \mathfrak{g}$ corresponds to a unique ideal $\tilde{\mathfrak{l}}$ in $\mathcal{F}[\partial, \int]$, the necessary refinement of $\mathcal{F}[\partial, \int]$ can now be defined as

$$\mathcal{F}^{\#}[\partial, \textstyle\int] = \mathcal{F}[\partial, \int]/\tilde{\mathfrak{l}} \cong \mathcal{F}\langle \partial, \int \rangle/(\mathfrak{l} + \mathfrak{g})$$

whose elements shall be called *expanded integro-differential operators*. Note that $\tilde{\mathfrak{l}}$ is really the "same" ideal as $\mathfrak{l}$ except that now $f, g, h, \tilde{f}, \tilde{g} \in \mathcal{F}[\partial, \int]$. By the isomorphism above, coming from the Third Isomorphism Theorem [31, Theorem 1.22], we can think of $\mathcal{F}^{\#}[\partial, \int]$ in two ways: Either we impose the linear relations on $\mathcal{F}[\partial, \int]$ or we merge them in with the Green's ideal – let us call these the a-posteriori and the combined approach, respectively.

For computational purposes, we need a *ground simplifier* on the free algebra [70, p. 525], which we define here as a K-linear canonical simplifier for $\mathcal{F}\langle \partial, \int \rangle/\mathfrak{l}$. Since all reduction rules of Table 1 are (bi)linear in $f, g \in \mathcal{F}$, any ground simplifier descends to a canonical simplifier σ on $\mathcal{F}^{\#}[\partial, \int]$. In our implementations, σ always operates by basis expansion (see below), but other strategies are conceivable. We can apply σ either a-posteriori or combined:

- In the first case we apply σ as a *postprocessing step* after computing the normal forms with respect to Table 1. We have chosen this approach in the upcoming Maple implementation [49].
- In the *combined approach*, σ may be used at any point during a reduction along the rules of Table 1. It may be more efficient, however, to use σ on the rules themselves to create a new reduction system on $\mathcal{F}\langle\partial, \int\rangle$; see below for an example. We have taken this approach in our earlier implementation [69, 70] and in the current implementation.

Generally the first approach seems to be superior, at least when σ tends to create large expressions that are not needed for the rewriting steps of Table 1; this is what typically happens if the ground simplifier operates by basis expansion.

Assume now we choose a K-basis $(b_i)_{i\in I}$ of $\mathcal{F}$. If $(\hat{b}_i)_{i\in I}$ is the dual basis, we can describe the linear ideal more economically as

$$\mathfrak{l} = \Big(f - \sum_{i\in I} \hat{b}_i(f)\, b_i \mid f \in \mathcal{F}\Big),$$

so the linear basis $(b_i)_{i\in I}$ gives rise to an ideal basis for $\mathfrak{l}$. Its generators $f - \sum_i \cdots$ can be oriented to create a ground simplifier $\sigma\colon f \;\mapsto\; \sum_i \cdots$ effecting *basis expansion*.

If one applies now such a ground simplifier coming from a basis $(b_i)_{i\in I}$ in the *combined approach*, one can restrict the generators of $\mathcal{F}\langle\partial, \int\rangle$ to basis elements $b_i \in \mathcal{F}$ rather than all $f \in \mathcal{F}$, and the reduction rules can be adapted accordingly. For example, when $\mathcal{F}$ contains the exponential polynomials, the Leibniz rule $\partial f \to f\partial + (\partial \cdot f)$ gets instantiated for $f = xe^x$ as $\partial(xe^x) \to (xe^x)\partial + e^x + x$, where the right-hand side now has three instead of two monomials! This is why the choice of basis was built into the definition of the precursor of $\mathcal{F}^\#[\partial, \int]$ as in [70].

Before leaving this section on integro-differential operators, let us mention some interesting *related work* on these objects, carried out from a more algebraic viewpoint. In his papers [5, 6], Bavula establishes an impressive list of various (notably ring-theoretic) properties for algebras of integro-differential operators. The setup considered in these papers is, on the one hand, in many respects more general since it deals with partial rather than ordinary differential operators but, on the other hand, the coefficients are restricted to polynomials.

Seen from the more applied viewpoint taken here, the most significant difference is the lack of multiple point evaluations (and thus boundary conditions). Apart from these obvious differences, there is also a somewhat *subtle difference* in the meaning of $\mathtt{E} = 1 - \int \circ\, \partial$ that we have tried to elucidate in a skew-polynomial setting [67]. The upshot is that while our approach views $\mathtt{E}$ as a specific evaluation (the prototypical example is given after Definition 2), it does not have a canonical action in V. Bavula's setting (and neither in our skew-polynomial approach). This is a subtle but far-reaching difference that deserves future investigation.

5 Applications for Boundary Problems

In this section we combine the tools developed in the previous sections to build an *algorithm* for solving linear boundary problems over an ordinary integro-differential algebra; see also [72] for further details. We also outline a *factorization method* along a given factorization of the defining differential operator applicable to boundary problems for both linear *ordinary* and *partial* differential equations; see [66] in an abstract linear algebra setting and [74] for an overview.

For motivating our algebraic setting of boundary problems, let us consider our standard example of an integro-differential algebra $(\mathcal{F}, \partial, \textstyle\int)$ with the integral operator

$$\textstyle\int : f \mapsto \int_0^x f(\xi)\, d\xi$$

for $\mathcal{F} = C^\infty[0, 1]$. The *simplest two-point boundary problem* reads then as follows: Given $f \in \mathcal{F}$, find $u \in \mathcal{F}$ such that

$$\boxed{\begin{array}{l} u'' = f, \\ u(0) = u(1) = 0. \end{array}} \tag{15}$$

In this and the subsequent examples, we let D and A denote respectively the derivation ∂ and the integral operator $\textstyle\int$. Moreover, we denote by L the corresponding evaluation $\mathtt{E}$, which is the character

$$L : f \mapsto f(0).$$

To express boundary problems we need additionally the evaluation at the endpoint of the interval

$$R : f \mapsto f(1).$$

Note that u is annihilated by any linear combination of these functionals so that problem (15) can be described by the pair $(D^2, [L, R])$, where $[L, R]$ is the subspace generated by L, R in the dual space $\mathcal{F}^*$.

The solution algorithm presupposes a constructive fundamental system for the underlying homogeneous equation but imposes no other conditions (in the literature one often restricts to self-adjoint and/or second-order boundary problems). This is always possible (relative to root computations) in the important special case of LODEs with constant coefficient.

5.1 The Solution Algorithm

In the following, we introduce the notion of *boundary problem* in the context of ordinary integro-differential algebras. Unless specified otherwise, all integro-differential algebras in this section are assumed to be ordinary and over a fixed field K.

Definition 27. Let $(\mathcal{F}, \partial, \int)$ be an ordinary integro-differential algebra. Then a *boundary problem* of order n is a pair $(T, \mathcal{B})$, where $T \in \mathcal{F}[\partial]$ is a regular differential operator of order n and $\mathcal{B} \subseteq |\mathcal{F}^\bullet)$ is an n-dimensional subspace of boundary conditions.

Thus a boundary problem is specified by a differential operator T and a *boundary space* $\mathcal{B} = [\beta_1, \ldots, \beta_n]$ generated by linearly independent boundary conditions $\beta_1, \ldots, \beta_n \in |\mathcal{F}^\bullet)$. In traditional notation, the boundary problem $(T, \mathcal{B})$ is then given by

$$\boxed{\begin{aligned} &Tu = f, \\ &\beta_1 u = \cdots = \beta_n u = 0. \end{aligned}} \tag{16}$$

For a given boundary problem, we can restrict to a finite subset $\Phi \subseteq \mathcal{F}^\bullet$, with the consequence that all subsequent calculations can be carried out in $\mathcal{F}_\Phi[\partial, \int]$ instead of $\mathcal{F}[\partial, \int]$. We will disregard this issue here for keeping the notation simpler.

Definition 28. A boundary problem $(T, \mathcal{B})$ is called *regular* if for each $f \in \mathcal{F}$ there exists a unique solution $u \in \mathcal{F}$ in the sense of (16).

The condition requiring T to have the same order as the dimension of $\mathcal{B}$ in Definition 27 is only necessary but not sufficient for ensuring regularity: the boundary conditions might collapse on *Ker*(T). A simple example of such a *singular boundary problem* is $(-D^2, [LD, RD])$ using the notation from before; see also [70, p. 191] for more details on this particular boundary problem.

For an *algorithmic* test of regularity, one may also apply the usual regularity criterion for two-point boundary problems, as described in [66]. Taking any fundamental system of solutions $u_1, \ldots, u_n$ for the homogeneous equation, one can show that a boundary problem $(T, \mathcal{B})$ is regular if and only if the *evaluation matrix*

$$\beta(u) = \begin{pmatrix} \beta_1(u_1) & \cdots & \beta_1(u_n) \\ \vdots & \ddots & \vdots \\ \beta_n(u_1) & \cdots & \beta_n(u_n) \end{pmatrix} \in K^{n \times n}$$

is regular.

For a regular boundary problem $(T, \mathcal{B})$, we can now define its *Green's operator* G as the linear operator mapping a given forcing function $f \in \mathcal{F}$ to the unique solution $u \in \mathcal{F}$ of (16). It is characterized by the identities

$$TG = 1 \quad \text{and} \quad \mathrm{Im}(G) = \mathcal{B}^\perp,$$

where $\mathcal{B}^\perp = \{u \in \mathcal{F} \mid \beta(u) = 0 \text{ for all } \beta \in \mathcal{B}\}$ is the subspace of all "functions" satisfying the boundary conditions. We also write

$$G = (T, \mathcal{B})^{-1}$$

for the Green's operator of $(T, \mathcal{B})$.

The investigation of *singular boundary problems* (i.e. non-regular ones) is very enlightening but leads us too far afield; we shall investigate it at another junction. Let us just mention that it involves so-called modified Green's operators and functions [80, p. 216] and that is paves the way to an interesting non-commutative analog of the classical Mikusiński calculus [60].

We will now recast Theorem 20 in the language of Green's operators of initial value problems. Given a regular differential operator T of order n, the theorem implies that the initial value problem $(T, [\mathrm{E}, \mathrm{E}\partial, \ldots, \mathrm{E}\partial^{n-1}])$ is regular. We call its Green's operator the *fundamental right inverse* of T and denote it by $T^{\blacklozenge}$.

Corollary 29. *Let* $(\mathcal{F}, \partial, \int)$ *be an ordinary integro-differential algebra and let* $T \in \mathcal{F}[\partial]$ *be a regular differential operator of order* n *with regular fundamental system* $u_1, \ldots, u_n$. *Then its fundamental right inverse is given by*

$$T^{\blacklozenge} = \sum_{i=1}^{n} u_i \textstyle\int d^{-1} d_i \in \mathcal{F}[\partial, \textstyle\int], \tag{17}$$

where $d, d_1, \ldots, d_n$ *are as in Theorem 20.*

Before turning to the solution algorithm for boundary problems, let us also mention the following practical formula for specializing Corollary 29 to the important special case of LODEs with *constant coefficients*, which could also be proved directly e.g. via the Lagrange interpolation formula. For simplicity, we restrict ourselves to the case where the characteristic polynomial is separable.

Corollary 30. *Let* $(\mathcal{F}, \partial, \int)$ *be an ordinary integro-differential algebra and consider the differential operator* $T = (\partial - \lambda_1) \cdots (\partial - \lambda_n) \in \mathcal{F}[\partial]$ *with* $\lambda_1, \ldots, \lambda_n \in K$ *mutually distinct. Assume each* $u' = \lambda_i u$, $\mathrm{E} \cdot u = 1$ *has a solution* $u = e^{\lambda_i x} \in \mathcal{F}$ *with reciprocal* $u^{-1} = e^{-\lambda_i x} \in \mathcal{F}$. *Then we have*

$$T^{\blacklozenge} = \sum_{i=1}^{n} \mu_i \, e^{\lambda_i x} \textstyle\int e^{-\lambda_i x},$$

where $\mu_i^{-1} = (\lambda_i - \lambda_1) \cdots (\lambda_i - \lambda_{i-1})(\lambda_i - \lambda_{i+1}) \cdots (\lambda_i - \lambda_n)$.

Proof. Let us write V for the $n \times n$ Vandermonde determinant in $\lambda_1, \ldots, \lambda_n$ and V_i for the $(n-1) \times (n-1)$ Vandermonde determinant in $\lambda_1, \ldots, \lambda_{i-1}, \lambda_{i+1}, \ldots, \lambda_n$. Evaluating the quantities of (17), one sees immediately that

$$d = e^{(\lambda_1 + \cdots \lambda_n)x} \, V \quad \text{and} \quad d_i = (-1)^{n+i} e^{(\lambda_1 + \cdots + \lambda_{i-1} + \lambda_{i+1} + \cdots + \lambda_n)x} \, V_i.$$

Hence we have $d_i/d = (-1)^{n+i}\, e^{-\lambda_i x}\, V_i/V$. Using the well-known formula for the Vandermonde determinant, one obtains $d_i/d = \mu_i\, e^{-\lambda_i x}$, and now the result follows from Corollary 29. □

Summarizing our earlier results, we can now give a *solution algorithm* for computing $G = (T, \mathscr{B})^{-1}$, provided we have a regular fundamental system $u_1, \ldots, u_n$ for $Tu = 0$ and a K-basis $\beta_1, \ldots, \beta_n$ for $\mathscr{B}$. The algorithm proceeds in three steps:

1. Construct the fundamental right inverse $T^{\blacklozenge} \in \mathscr{F}[\partial, \textstyle\int]$ as in Corollary 29.
2. Determine the projector $P = \sum_{i=1}^{n} u_i \tilde{\beta}_i \in \mathscr{F}[\partial, \textstyle\int]$ as in (14).
3. Compute $G = (1 - P)T^{\blacklozenge} \in \mathscr{F}[\partial, \textstyle\int]$.

Theorem 31. *The above algorithm computes the Green's operator $G \in \mathscr{F}[\partial, \textstyle\int]$ for any regular boundary problem $(T, \mathscr{B})$.*

Proof. See [72]. □

The computation of *Green's operators* for boundary problems for ODEs using the above algorithm takes on the following concrete form in TH∃OREM∀ code.

$$\text{GreensOp}_{\mathrm{P}}[\mathrm{F}, \mathscr{B}] = \left(1 \underset{\mathfrak{A}}{-} \underset{\mathfrak{A}}{}\ \text{Proj}_{\mathrm{P}}[\mathscr{B}, \mathrm{F}]\right) \underset{\mathfrak{A}}{\star} \text{RightInv}_{\mathrm{P}}[\mathrm{F}]$$

Here $\mathscr{B}$ is a basis for the boundary space and F a regular fundamental system.

Let us consider again example (15): Given $f \in \mathscr{F}$, find $u \in \mathscr{F}$ such that

$$\boxed{\begin{aligned} &u'' = f, \\ &u(0) = u(1) = 0. \end{aligned}}$$

The Green's operator G of the boundary problem can be obtained by our implementation via the following computation

```
G = GreensOp[D², ⟨L, R⟩]

-A x - x B + x A x + x B x
```

where we use the notation from before: $Au = \int_0^x u(\xi)\, d\xi$, $Lu = u(0)$, $Ru = u(1)$ and in addition, $Bu = \int_x^1 u(\xi)\, d\xi$. The corresponding Green's function is computed in an immediate postprocessing step:

```
GreensFct[G]

g[x, ξ] = { -ξ + x ξ  ⇐  ξ ≤ x
          { -x + x ξ  ⇐  x < ξ
```

As noted in [70], the Green's function provides a canonical form for the Green's operator. Moreover, one can obtain the function $u(x)$ and thus solve the boundary problem through knowledge of the Green's function in the following identity:

$$u(x) = Gf(x) = \int_0^1 g(x,\xi) f(\xi)\, d\xi.$$

By replacing the Green's function obtained above in the latter integral we obtain

$$u(x) = (x-1)\int_0^x \xi f(\xi)\, d\xi + x\int_x^1 (\xi - 1) f(\xi)\, d\xi.$$

Furthermore, we can look at some specific instances of the forcing function $f(x)$. Let us first consider the simple example $f(x) = x$. By an immediate calculation, we obtain the expression for the action of G on $f(x)$, which is $u(x)$:

GreensOpAct[G, x]

$$\frac{x}{6} + \frac{2x^4}{3} + x^3\left(-\frac{5}{6}\right)$$

The expression for the solution function $u(x)$ can easily become more complicated, as we can see in the next example, where we consider the instance

$$f(x) = e^{2x} + 3x^2 \sin x^3.$$

Relying on Mathematica for handling symbolic integration, we obtain:

GreensOpAct[G, e^(2x) + 3x^2 Sin[x]^3]

$$\frac{-1}{4} + \frac{e^{2x}}{4} + \frac{x}{4} + \frac{e^{2}x}{4} + \left(\frac{-3}{2}\right) e^{2x}x + e^{2x}x^2 + (-9)\,x\mathrm{Cos}[1] + \frac{1}{9}x\mathrm{Cos}[3] + (-18)\,x\mathrm{Cos}[x] + 27x^2\mathrm{Cos}[x]$$
$$+ \frac{9}{2}x^3\mathrm{Cos}[x] + \left(\frac{-9}{2}\right)x^4\mathrm{Cos}[x] + \frac{2}{9}x\mathrm{Cos}[3x] + \left(\frac{-1}{3}\right)x^2\mathrm{Cos}[3x] + \left(\frac{-1}{2}\right)x^3\mathrm{Cos}[3x]$$
$$+ \frac{1}{2}x^4\mathrm{Cos}[3x] + \frac{45}{4}x\mathrm{Sin}[1] + \frac{1}{36}x\mathrm{Sin}\ 3 + \frac{27\mathrm{Sin}[x]}{2} + (-27)\,x\mathrm{Sin}[x] + \left(\frac{-45}{4}\right)x^2\mathrm{Sin}[x]$$
$$+ \frac{27}{2}x^3\mathrm{Sin}[x] + \left(\frac{-1}{18}\right)\mathrm{Sin}[3x] + \frac{1}{9}x\mathrm{Sin}[3x] + \frac{5}{12}x^2\mathrm{Sin}[3x] + \left(\frac{-1}{2}\right)x^3\mathrm{Sin}[3x]$$

As a last example, let us consider $f(x) = \sin\sin x$. As we can notice below, it cannot be integrated with Mathematica:

GreensOpAct[G, Sin[Sin[x]]]

$$\int_1^x (-x + x\xi)\,\mathrm{Sin}[\mathrm{Sin}[\xi]]\, d\xi + \int_0^x (-\xi + x\xi)\,\mathrm{Sin}[\mathrm{Sin}[\xi]]\, d\xi$$

In order to carry out the integrals involved in the application of the Green's operator to a forcing function, one can use any numerical quadrature method (as also available in many computer algebra systems).

5.2 *Composing and Factoring Boundary Problems*

In the following, we discuss the *composition* of boundary problems corresponding to their Green's operators. We also describe how *factorizations* of a boundary problem along a given factorization of the defining operator can be characterized and constructed. We refer again to [66, 72] for further details. We assume that all operators are defined on suitable spaces such that the composition is well-defined. It is worth mentioning that the following approach works in an *abstract setting*, which includes in particular boundary problems for linear partial differential equations (LPDEs) and systems thereof; for simplicity, we will restrict ourselves in the examples to the LODE setting.

Definition 32. We define the *composition* of boundary problems $(T_1, \mathscr{B}_1)$ and $(T_2, \mathscr{B}_2)$ by

$$(T_1, \mathscr{B}_1) \circ (T_2, \mathscr{B}_2) = (T_1 T_2, \mathscr{B}_1 \cdot T_2 + \mathscr{B}_2).$$

So the boundary conditions from the first boundary problem are "translated" by the operator from the second problem. The composition of boundary problems is associative but in general not commutative. The next proposition tells us that the composition of boundary problems preserves *regularity*.

Proposition 33. *Let $(T_1, \mathscr{B}_1)$ and $(T_2, \mathscr{B}_2)$ be regular boundary problems with Green's operators G_1 and G_2. Then $(T_1, \mathscr{B}_1) \circ (T_2, \mathscr{B}_2)$ is regular with Green's operator $G_2 G_1$ so that*

$$((T_1, \mathscr{B}_1) \circ (T_2, \mathscr{B}_2))^{-1} = (T_2, \mathscr{B}_2)^{-1} \circ (T_1, \mathscr{B}_1)^{-1}.$$

The simplest example of composing two boundary (more specifically, initial value) problems for ODEs is the following. Using the notation from before, one sees that

$$(D, [L]) \circ (D, [L]) = (D^2, [LD] + [L]) = (D^2, [L, LD]).$$

Let now $(T, \mathscr{B})$ be a boundary problem and assume that we have a factorization $T = T_1 T_2$ of the defining differential operator. We refer to [66, 72] for a characterization and construction of all factorizations

$$(T, \mathscr{B}) = (T_1, \mathscr{B}_1) \circ (T_2, \mathscr{B}_2)$$

into boundary problems. In particular, if $(T, \mathscr{B})$ is regular, it can be factored into regular boundary problems: the left factor $(T_1, \mathscr{B}_1)$ is *unique*, while for the right factor $(T_2, \mathscr{B}_2)$ we can choose any subspace $\mathscr{B}_2 \leq \mathscr{B}$ that makes the problem regular. We can compute the uniquely determined boundary conditions for the left factor by $\mathscr{B}_1 = \mathscr{B} \cdot G_2$, where G_2 is the Green's operator for some regular right factor $(T_2, \mathscr{B}_2)$. By factorization, one can split a problem of higher order into subproblems of *lower* order, given a factorization of the defining operator. For algorithms and results about factoring ordinary differential operators we refer to [64, 78, 84].

Given a fundamental system of the differential operator T and a right inverse of T_2, one can factor boundary problems in an algorithmic way as shown in [66] and in an integro-differential algebra [72]. As described in [74], we can also compute boundary conditions $\mathscr{B}_2 \leq \mathscr{B}$ such that $(T_2, \mathscr{B}_2)$ is a regular right factor, given only a fundamental system of T_2. The unique left factor can be then computed as explained above. This allows us to factor a boundary problem if we can factor the differential operator and compute a fundamental system of only one factor. The remaining lower order problems can then also be solved by numerical methods.

Here is how we can compute the boundary conditions of the left and right factor problems for the boundary problem $(D^2, [L, R])$ from previous example (15), along the trivial factorization with $T_1 = T_2 = D$. The indefinite integral $A = \int_0^x$ is the Green's operator for the regular right factor $(D, [L])$.

```
Fact[D, D, ⟨L, R⟩, ⟨R⟩]
⟨⟨A + B⟩, ⟨L⟩⟩
```

This factorization reads in traditional notation as

$$\boxed{\begin{array}{l} u' = f \\ \int_0^1 u(\xi)\, d\xi = 0 \end{array}} \circ \boxed{\begin{array}{l} u' = f \\ u(0) = 0 \end{array}} = \boxed{\begin{array}{l} u'' = f \\ u(0) = u(1) = 0 \end{array}}.$$

Note that the boundary condition for the unique left factor is an integral (Stieltjes) boundary condition.

We consider as a second example the fourth order boundary problem [72, Example 33]:

$$\boxed{\begin{array}{l} u'''' + 4u = f, \\ u(0) = u(1) = u'(0) = u'(1) = 0. \end{array}} \tag{18}$$

Factoring the boundary problem along $D^4 + 4 = (D^2 - 2i)(D^2 + 2i)$, we obtain the following boundary conditions for the factor problems.

```
Fact[D² - 2 i, D² + 2 i, ⟨L, R, L D, R D⟩, ⟨e^((-1+i) x), e^((1-i) x)⟩]
⟨⟨A e^((Complex[-1,1]) x) + B e^((Complex[-1,1]) x), A e^((Complex[1,-1]) x) + B e^((Complex[1,-1]) x)⟩, ⟨L, R⟩⟩
```

6 Integro-Differential Polynomials

In this section, we describe the algebra of *integro-differential polynomials* [73] obtained by adjoining an indeterminate function to a given integro-differential algebra $(\mathcal{F}, \partial, \int)$. Intuitively, these are all terms that one can build using an indeterminate u, coefficient functions $f \in \mathcal{F}$ and the operations $+, \cdot, \partial, \int$, identifying two terms if one can be derived from the other by the axioms of integro-differential algebras and the operations in $\mathcal{F}$. A typical term for $(K[x], \partial, \int)$ looks like this:

$$(4uu' \textstyle\int (x+3)u'^3)(u' \textstyle\int u''^2) + \textstyle\int x^6 uu''^5 \textstyle\int (x^2+5x) u^3 u'^2 \textstyle\int u$$

From the computational point of view, a fundamental problem is to find a canonical simplifier (see Sect. 2) on these objects. For example, the above term can be transformed to

$$4uu'^2 \textstyle\int xu'^3 \textstyle\int u''^2 + 4uu'^2 \textstyle\int u''^2 \textstyle\int xu'^3 + 12uu'^2 \textstyle\int u'^3 \textstyle\int u''^2 + 12uu'^2 \textstyle\int u''^2 \textstyle\int u'^3$$
$$+ \textstyle\int x^6 uu''^5 \textstyle\int x^2 u^3 u'^2 \textstyle\int u + 5 \textstyle\int x^6 uu''^5 \textstyle\int x u^3 u'^2 \textstyle\int u.$$

by the Baxter axiom and the K-linearity of the integral.

As outlined in the next subsection, a notion of polynomial can be constructed for any variety in the sense of *universal algebra*. (In this general sense, an algebra is a set with an arbitrary number of operations, and a variety is a collection of such algebras satisfying a fixed set of identities. Typical examples are the varieties of groups, rings, and lattices.)

For sample computations in the algebra of integro-differential polynomials, we use a *prototype implementation* of integro-differential polynomials, based on the TH∃OREM∀ functor mechanism (see Sect. 2).

6.1 Polynomials in Universal Algebra

In this subsection, we describe the idea of the general construction of polynomials in universal algebra [45]. We refer to [54] for a comprehensive treatment; see also the surveys [1, 27]. For the basic notions in universal algebra used below, see for example [4] or [32, Chap. 1].

Let $\mathcal{V}$ be a variety defined by a set $\mathcal{E}$ of identities over a signature Σ. Let A be a fixed "coefficient domain" from the variety $\mathcal{V}$, and let X be a set of indeterminates (also called "variables"). Then all terms in the signature Σ with constants (henceforth referred to as coefficients) in A and indeterminates in X represent the same polynomial if their equality can be derived in finitely many steps from the identities in $\mathcal{E}$ and the operations in A. The set of all such terms $\mathcal{T}_\Sigma(A \cup X)$

modulo this congruence $\equiv$ is an algebra in $\mathscr{V}$, called the *polynomial algebra* for $\mathscr{V}$ in X over A and denoted by $A_{\mathscr{V}}[X]$.

The polynomial algebra $A_{\mathscr{V}}[X]$ contains A as a subalgebra, and $A \cup X$ is a generating set. As in the case of polynomials for commutative rings, we have the *substitution homomorphism* in general polynomial algebras. Let B be an algebra in $\mathscr{V}$. Then given a homomorphism $\varphi_1 \colon A \to B$ and a map $\varphi_2 \colon X \to B$, there exists a unique homomorphism

$$\varphi \colon A_{\mathscr{V}}[X] \to B$$

such that $\varphi(a) = \varphi_1(a)$ for all $a \in A$ and $\varphi(x) = \varphi_2(x)$ for all $x \in X$. So in a categorical sense the polynomial algebra $A_{\mathscr{V}}[X]$ is a free product of the coefficient algebra A and the free algebra over X in $\mathscr{V}$; see also [1].

For computing with polynomials, we will construct a *canonical simplifier* on $A_{\mathscr{V}}[X]$ with associated system of canonical forms $\mathscr{R}$. As explained before (Sect. 2), the canonical simplifier provides for every polynomial in $A_{\mathscr{V}}[X]$, represented by some term T, a canonical form $R \in \mathscr{R}$ that represents the same polynomial, with different terms in $\mathscr{R}$ representing different polynomials; see also [54, p. 23].

The set $\mathscr{R}$ must be large enough to generate all of $A_{\mathscr{V}}[X]$ but small enough to ensure *unique representatives*. The latter requirement can be ensured by endowing a given set of terms with the structure of an algebra in the underlying variety.

Proposition 34. *Let $\mathscr{V}$ be a variety over a signature Σ, let A be an algebra in $\mathscr{V}$ and X a set of indeterminates. If $\mathscr{R} \subseteq \mathscr{T}_{\Sigma}(A \cup X)$ is a set of terms with $A \cup X \subseteq \mathscr{R}$ that can be endowed with the structure of an algebra in $\mathscr{V}$, then different terms in $\mathscr{R}$ represent different polynomials in $A_{\mathscr{V}}[X]$.*

Proof. Since $\mathscr{R}$ can be endowed with the structure of an algebra in the variety $\mathscr{V}$ and $A \cup X \subseteq \mathscr{R}$, there exists a unique substitution homomorphism

$$\varphi \colon A_{\mathscr{V}}[X] \to \mathscr{R}$$

such that $\varphi(a) = a$ for all $a \in A$ and $\varphi(x) = x$ for all $x \in X$. Let

$$\pi \colon \mathscr{R} \to A_{\mathscr{V}}[X]$$

denote the restriction of the canonical map associated with $\equiv$. Then we have $\varphi \circ \pi(R) = R$ for all $R \in \mathscr{R}$, so π is injective, and different terms in $\mathscr{R}$ indeed represent different polynomials. □

As a well-known example, take the *polynomial ring* $R[x]$ in one indeterminate x over a commutative ring R, which is $A_{\mathscr{V}}[X]$ for $A = R$ and $X = \{x\}$ with $\mathscr{V}$ being the variety of commutative unital rings. The set of all terms of the form $a_n x^n + \cdots + a_0$ with coefficients $a_i \in R$ and $a_n \neq 0$ together with 0 is a system of canonical forms for $R[x]$. One usually defines the polynomial ring directly in terms of these canonical forms. Polynomials for groups, bounded lattices and Boolean algebras are discussed in [54] along with systems of canonical forms.

6.2 Differential Polynomials

For illustrating the general construction described above, consider the algebra of *differential polynomials* over a commutative differential K-algebra $(\mathscr{F}, \partial)$ in one indeterminate function u, usually denoted by $\mathscr{F}\{u\}$. Clearly this is $A_{\mathscr{V}}[X]$ for $A = \mathscr{F}$ and $X = \{u\}$ with $\mathscr{V}$ being the variety of differential K-algebras. Terms are thus built up with the indeterminate u, coefficients from $\mathscr{F}$ and the operations $+, \cdot, \partial$; a typical example being

$$\partial^2(f_1 u^2 + u)\partial(f_2 u^3) + \partial^3(f_3 u).$$

By applying the Leibniz rule and the linearity of the derivation, it is clear that every polynomial is congruent to a K-linear combination of terms of the form

$$f \prod_{i=0}^{\infty} u_i^{\beta_i}, \tag{19}$$

where $f \in \mathscr{F}$, the notation u_n is short for $\partial^n(u)$, and only finitely many $\beta_i \in \mathbb{N}$ are nonzero. In the following, we use the multi-index notation $f u^\beta$ for terms of this form. For instance, $u^{(1,0,3,2)}$ is the multi-index notation for $u(u'')^3(u''')^2$. The order of a differential monomial u^β is given by the highest derivative appearing in u^β or $-\infty$ if $\beta = 0$.

Writing $\mathscr{R}$ for the set of all K-linear combinations of terms of the form (19), we already know that every polynomial is congruent to a term in $\mathscr{R}$. When $\mathscr{F} = K[x]$, a typical element of $\mathscr{R}$ is given by

$$(3x^3 + 5x)\, u^{(1,0,3,2)} + 7x^5 u^{(2,0,1)} + 2xu^{(1,1)}.$$

To show that $\mathscr{R}$ is a system of canonical forms for $\mathscr{F}\{u\}$, by Proposition 34 it suffices to endow $\mathscr{R}$ with the structure of a commutative differential algebra. As a commutative algebra, $\mathscr{R}$ is just the polynomial algebra in the infinite set of indeterminates $u_0, u_1, u_2, \ldots$. For defining a derivation in a commutative algebra, by the Leibniz rule and K-linearity, it suffices to specify it on the generators. Thus $\mathscr{R}$ becomes a differential algebra by setting $\partial(u_k) = u_{k+1}$. One usually defines the differential polynomials directly in terms of these canonical forms, see for example [48].

6.3 Integro Polynomials

We outline the *integro polynomials* over a Rota-Baxter algebra as in Definition 8. This is related to the construction of free objects in general Rota-Baxter algebras; we refer to [41] for details and references. By iterating the Baxter axiom (8), one obtains a generalization that is called the *shuffle identity* on $\mathcal{F}$:

$$(\textstyle\int f_1 \int \cdots \int f_m) \cdot (\int g_1 \int \cdots \int g_n) = \sum \int h_1 \int \cdots \int h_{m+n} \tag{20}$$

Here the sum ranges over all shuffles of $(f_1, \ldots, f_m)$ and $(g_1, \ldots, g_n)$; see [65, 68, 76] for details. The sum consists of $\binom{m+n}{n}$ shuffles, obtained by "shuffling" together $\int f_1 \int \cdots \int f_m$ and $\int g_1 \int \cdots \int g_n$ as words over the letters $\int f_i$ and $\int g_j$, such that the inner order in the words is preserved. For instance, we have

$$(\textstyle\int f_1 \int f_2) \cdot (\int g_1) = \int f_1 \int f_2 \int g_1 + \int f_1 \int g_1 \int f_2 + \int g_1 \int f_1 \int f_2.$$

for the simple $m = 2, n = 1$ case.

The integro polynomials over $\mathcal{F}$ are defined as $A_{\mathcal{V}}[X]$ for $A = \mathcal{F}$ and $X = \{u\}$ with $\mathcal{V}$ being the variety of Rota-Baxter algebras over K. The full construction of the canonical forms for integro polynomials is included in the following subsection. But it is clear that by *expanding products of integrals* by the shuffle identity, every integro polynomial is congruent to a K-linear combination of terms of the form

$$f u^k \textstyle\int f_1 u^{k_1} \int \cdots \int f_m u^{k_m} \tag{21}$$

with $f, f_1, \ldots, f_m \in \mathcal{F}$ and $k, k_1, \ldots, k_m \in \mathbb{N}$. However, they cannot be canonical forms, since terms like $\int (f + g)u$ and $\int fu + \int gu$ or $\int \lambda f u$ and $\lambda \int f u$ represent the same polynomials.

Writing $\mathcal{R}$ for the set of all K-linear combinations of terms of the form (21), the multiplication of two elements of $\mathcal{R}$ can now be defined via (20) as follows. Since the product of (21) with another term $gu^l \int g_1 u^{l_1} \int \cdots \int g_n u^{l_n}$ should clearly be given by $fg\, u^{k+l} (\int f_1 u^{k_1} \int \cdots \int f_m u^{k_m}) (\int gu^l g_1 u^{l_1} \int \cdots f_n u^{l_n})$, it remains to define the so-called *shuffle product* on integral terms (those having the form (21) with $f = 1$ and $k = 0$). This can be achieved immediately by using (20) with $f_i u^{k_i}$ and $g_j u^{l_j}$ in place of f_i and g_j, respectively. It is easy to see that the shuffle product is commutative and distributive with respect to addition.

The shuffle product can also be defined *recursively* [68]. Let J and $\tilde{J}$ range over integral terms (note that 1 is included as the special case of zero nested integrals). Then we have

$$(\textstyle\int f u^k J) \cdot (\int \tilde{f} u^{\tilde{k}} \tilde{J}) = (\int f u^k) \sqcup J \cdot (\int \tilde{f} u^{\tilde{k}} \tilde{J}) + (\int \tilde{f} u^{\tilde{k}}) \sqcup (\int f u^k J) \cdot \tilde{J}, \tag{22}$$

where $\sqcup \colon \mathcal{R} \times \mathcal{R} \to \mathcal{R}$ denotes the operation of nesting integrals (with the understanding that $\cdot$ binds stronger than $\sqcup$), defined on basis vectors by

$$\textstyle\int F_1 \int \cdots \int F_m \sqcup \int G_1 \int \cdots \int G_n = \int F_1 \int \cdots \int F_m \int G_1 \int \cdots \int G_n, \tag{23}$$

and extended bilinearly to all of $\mathcal{R}$. Here F_i and G_j stand for $f_i u^{k_i}$ and $g_j u^{l_j}$, respectively. For example, $\int F_1 \int F_2$ and $\int G_1$ can be multiplied as

$$(\textstyle\int F_1) \sqcup (\int F_2) \cdot (\int G_1) + (\int G_1) \sqcup 1 \cdot (\int F_1 \int F_2) = (\int F_1) \sqcup (\int F_2 \int G_1 + \int G_1 \int F_2)$$
$$+ (\textstyle\int G_1) \sqcup (\int F_1 \int F_2) = \int F_1 \int F_2 \int G_1 + \int F_1 \int G_1 \int F_2 + \int G_1 \int F_1 \int F_2,$$

analogous to the previous computation.

6.4 *Representing Integro-Differential Polynomials*

In the following, we describe in detail the universal algebra construction of the integro-differential polynomials and their canonical forms. We refer to [39, 40] for the related problem of constructing free objects in differential Rota-Baxter algebras. We consider the variety of integro-differential algebras. Its *signature* Σ contains: the ring operations, the derivation ∂, the integral $\int$, the family of unary "scalar multiplications" $(\cdot_\lambda)_{\lambda \in K}$, and for convenience we also include the evaluation $\mathtt{E}$. The *identities* $\mathscr{E}$ are those of a K-algebra, then K-linearity of the three operators ∂, $\int$, $\mathtt{E}$, the Leibniz rule (2), the section axiom (3), the Definition 2 of the evaluation, and the differential Baxter axiom (6).

Definition 35. Let $(\mathscr{F}, \partial, \int)$ be an integro-differential algebra. Then the algebra of *integro-differential polynomials* in u over $\mathscr{F}$, denoted by $\mathscr{F}\{u\}$ in analogy to the differential polynomials, is the polynomial algebra $A_{\mathscr{V}}[X]$ for $A = \mathscr{F}$ and $X = \{u\}$ with $\mathscr{V}$ being the variety of integro-differential algebras over K.

Some identities following from $\mathscr{E}$ describe *basic interactions* between operations in $\mathscr{F}$: the pure Baxter axiom (8), multiplicativity of the evaluation (5), the identities

$$\mathtt{E}^2 = \mathtt{E}, \quad \partial\mathtt{E} = 0, \quad \mathtt{E}\textstyle\int = 0, \quad \int(\mathtt{E}f)g = (\mathtt{E}f)\int g, \tag{24}$$

and the variant (7) of the differential Baxter axiom connecting all three operations.

We need to introduce some *notational conventions*. We use f, g for coefficients in $\mathscr{F}$, and V for terms in $\mathscr{T}_\Sigma(\mathscr{F} \cup \{u\})$. As for differential polynomials, we write u_n for the nth derivative of u. Moreover, we write

$$V(0) \text{ for } \mathtt{E}(V) \quad \text{and} \quad u(0)^\alpha \text{ for } \prod_{i=0}^{\infty} u_i(0)^{\alpha_i},$$

where α is a multi-index.

As a first step towards canonical forms, we describe below a system of terms that is *sufficient for representing* every integro-differential polynomial (albeit not uniquely as we shall see presently).

Lemma 36. *Every polynomial in $\mathscr{F}\{u\}$ can be represented by a finite K-linear combination of terms of the form*

$$f\,u(0)^{\alpha}u^{\beta}\textstyle\int f_1 u^{\gamma_1}\int\cdots\int f_n u^{\gamma_n}, \tag{25}$$

where $f, f_1, \ldots, f_n \in \mathscr{F}$, and each multi-index as well as n may be zero.

Proof. The proof is done by induction on the structure of terms, using the above identities (8), (5), (20) and (24) of integro-differential algebras. □

With the aid of the previous lemma we can determine the constants of $\mathscr{F}\{u\}$.

Proposition 37. *Every constant in $\mathscr{F}\{u\}$ is represented as a finite sum $\sum_{\alpha} c_{\alpha} u(0)^{\alpha}$ with constants c_{α} in $\mathscr{F}$.*

Proof. By the identity $\int\partial = 1 - \mathrm{E}$, a term V represents a constant in $\mathscr{F}\{u\}$ if and only if $\mathrm{E}(V) \equiv V$. Since V is congruent to a finite sum of terms of the form (25) and since $\mathrm{Im}(\mathrm{E}) = \mathscr{C}$, the identities for E imply that V is congruent to a finite sum of terms of the form $c_{\alpha}u(0)^{\alpha}$. □

The above representation (25) of the integro-differential polynomials is *not unique* since for example when trying to integrate differential polynomials by using integration by parts, terms like

$$\textstyle\int f u' \quad \text{and} \quad f u - \int f' u - f(0)\, u(0)$$

are equivalent. It becomes even more tedious to decide that, for instance,

$$2x\, u(0)^{(3,1)} u^{(1,3,0,4)} \textstyle\int (2x^3 + 3x)\, u^{(1,2,3)} \int (x+2)\, u^{(2)}$$

and

$$4x\, u(0)^{(3,1)} u^{(1,3,0,4)} \textstyle\int x^3 u^{(1,2,3)} \int x\, u^{(2)} + 6x\, u(0)^{(3,1)} u^{(1,3,0,4)} \int x\, u^{(1,2,3)} \int (x+2)\, u^{(2)}$$
$$+12x\, u(0)^{(3,1)} u^{(1,3,0,4)} \textstyle\int x\, u^{(1,2,3)} \int u^{(2)}$$

represent the same polynomial. In general, the following identity holds:

Lemma 38. *We have*

$$\textstyle\int V u_k^{\beta_k} u_{k+1} \equiv \dfrac{1}{\beta_k + 1}\left(V u_k^{\beta_k+1} - \int V' u_k^{\beta_k+1} - V(0)\, u_k(0)^{\beta_k+1}\right) \tag{26}$$

where $k, \beta_k \geq 0$.

Proof. Using (7) and the Leibniz rule, the left-hand side becomes

$$\textstyle\int (V u_k^{\beta_k})(u_k)' \equiv V u_k^{\beta_k+1} - \int V' u_k^{\beta_k+1} - \beta_k \int V u_k^{\beta_k} u_{k+1} - V(0)\, u_k(0)^{\beta_k+1},$$

and the equation follows by collecting the $\int V u_k^{\beta_k} u_{k+1}$ terms. □

The important point to note here is that if the highest derivative in the differential monomial u^β of order $k+1$ appears *linearly*, then the term $\int f u^\beta$ is congruent to a sum of terms involving differential monomials of order at most k. This observation leads us to the following classification of monomials; confer also [10, 35].

Definition 39. A differential monomial u^β is called *quasiconstant* if $\beta = 0$, *quasilinear* if $\beta \neq 0$ and the highest derivative appears linearly; otherwise it is called *functional*. An integro-differential monomial (25) is classified according to its outer differential monomial u^β, and its order is defined to be that of u^β.

Proposition 40. *Every polynomial in $\mathcal{F}\{u\}$ can be represented by a K-linear combination of terms of the form*

$$f\,u(0)^\alpha u^\beta \textstyle\int f_1 u^{\gamma_1} \int \cdots \int f_n u^{\gamma_n}, \tag{27}$$

where $f, f_1, \ldots, f_n \in \mathcal{F}$, the multi-indices α, β as well as n may be zero and the $u^{\gamma_1}, \ldots, u^{\gamma_n}$ are functional.

Proof. By Lemma 36 we can represent every polynomial in $\mathcal{F}\{u\}$ as a K-linear combination of terms of the form

$$f\,u(0)^\alpha u^\beta \textstyle\int f_1 u^{\gamma_1} \int \cdots \int f_n u^{\gamma_n}, \tag{28}$$

where the multi-indices and n can also be zero. Let us first prove by induction on depth that every term can be written as in (28) but with nonzero multi-indices γ_k. The base case $n = 1$ is trivial since $\int f_1$ can be pulled to the front. For the induction step we proceed from right to left, using the identity

$$\textstyle\int f \int V \equiv \int f \cdot \int V - \int V \int f$$

implied by the pure Baxter axiom (8).

For proving that every multi-index γ_k in (28) can be made functional, we use noetherian induction with respect to the preorder on $J = \int f_1 u^{\gamma_1} \int \cdots \int f_n u^{\gamma_n}$ that first compares depth and then the order of u^{γ_1}. One readily checks that the left-hand side of (26) is greater than the right-hand side with respect to this preorder, provided that V is of this form.

Applying Lemma 38 inductively, a term $\int f_1 u^{\gamma_1}$ is transformed to a sum of terms involving only integral terms with functional differential monomials, and the base case $n = 1$ follows. As induction hypothesis, we assume that all terms that are smaller than $J = f\,u(0)^\alpha u^\beta \int f_1 u^{\gamma_1} \int \cdots \int f_n u^{\gamma_n}$ can be written as a sum of terms involving only functional monomials. Since $\int f_2 u^{\gamma_2} \int \cdots \int f_n u^{\gamma_n}$ is smaller than J, it can be written as sum of terms involving only functional monomials; we may thus assume that $u^{\gamma_2}, \ldots, u^{\gamma_n}$ are all functional. Since γ_1 is nonzero, we are left with the case when u^{γ_1} is quasilinear. Applying again Lemma 38 inductively, we can replace u^{γ_1} in J by a sum of terms involving only integral terms with functional differential monomials. The induction step follows then by the linearity of $\int$. □

For *implementing* the integro-differential polynomials in TH∃OREM∀ we use the functor hierarchy described in Sect. 2. The multi-index representation u^β for terms of the form (19) is realized by the monoid $\mathbb{N}^*$ of natural tuples with finitely many nonzero entries, generated by a functor named TuplesMonoid. The nested integrals $\int f_1 u^{\gamma_1} \int \cdots \int f^n u^{\gamma_n}$ are represented as lists of pairs of the form $\langle f_k, \gamma_k \rangle$, with $f_k \in \mathcal{F}$ and $\gamma_k \in \mathbb{N}^*$. The terms of the form (25) are then constructed via a cartesian product of monoids as follows:

```
Definition["Term Monoid for IDP", any[𝓕, N],
 TermMonoid[𝓕, N] = TuplesMonoid[N] × TuplesMonoid[N] × TuplesMonoid[𝓕 × TuplesMonoid[N]]]
```

Using this construction, the integro-differential polynomials are built up by the functor FreeModule[$\mathcal{F}$,B] that constructs the $\mathcal{F}$-module with basis B. It is instantiated with $\mathcal{F}$ being a given integro-differential algebra and B the term monoid just described. We will equip this domain with the operations defined as below, using a functor named IntDiffPol[$\mathcal{F}$,K]. Later in this section we will present some sample computations.

6.5 *Canonical Forms for Integro-Differential Polynomials*

It is clear that K-linear combinations of terms of the form (27) are still not canonical forms for the integro-differential polynomials since by the linearity of the integral, terms like

$$f \textstyle\int (g+h)u \quad \text{and} \quad f \int gu + f \int hu$$

or terms like

$$f \textstyle\int \lambda gu \quad \text{and} \quad \lambda f \int gu$$

with $f, g, h \in \mathcal{F}$ and $\lambda \in K$ represent the same polynomial. To solve this problem, we can consider terms of the form (27) modulo these identities coming from *linearity* in the "coefficient" f and the integral, in analogy to the ideal $\mathfrak{l}$ introduced in Sect. 4 for $\mathcal{F}^\#[\partial, \int]$. Confer also [39], where the tensor product is employed for constructing free objects in differential Rota-Baxter algebras. In the following, we assume for simplicity that $\mathcal{F}$ is an ordinary integro-differential algebra.

More precisely, let $\mathcal{R}$ denote the set of terms of the form (27) and consider the free K-vector space generated by $\mathcal{R}$. We identify terms

$$f u(0)^\alpha u^\beta \textstyle\int f_1 u^{\gamma_1} \int \cdots \int f_n u^{\gamma_n}$$

with the corresponding basis elements in this vector space. Then we factor out the subspace generated by the following identities (analogous to the construction of the tensor product):

$$
\begin{aligned}
&f U \textstyle\int f_1 u^{\gamma_1} \int \cdots \int (f_k + \tilde{f}_k) u^{\gamma_k} \int \cdots \int f_n u^{\gamma_n} \\
&\quad = f U \textstyle\int f_1 u^{\gamma_1} \int \cdots \int f_k u^{\gamma_k} \int \cdots \int f_n u^{\gamma_n} + f U \int f_1 u^{\gamma_1} \int \cdots \int \tilde{f}_k u^{\gamma_k} \int \cdots \int f_n u^{\gamma_n} \\
&f U \textstyle\int f_1 u^{\gamma_1} \int \cdots \int (\lambda f_k) u^{\gamma_k} \int \cdots \int f_n u^{\gamma_n} = \lambda f U \int f_1 u^{\gamma_1} \int \cdots \int f_k u^{\gamma_k} \int \cdots \int f_n u^{\gamma_n}
\end{aligned}
$$

Here U is short for $u(0)^\alpha u^\beta$, and there are actually two more identities of the same type for ensuring K-linearity in f. We write $[\mathscr{R}]$ for this quotient space and denote the corresponding equivalence classes by

$$
[f u(0)^\alpha u^\beta \textstyle\int f_1 u^{\gamma_1} \int \cdots \int f_n u^{\gamma_n}]. \tag{29}
$$

By construction, the quotient module $[\mathscr{R}]$ now respects the *linearity relations*

$$
\begin{aligned}
&[f U \textstyle\int f_1 u^{\gamma_1} \int \cdots \int (f_k + \tilde{f}_k) u^{\gamma_k} \int \cdots \int f_n u^{\gamma_n}] \\
&\quad = [f U \textstyle\int f_1 u^{\gamma_1} \int \cdots \int f_k \int \cdots \int f_n u^{\gamma_n}] + [f U \int f_1 u^{\gamma_1} \int \cdots \int \tilde{f}_k \int \cdots \int f_n u^{\gamma_n}] \\
&[f U \textstyle\int f_1 u^{\gamma_1} \int \cdots \int (\lambda f_k) u^{\gamma_k} \int \cdots \int f_n u^{\gamma_n}] = \lambda [f U \int f_1 u^{\gamma_1} \int \cdots \int f_k u^{\gamma_k} \int \cdots \int f_n u^{\gamma_n}].
\end{aligned}
$$

together with the ones for linearity in f.

As for the tensor product, we have canonical forms for the factor space by expanding the "coefficient" f and all the f_k in (29) with respect to a *K-basis* $\mathscr{B}$ for $\mathscr{F}$, assuming $\mathscr{B}$ contains 1. Then every polynomial can be written as a K-linear combination of terms of the form

$$
b u(0)^\alpha u^\beta \textstyle\int b_1 u^{\gamma_1} \int \cdots \int b_n u^{\gamma_n}, \tag{30}
$$

where $b, b_1, \ldots, b_n \in \mathscr{B}$ with the condition on multi-indices as in Proposition 40.

To show that terms of the form (30) are canonical forms for the integro-differential polynomials, we endow the quotient space $[\mathscr{R}]$ with an *integro-differential structure* and invoke Proposition 34. For this we define the operations on the generators (29) and check that they respect the above linearity relations on $[\mathscr{R}]$.

First, we define a *multiplication* on $[\mathscr{R}]$. Let $\mathscr{R}_0 \subseteq \mathscr{R}$ denote the K-subspace generated by integral terms $\int f_1 u^{\gamma_1} \int \cdots \int f_n u^{\gamma_n}$, including $1 \in \mathscr{R}$ as the case $n = 0$. Clearly, the nesting operation (23) can be defined in a completely analogous manner on such integral terms (the only difference being that we have now derivatives of the indeterminate). Since it is clearly K-linear, it induces an operation $\sqcup\colon [\mathscr{R}_0] \times [\mathscr{R}_0] \to [\mathscr{R}_0]$. The next step is to define the shuffle product on $\mathscr{R}_0$ just as in (22), again with obvious modifications. Passing to the quotient yields the shuffle product $\cdot\colon [\mathscr{R}_0] \times [\mathscr{R}_0] \to [\mathscr{R}_0]$. This product is finally extending to a multiplication on all of $[\mathscr{R}]$ by setting

$$
[f u(0)^\alpha u^\beta J][\tilde{f} u(0)^{\tilde{\alpha}} u^{\tilde{\beta}} \tilde{J}] = [f \tilde{f} u(0)^{\alpha+\tilde{\alpha}} u^{\beta+\tilde{\beta}} (J \cdot \tilde{J})]
$$

where J and $\tilde{J}$ range over $\mathscr{R}_0$. Let us compute an example:

$$\mathbf{MultIDP}[\mathbf{u}[0]^{\langle 1\rangle}\,\mathbf{u}^{\langle 2\rangle}\,\text{"}\!\int\!\text{"}\,(3\,\mathbf{x})\,\mathbf{u}^{\langle 1,1\rangle}\,\text{"}\!\int\!\text{"}\,\mathbf{x}^2\,\mathbf{u}^{\langle 0,2\rangle},\ 3\,\mathbf{u}[0]^{\langle 2,3\rangle}\,\mathbf{u}^{\langle 3,1\rangle}\,\text{"}\!\int\!\text{"}\,\mathbf{x}\,\mathbf{u}^{\langle 1,0,1\rangle}]$$

$$\begin{aligned}
&-18\,u[0]^{\langle 3,3\rangle}\,u^{\langle 5,1\rangle}\int x\,u^{\langle 1,1\rangle}\int u^{\langle 0,2\rangle}\int x^2\,u^{\langle 0,2\rangle} - 27\,u[0]^{\langle 3,3\rangle}\,u^{\langle 5,1\rangle}\int x\,u^{\langle 1,1\rangle}\int x^2\,u^{\langle 0,2\rangle}\int u^{\langle 0,2\rangle}\\
&\quad + 27\,x\,u[0]^{\langle 3,3\rangle}\,u^{\langle 6,2\rangle}\int x\,u^{\langle 1,1\rangle}\int x^2\,u^{\langle 0,2\rangle} - \frac{9}{2}\,x\,u[0]^{\langle 3,3\rangle}\,u^{\langle 7,1\rangle}\int u^{\langle 0,2\rangle}\int x^2\,u^{\langle 0,2\rangle}\\
&- 27\,u[0]^{\langle 4,4\rangle}\,u^{\langle 5,1\rangle}\int x\,u^{\langle 1,1\rangle}\int x^2\,u^{\langle 0,2\rangle} + \frac{3}{2}\,u[0]^{\langle 5,3\rangle}\,u^{\langle 5,1\rangle}\int u^{\langle 0,2\rangle}\int x^2\,u^{\langle 0,2\rangle}
\end{aligned}$$

Since the multiplication on $\mathscr{F}$ and the shuffle product are commutative, associative, and distributive over addition, the multiplication on $[\mathscr{R}]$ is well-defined and gives $[\mathscr{R}]$ the structure of a commutative K-algebra.

The definition of a *derivation* ∂ on this algebra is straightforward, using the fact that it should respect K-linearity and the Leibniz rule (treating also the $u(0)^\alpha$ as constants), that it should restrict to the derivation on differential polynomials (which in turn restricts to the derivation on $\mathscr{F}$), and finally that it should also satisfy the section axiom (3). Here is a sample computation:

$$\mathbf{DiffIDP}[\mathbf{u}[0]^{\langle 1\rangle}\,\mathbf{u}^{\langle 2,1\rangle}\,\text{"}\!\int\!\text{"}\,\mathbf{x}\,\mathbf{u}^{\langle 1,0,2\rangle}\,\text{"}\!\int\!\text{"}\,(3\,\mathbf{x}^2)\,\mathbf{u}^{\langle 0,2\rangle} + 3\,\mathbf{u}[0]^{\langle 2,3\rangle}\,\mathbf{u}^{\langle 3,2\rangle}\,\text{"}\!\int\!\text{"}\,(2\,\mathbf{x}^3 + 4\,\mathbf{x})\,\mathbf{u}^{\langle 2,1\rangle}]$$

$$\begin{aligned}
&6\,u[0]^{\langle 1\rangle}\,u^{\langle 1,2\rangle}\int x\,u^{\langle 1,0,2\rangle}\int x^2\,u^{\langle 0,2\rangle} + 3\,u[0]^{\langle 1\rangle}\,u^{\langle 2,0,1\rangle}\int x\,u^{\langle 1,0,2\rangle}\int x^2\,u^{\langle 0,2\rangle}\\
&\quad + 3\,x\,u[0]^{\langle 1\rangle}\,u^{\langle 3,1,2\rangle}\int x^2\,u^{\langle 0,2\rangle} + 18\,u[0]^{\langle 2,3\rangle}\,u^{\langle 2,3\rangle}\int x^3\,u^{\langle 2,1\rangle} + (24\,x + 6\,x^3)\,u[0]^{\langle 2,3\rangle}\,u^{\langle 5,3\rangle}\\
&\quad + 12\,u[0]^{\langle 2,3\rangle}\,u^{\langle 3,1,1\rangle}\int x^3\,u^{\langle 2,1\rangle} + 8\,x\,u[0]^{\langle 2,3\rangle}\,u^{\langle 6,1,1\rangle} - 3\,u[0]^{\langle 5,3\rangle}\,u^{\langle 2,3\rangle} + 2\,u[0]^{\langle 5,3\rangle}\,u^{\langle 3,1,1\rangle}
\end{aligned}$$

Using the K-linearity of this derivation, one verifies immediately that it is well-defined. From the definition it is clear that K-linear combinations of generators of the form $[u(0)^\alpha]$ are constants for ∂, and one can also check that all constants are actually of this form.

Finally, we define a K-linear *integral* on the differential K-algebra $([\mathscr{R}], \partial)$. Since we have to distinguish three different types of integrals, here and subsequently we will use the following notation: the usual big integral sign $\displaystyle\int$ for the integration to be defined, the small integral sign $\textstyle\int$ for the elements of $\mathscr{R}$ as we have used it before, and $\int_{\mathscr{F}}$ for the integral on $\mathscr{F}$.

The definition of the integral on $[\mathscr{R}]$ is recursive, first by depth and then by order of u^β, following the classification of monomials from Definition 39. In the *base case* of zero depth and order, we put

$$\int [f\,u(0)^\alpha] = [\textstyle\int_{\mathscr{F}} f][u(0)^\alpha]. \tag{31}$$

Turning to *quasiconstant* monomials, we use the following definition (which actually includes the base case when $J = 1$):

$$\int [f\,u(0)^\alpha J] = [u(0)^\alpha (\textstyle\int_{\mathscr{F}} f) J] - [u(0)^\alpha \textstyle\int (\int_{\mathscr{F}} f) J']. \tag{32}$$

In the *quasilinear* case we write the generators in form

$$[f\,u(0)^\alpha V u_k^{\beta_k} u_{k+1} J] \quad \text{with} \quad V = u_0^{\beta_0} \cdots u_{k-1}^{\beta_{k-1}}$$

and construct the integral via (26). Writing $s = \beta_k + 1$, we have $u_k^{\beta_k} u_{k+1} = (u_k^s)'/s$, so we can define

$$\int [f\,u(0)^\alpha V (u_k^s)' J] = [f\,u(0)^\alpha V u_k^s J] - [u(0)^\alpha] \int [f\,VJ]'[u_k^s] - f(0)\,[u(0)^{\alpha+\beta} \hat{J}], \tag{33}$$

where we write $f(0)$ for $\mathtt{E}(f)$ and $\hat{J}$ is 1 for $J = 1$ and zero otherwise. In the *functional* case, we set

$$\int [f\,u(0)^\alpha u^\beta J] = [u(0)^\alpha \textstyle\int f u^\beta J], \tag{34}$$

so here we can just let the integral sign slip into the equivalence class. One may check that the integral is well-defined in all the cases by an easy induction proof, using K-linearity of the integral, the evaluation on $\mathscr{F}$, and the derivation on $[\mathscr{R}]$.

Here is a small *example* of an integral computed in the quasiconstant case (note that `IntIDP` corresponds to the big integral and "$\int$" to $\int$ in our notation):

```
IntIDP[u[0]^(1) "∫" x u^(1,0,2) "∫" (x^2 + 2) u^(1,2)]
```

$$2\,x\,u[0]^{\langle 1\rangle} \int x\,u^{\langle 1,0,2\rangle} \int u^{\langle 1,2\rangle} + x\,u[0]^{\langle 1\rangle} \int x\,u^{\langle 1,0,2\rangle} \int x^2\,u^{\langle 1,2\rangle} - 2\,u[0]^{\langle 1\rangle} \int x^2\,u^{\langle 1,0,2\rangle} \int u^{\langle 1,2\rangle}$$
$$- u[0]^{\langle 1\rangle} \int x^2\,u^{\langle 1,0,2\rangle} \int x^2\,u^{\langle 1,2\rangle}$$

The next example computes an integral in the quasilinear case:

```
IntIDP[u[0]^(3,2) u^(2,1) "∫" x u^(1,0,2) "∫" x^2 u^(1,1)]
```

$$\frac{1}{6}\,u[0]^{\langle 5,2\rangle} \int x\,u^{\langle 4,0,2\rangle} - \frac{1}{6}\,x^2\,u[0]^{\langle 3,2\rangle}\,u^{\langle 2\rangle} \int x\,u^{\langle 4,0,2\rangle} + \frac{1}{6}\,x^2\,u[0]^{\langle 3,2\rangle}\,u^{\langle 5\rangle} \int x\,u^{\langle 1,0,2\rangle}$$
$$- \frac{1}{6}\,u[0]^{\langle 5,2\rangle}\,u^{\langle 3\rangle} \int x\,u^{\langle 1,0,2\rangle}$$

Note that all differential monomials within integrals are functional again, as it must be by our definition of $[\mathscr{R}]$.

By construction the integral defined above is a section of the derivation on $[\mathscr{R}]$. So for showing that $[\mathscr{R}]$ is an *integro-differential algebra* with operations, it remains only to prove the differential Baxter axiom (4). Equivalently, we can show that the evaluation

$$\mathtt{E} = 1 - \int \partial$$

is multiplicative by Corollary 4.

Recall that the algebra of constants $\mathscr{C}$ in $([\mathscr{R}], \partial)$ consists of K-linear combinations of generators of the form $[u(0)^\alpha]$. By a short induction proof, we see that

$$\int [u(0)^\alpha][R] = [u(0)^\alpha] \int [R]. \tag{35}$$

Hence the integral is homogeneous over the constants.

For showing that the *evaluation* is multiplicative, we first reassure ourselves that it operates in the expected way on integro-differential monomials.

Lemma 41. *We have*

$$\mathrm{E}\,[f\,u(0)^\alpha u^\beta J] = f(0)\,[u(0)^{\alpha+\beta}\hat{J}],$$

where $\hat{J}$ *is* 1 *for* $J = 1$ *and zero otherwise as in* (33).

Proof. Note that E is $\mathscr{C}$-linear by (35), so we can omit the factor $u(0)^\alpha$. Assume first $\beta = 0$. Then by the quasiconstant case (32) of the definition of the integral, we have

$$\mathrm{E}\,[fJ] = [fJ] - \int [fJ]' = [fJ] - [(\textstyle\int_{\mathscr{F}} f')J] + \displaystyle\int [(\textstyle\int_{\mathscr{F}} f')J'] - \displaystyle\int [fJ'],$$

which by $\int_{\mathscr{F}} f' = f - f(0)$ gives

$$f(0)\,[J] - f(0) \int [J]' = f(0)[\hat{J}]$$

because

$$\int [J]' = [J] \quad \text{for} \quad J \neq 1$$

by the functional case (34) and zero for $J = 1$. If $\beta \neq 0$ is of order k, we write $u^\beta = V u_k^s$ with $s \neq 0$, and we compute

$$\mathrm{E}\,[f\,u^\beta J] = [f\,V u_k^s J] - \int [f\,V u_k^s J]' = f(0)\,[u(0)^\beta \hat{J}]$$

by the quasilinear case (33) and the Leibniz rule. □

Theorem 42. *With the operations defined as above,* $([\mathscr{R}], \partial, \int)$ *has the structure of an integro-differential algebra.*

Proof. As mentioned above, it suffices to prove that E is multiplicative, and we need only do this on the generators. Again omitting the $u(0)^\alpha$, we have to check that

$$\mathrm{E}\,[f u^\beta J][\tilde{f} u^{\tilde{\beta}} \tilde{J}] = \mathrm{E}[f\,\tilde{f}\,u^{\beta+\tilde{\beta}}\,(J \cdot \tilde{J})] = \mathrm{E}[f u^\beta J] \cdot \mathrm{E}[\tilde{f} u^{\tilde{\beta}} \tilde{J}].$$

The case $J = \tilde{J} = 1$ follows directly from Lemma 41 and the multiplicativity of $\mathtt{E}$ in $\mathscr{F}$. Otherwise the shuffle product $J \cdot \tilde{J}$ is a sum of integral terms, each of them unequal one. Using again Lemma 41 and the linearity of $\mathtt{E}$, the evaluation of this sum vanishes, as does $\mathtt{E}[f u^{\beta} J] \cdot \mathtt{E}[\tilde{f} u^{\tilde{\beta}} \tilde{J}]$. □

Since $[\mathscr{R}]$ is an integro-differential algebra, we can conclude by Proposition 40 and Proposition 34 that $[\mathscr{R}]$ leads to canonical forms for integro-differential polynomials, up to the linearity relations: After a choice of basis, terms of the form (30) constitute a system of *canonical forms* for $\mathscr{F}\{u\}$. In the THEOREMA implementation, we actually compute in $[\mathscr{R}]$ and do basis expansions only for deciding equality.

7 From Rewriting to Parametrized Gröbner Bases

Equipped with the integro-differential polynomials, we can now tackle the task of proving the convergence of the reduction rules in Table 1. As explained in Sect. 4, we will use the *Diamond Lemma* (Theorem 22) for this purpose. First of all we must therefore construct a noetherian monoid ordering $>$ on $\mathscr{F}\langle\partial, \textstyle\int\rangle$ that is compatible with the reduction rules. In fact, there is a lot of freedom in defining such a $>$. It is sufficient to put $\partial > f$ for all $f \in \mathscr{F}$ and extend this to words by the graded lexicographic construction. The resulting partial ordering is clearly noetherian (since it is on the generators) and compatible with the monoid structure (by its grading). It is also compatible with the rewrite system because all rules reduce the word length except for the Leibniz rule, which is compatible because $\partial > f$.

Thus it remains to prove that all ambiguities of Table 1 are resolvable, and we have to compute the corresponding S-polynomials and reduce them to zero. On the face of it, there are of course *infinitely many* of these, suitably parametrized by $f, g \in \mathscr{F}$ and $\varphi, \psi \in \Phi$. For example, let us look at the minimal fork generated by $\textstyle\int u \textstyle\int v \textstyle\int$. In this case, the rule $\textstyle\int f \textstyle\int$ may be applied either with $f = u$ or with $f = v$ yielding the reductions

$$\begin{array}{ccc} & \textstyle\int u \textstyle\int v \textstyle\int & \\ \swarrow & & \searrow \\ (\textstyle\int \cdot u) \textstyle\int v \textstyle\int - \textstyle\int (\textstyle\int \cdot u) v \textstyle\int & & \textstyle\int u (\textstyle\int \cdot v) \textstyle\int - \textstyle\int u \textstyle\int (\textstyle\int \cdot v) \end{array}$$

with the S-polynomial $p = (\textstyle\int \cdot u) \textstyle\int v \textstyle\int - \textstyle\int (\textstyle\int \cdot u) v \textstyle\int - \textstyle\int u (\textstyle\int \cdot v) \textstyle\int + \textstyle\int u \textstyle\int (\textstyle\int \cdot v)$. But actually we should not call p *an* S-polynomial since it represents infinitely many: one for each choice of $u, v \in \mathscr{F}$.

How should one handle this infinitude of S-polynomials? The problem is that for reducing S-polynomials like p one needs not only the relations embodied in the reduction of Table 1 but also properties of the operations $\partial, \textstyle\int : \mathscr{F} \to \mathscr{F}$ acting on

$u, v \in \mathscr{F}$. Since these computations can soon become unwieldy, one should prefer a method that can be automated. There are two options that may be pursued:

- Either one retreats to the viewpoint of *rewriting*, thinking of Table 1 as a two-level rewrite system. On the upper level, it applies the nine parametrized rules with $f, g \in \mathscr{F}$ and $\varphi, \psi \in \Phi$ being arbitrary expressions. After each such step, however, there are additional reductions on the lower level for applying the properties of $\partial, \int : \mathscr{F} \to \mathscr{F}$ on these expressions. Using a custom-tailored reduction system for the lower level, this approach was used in the old implementation for generated an automated confluence proof [70].
- Or one views an S-polynomial like p nevertheless as a *single* element, not in $\mathscr{F}\langle\partial, \int\rangle$ but in $\hat{\mathscr{F}}\langle\partial, \int\rangle$ with $\hat{\mathscr{F}} = \mathscr{F}\{u, v\}$. With this approach, one remains within the paradigm of *parametrized Gröbner bases*, and the interlocked operation of the two levels of reduction is clarified in a very coherent way: The upper level is driven by the canonical simplifier on $\hat{\mathscr{F}}[\partial, \int]$, the lower level by that on $\mathscr{F}\{u, v\}$.

It is the second approach that we will explain in what follows.

Using $\hat{\mathscr{F}}\langle\partial, \int\rangle$ instead of $\mathscr{F}\langle\partial, \int\rangle$ takes care of the parameters $f, g \in \mathscr{F}$ but then there are also the *characters* $\varphi, \psi \in \Phi$. The proper solution to this problem would be to use a refined version of integro-differential polynomials that starts from a whole family $(\int_\varphi)_{\varphi\in\Phi}$ of integrals instead of the single integral $\int$, thus leading to a corresponding family of evaluations $u(\varphi)$ instead of the single evaluation $u(0)$. We plan to pursue this approach in a forthcoming paper. For our present purposes, we can take either of the following positions:

- The characters φ, ψ may range over an *infinite* set Φ, but they are harmless since unlike the $f, g \in \mathscr{F}$ they do not come with any operations (whose properties must be accounted for by an additional level of reduction). In this case, Table 1 is still an infinitely generated ideal in $\hat{\mathscr{F}}\langle\partial, \int\rangle$, and we have to reduce infinitely many S-polynomials. But the ambiguities involving characters are all of a very simple nature, and their reduction of their S-polynomials is straightforward.
- Alternatively, we may restrict ourselves to a *finite* set of characters (as in most applications!) so that Table 1 actually describes a finitely generated ideal in $\hat{\mathscr{F}}\langle\partial, \int\rangle$, and we need only consider finitely many S-polynomials.

The second alternative is somewhat inelegant due to the proliferation of instances for rules like $\varphi\psi \to \psi$. In our implementation, we have thus followed the first alternative with a straightforward treatment of parametrization in φ, ψ but we will ignore this issue in what follows.

We can now use the new TH∃OREM∀ implementation for checking that the nine rules in Table 1 form a Gröbner basis in $\hat{\mathscr{F}}\langle\partial, \int\rangle$. As explained before, we use the Diamond Lemma for this purpose (note that the noetherian monoid ordering $>$ applies also to $\hat{\mathscr{F}}\langle\partial, \int\rangle$ except that we have now just two generators $u, v \in \hat{\mathscr{F}} = \mathscr{F}\{u, v\}$ instead of all $f \in \mathscr{F}$). Hence it remains to check that all S-polynomials reduce to zero. We realize this by using the appropriate functor hierarchy, as follows.

We first build up the algebra of the integro-differential polynomials having, in turn, integro-differential polynomials as coefficients, via the functor

$$\texttt{IntDiffPolys[IntDiffPolys[}\mathscr{F}\texttt{, K], K]}$$

and we denote the resulting domain by $\mathbb{P}$. Then we consider an instance of the functor constructing the integro-differential operators over $\mathbb{P}$. Finally, the computations are carried out over the algebra created by the `GroebnerExtension` functor taking the latter instance as input domain, that allows to perform polynomial reduction, S-polynomials and the Gröbner basis procedure.

Of course, the S-polynomials are generated automatically, but as a concrete example we check the minimal fork considered above:

$$\mathbf{ReducePol}_{\mathbb{P}}[((\text{"}\int\text{"}\cdot u^{\langle 1\rangle})\ \text{"}\int\text{"}\cdot v^{\langle 1\rangle}\ \text{"}\int\text{"} - \text{"}\int\text{"}\ (\text{"}\int\text{"}\cdot u^{\langle 1\rangle})\ v^{\langle 1\rangle}\ \text{"}\int\text{"})$$
$$-(\text{"}\int\text{"}\,u^{\langle 1\rangle}\ (\text{"}\int\text{"}\cdot v^{\langle 1\rangle})\ \text{"}\int\text{"} - \text{"}\int\text{"}\,u^{\langle 1\rangle}\ \text{"}\int\text{"}\ (\text{"}\int\text{"}\cdot v^{\langle 1\rangle})))]$$

0

As it turns out, there are 17 nontrivial S-polynomials, and they all reduce to zero. This leads us finally to the desired convergence result for $\mathscr{F}[\partial, \int]$.

Theorem 43. *The system of Table 1 represents a noncommutative Gröbner basis in* $\mathscr{F}\langle\partial, \int\rangle$ *for any graded lexicographic ordering satisfying* $\partial > f$ *for all* $f \in \mathscr{F}$.

Proof. By the Diamond Lemma we must show that all S-polynomials $p \in \mathscr{F}\langle\partial, \int\rangle$ reduce to zero. Since they may contain at most two parameters $f, g \in \mathscr{F}$, let us write them as $p(f, g)$. But we have just observed that the corresponding S-polynomials $p(u, v) \in \hat{\mathscr{F}}\langle\partial, \int\rangle$ with $\hat{\mathscr{F}} = \mathscr{F}\{u, v\}$ reduce to zero. Using the substitution homomorphism

$$\varphi\colon \hat{\mathscr{F}} \to \mathscr{F},\ (u, v) \mapsto (f, g),$$

lifted to $\hat{\mathscr{F}}[\partial, \int] \to \mathscr{F}[\partial, \int]$ in the obvious way, we see that $p(f, g) = \varphi\, p(u, v)$ reduces to zero as well. □

From the conclusion of the Diamond Lemma, we can now infer that Table 1 indeed establishes a canonical simplifier for $\mathscr{F}[\partial, \int]$.

8 Conclusion

The *algebraic treatment of boundary problems* is a new development in Symbolic Analysis that takes its starting point in differential algebra and enriches its structures by introducing an explicit notion of integration and boundary evaluations. Recall the three basic tools that we have introduced for this purpose:

- The category of *integro-differential algebras* $(\mathcal{F}, \partial, \int)$ for providing a suitable notion of "functions". (As explained in Sect. 2, here we do not think of categories and functors in the sense of Eilenberg and Maclane – this is also possible and highly interesting but must be deferred to another paper.)
- The functor creating the associated *integro-differential operators* $\mathcal{F}[\partial, \int]$ as a convenient language for expressing boundary problems (differential operators, boundary operators) and their solving Green's operators (integral operators).
- The functor creating the associated *integro-differential polynomials* $\mathcal{F}\{u\}$, which describe the extension of an integro-differential algebra by a generic function u.

In each of these three cases, the differential algebra counterpart (i.e. without the "integro-") is well-known, and it appears as a rather simple substructure in the full structure. For example, the differential polynomials $\mathcal{F}\{u\}$ over a differential algebra $(\mathcal{F}, \partial)$ are simple to construct since the Leibniz rule effectively flattens out compound terms. This is in stark contrast to an integro-differential algebra $(\mathcal{F}, \partial, \int)$, where the Baxter rule forces the presence of nested integrals for ensuring closure under integration.

The interplay between these three basic tools is illustrated in a *new confluence proof*: For an arbitrary integro-differential algebra $(\mathcal{F}, \partial, \int)$, the rewrite system for the integro-differential operators $\mathcal{F}[\partial, \int]$ is shown to be a noncommutative Gröbner basis by the aid of the integro-differential polynomials $\mathcal{F}\{u, v\}$. Having a confluent rewrite system leads to a canonical simplifier, which is crucial for the algorithmic treatment as expounded in Sect. 2.

Regarding our overall mission – the algebraic treatment of boundary problems and integral operators – we have only scratched the surface, and much is left to be done. We have given a brief overview of solving, multiplying and factoring boundary problems in Sect. 5. But the real challenge lies ahead, namely how to *extend our framework* to:

- *Linear Boundary Problems for LPDEs*: As mentioned at the start of Sect. 5, the algebraic framework for multiplying and factoring boundary problems is set up to allow for LPDEs; see [66] for more details. But the problematic issue is how to design a suitable analog of $\mathcal{F}[\partial, \int]$ for describing integral and boundary operators (again the differential operators are obvious). This involves more than replacing ∂ by $\partial/\partial_x, \partial/\partial_y$ and $\int$ by $\int_0^x, \int_0^y$ because even the simplest Green's operators employ one additional feature: the transformation of variables, along with the corresponding interaction rules for differentiation (chain rule) and integration (substitution rule); see [74] for some first steps in this direction.
- *Nonlinear Boundary Problems:* A radically new approach is needed for that, so it seems appropriate to concentrate first on boundary problems for nonlinear ODEs and systems thereof. A natural starting point for such an investigation is the differential algebra setting, i.e. the theory of differential elimination [11, 12, 44]. By incorporating initial or boundary conditions, we can use explicit integral operators on equations, in addition to the usual differential operators (prolongations).

As a consequence, the natural objects of study would no longer be differential but integro-differential polynomials.

We are well aware that such an approach will meet with many difficulties that will become manifest only as we progress. Nevertheless, we are confident that an algebraic – and indeed symbolic – treatment along these lines is possible.

Acknowledgements We acknowledge gratefully the support received from the *SFB F013* in Subproject F1322 (principal investigators Bruno Buchberger and Heinz W. Engl), in earlier stages also Subproject F1302 (Buchberger) and Subproject F1308 (Engl). This support from the Austrian Science Fund (FWF) was not only effective in its financial dimension (clearly a necessary but not a sufficient condition for success), but also in a "moral" dimension: The stimulating atmosphere created by the unique blend of symbolic and numerical communities in this SFB – in particular the Hilbert Seminar mentioned in Sect. 1 – has been a key factor in building up the raw material for our studies.

Over and above his general role in the genesis and evolution of the SFB F1322, we would like to thank *Heinz W. Engl* for encouragement, critical comments and helpful suggestions, not only but especially in the early stages of this project.

Loredana Tec is a recipient of a DOC-fFORTE-fellowship of the Austrian Academy of Sciences at the Research Institute for Symbolic Computation (RISC), Johannes Kepler University Linz. Georg Regensburger was partially supported by the Austrian Science Fund (FWF): J 3030-N18.

We would also like to thank an anonymous referee for giving us plenty of helpful suggestions and references that certainly increased the value of this article.

References

1. Aichinger, E., Pilz, G.F.: A survey on polynomials and polynomial and compatible functions. In: Proceedings of the Third International Algebra Conference, pp. 1–16. Kluwer, Acad. Publ., Dordrecht (2003)
2. Albrecher, H., Constantinescu, C., Pirsic, G., Regensburger, G., Rosenkranz, M.: An algebraic operator approach to the analysis of Gerber-Shiu functions. Insurance Math. Econom. **46**, 42–51 (2010)
3. Aschenbrenner, M., Hillar, C.J.: An algorithm for finding symmetric Gröbner bases in infinite dimensional rings. In: D. Jeffrey (ed.) Proceedings of ISSAC '08, pp. 117–123. ACM, New York NY, USA(2008)
4. Baader, F., Nipkow, T.: Term Rewriting and all that. Cambridge University Press, Cambridge (1998)
5. Bavula, V.V.: The group of automorphisms of the algebra of polynomial integro-differential operators (2009). http://arxiv.org/abs/0912.2537
6. Bavula, V.V.: The algebra of integro-differential operators on a polynomial algebra (2009). http://arxiv.org/abs/0912.0723
7. Baxter, G.: An analytic problem whose solution follows from a simple algebraic identity. Pacific J. Math. **10**, 731–742 (1960)
8. Becker, T., Weispfenning, V.: Gröbner bases, *Graduate Texts in Mathematics*, vol. 141. Springer, New York (1993). A computational approach to commutative algebra, In cooperation with Heinz Kredel
9. Bergman, G.M.: The diamond lemma for ring theory. Adv. Math. **29**(2), 178–218 (1978)
10. Bilge, A.H.: A REDUCE program for the integration of differential polynomials. Comput. Phys. Comm. **71**(3), 263–268 (1992)

11. Boulier, F., Lazard, D., Ollivier, F., Petitot, M.: Representation for the radical of a finitely generated differential ideal. In: Proceedings of ISSAC '95, pp. 158–166. ACM, New York (1995)
12. Boulier, F., Ollivier, F., Lazard, D., Petitot, M.: Computing representations for radicals of finitely generated differential ideals. Appl. Algebra Engrg. Comm. Comput. **20**(1), 73–121 (2009)
13. Bourbaki, N.: Algebra I. Chapters 1–3. Elements of Mathematics (Berlin). Springer-Verlag, Berlin (1998)
14. Brouwer, A.E., Draisma, J.: Equivariant Gröbner bases and the Gaussian two-factor model (2009). http://arxiv.org/abs/0908.1530
15. Brown, R.C., Krall, A.M.: Ordinary differential operators under Stieltjes boundary conditions. Trans. Amer. Math. Soc. **198**, 73–92 (1974)
16. Brown, R.C., Krall, A.M.: n-th order ordinary differential systems under Stieltjes boundary conditions. Czechoslovak Math. J. **27**(1), 119–131 (1977)
17. Buchberger, B.: A Critical-Pair/Completion Algorithm for Finitely Generated Ideals in Rings. In: E. Boerger, G. Hasenjaeger, D. Roedding (eds.) Logic and Machines: Decision Problems and Complexity, LNCS, vol. 171, pp. 137–161 (1984)
18. Buchberger, B.: History and basic features of the critical-pair/completion procedure. J. Symbolic Comput. **3**(1-2), 3–38 (1987)
19. Buchberger, B.: Ein algorithmisches Kriterium für die Lösbarkeit eines algebraischen Gleichungssystems. Aequationes Math. **4**, 374–383 (1970). English translation: An algorithmical criterion for the solvability of a system of algebraic equations. In: B. Buchberger, F. Winkler (eds.) Gröbner bases and applications, Cambridge University Press (1998)
20. Buchberger, B.: Introduction to Gröbner bases. In: B. Buchberger, F. Winkler (eds.) Gröbner bases and applications. Cambridge University Press (1998)
21. Buchberger, B.: Groebner Rings. Contributed talk at International Conference on Computational Algebraic Geometry, University of Hyderabad, India (2001)
22. Buchberger, B.: Groebner rings and modules. In: S. Maruster, B. Buchberger, V. Negru, T. Jebelean (eds.) Proceedings of SYNASC 2001, pp. 22–25 (2001)
23. Buchberger, B.: Groebner Rings in Theorema: A Case Study in Functors and Categories. Tech. Rep. 2003-49, Johannes Kepler University Linz, Spezialforschungsbereich F013 (2003)
24. Buchberger, B.: An algorithm for finding the bases elements of the residue class ring modulo a zero dimensional polynomial ideal (German). Ph.D. thesis, Univ. of Innsbruck (1965). English translation published in J. Symbolic Comput. **41**(3-4), 475–511 (2006)
25. Buchberger, B.: Groebner bases in Theorema using functors. In: J. Faugere, D. Wang (eds.) Proceedings of SCC '08, pp. 1–15. LMIB Beihang University Press (2008)
26. Buchberger, B., Craciun, A., Jebelean, T., Kovacs, L., Kutsia, T., Nakagawa, K., Piroi, F., Popov, N., Robu, J., Rosenkranz, M., Windsteiger, W.: Theorema: Towards computer-aided mathematical theory exploration. J. Appl. Log. **4**(4), 359–652 (2006)
27. Buchberger, B., Loos, R.: Algebraic simplification. In: Computer algebra, pp. 11–43. Springer, Vienna (1983)
28. Bueso, J., Gómez Torrecillas, J., Verschoren, A.: Algorithmic Methods in Non-Commutative Algebra: Applications to Quantum Groups. Springer (2003)
29. Chyzak, F., Salvy, B.: Non-commutative elimination in Ore algebras proves multivariate identities. J. Symbolic Comput. **26**(2), 187–227 (1998)
30. Coddington, E.A., Levinson, N.: Theory of ordinary differential equations. McGraw-Hill Book Company, Inc., New York-Toronoto-London (1955)
31. Cohn, P.M.: Introduction to Ring Theory. Springer, London (2000)
32. Cohn, P.M.: Further Algebra and Applications. Springer-Verlag, London (2003)
33. Cohn, P.M.: Basic Algebra: Groups, Rings and Fields. Springer, London (2003)
34. Cucker, F., Shub, M. (eds.): Foundations of Computational Mathematics. Springer (1997). See http://www.focm.net/ for other FoCM based publications

35. Gelfand, I.M., Dikiĭ, L.A.: Fractional powers of operators, and Hamiltonian systems. Funkcional. Anal. i Priložen. **10**(4), 13–29 (1976). English translation: Functional Anal. Appl. 10 (1976), no. 4, 259–273 (1977)
36. Grabmeier, J., Kaltofen, E., Weispfenning, V. (eds.): Computer algebra handbook. Springer-Verlag, Berlin (2003)
37. Guo, L.: Baxter algebras and differential algebras. In: Differential algebra and related topics (Newark, NJ, 2000), pp. 281–305. World Sci. Publ., River Edge, NJ (2002)
38. Guo, L.: What is...a Rota-Baxter algebra? Notices Amer. Math. Soc. **56**(11), 1436–1437 (2009)
39. Guo, L., Keigher, W.: On differential Rota-Baxter algebras. J. Pure Appl. Algebra **212**(3), 522–540 (2008)
40. Guo, L., Sit, W.Y.: Enumeration and generating functions of differential Rota-Baxter words. Math. Comput. Sci. (2011). http://dx.doi.org/10.1007/s11786-010-0062-1
41. Guo, L., Sit, W.Y.: Enumeration and generating functions of Rota-Baxter words. Math. Comput. Sci. (2011). http://dx.doi.org/10.1007/s11786-010-0061-2
42. Helton, J., Stankus, M.: NCGB 4.0: A noncommutative Gröbner basis package for mathematica (2010). http://www.math.ucsd.edu/~ncalg/
43. Hillar, C.J., Sullivant, S.: Finite Gröbner bases in infinite dimensional polynomial rings and applications (2009). http://arxiv.org/abs/0908.1777
44. Hubert, E.: Notes on triangular sets and triangulation-decomposition algorithms ii: Differential systems. In: U. Langer, F. Winkler (eds.) Symbolic and Numerical Scientific Computations, *Lecture Notes in Computer Science*, vol. 2630. Springer (2003)
45. Hule, H.: Polynome über universalen Algebren. Monatsh. Math. **73**, 329–340 (1969)
46. Keigher, W.F.: On the ring of Hurwitz series. Comm. Algebra **25**(6), 1845–1859 (1997)
47. Keigher, W.F., Pritchard, F.L.: Hurwitz series as formal functions. J. Pure Appl. Algebra **146**(3), 291–304 (2000)
48. Kolchin, E.: Differential algebra and algebraic groups, *Pure and Applied Mathematics*, vol. 54. Academic Press, New York (1973)
49. Korporal, A., Regensburger, G., Rosenkranz, M.: A Maple package for integro-differential operators and boundary problems. ACM Commun. Comput. Algebra **44**(3), 120–122 (2010). Also presented as a poster at ISSAC '10
50. Köthe, G.: Topological Vector Spaces (Volume I). Springer, New York (1969)
51. La Scala, R., Levandovskyy, V.: Letterplace ideals and non-commutative Gröbner bases. J. Symbolic Comput. **44**(10), 1374–1393 (2009)
52. Lang, S.: Real and Functional Analysis, *Graduate Texts in Mathematics*, vol. 142. Springer, New York (1993)
53. Lang, S.: Algebra, *Graduate Texts in Mathematics*, vol. 211, 3rd edn. Springer, New York (2002)
54. Lausch, H., Nöbauer, W.: Algebra of Polynomials, *North-Holland Mathematical Library*, vol. 5. North-Holland Publishing Co., Amsterdam (1973)
55. Levandovskyy, V.: PLURAL, a non-commutative extension of SINGULAR: past, present and future. In: Mathematical software—ICMS 2006, *LNCS*, vol. 4151, pp. 144–157. Springer, Berlin (2006)
56. Levandovskyy, V.: Gröbner basis implementations: Functionality check and comparison. Website (2008). http://www.ricam.oeaw.ac.at/Groebner-Bases-Implementations/
57. Madlener, K., Reinert, B.: String rewriting and Gröbner bases—a general approach to monoid and group rings. In: Symbolic rewriting techniques, Progr. Comput. Sci. Appl. Logic, vol. 15, pp. 127–180. Birkhäuser, Basel (1998)
58. Madlener, K., Reinert, B.: Gröbner bases in non-commutative reduction rings. In: B. Buchberger, F. Winkler (eds.) Gröbner Bases and Applications, pp. 408–420. Cambridge University Press, Cambridge (1998)
59. Madlener, K., Reinert, B.: Non-commutative reduction rings. Rev. Colombiana Mat. **33**(1), 27–49 (1999)
60. Mikusiński, J.: Operational Calculus. Pergamon Press, New York (1959)

61. Mora, F.: Groebner bases for non-commutative polynomial rings. In: AAECC-3: Proceedings of the 3rd International Conference on Algebraic Algorithms and Error-Correcting Codes, pp. 353–362. Springer, London, UK (1986)
62. Mora, T.: An introduction to commutative and noncommutative Gröbner bases. Theoret. Comput. Sci. **134**(1), 131–173 (1994)
63. Nashed, M.Z., Votruba, G.F.: A unified operator theory of generalized inverses. In: M.Z. Nashed (ed.) Generalized Inverses and Applications (Proc. Sem., Math. Res. Center, Univ. Wisconsin, Madison, Wis., 1973), pp. 1–109. Academic Press, New York (1976)
64. van der Put, M., Singer, M.F.: Galois Theory of linear differential equations, *Grundlehren der Mathematischen Wissenschaften*, vol. 328. Springer, Berlin (2003)
65. Ree, R.: Lie elements and an algebra associated with shuffles. Ann. Math. (2) **68**, 210–220 (1958)
66. Regensburger, G., Rosenkranz, M.: An algebraic foundation for factoring linear boundary problems. Ann. Mat. Pura Appl. (4) **188**(1), 123–151 (2009)
67. Regensburger, G., Rosenkranz, M., Middeke, J.: A skew polynomial approach to integro-differential operators. In: J.P. May (ed.) Proceedings of ISSAC '09, pp. 287–294. ACM, New York, NY, USA (2009)
68. Reutenauer, C.: Free Lie Algebras, vol. 7. The Clarendon Press Oxford University Press, New York (1993)
69. Rosenkranz, M.: The Green's algebra: A polynomial approach to boundary value problems. Phd thesis, Johannes Kepler University, Research Institute for Symbolic Computation (2003). Also available as RISC Technical Report 03-05, July 2003
70. Rosenkranz, M.: A new symbolic method for solving linear two-point boundary value problems on the level of operators. J. Symbolic Comput. **39**(2), 171–199 (2005)
71. Rosenkranz, M., Buchberger, B., Engl, H.W.: Solving linear boundary value problems via non-commutative Gröbner bases. Appl. Anal. **82**, 655–675 (2003)
72. Rosenkranz, M., Regensburger, G.: Solving and factoring boundary problems for linear ordinary differential equations in differential algebras. J. Symbolic Comput. **43**(8), 515–544 (2008)
73. Rosenkranz, M., Regensburger, G.: Integro-differential polynomials and operators. In: D. Jeffrey (ed.) Proceedings of ISSAC '08, pp. 261–268. ACM, New York (2008)
74. Rosenkranz, M., Regensburger, G., Tec, L., Buchberger, B.: A symbolic framework for operations on linear boundary problems. In: V.P. Gerdt, E.W. Mayr, E.H. Vorozhtsov (eds.) Computer Algebra in Scientific Computing. Proceedings of the 11th International Workshop (CASC 2009), *LNCS*, vol. 5743, pp. 269–283. Springer, Berlin (2009)
75. Rota, G.C.: Baxter algebras and combinatorial identities (I, II). Bull. Amer. Math. Soc. **75**, 325–334 (1969)
76. Rota, G.C.: Ten mathematics problems I will never solve. Mitt. Dtsch. Math.-Ver. (2), 45–52 (1998)
77. Salvy, B., Zimmerman, P.: Gfun: a maple package for the manipulation of generating and holonomic functions in one variable. ACM Trans. Math. Softw. **20**(2), 163–177 (1994)
78. Schwarz, F.: A factorization algorithm for linear ordinary differential equations. In: Proceedings of ISSAC '89, pp. 17–25. ACM, New York (1989)
79. Seiler, W.: Computer algebra and differential equations: An overview. mathPAD **7**, 34–49 (1997)
80. Stakgold, I.: Green's Functions and Boundary Value Problems. John Wiley & Sons, New York (1979)
81. Stifter, S.: A generalization of reduction rings. J. Symbolic Comput. **4**(3), 351–364 (1987)
82. Stifter, S.: Gröbner bases of modules over reduction rings. J. Algebra **159**(1), 54–63 (1993)
83. Tec, L., Regensburger, G., Rosenkranz, M., Buchberger, B.: An automated confluence proof for an infinite rewrite system parametrized over an integro-differential algebra. In: K. Fukuda, J. van der Hoeven, M. Joswig, N. Takayama (eds.) Mathematical Software - Proceedings of ICMS 2010., *LNCS*, vol. 6327, pp. 245–248. Springer (2010)

84. Tsarev, S.P.: An algorithm for complete enumeration of all factorizations of a linear ordinary differential operator. In: Proceedings of ISSAC '96, pp. 226–231. ACM, New York (1996)
85. Ufnarovski, V.: Introduction to noncommutative Gröbner bases theory. In: B. Buchberger, F. Winkler (eds.) Gröbner bases and applications, pp. 259–280. Cambridge University Press (1998)
86. Ufnarovskij, V.A.: Combinatorial and asymptotic methods in algebra. In: Algebra, VI, *Encyclopaedia Math. Sci.*, vol. 57, pp. 1–196. Springer, Berlin (1995)
87. Windsteiger, W.: Building up hierarchical mathematical domains using functors in Theorema. Electr. Notes Theor. Comput. Sci. **23**(3), 401–419 (1999)

Linear Partial Differential Equations and Linear Partial Differential Operators in Computer Algebra

Ekaterina Shemyakova and Franz Winkler

Abstract In this survey paper we describe our recent contributions to symbolic algorithmic problems in the theory of Linear Partial Differential Operators (LPDOs). Such operators are derived from Linear Partial Differential Equations in the usual way. The theory of LPDOs has a long history, dealing with problems such as the determination of differential invariants, factorization, and exact methods of integration. The study of constructive factorization have led us to the notion of obstacles to factorization, to the construction of a full generating set of invariants for bivariate LPDOs of order 3, to necessary and sufficient conditions for the existence of a factorization in terms of generating invariants, and a result concerning multiple factorizations of LPDOs. We give links to our further work on generalizations of these results to n-variate LPDOs of arbitrary order.

1 Introduction

The solution of Partial Differential Equations (PDEs) is one of the most important problems of mathematics, and has an enormous area of applications in science, engineering, and even finance. The study of PDEs started in the eighteenth century with the work of Euler, d'Alembert, Lagrange and Laplace as a central tool in the description of the mechanics of continua and, more generally, as the principal mode of analytical study of models in the physical sciences. The analysis of physical models has remained to the present day one of the fundamental concerns of the development of PDEs. However, beginning in the middle of the nineteenth century, particularly with the work of Riemann, PDEs also became an essential tool in other branches of mathematics.

E. Shemyakova · F. Winkler (✉)
RISC, Johannes Kepler University, Linz, Austria
e-mail: ekaterina.shemyakova@risc.jku.at; franz.winkler@risc.jku.at

U. Langer and P. Paule (eds.), *Numerical and Symbolic Scientific Computing*, Texts and Monographs in Symbolic Computation, DOI 10.1007/978-3-7091-0794-2_14,

As is the case for many other types of mathematical problems (for example integration), solution methods for PDEs can be classified into symbolic (or analytical) and numerical methods. Of course, an analytical solution is to be preferred, if it can be found.

Whereas some simple Ordinary Differential Equations (ODEs) can still be solved analytically, this happens more and more rarely as the complexity of the equations increases. Only few very special PDEs can be solved analytically. Such solutions are often expressions in quadratures.

Algebraic methods for the solution of PDEs originate, e.g., from the work of Galois, whose theory of transformation groups of solutions of algebraic equations was applied to differential equations by Kolchin; Lie, who introduced continuous transformation groups; Cartan, whose theory makes use of the equivalence method of differential geometry, which determines whether two geometrical structures are the same up to a diffeomorphism; Ritt, who studied the integration of algebraic differential systems of equations; Weyl, who introduced the famous Weyl algebra of differential operators with polynomial coefficients; Laplace and Darboux, who developed exact integration methods for Linear Partial Differential Equations (LPDEs).

In this survey paper we describe our recent contributions to some classical problems appearing in the work of Laplace and Darboux on exact integration methods for LPDEs. Linear Partial Differential Operators (LPDOs) are derived from LPDEs in the usual way. Algebraic methods are in fact well suited for the investigation of LPDEs and LPDOs, due to the nice algebraic structure of these objects. Besides the problem of actually determining solutions to LPDEs, we are also interested in understanding their structure and properties such as factorization of the corresponding LPDOs and determination of generating sets of invariants. Here, the coefficient field of the factors and of the invariants are considered to be from universal [17] field. However, in most of the presented results one can have factorization or invariants over the base field of a given operator.

The Laplace transformation method has been known for approximately two centuries, and it has been generalized by Darboux about a century ago. It has served as a basis for many modern exact integration algorithms. The original methods interlace very elegantly ideas of generating sets of invariants of LPDOs with respect to gauge transformations and of factoring these LPDOs. The transformations of second-order bivariate LPDOs are described in terms of the values of their generating invariants, and the test for the transformed operator to be factorable is also described in terms of invariants.

Invariant descriptions of invariant properties and algorithms are actively used in science, and the search for generating sets of invariants is an important problem in itself. For the purpose of generalizing the results of Laplace, we need generating sets of invariants for all types of LPDOs with respect to gauge transformations. For bivariate operators of orders three and higher this problem had remained unsolved for two centuries. The case of order two has been considered by Laplace; he discovered a generating set of invariants for hyperbolic LPDOs of order two consisting of two expressions in the coefficients of the LPDO, generally referred to as the Laplace invariants h and k. We have been able to find a generating set of invariants for bivariate LPDOs of order 3. Recently this result has been generalized to n-variate operators of arbitrary order.

The next section, Sect. 2, contains an outline of the relevant results and notations. Our own contributions are described in subsequent sections.

In Sect. 3 we generalize a phenomenon of second-order hyperbolic LPDOs noticed by Laplace: there can be only two different types of incomplete factorization and the corresponding remainders are the Laplace invariants h and k. For LPDOs of order greater than 2 the remainders are not invariant and in fact there are infinitely many of them. Thus, in Sect. 3, we consider instead incomplete factorizations in a specially defined ring, which we call the ring of obstacles. Then the obstacle to a factorization is a uniquely defined element of this ring and is invariant. Obstacles to factorization have other interesting properties, some of which are described here. Within this study we have generalized the factorization algorithm of Grigoriev and Schwarz for LPDOs (see Theorems 3 and 4).

Factorization of LPDOs and many other related problems are invariant with respect to gauge transformations. Thus, they can be defined in terms of differential invariants with respect to the considered transformations. As there are infinitely many such invariants, we look for a generating set of invariants. A generating set of invariants for second-order hyperbolic LPDOs has been found by Laplace, and it is a prerequisite for the integration method based on Laplace transformations. For LPDOs of orders three and higher some individual invariants were known before, but they were not enough to form a generating set of invariants. In Sect. 4 we show how to determine 4 independent invariants for hyperbolic LPDOs of order 3 in the plane using the invariability of obstacles to factorizations. A different method gave us the fifth invariant. We have proved that these five invariants, together with the trivial ones (the coefficients of the symbol, i.e. the highest order component, of an LPDO), form a generating set of invariants.

Since the property of the existence of a factorization extending a given factorization of the symbol of an LPDO is invariant under gauge transformations, we can give an invariant description of this property, i.e. in terms of generating invariants. In Sect. 5 we solve this problem for bivariate, hyperbolic third-order LPDOs. The operation of taking the formal adjoint can also be defined for equivalence classes of LPDOs, and explicit formulae defining this operation in terms of invariants are obtained.

In Sect. 6 we outline a result which we have discovered unexpectedly while studying generating sets of invariants and the invariant conditions for the existence of factorizations extending a given factorization of the symbol of an LPDO. Namely, a third-order bivariate LPDO has both first-order left and right factors with co-prime symbols if and only if the operator has a factorization into three factors, the left one of which is exactly the initial left factor and the right one is exactly the initial right factor. For this property to hold, co-primality of the symbols and restriction to operators of order 3 are of the essence.

In Sect. 7 we briefly describe our software package `LPDOs`, containing symbolic algorithms for the investigation of LPDOs. The package `LPDOs` is based on Maple.

2 Algebraic Methods for LPDEs and LPDOs

2.1 Notations

Consider a field K of characteristic zero with commuting derivations $\partial_1, \ldots, \partial_n$, and the corresponding non-commutative ring of linear partial differential operators (LPDOs) $K[D] = K[D_1, \ldots, D_n]$, where D_i corresponds to the derivation ∂_i for all $i \in \{1, \ldots, n\}$. In $K[D]$ the variables $D_1, \ldots, D_n$ commute with each other, but not with elements of K. We write multiplication in $K[D]$ as "$\circ$"; i.e. $L_1 \circ L_2$ for $L_1, L_2 \in K[D]$. As usual, the sign "$\circ$" is often omitted, and we simply write $L_1 L_2$. An operator $L \in K[D]$ is *applied* to an element $c \in K$ as follows:

- if $L = L_1 + L_2$, then $L(c) = L_1(c) + L_2(c)$,
- if $L = L_1 \circ L_2$, then $L(c) = L_1(L_2(c))$,
- if $L = D_i$, then $L(c) = \partial_i(c)$,
- if $L = a$, for $a \in K$, then $L(c) = a \cdot c$.

In particular, for $a \in K$, because of the Leibniz rule this means:

$$D_i \circ a(c) = \partial_i(a \cdot c) = (a \circ D_i + \partial_i(a))(c).$$

i.e., for $a \in K$ we have $D_i a = a D_i + \partial_i(a)$. Note that the multiplication in $K[D]$, $L \circ a$, and the application of L to an element $a \in K$, $L(a)$, are strictly different operations.

Any operator $L \in K[D]$ has the form

$$L = \sum_{|J|=0}^{d} a_J D^J \ , a_J \in K, \tag{1}$$

where $J = (j_1, \ldots, j_n)$ is a multi-index in $\mathbb{N}^n$, $|J| = j_1 + \cdots + j_n$, and $D^J = D_1^{j_1} \ldots D_n^{j_n}$. The equation $L(u) = 0$, where u ranges over K, is the linear partial differential equation (LPDE) corresponding to the LPDO (1).

When considering the bivariate case $n = 2$, we use the following formal notations: $\partial_1 = \partial_x$, $\partial_2 = \partial_y$, $\partial_1(f) = f_x$, $\partial_2(f) = f_y$, where $f \in K$, and correspondingly $D_1 \equiv D_x$, and $D_2 \equiv D_y$. Note that this does in no way insinuate that K is a field of functions of x und y, i.e. $f = f(x, y)$. The notations ∂_x, ∂_y etc. are introduced solely for ease of notation. Thus, instead of $\partial_1(a_{11})$, for $a_{11} \in K$, we can write a_{11x}.

The homogeneous commutative polynomial

$$\mathrm{Sym}(L) = \sum_{|J|=d} a_J X^J \tag{2}$$

in formal variables $X_1, \ldots, X_n$ is called the (*principal*) *symbol* of L. Vice versa, given a homogeneous polynomial $S \in K[X]$, we define the operator $\widehat{S} \in K[D]$ as the result of substituting D_i for each variable X_i; that is, for S defined by (2) the corresponding operator is $\widehat{S} = L$ defined by (1). If there is no danger of misunderstanding we use just S to denote the operator $\widehat{S}$.

The operator L is called *hyperbolic* if its symbol $\mathrm{Sym}(L)$ can be factored into first-order factors of multiplicity one each.

We will also have occasion to collect the summands of a given order i, for $0 \leq i \leq d$, in (1) and call this sum L_i the *component of order i* of the operator L. With this notation, the operator $L \in K[D]$ can be written as $L = \sum_{i=0}^{d} L_i$.

Below we assume, unless stated otherwise, that the field K is *differentially closed*, i.e. it contains solutions of (non-linear in the generic case) differential equations with coefficients in K.

Let K^* denote the set of invertible elements in K. For $L \in K[D]$ and every $g \in K^*$ consider *gauge transformation*

$$L \to L^g = g^{-1} \circ L \circ g.$$

Then an algebraic differential expression I in the coefficients of L is (*differentially*) *invariant* under gauge transformations (we consider only these here) if it is unaltered by these transformations. Trivial examples of invariants are the coefficients of the symbol of an operator. We denote the set of all invariants of L by $\mathrm{Inv}(L)$.

A set of algebraic expressions G in the coefficients of L is *invariationally closed* iff

- $c \cdot I \in G$, for all $c \in K$ and $I \in G$,
- $I_1 + I_2 \in G$, for all $I_1, I_2 \in G$,
- $I_1 \cdot I_2 \in G$, for all $I_1, I_2 \in G$,
- $D_i(I) \in G$, for all $I \in G$ and $1 \leq i \leq n$.

Given a set of invariants G of L, the set of invariants *generated* by G is the smallest set of algebraic expressions in the coefficients of L, which contains G and is invariationally closed. G is called a *generating set for the invariants* of L iff the set of invariants generated by G is equal to $\mathrm{Inv}(L)$, the set of all invariants of L.

Another important transformation of LPDOs is *transposition*. Given $L \in K[D]$, the transposed operator is defined as

$$L^t(f) = \sum_{|J| \leq d} (-1)^{|J|} D^J(a_J f),$$

where $f \in K$. If $L = L_1 \circ L_2$, then $L^t = L_2^t \circ L_1^t$, that is if there is a factorization for L then there is an analogous one for L^t. For this reason transposition is important in studying factorizations of LPDOs.

2.2 Laplace Transformation Method and its Generalizations

Here we outline an old method, which illustrates several important ideas of exact methods for LPDOs. The method was suggested by Laplace (1749–1827), and then has been developed further and studied in depth by Darboux (1842–1917) [5]. The classical *Laplace transformation method* is applied to a second-order linear hyperbolic PDE in the plane in its normalized form

$$z_{xy} + az_x + bz_y + cz = 0, \tag{3}$$

where $a = a(x, y), b = b(x, y), c = c(x, y)$. Consider the corresponding LPDO,

$$L = D_x D_y + a D_x + b D_y + c, \tag{4}$$

and notice that there are at most two incomplete factorizations of this LPDO, namely,

$$L = (D_x + b) \circ (D_y + a) + h = (D_y + a) \circ (D_x + b) + k, \tag{5}$$

where the expressions for the "remainders" h and k can be computed explicitly:

$$h = c - a_x - ab, \quad k = c - b_y - ab. \tag{6}$$

The differential algebraic expressions h and k in the coefficients of L are known as the *Laplace invariants*. Indeed, Laplace proved the following theorem.

Theorem 1. *The Laplace invariants are invariants of the operator L in (4) with respect to gauge transformations. Both of them together form a generating set of all (differential) invariants of the operator L.*

The operator L is factorable if and only if h or k is zero.

2.2.1 The Laplace Transformation Method

1. If h or k is equal to zero, L is factorable, and hence the (3) is integrable. For example, if $h = 0$, we have $L = (D_x + b)(D_y + a)$, and the problem of integration of the equation (3) is reduced to the problem of integration of the two first order equations:

$$\begin{cases} (D_x + b)(z_1) = 0, \\ (D_y + a)(z) = z_1. \end{cases}$$

Accordingly one gets the general solution of the initial equation (3) as

$$z = \left(A(x) + \int B(y) \exp\left(\int ady - bdx\right) dy\right) e^{-\int ady} \tag{7}$$

with two arbitrary functions $A(x)$ and $B(y)$.

2. If neither h nor k are zero, one can apply the two Laplace transformations $L \to L_1$ and $L \to L_{-1}$, which are defined by the substitutions

$$z_1 = (D_y + a)(z), \quad z_{-1} = (D_x + b)(z). \tag{8}$$

Such transformations preserve the form of the equation. For example, $L \to L_1 = D_x \circ D_y + a_1 D_x + b_1 D_y + c_1$, where $a_1 = a - \partial_y(ln|h|)$, $b_1 = b$, $c_1 = c + b_y - a_x - b\partial_y(ln|h|)$.

The Laplace invariants of the new operators can be expressed in terms of the invariants of the initial operator. Thus, for the operators L_1 and L_{-1}, we have

$$\begin{aligned} h_1 &= 2h - k - \partial_{xy}(ln|h|), & k_1 &= h, \\ h_{-1} &= k, & k_{-1} &= 2k - h - \partial_{xy}(ln|k|). \end{aligned}$$

The invariants k_1, and h_{-1} are non-zero. We check whether the invariants h_1 and k_{-1} are zero. If for example $h_1 = 0$, we solve the new equation $L_1(z_1) = 0$ in quadratures as described above. Then, using the inverse substitution

$$z = \frac{1}{h}(z_1)_{-1}, \tag{9}$$

we obtain the complete solution of the original equation $L(z) = 0$. The case $k_{-1} = 0$ is treated analogously. If neither h_1 nor k_{-1} are equal to zero, we apply the Laplace transformations again.

Thus, in the generic case, we obtain two sequences: $\cdots \to L_{-2} \to L_{-1} \to L$ and $L \to L_1 \to L_2 \to \ldots$. The inverse substitution (9) implies $L = h^{-1}(L_1)_{-1}h$, and one can prove that the Laplace invariants do not change under such substitution. This means that essentially we have one chain

$$\cdots \leftrightarrow L_{-2} \leftrightarrow L_{-1} \leftrightarrow L \leftrightarrow L_1 \leftrightarrow L_2 \leftrightarrow \ldots,$$

and the corresponding chain of invariants

$$\cdots \leftrightarrow k_{-2} \leftrightarrow k_{-1} \leftrightarrow k \leftrightarrow h \leftrightarrow h_1 \leftrightarrow h_2 \leftrightarrow \ldots. \tag{10}$$

In that way one iterates the Laplace transformations until one of the Laplace invariants in the sequence (10) vanishes. In this case, one can solve the corresponding transformed equation in quadratures. Using the same differential substitution (8) one obtains the complete solution of the original equation.

One can also prove, cf. [5,10,11], that if the chain (10) is finite in both directions, then one may obtain a quadrature free expression of the general solution of the original equation.

2.2.2 Generalization to Non-Linear Case

Darboux [5] has also suggested an explicit integration method of non-linear second-order scalar equations of the form $F(x, y, z, z_x, z_y, z_{xx}, z_{xy}, z_{yy}) = 0$. The idea is to consider a linearization of the equation, and then to apply the Laplace method. The relationship between the Laplace invariants of the linearized operator P and Darboux integrability of the initial equation was established by Sokolov, Ziber, Startsev [29, 35], who proved that a second order hyperbolic non-linear equation is Darboux integrable if and only if both Laplace sequences are finite. Later Anderson, Juras, and Kamran [1, 2, 15] generalized this to the case of the equations of the general form as a consequence of their analysis of higher degree conservation laws for different types of partial differential equations.

2.2.3 Generalization to Multivariate Case

Dini [6, 7] suggested a generalization of the Laplace transformations for a certain class of second-order operators in the space of arbitrary dimension. But no general statement was given on the range of applicability of his trick. Recently Tsarev [39] proved that for a generic second-order linear partial differential operator in three-dimensional space,

$$L = \sum_{i+j+k \leq 2} a_{ijk}(x, y, z) D_x D_y D_z$$

there exist two Dini transformations under the assumption that its principal symbol factors.

2.2.4 Generalization to Systems of LPDOs

There were also several attempts to generalize the Laplace method to some systems of equations. Athorne and Yilmaz [3, 14] proposed a special transformation, which is applicable to systems whose order coincides with the number of independent variables. A serious effort to generalize the classical theory to operators of higher order (in two independent variables) was undertaken by Roux [22]. Recently Tsarev [38] described another procedure, which generalizes the Cascade Method to the case of arbitrary order hyperbolic operators.

2.3 *Factorization of LPDOs*

Thus, the factorization of LPDOs, the theory of invariants for LPDOs, and exact integration algorithms of Partial Differential Equations (PDEs) are closely related.

The Loewy decomposition method for exact integration, cf. [23], requires various generalizations of the Laplace transformation method and factorization of LPDOs.

For an ordinary linear differential operator L the Loewy uniqueness theorem [19] states that if $L = P_1 \circ \cdots \circ P_k = \widetilde{P}_1 \circ \cdots \circ \widetilde{P}_l$ are two different irreducible factorizations, then they have the same number of factors (that is $k = l$) and the factors are pairwise "similar" in some transposed order. In contrast the factoring of LPDOs can lead to very different factorizations: even different numbers of irreducible factors are possible [4]: the operator $L = D_x^3 + xD_x^2 D_y + 2D_x^2 + (2x+2)D_x D_y + D_x + (2+x)D_y$ has two factorizations into different numbers of irreducible factors:

$$L = Q \circ Q \circ P = R \circ Q, \tag{11}$$

for the operators $P = D_x + xD_y$, $Q = D_x + 1$, $R = D_x^2 + xD_x D_y + D_x + (2 + x)D_y$, where operator R is absolutely irreducible, that is one cannot factor it into a product of first-order operators with coefficients in any extension of $\mathbb{Q}(x, y)$.

2.3.1 Generalized Factorizations of LPDOs

One important direction of the development has consisted in attacking the non-uniqueness of factorizations by inventing new definitions of factorizations [13, 18, 37]. Then the conventional factorization of ordinary operators becomes a special case of the generalized factorization, and some analogues of the Loewy–Ore uniqueness theorem can be proved. In one of the earliest attempts [18] the factoring of a linear homogeneous partial differential system is treated as finding superideals of a left ideal in the ring of LPDOs rather than factoring a single LPDO, and a generalization of the Beke–Schlesinger algorithm for factoring LODOs with coefficients in $\overline{\mathbb{Q}}(x, y)$ has been given. This approach is based on an algorithm for finding hyperexponental solutions of such ideals. In [13] a given LPDO is considered as a generator of a left D-module over an appropriate ring of differential operators. In this algebraic approach decomposing a D-module means finding supermodules which describe various parts of the solution of the original problem.

2.3.2 Factorization Algorithm of Grigoriev and Schwarz

Another direction was initiated by Miller [21], who suggested to use some analogue of the well-known method of Hensel lifting in factorization of polynomials. Miller considered LPDOs of order two and three only. Grigoriev and Schwarz [12] have generalized Miller's results to LPDOs of arbitrary order.

The result of Grigoriev and Schwarz is important because it determines a class of LPDOs which have at most one factorization extending a given factorization of the symbol. Here we sketch the suggested proof, because it is constructive and it is also a description of the algorithm.

Theorem 2. *Let $L \in K[D]$, and*

$$\mathrm{Sym}_L = S_1 \cdot S_2, \quad \textit{with} \quad \gcd(S_1, S_2) = 1. \tag{12}$$

Then there exists at most one factorization of the form

$$L = F_1 \circ F_2\,, \quad \textit{with} \quad \mathrm{Sym}(F_1) = S_1\,, \quad \mathrm{Sym}(F_2) = S_2.$$

Proof. Consider L, F_1 and F_2 as the sums of their components:

$$L = \sum_{i=0}^{d} L_i\,, \quad F_1 = \widehat{S}_1 + \sum_{i=0}^{k_1-1} G_i\,, \quad F_2 = \widehat{S}_2 + \sum_{i=0}^{k_2-1} H_i$$

where $d = \mathrm{ord}(L)$, $k_1 = \mathrm{ord}(\widehat{S}_1)$, $k_2 = \mathrm{ord}(\widehat{S}_2)$. Then the considered factorization has the form

$$\sum_{i=0}^{d} L_i = \left(\widehat{S}_1 + G_{k_1-1} + \cdots + G_0\right) \circ \left(\widehat{S}_2 + H_{k_2-1} + \cdots + H_0\right).$$

By equating the components of the two sides of this equality, one gets the following system in the homogeneous polynomials corresponding to the operators H_i and G_j, which we denote by the same letters:

$$\begin{cases} L_{d-1} = S_1 \cdot H_{k_2-1} + G_{k_1-1} \cdot S_2\,, \\ L_{d-2} = S_1 \cdot H_{k_2-2} + G_{k_1-2} \cdot S_2 \ + P_{d-2}, \\ \dots \\ L_i \quad = S_1 \cdot H_{i-k_1} + G_{i-k_2} \cdot S_2 \ + P_i\,, \\ \dots \end{cases}$$

where P_i are some expressions of derivatives of H_{k_2-j}, G_{k_1-j} with $j < i$. Thus, if one solves the system in descendent order, the polynomials P_i can be considered as known. Also here L_i stands for the homogeneous polynomial corresponding to the component L_i of L.

Consider one equation of the system:

$$L_i = S_1 \cdot H_{i-k_1} + G_{i-k_2} \cdot S_2 + P_i.$$

It is equivalent to a linear algebraic system in the coefficients of the polynomials H_{i-k_1} and G_{i-k_2}. Since S_1 and S_2 are coprime, there is at most one solution of the system, and likewise for the equation. Thus, at every step one either gets the next components of H and G, or (in case the linear algebraic system is inconsistent) concludes that there is no factorization of the operator L extending the polynomial factorization of the symbol (12). □

By induction on the number of factors one proves the following theorem.

Theorem 3 (Grigoriev and Schwarz). *Let $L \in K[D]$, and*

$$\mathrm{Sym}_L = S_1 \cdot S_2 \ldots S_k, \quad \textit{with} \quad \gcd(S_i, S_j) = 1\ \forall i \neq j.$$

Then there exists at most one factorization $L = F_1 \circ \cdots \circ F_k$*, such that* $\mathrm{Sym}_{F_i} = S_i$ *for* $i = 1, \dots k$.

Despite all these results, and many others (for example [3, 38, 40]), the general factorization problem remains wide open.

3 Obstacles to Factorizations

In this section we outline our results on the generalization of the incomplete factorization idea of Laplace.

The commutativity of the symbols of LPDOs implies that any factorization of an LPDO extends some factorization of its symbol. In general, if $L \in K[D]$ and $\mathrm{Sym}(L) = S_1 \cdot \cdots \cdot S_k$, then we say that the factorization $L = F_1 \circ \cdots \circ F_k$, with $\mathrm{Sym}(F_i) = S_i$ for all $i \in \{1, \dots, k\}$, is of the *factorization type* $(S_1) \dots (S_k)$.

In Sect. 2.2 we have seen that for second-order hyperbolic LPDOs the remainders of the incomplete factorizations (5) of the factorization types $(X)(Y)$ and $(Y)(X)$ are h and k, respectively, and h and k are invariants with respect to gauge transformations of LPDOs. Note that any incomplete factorization of the type $(X)(Y)$ has remainder h, and any of the type $(Y)(X)$ has remainder k. Unfortunately, for LPDOs of order greater than two this is not true any more.

In [30, 32] we suggested instead the consideration of incomplete factorizations in a specially defined ring, which we called the ring of obstacles. Then the obstacle to a factorization is a uniquely defined element of this ring and is invariant. Other interesting properties are also proved, among which is a generalization of the Grigoriev–Schwarz Theorem to the case of non-coprime symbols of the factors. Below we outline these results.

3.1 Generalization of Grigoriev–Schwarz Theorem to Non-Coprime Case

In view of the Grigoriev–Schwarz factorization algorithm (see the proof of Theorem 3) we introduce the following notation.

Definition 1. Let $L, F_i \in K[D]$, $i = 1, \dots, k$ and assume that for some $t \in \{0, \dots, \mathrm{ord}(L)\}$

$$\mathrm{ord}(L - F_1 \circ \cdots \circ F_k) < t \tag{13}$$

holds. Then we say that $F_1 \circ \cdots \circ F_k$ is a *partial factorization (incomplete factorization) of order* t of the operator L. If, in addition, $S_i = \mathrm{Sym}_{F_i}$, for $i = 1, \dots, k$ (so $\mathrm{Sym}_L = S_1 \dots S_k$), then this partial factorization is of the *factorization type* $(S_1) \dots (S_k)$.

Every factorization of $L \in K[D]$ is a partial factorization of order 0, and if $\mathrm{Sym}_L = S_1 \dots S_k$, then the corresponding composition of operators $\widehat{S}_1 \circ \dots \circ \widehat{S}_k$ is a partial factorization of order d.

Definition 2. Let $L \in K[D]$, $\mathrm{Sym}_L = S_1 \cdots S_k$ with $\mathrm{ord}(S_i) = d_i$, $i = 1, \dots, k$, $d = d_1 + \dots + d_k$. Let

$$F_1 \circ \dots \circ F_k, \quad F_1' \circ \dots \circ F_k'$$

be two partial factorizations of orders t and t', respectively, where $t' < t$. Then $F_1' \circ \dots \circ F_k'$ is an *extension* of $F_1 \circ \dots \circ F_k$ iff

$$\mathrm{ord}(F_i - F_i') < t - (d - d_i), \ \forall i \in \{1, \dots, k\}.$$

Example 1. Consider the fifth-order operator $L = (D_x^2 + D_y + 1) \circ (D_x^2 D_y + D_x D_y + D_x + 1)$. Factorizations of the form $(D_x^2 + \dots) \circ (D_x^2 D_y + \dots)$ are partial factorizations of order 5. Their extensions are the following fourth-order partial factorizations of the type $(D_x^2 + D_y + \dots) \circ (D_x^2 D_y + D_x D_y + \dots)$.

Consider an operator $L \in K[D]$ of some order d, and some factorization of its symbol $\mathrm{Sym}_L = S_1 \cdot S_2$. If $\gcd(S_1, S_2) = 1$, then by the Grigoriev–Schwarz Theorem, there exists at most one extension of this partial factorization to a factorization of the whole operator L .

Suppose now that there exists a nontrivial common divisor of S_1 and S_2. Then extensions might not be unique (see, e.g., (11)). However we can prove the following theorem:

Theorem 4 (Generalization of Grigoriev–Schwarz theorem to non-coprime case). *Let $L \in K[D]$ be an operator of order d, and let*

$$\mathrm{Sym}_L = S_1 \cdot S_2 \ , \ \gcd(S_1, S_2) = S_0 \ , \ \mathrm{ord}(S_0) = d_0$$

be a factorization of the symbol of L. Then for every partial factorization of order $d - d_0$ of the type $(S_1)(S_2)$, there is at most one extension to a complete factorization of L of the same type.

3.1.1 Obstacles to Factorizations

Definition 3. Let $L \in K[D]$, $\mathrm{Sym}_L = S_1 \dots S_k$. An operator $R \in K[D]$ is called a *common obstacle* to factorization of the type $(S_1)(S_2)\dots(S_k)$ if there exists a factorization of this type for the operator $L - R$ and R has minimal possible order.

Common obstacles are closely related to partial factorizations.

Proposition 1. *Let $L \in K[D]$, $\mathrm{Sym}_L = S_1 \cdots S_k$. A common obstacle to a factorization of the type $(S_1)\dots(S_k)$ is of order t if and only if the minimal order of a partial factorization of this type is $t + 1$.*

Though common obstacles are the natural generalization of the Laplace invariants, they do not preserve the important properties of those. Neither common obstacles nor their symbols are unique in general, or invariant. In order to describe all factorable (or unfactorable) LPDOs in some algebraic terms, and understand what actually prevents an LPDO to be factorable, we suggest to consider certain factor rings.

Definition 4. Let $L \in K[D]$ and $\mathrm{Sym}_L = S_1 \cdots S_k$. Then

$$K(S_1, \ldots, S_k) \;:=\; K[X]/I,$$

where I is the homogeneous ideal

$$I = \left(\frac{\mathrm{Sym}_L}{S_1}, \ldots, \frac{\mathrm{Sym}_L}{S_k} \right),$$

is the *ring of obstacles* to factorization of L *of the type* $(S_1) \ldots (S_k)$.

The following property justifies the introduction of this new notion, the ring of obstacles.

Theorem 5. *Let* $L \in K[D]$ *and* $\mathrm{Sym}_L = S_1 \cdots S_k$, *where the* S_i *are pairwise coprime. Then the symbols of all common obstacles to factorization of the type* $(S_1) \ldots (S_k)$ *belong to the same class in the ring of obstacles* $K(S_1, \ldots, S_k)$.

Definition 5. The class of common obstacles in the ring of obstacles is called the *obstacle to factorization*.

Remark 1. Every element in the obstacle to factorization is a common obstacle.

The following theorems describes the major properties of obstacles to factorizations.

Theorem 6. *Obstacles are invariant under gauge transformations.*

Theorem 7. *Let* $n = 2$, $L \in K[D]$, $\mathrm{ord}(L) = d$, *and let* $\mathrm{Sym}_L = S_1 \cdots S_k$, *where* S_i, $i \in \{1, \ldots, k\}$ *are pairwise coprime. Then common obstacles to the factorization of* L *of this type are of order equal to or less than* $d - 2$.

Theorem 8. *Let* $L \in K[D]$ *be a bivariate hyperbolic operator of order* d. *Then for each type of factorization there is a unique common obstacle.*

4 Generating Sets of Invariants for Hyperbolic Third-Order LPDOs

The Laplace transformation method of Sect. 2.2 essentially uses the generating set of invariants $\{h, k\}$ for the considered class of operators. For generalizations of the method we need generating sets of invariants with respect to gauge transformations for LPDOs of order higher than two. The generating invariants are also used as convenient terms in which all invariant properties can be expressed (see for example Sect. 5). Often this leads to simplification of the corresponding problem. In addition, with a generating set of invariants in hand we may be able to classify uniquely a number of simple equations. For example, an equation of the form $z_{xy} + az_x + bz_y + cz = 0$, where $a, b, c \in K$, is gauge equivalent to the wave equation $z_{xy} = 0$ whenever $h = k = 0$.

Although the generating set of invariants $\{h, k\}$ for hyperbolic bivariate second-order LPDOs has been discovered two hundred years ago, for hyperbolic operators of higher orders even in the plane not much has been known. Individual invariants were discovered and rediscovered several times (see [39] and [16]), generating sets of invariants were not found.

Following our generalization of the Laplace idea, we employ Theorem 6 to obtain invariants for bivariate third-order hyperbolic operators. It is easy to see that neither these invariants, nor individual invariants found in [39] and [16], nor they all together can generate all differential invariants. Specifically, if we consider normalized form (14)) of hyperbolic third-order LPDOs in the plane, we see that neither of those individual invariants can generate any invariant depending on the coefficient a_{00}. Thus, some new invariants needed to be found and a minimal system needed to be extracted. In [24, 31] we demonstrate a set of five invariants forming a generating set of invariants for third-order hyperbolic LPDO in the plane given in a normalized form.

Theorem 9. *For some non-zero $q \in K$ consider operators of the form*

$$L = (D_x + qD_y)D_xD_y + a_{20}D_x^2 + a_{11}D_xD_y + a_{02}D_y^2 + a_{10}D_x + a_{01}D_y + a_{00}, \quad (14)$$

where the coefficients belong to K, $q \neq 0$. Then the following is a generating set of invariants with respect to gauge transformations:

$$I_q = q,$$

$$I_1 = 2a_{20}q^2 - a_{11}q + 2a_{02},$$

$$I_2 = \partial_x(a_{20})q^2 - \partial_y(a_{02})q + a_{02}q_y,$$

$$I_3 = a_{10} + a_{20}(qa_{20} - a_{11}) + \partial_y(a_{20})q - \partial_y(a_{11}) + 2a_{20}q_y,$$

$$I_4 = a_{01}q^2 + a_{02}2 - (3q_x + a_{11}q)a_{02} + q_xqa_{11} - \partial_x(a_{11})q2 + q\partial_x(a_{02}),$$

$$I_5 = a_{00} - \frac{1}{2}\partial_{xy}(a_{11}) + q_x\partial_y(a_{20}) + q_{xy}a_{20}$$

$$+ \left(2qa_{20} + \frac{2}{q}a_{02} - a_{11} + q_y\right)\partial_x(a_{20}) - \frac{1}{q}a_{02}a_{10} - a_{01}a_{20} + \frac{1}{q}a_{20}a_{11}a_{02}.$$

Thus, an operator $L' \in K[D]$

$$L' = (D_x+qD_y)D_xD_y+b_{20}D_x^2+b_{11}D_xD_y+b_{02}D_y^2+b_{10}D_x+b_{01}D_y+b_{00} \quad (15)$$

is equivalent to L (w.r.t. gauge transformations $L \to g^{-1}Lg$) if and only if their corresponding invariants I_1, I_2, I_3, I_4, I_5 are equal.

Example 2. An interesting case is when all the coefficients are constants in the differential field K. Then for the class of (14) the following is a generating set of differential invariants:

$$\begin{aligned}
I_q &= q,\\
I_1 &= 2q2a_{20} - qa_{11} + 2a_{02},\\
I_2 &= 0,\\
I_3 &= -a_{20}a_{11} + a_{10} + a_{20}^2 q,\\
I_4 &= a_{01}q2 + a_{02}2 - a_{02}a_{11}q,\\
I_5 &= a_{00}q + a_{20}a_{11}a_{02} - a_{02}a_{10} - a_{01}a_{20}q.
\end{aligned}$$

Notice that here the generating invariants are constant-valued functions. Also we conclude that $I_2 = 0$ is a necessary condition for an LPDO of the form (14) to have constant coefficients only.

Example 3. Let us consider an LPDO R for which we know that it has factorization

$$R = R_1 \circ R_2 \circ R_3,$$

where $R_1 = D_x + r_1$, $R_2 = D_y + r_2$, $R_3 = D_x + D_y + r_3$ and $r_1, r_2, r_3 \in K$. Then

$$\left.\begin{aligned}
R = {} & D_xD_y(D_x + D_y) + r_2D_x2 + (r_1 + r_2 + r_3)D_xD_y + r_1D_y2\\
& + (r_1r_2 + r_2r_3 + r_{3y} + r_{2x})D_x + (r_1r_2 + r_1r_3 + r_{2x} + r_{3x})D_y\\
& + r_1r_2r_3 + r_1r_{3y} + r_{2x}r_3 + r_2r_{3x} + r_{3xy},
\end{aligned}\right\} \quad (16)$$

R has the following generating set of differential invariants:

$$\left.\begin{aligned}
I_q &= 1,\\
I_1 &= r_2 + r_1 - r_3,\\
I_2 &= I_3 = -r_{1y} + r_{2x},\\
I_4 &= 0,\\
I_5 &= (r_{3xy} - r_{1xy} - r_{2xy})/2.
\end{aligned}\right\} \quad (17)$$

One can see that $I_4 = 0$ is a necessary condition for the existence of a factorization of the factorization type $(X)(Y)(X + Y)$ for an LPDO with the symbol $D_xD_y(D_x + D_y)$.

Let us consider a subclass of the class of LPDOs (14) that contains only operators of the form (16). Then every LPDO within this class has $I_4 = 0$, and within this new class gauge transformations form new equivalence classes, and the following is a generating set of invariants:

$$r_2 + r_1 - r_3\,, -r_{1y} + r_{2x}.$$

Applying gauge transformation with $\exp(x - y + x2 + y3)$, we get the LPDO $R^{\exp(x-y+x2+y3)}$. It has complicated coefficients, for example, the component of order zero is

$$-2 + 2r_2r_{3x} + 2r_1r_{2x} + r_1r_2r_3 + 2r_2 - r_3 - 2x + 6y + 2r_2x - 4x2 + 3r_1r_2y2$$
$$+ 3r_1r_3y2 + 6r_1y^2x + 6r_2y^2x + 6r_3y^2x + 9y4 - r_{3x} + r_{3xy} + 12y^2x2 + 9r_1y4 + 6r_1y$$
$$+ 3r_{2x}y2 + 3r_{3x}y2 + 3r_2y2 - 3r_1y2 + 3r_3y2 + 3y2 + r_2r_3 + 4r_2x2 - r_1r_3 + 2r_{2x}x$$
$$+ r_2r_{3x} + r_{2x}r_3 + r_1r_{3y} - 2r_1x + 2r_{3y}x + 12xy + 18xy4 + r_{3y} - 2r_3x.$$

Given an LPDO F for which no connection to R is known, we can compute the values of the six invariants of the generating set of invariants from Theorem 9. If they happen to be exactly the same as those for R, we conclude that $F = R^f$ for some $f \neq 0$, $f \in K$, and that F is factorable into first-order factors, as the property of the existence of a factorization is invariant under gauge transformations (see further details in Sect. 5).

5 Existence of Factorization in Terms of Invariants

In the case considered by Laplace, the invariants h and k can be simply obtained from the incomplete factorizations, $L = (D_x + b) \circ (D_y + a) + h = (D_y + a) \circ (D_x + b) + k$. That is why the invariant necessary and sufficient condition for factorizability becomes so simple: $h = 0$ or $k = 0$. For hyperbolic operators of the next higher order – three – the situation is much more difficult: the "remainder" of an incomplete factorization is not invariant in the generic case (see Sect. 3).

However, the properties of the existence of a factorization and of the existence of a factorization with given higher order terms of the factors are invariant with respect to gauge transformations. Indeed, if $L = F_1 \circ F_2 \circ \cdots \circ F_k$, for some LPDOs $L, F_i \in K[D]$, then for every $g \in K^*$ we have $g^{-1} \circ L \circ g = (g^{-1} \circ F_1 \circ g) \circ (g^{-1} \circ F_2 \circ g) \circ \cdots \circ (g^{-1} \circ F_k \circ g)$. This means that theoretically a description in terms of invariants for the existence of such factorizations can be given. In [33] invariant expressions (that is algebraic expressions in terms of generating invariants and their derivations) defining all these properties for hyperbolic and non-hyperbolic LPDOs in the plane were found. Also we showed that the operation of taking the formal adjoint can be defined in terms of invariants, that is for equivalence classes of LPDOs. Explicit formulae defining this operation in terms of invariants were obtained. The operation of formal adjoint is highly interesting for factorization of LPDOs; for if the initial operator has a factorization, then so does its adjoint, and these factorizations are closely related.

5.1 Existence of Factorization for Hyperbolic Bivariate LPDOs of Order Three in Terms of Invariants

Theorem 10. *[33] Given the values of the invariants $q, I_1, I_2, I_3, I_4, I_5$ (from Theorem 9) for an equivalence class of operators of the form (14), the LPDOs of this class have a factorization of factorization type (here we denote $S = X + qY$) $(S)(XY)$ if and only if*

$$\left.\begin{array}{l} I_3q3 - I_{1y}q2 + q_yI_1q - I_4 + qI_{1x} - 2q_xI_1 - 3qI2 = 0, \\ -q^2I_{4y} + 1/2q^3I_{1xy} - qI_{4x} - 3/2q^2q_xI_{1y} + q^3I_5 + q^2I_{1xx} \\ -3/2I_1q^2q_{xy} - 2I_1qq_{xx} + 5I_1qq_xq_y + 6I_1q_x2 + 3I_4q_x \\ +3I_4qq_y - qI_1I_{1x} + I_1I_4 + 2q_xI_12 - 4I_{1x}qq_x - 3/2I_{1x}q^2q_y \\ -2q^2I_{2x} - q^3I_{2y} + I_2qI_1 + 4I_2qq_x + 2I_2q^2q_y = 0; \end{array}\right\} \tag{18}$$

$(S)(X)(Y)$ if and only if (18) & $-I_4 + qI_{1x} - 2q_xI_1 - qI_2 = 0$;
$(X)(SY)$ if and only if

$$\left.\begin{array}{l} (D_1): \quad qq_{xx} - I_4 - 2q_x = 0, \\ -3/2q_xqI_{1y} - q^3I_{3x} + I_5q2 + 1/2q^2I_{1xy} - 1/2qq_yI_{1x} \\ +q_xq^2I_3 + 2I_1q_xq_y - 1/2I_1q_{xy}q - 4q_xI_2 + qI_{2x} = 0; \end{array}\right\} \tag{19}$$

$(X)(Y)(S)$ if and only if (19) & (D_1);
$(XY)(S)$ if and only if

$$\left.\begin{array}{l} -qI_2 + qq_xq_y + q_{yy}q3 - q^2q_{xy} + qq_{xx} + I_3q3 - I_4 - 2q_x2 = 0, \\ q^3I_5 + qI_{4x} + 1/2q^3I_{1xy} - 3/2q^2q_xI_{1y} + I_1I_4 + q^2I_{2x} + 2I_1qq_xq_y \\ +2I_1q_x2 - 5I_4q_x - 1/2I_1q^2q_{xy} - I_1qq_{xx} + I_4qq_y - 1/2I_{1x}q^2q_y - 4I_2qq_x \\ -10q_x3 - q^2q_{xxx} - q^4I_{3x} + I_3q^3q_x + 2qq_x^2q_y - q^2q_yq_{xx} + 8qq_xq_{xx} = 0; \end{array}\right\} \tag{20}$$

$(YS)(X)$ if and only if

$$\left.\begin{array}{l} (C_1): \quad -2q_xI_1 + qI_{1x} - I_4 - 2qI_2 = 0, \\ -qI_{4y} + 1/2q^2I_{1xy} + I_5q2 - I_2I_1 - q^2I_{2y} + 2q_yI_4 + 3I_1q_xq_y - \\ 3/2I_1q_{xy}q - 1/2qq_yI_{1x} - 3/2q_xqI_{1y} = 0: \end{array}\right\} \tag{21}$$

$(S)(Y)(X)$ *if and only if* (18) & (C_1);
$(XS)(Y)$ *if and only if*

$$\left.\begin{array}{l}(B_1): \quad I_3q2 - qI_{1y} + q_yI_1 - 2I_2 = 0 \\ -1/2qq_yI_{1x} - 3/2q_xqI_{1y} + I_5q2 - q^3I_{3x} + 1/2q^2I_{1xy} + q_xq^2I_3 \\ -2q_xI_2 + qI_{2x} + 3I_1q_xq_y - 3/2I_1q_{xy}q - I_2I_1 + 2q_y^2q_xq - 2q_yqI_2 \\ -2q_{xy}q^2q_y + 2q_{xy}qq_x - 2q_yq_x2;\end{array}\right\} \quad (22)$$

$(Y)(SX)$ *if and only if*

$$\left.\begin{array}{l}(A_1): \quad I_3 + q_{yy} = 0, \\ -qI_{4y} - 3/2q_xqI_{1y} + I_5q2 + 1/2q^2I_{1xy} + 2q_yI_4 - 1/2qq_yI_{1x} + \\ 3I_1q_xq_y - 3/2I_1q_{xy}q - q^2I_{2y} = 0;\end{array}\right\} \quad (23)$$

$(Y)(X)(S)$ *if and only if* (20)& $-qq_{xx} + I_4 + 2q_x2 + qI_2 - qq_xq_y + q^2q_{xy} = 0$;
$(Y)(S)(X)$ *if and only if* (21) & (A_1).
$(X)(S)(Y)$ *if and only if* (19) & (B_1);

Consider the important case, where all coefficients of LPDOs in question are constants. Example 2 implies that the generating invariants are constant-valued functions, and the formulae of Theorem 10 can be simplified drastically.

Corollary 1. *Given the values of the invariants* $q, I_1, I_2, I_3, I_4, I_5$ *(from Theorem 9) for an equivalence class of operators of the form (14), in which all coefficients are constants. Then the LPDOs of this class have a factorization of factorization type* $(S)(XY)$ *if and only if*

$$\left.\begin{array}{l}I_3 - I_4 - 3I2 = 0, \\ I_5 + I_1I_4 + I_2I_1 = 0;\end{array}\right\} \quad (24)$$

$(S)(X)(Y)$ *if and only if* (24) & $-I_4 - I_2 = 0$;
$(S)(Y)(X)$ *if and only if* (24) & (C_1);
$(X)(SY)$ *if and only if*

$$\left.\begin{array}{l}(D_1): \quad I_4 = 0, \\ \qquad\qquad I_5 = 0;\end{array}\right\} \quad (25)$$

$(X)(S)(Y)$ *if and only if* (25) & (B_1);
$(X)(Y)(S)$ *if and only if* (25) & (D_1);
$(XY)(S)$ *if and only if*

$$\left.\begin{array}{l}-I_2 + I_3 - I_4 = 0, \\ I_5 + I_1I_4 = 0;\end{array}\right\} \quad (26)$$

$(YS)(X)$ *if and only if*

$$\left.\begin{array}{l}(C_1): \quad I_4 + 2I_2 = 0, \\ \qquad\qquad I_5 - I_2I_1 = 0;\end{array}\right\} \quad (27)$$

$(XS)(Y)$ *if and only if*

$$(B_1): \quad \left.\begin{aligned} I_3 - I_{1y} - 2I_2 = 0 \\ I_5 - I_2 I_1 = 0; \end{aligned}\right\} \tag{28}$$

$(Y)(SX)$ *if and only if*

$$(A_1): \quad \left.\begin{aligned} I_3 = 0, \\ I_5 = 0; \end{aligned}\right\} \tag{29}$$

$(Y)(X)(S)$ *if and only if* (26) & $I_4 + I_2 = 0$;
$(Y)(S)(X)$ *if and only if* (27) & (A_1).

Example 4. Consider R from Example 3 in the form (16). Substituting the generating invariants computed in Example 3, we get in particular that L has a factorization of the factorization type $(X(X+Y))(Y)$ if and only if

$$r_{2x} + r_{2y} - r_{3y} = 0 \quad \& \quad (r_2 + r_1 - r_3)(-r_{1y} + r_{2x}) = 0;$$

and that L has a factorization of the factorization type $(Y)((X+Y)X)$ if and only if

$$-r_{1y} + r_{2x} = 0 \quad \& \quad r_{1yy} - r_{2xy} = 0.$$

So if r_1, r_2, r_3 are constants, then L has both a factorization of factorization type $(Y)((X+Y)X)$ and also of factorization type $(X)(Y)(X+Y)$. This observation corresponds to the fact that factorizations with constant coefficients commute.

5.2 *Formal Adjoint in Terms of Invariants*

Below we consider the operation of taking the formal adjoint of an LPDO. We also define this operation on the equivalence classes of third-order bivariate non-hyperbolic LPDO. The original aim of this investigation was to reduce the number of cases in the proof of Thereom 10, however we find this result interesting in itself.

For an operator $L = \sum_{|J|\le d} a_J D^J$, where $a_J \in K$, $J \in \mathbf{N}^n$ and $|J|$ is the sum of the components of J, the *formal adjoint* is defined as

$$L^\dagger(f) = \sum_{|J|\le d} (-1)^{|J|} D^J(a_J f), \ \forall f \in K.$$

The formal adjoint possesses the following properties which are useful in the theory of factorization:

$$(L^\dagger)^\dagger = L, \quad (L_1 \circ L_2)^\dagger = L_2^\dagger \circ L_1^\dagger, \quad \mathrm{Sym}_L = (-1)^{\mathrm{ord}(L)} \mathrm{Sym}_{L^\dagger}.$$

The property of having a factorization is invariant under the operation of taking the formal adjoint, while the property of having a factorization of certain factorization type is not invariant under the operation of taking the formal adjoint. Indeed, an operator L has a factorization of some factorization type $(S_1)(S_2)$ (where $\mathrm{Sym}_L = S_1 S_2$) if and only if $L^\dagger$ has a corresponding factorization of factorization type $(S_2)(S_1)$.

The operation of taking the formal adjoint can be defined on equivalence classes of LPDOs. Obviously the operations of taking the formal adjoint and applying a gauge transformation commute: for every $g \in K^*$, and $f = g^{-1}$ we have

$$(g^{-1} \circ L \circ g)^\dagger = g^\dagger \circ L^\dagger \circ (g^{-1})^\dagger = g \circ L^\dagger \circ g^{-1} = f^{-1} \circ L^\dagger \circ f.$$

Example 5 (LPDOs of order 2). Consider an operator $L \in K[D]$ of the form $L = D_{xy} + aD_x + bD_y + c$, and its generating set of invariants $h = c - a_x - ab$ and $k = c - b_y - ab$. The formal adjoint is

$$L^\dagger = D_{xy} - aD_x - bD_x + c - a_x - b_y$$

and $h^\dagger = c - b_y - ab$ and $k^\dagger = c - a_x - ab$; and so $h^\dagger = k$, $k^\dagger = h$.

Remark 2. Given an LPDO L of the form (14) the adjoint operator $L^\dagger$ has symbol $-\mathrm{Sym}(L)$, and therefore, formulae (9) are not suitable for defining a generating system of invariants. To obtain a generating set of invariants for the class of LPDOs of the form

$$L = -(D_x + qD_y)D_xD_y + a_{20}D_x 2 + a_{11}D_{xy} + a_{02}D_y 2 + a_{10}D_x + a_{01}D_y + a_{00}, \tag{30}$$

we substitute every occurrence of a_{ij} with $-a_{ij}$ for all i, j in the formulae for invariants in Theorem 9. The resulting invariants we denote by $I_1^\dagger, \ldots, I_5^\dagger$.

Example 6. Consider again the operator LPDO R from Example 3 in the form (16). The adjoint LPDO for R is

$$\left.\begin{aligned} R^\dagger = &-(D_x + qD_y)D_xD_y + r_2D_x 2 + (r_1 + r_2 + r_3)D_{xy} + r_1D_y 2 \\ &+ (-r_1r_2 - r_2r_3 + r_{3y} + r_{2x} + r_{1y} + r_{2y})D_x \\ &+ (-r_1r_2 - r_1r_3 + r_{1x} + 2r_{1y})D_y \\ &+ r_1r_2r_3 - r_{1x}r_2 - r_1r_{2x} - r_{1y}r_2 - r_1r_{2y} + r_{1y}r_3 + r_{1xy} + r_{1yy}. \end{aligned}\right\} \tag{31}$$

Theorem 11 (formal adjoint for equivalence classes). *Consider the equivalence classes of (14) given by the values of the invariants $q, I_1, I_2, I_3, I_4, I_5$ (from Theorem 9). Then the operation of taking the formal adjoint is defined by the following formulae*

$$I_1^\dagger = I_1 - 2q_x + 2q_yq,$$

$$I_2^\dagger = -I_2 - qq_{xy} + q_yq_x,$$

$$I_3^\dagger = -I_3 + \frac{1}{q2}\Big(2I_2 - q_yI_1 + qI_{1y} + q_{yy}q2\Big),$$

$$I_4^\dagger = -I_4 + \Big(-2qI_2 - 2q_xI_1 + qI_{1x} + 2q_x2 - qq_{xx} \Big),$$

$$I_5^\dagger = I_5 + (4q_xq_y/q2 - 2q_{xy}/q)I_1 + q_xI_3 + 2q_y/q^2I_4 - 2q_x/qI_{1y} - q_y/qI_{1x} + I_{1xy}$$
$$-qI_{3x} - \frac{1}{q}I_{4y} + q_xq_{yy} - I_{2y} + \frac{1}{q}I_{2x} + (-2q^2q_x - q2I_1)/(q4)I_2.$$

6 Multiple Factorizations of an LPDO

Studying generating sets of invariants and the invariant conditions for the existence of factorization extending a given factorization of highest order terms, we have discovered [27] unexpectedly an interesting result, which can be formulated very simply: a third-order bivariate LPDO has first-order left and right factors such that their symbols are co-prime if and only if the operator has a factorization into three factors, the left one of which is exactly the initial left factor and the right one is exactly the initial right factor. It has been also proved that the condition that the symbols of the initial left and right factors are co-prime is essential, and that the analogous statement "as it is" is not true for LPDOs of order four.

Theorem 12. *A third-order bivariate operator L has a first-order left factor F_1 and a first-order right factor F_2 with $\gcd(\mathrm{Sym}(F_1), \mathrm{Sym}(F_2)) = 1$ if and only if L has a factorization into three factors, the left one of which is exactly F_1 and the right one is exactly F_2.*

The following diagram is an informal illustration of the statement of the theorem:

$$(L = F_1 \circ \dots \quad \wedge \quad L = \dots \circ F_2) \Longleftrightarrow L = F_1 \circ \dots \circ F_2.$$

Example 7 (Symbol $(X+Y)XY$, $S_1 = X+Y$, $S_2 = Y$). We found an operator with two factorizations $L = (D_x + D_y + x) \circ (D_{xy} + yD_x + y2D_y + y3)$ and $L = (D_{xx} + D_{xy} + (x + y2)D_x + y2D_y + xy2 + 2y) \circ (D_y + y)$. Then L has the factorization $L = (D_x + D_y + x) \circ (D_x + y2) \circ (D_y + y)$.

Example 8 (Symbol X^2Y, $S_1 = X$, $S_2 = Y$). We found an operator with two factorizations $L = (D_x+x) \circ (D_{xy}+yD_x+y2D_y+y3)$ and $L = (D_{xx}+(x+y2)D_x+xy2) \circ (D_y+y)$. Then L has the factorization $L = (D_x+x) \circ (D_x+y2) \circ (D_y+y)$.

Example 9 (Symbol X^2Y, $S_1 = Y$, $S_2 = X$). We found an operator with two factorizations $L = (D_y + x) \circ (D_{xx} + yD_x + y3 - y4)$ and $L = (D_xD_y + xD_x + y^2D_y + xy2 + 2y) \circ (D_x + y - y2)$. Then L has the factorization $L = (D_y + x) \circ (D_x + y2) \circ (D_x + y - y2)$.

Proposition 2. *The condition $\gcd(S_1, S_2) = 1$ in Theorem 12 cannot be omitted.*

Proof. Hyperbolic case: Consider the equivalence class of (14) defined by the following values of the invariants from Thereom 9: $q = 1$, $I_1 = I_2 = I_5 = 0$,

$I_3 = I_4 = x - y$. Using Theorem 10 one can verify that the operators of this class have factorizations of the types $(S)(XY)$ and $(XY)(S)$ only.

This equivalence class is not empty. For example, the operator $A_3 = D_x^2 D_y + D_x D_y^2 + (x-y)(D_x + D_y)$ belongs to this equivalence class. Only the following two factorizations exist for A_3: $A_3 = (D_x D_y + x - y)(D_x + D_y) = (D_x + D_y)(D_x D_y + x - y)$.

Non-hyperbolic case: Consider the operator of Landau

$$D_x^3 + xD_x^2 D_y + 2D_x^2 + (2x+2)D_x D_y + D_x + (2+x)D_y,$$

which has two factorizations into different numbers of irreducible factors:

$$L = Q \circ Q \circ P = R \circ Q,$$

for the operators $P = D_x + xD_y,\quad Q = D_x + 1,\quad R = D_x^2 + xD_x D_y + D_x + (2+x)D_y$. That is, factorizations of the types $(X)(SX)$, $(SX)(X)$ exist, while those of the type $(X)(S)(X)$ do not. Here we denote $S = X + xY$. □

Proposition 3. *The statement of Theorem 12 is not always true for a general fourth-order hyperbolic operator.*

Proof. For example, consider the operator

$$\begin{aligned} L &= (D_x + D_y) \circ (D_x D_y (D_x + D_y) + xD_x^2 + (2 - x2)D_x + xD_y - 2x + x2) \\ &= (D_x(D_x + D_y)2 - xD_x(D_x + D_y) + (x-2)D_x + (x-1)D_y + 1) \circ (D_y + x). \end{aligned}$$

The second factor in the first factorization is irreducible. □

Though there is no straightforward generalization, Theorem 12 been generalized to LPDOs of higher orders. However, this result is beyond the scope of this survey.

7 Software Development: The Maple Package `LPDOs`

Our Maple based package `LPDOs` serves as a tool for investigations in the area of symbolic algorithms for LPDOs. Since none of the existing packages for LPDOs and PDEs in Maple suits our problems properly, we have started from scratch, and introduced our own code for LPDOs and their basic manipulations. For example, the well-known package

```
Ore_Algebra
```

requires us to declare all the parameters at the very beginning. It is not always possible, because new parameters with names unknown in advance may appear during computation. For example, the output of *pdsolve* may contain new parameters

```
_C1, _F1(x+y)
```

and so on.

Besides basic manipulations of LPDOs, e.g., computing the conjugation of an operator for a given non-zero function, the taking of the formal adjoint, our package LPDOs can compute Laplace invariants, transformations, invariant chains (for the classical case and for the Schrödinger case), as well as invariants for third-order bivariate hyperbolic and non-hyperbolic LPDOs, obstacles to factorizations and factorizations of LPDOs of orders two and three (Grigoriev–Schwarz algorithm). Invariant definition of the existence of a factorization, and the operation of taking the formal adjoint have been implemented. The package is compatible with the existing internal MAPLE packages for LPDOs.

The package works with an arbitrary number of variables, arbitrary parameters, and coefficients. The only restriction is that the number of independent variables, and their names should be declared at the very beginning. For example, declare the independent variables x, y:

```
>> > > LPDO__set_vars([x,y]):
```

To work with an operator, we declare its name:

```
>> > > L3:=LPDO__create():
```

At this stage, $L3$ is the zero-operator, that is the operator, that multiplies by zero. Then one can change the operator, describing its coefficients. For example, define $L3 = D_x D_y (x D_x + y D_y) + D_x + 1$:

```
>> > > LPDO__add_value(L3, x,  [2,1]):
>> > > LPDO__add_value(L3, y,  [1,2]):
>> > > LPDO__add_value(L3, 1,  [1,0]):
>> > > LPDO__add_value(L3, 1,  [0,0]):
```

One can also create the same LPDO by specifying the list of its coefficient:

```
Create_LPDO_from_coeff([0,1,0,0,0,0,0,x,y,0]):
```

Compute, for example, invariants from Theorem 9 for LPDO

```
>> > > I1 := LPDO__Inv(1,L3);
                                    I1 := 0

>> > > I2 := LPDO__Inv(2,L3);
                                    I2 := 0

>> > > I3 := LPDO__Inv(3,L3);
                                            2
                                    I3 := x

>> > > I4 := LPDO__Inv(4,L3);
                                    I4 := 0

>> > > I5 := LPDO__Inv(5,L3);
                                          3
                                  I5 := x  y
```

Now, if $C3$ is the result of the gauge transformation of $L3$ with some $f(x, y)$, then the invariants must be the same. Here $f[x, x, y]$ stands for f_{xxy}, and so on.

```
>> > > C3:=LPDO__conj(L3,f(x,y)): LPDO__print(C3);
                      f + f[x] + x f[x, x, y] + y f[x, y, y]
              [0, 0], --------------------------------------
                                        f
                          f + 2 x f[x, y] + y f[y, y]
                  [1, 0], ---------------------------
                                        f
                            x f[x, x] + 2 y f[x, y]
                     [0, 1], -----------------------
                                        f
                                     x f[y]
                             [2, 0], ------
                                        f
                               2 (x f[x] + y f[y])
                        [1, 1], -------------------
                                        f
                                     y f[x]
                             [0, 2], ------
                                        f
                                [2, 1], x
                                [1, 2], y
>> > > for i from 1 to 5 do print(i, LPDO__Inv(i,C3)): end do;
                                    1, 0
                                    2, 0
                                        2
                                   3, x
                                    4, 0
                                      3
                                  5, x  y
```

8 Summary and Conclusions

We have given a survey of classical and recent results in the theory of factorization and invariants for LPDOs. In particular, we have focussed on the results obtained in our research group. We have been able to extend the work of Laplace to operators of order three; after this breakthrough we have realized that regularized moving frames (cf. [8, 9]), a method in differential geometry, allow reformulation of our results. Using moving frames, we have been able to find a method for determining generating sets of invariants for LPDOs of arbitrary order and in arbitrarily many variables [28]. Recently this result was also repeated and verified in [20], where Vessiot's methods were applied.

With generating sets of invariants in hand, we succeeded in finding algebraic-differential invariant descriptions of LPDOs having a factorization for every factorization type for order three [26]. Parametric factorization of LPDOs in the plane up to order four has been studied in [25]. Additionally, we succeeded in constructing differential transformations for parabolic LPDOs in the plane [36], which essentially generalize those invented by Laplace. The result about multiple factorizations of LPDOs has been generalized recently.

The area is actively developing. Thus, the very recent [34] theory of Hölden–Cassidy sequences for factorization of LPDOs opens many new avenues.

Acknowledgements This work was supported by the Austrian Science Foundation (FWF) in the projects SFB F013/F1304 and DIFFOP, Nr. P20336-N18.

References

1. Anderson, I., Juras, M.: Generalized Laplace invariants and the method of Darboux. Duke J. Math. **89**, 351–375 (1997)
2. Anderson, I., Kamran, N.: The variational bicomplex for hyperbolic second-order scalar partial differential equations in the plane. Duke J. Math. **87**, 265–319 (1997)
3. Athorne, C.: A $z \times r$ toda system. Phys. Lett. A. **206**, 162–166 (1995)
4. Blumberg, H.: Über algebraische Eigenschaften von linearen homogenen Differentialausdrücken. Ph.D. thesis, Göttingen (1912)
5. Darboux, G.: Leçons sur la théorie générale des surfaces et les applications géométriques du calcul infinitésimal, vol 2. Gauthier-Villars, Paris (1889)
6. Dini, U.: Sopra una classe di equazioni a derivate parziali di second ordine con un numero qualunque di variabili. Atti Acc. Naz. dei Lincei. Mem. Classe fis., mat., nat. **4**(5), 121–178 (1901) (also Opere III (1901), 489–566)
7. Dini, U.: Sopra una classe di equazioni a derivate parziali di second ordine. Atti Acc. Naz. dei Lincei. Mem. Classe fis., mat., nat., **4**(5), 431–467 (1902) (also Opere III (1902), 613–660)
8. Fels, M., Olver, P.J.: Moving coframes. I. A practical algorithm. Acta Appl. Math. **51**(2), 161–213 (1998)
9. Fels, M., Olver, P.J.: Moving coframes. II. Regularization and theoretical foundations. Acta Appl. Math. **55**(2), 127–208 (1999)
10. Forsyth, A.R.: Theory of Differential Equations, vol. VI. Cambridge University Press, Cambridge (1906)
11. Goursat, E.: Leçons sur l'intégration des équations aux dérivées partielles du seconde ordre a deux variables indépendants, vol. 2. Paris (1898)
12. Grigoriev, D., Schwarz, F.: Factoring and solving linear partial differential equations. Computing **73**(2), 179–197 (2004)
13. Grigoriev, D., Schwarz, F.: Generalized loewy-decomposition of d-modules. In: ISSAC '05: Proceedings of the 2005 International Symposium on Symbolic and Algebraic Computation, New York, NY, USA, pp. 163–170. ACM (2005)
14. Athorne, C., Yilmaz, H.: The geometrically invariant form of evolution equations. J. Phys. A. **35**, 2619–2625 (2002)
15. Juras, M.: Generalized laplace invariants and classical integration methods for second order scalar hyperbolic partial differential equations in the plane. Proceedings of the Conference Differential Geometry and Applications, pp. 275–284. Brno, Czech Republic (1996)
16. Kartashova, E.: A hierarchy of generalized invariants for linear partial differential operators. TMPh (Journal of Theoretical and Mathematical Physics), **147**(3), 839–846 (2006)
17. Kolchin, E.: Differential Algebra and Algebraic Groups. Academic, New York (1973)
18. Li, Z., Schwarz, F., Tsarev, S.P.: Factoring systems of linear pdes with finite-dimensional solution spaces. J. Symb. Comput. **36**(3–4), 443–471 (2003)
19. Loewy, A.: Über reduzible lineare homogene Differentialgleichungen. Math. Ann. **56**, 549—584 (1903)
20. Lorenz, A.: Jet Groupoids, Natural Bundles and the Vessiot Equivalence Method. Ph.D. thesis. RWTH Aachen University (2009)

21. Miller, F.H.: Reducible and Irreducible Linear Differential Operators. Ph.D. thesis, Columbia University (1932)
22. Le Roux, J.: Extensions de la méthode de laplace aux équations linéaires aux derivées partielles d'ordre supérieur au second. Bull. Soc. Math. de France **27**, 237–262 (1899) A digitized copy is obtainable from http://www.numdam.org/
23. Schwarz, F.: Algorithmic Lie Theory for Solving Ordinary Differential Equations. Pure and Applied Mathematics. Chapman & Hall/CRC, Boca Raton (2008)
24. Shemyakova, E.: A full system of invariants for third-order linear partial differential operators. Lect. Notes Comput. Sci. **4120**, 360–369 (2006)
25. Shemyakova, E.: The parametric factorizations of second-, third- and fourth-order linear partial differential operators on the plane. Math. Comput. Sci. **1**(2), 225–237 (2007)
26. Shemyakova, E.: Invariant properties of third-order non-hyperbolic linear partial differential operators. Lect. Notes Comput. Sci. **5625**, 154–169 (2009)
27. Shemyakova, E.: Multiple factorizations of bivariate linear partial differential operators. LNCS **5743**, 299–309 (2009)
28. Shemyakova, E., Mansfield, E.: Moving frames for Laplace invariants. Proceedings of ISSAC'08 (The International Symposium on Symbolic and Algebraic Computation), pp. 295–302 (2008)
29. Sokolov, V.V., Startsev, S.Ya., Zhiber, A.V.: On non-linear darboux integrable hyperbolic equations. Dokl. Acad. Nauk **343**(6), 746–748 (1995)
30. Shemyakova, E., Winkler, F.: Obstacle to factorization of LPDOs. In: Dumas J.-G. (ed.) Proceedings of Transgressive Computing 2006, Conference in Granada Spain, pp. 435–441, Grenoble, France (2006) Universite J. Fourier
31. Shemyakova, E., Winkler, F.: A full system of invariants for third-order linear partial differential operators in general form. Lect. Notes Comput. Sci. **4770**, 360–369 (2007)
32. Shemyakova, E., Winkler, F.: Obstacles to the factorization of linear partial differential operators into several factors. Program. Comput. Softw. **33**(2), 67–73 (2007)
33. Shemyakova, E., Winkler, F.: On the invariant properties of hyperbolic bivariate third-order linear partial differential operators. LNAI 5081 (2007)
34. Cassidy, P.J., Singer, M.F.: A Jordan-Höder theorem for differential algebraic groups. Elsevier, Amsterdam (2010)
35. Sokolov, V.V., Zhiber, A.V.: On the darboux integrable hyperbolic equations. Phys. Lett. A **208**, 303–308 (1995)
36. Tsarev, S.P., Shemyakova, E.: Differential transformations of parabolic second-order operators in the plane. Proc. Steklov Inst. Math. (Moscow) **266**, 219–227 (2009). http://arxiv.org/abs/0811.1492, submitted on 10 Nov 2008
37. Tsarev, S.P.: Factorization of linear partial differential operators and darboux integrability of nonlinear pdes. SIGSAM Bull. **32**(4), 21–28 (1998). also Computer Science e-print cs.SC/9811002 at http://arxiv.org/abs/cs/9811002, submitted on 31 Oct 1998
38. Tsarev, S.P.: Generalized laplace transformations and integration of hyperbolic systems of linear partial differential equations. In: ISSAC '05: Proceedings of the 2005 international symposium on Symbolic and algebraic computation, New York, NY, USA, pp. 325–331. ACM (2005)
39. Tsarev, S.P.: On factorization and solution of multidimensional linear partial differential equations. In: Computer Algebra 2006: Latest Advances in Symbolic Algorithms: Proceedings of the Waterloo Workshop in Computer Algebra 2006, Ontario, Canada, 10–12 April 2006, p. 181. World Scientific, Singapore (2007)
40. Wu, M.: On Solutions of Linear Functional Systems and Factorization of Modules over Laurent-Ore Algebras. Ph.D. thesis, Beijing (2005)